KU-218-642

MATHEMATICAL METHODS FOR THE PHYSICAL SCIENCES

MATHEMATICAL METHODS FOR THE PHYSICAL SCIENCES

An informal treatment for
students of physics and engineering

K. F. RILEY

Lecturer in physics, Cavendish Laboratory
Fellow of Clare College, Cambridge

CAMBRIDGE UNIVERSITY PRESS

Cambridge
London   New York   New Rochelle
Melbourne   Sydney

Published by the Press Syndicate of the University of Cambridge
The Pitt Building, Trumpington Street, Cambridge CB2 1RP
32 East 57th Street, New York, NY 10022, USA
296 Beaconsfield Parade, Middle Park, Melbourne 3206, Australia

© Cambridge University Press 1974

Library of congress catalogue card number: 73–89765

ISBN 0 521 20390 2 hard covers
ISBN 0 521 09839 4 paperback

First published 1974
Reprinted 1980, 1983

Transferred to digital printing 2004

TO MY PARENTS

Contents

xii   Contents

Preface

This book is intended for students of physical science, applied science and engineering, who, for the understanding and practice of their principal subjects, need a working knowledge of applicable mathematics.

Since it is not possible in a single text to cater for all degrees of mathematical facility, nor for all tastes in abstraction, a broad middle course has been adopted, set at the level of what, at the risk of being misunderstood, I describe as the 'average student'. It is hoped, however, that what is presented will also be of value to those who fall on either side of this central band, either as a less than rigorous introduction to the subject for the one group, or as an explanatory and illustrative text for the other.

The ground covered is roughly those areas of applied mathematics usually met by students of the physical sciences in their first and second years at university or technical college. Naturally much of it also forms parts of courses for mathematics students.

In any book of modest size it is impossible to cover all topics fully, and any one of the areas mentioned in this book can be, and has been, the subject of larger and more thorough works. My aim has been to take a 'horizontal slice' through the subject centred on the level of an average second-year student.

The preliminary knowledge assumed is that generally acquired by any student prior to entering university or college. In the United Kingdom, for example, it is that appropriate to the Advanced Level examination in Mathematics for Natural Science of one of the British Schools' Examination Boards. In the United States the material assumed is that which would normally be covered at junior college. Starting from this level, the first chapter of the book, consisting of a collection of topics mostly from the area of calculus, is aimed at providing a common base of the general techniques used in the development of the remaining chapters. Students who have had additional preparation, such as having offered Mathematics as a main A-Level subject, will find much of chapter 1 already familiar.

After the opening chapter, the contents of the remainder of the book fall under about half a dozen main headings. Chapters 2–4 deal with vectors and their uses, 5–8 with ordinary differential equations and 9–10 with partial differential equations. Stationary value problems are discussed in chapters 12–13 and matrices and tensors in 14–15, whilst chapter 11 on numerical methods and 16 on complex variables stand more or less alone, although both have connections with several other chapters.

The guiding principle for presenting the material has been that of introducing it wherever possible from a heuristic, physical point of view, and of deliberately avoiding strictly mathematical questions, such as the existence of limits, uniform convergence, interchanging integration and summation orders, etc., roughly speaking on the grounds that 'this is the real world; it must behave reasonably'. Free use has therefore been made of pictorial mathematics and sometimes of qualitative verbal descriptions instead of more compact mathematical symbolism. This has the effect of lengthening the book somewhat but makes the arguments less terse and tightly knit and therefore, for the average student, easier to unravel. In this same spirit, liberal use has also been made of parenthetical words of clarification [enclosed in square brackets]. They are certainly detrimental to the style and appearance of the text, but the author has found such asides to be appreciated by students and this is a quite sufficient justification for their use.

The general aim has been to present a topic in three stages, a qualitative introduction, the more formal presentation, and an explicit check or worked example to 'close the circle'. In the author's experience this last step is of great value to many students, who prefer to see the developed methods in action, rather than left as correct but abstract solutions. Such introductions and examples are based upon familiar situations and ideas in physical science and engineering, and it is such a general background which is, at the same time, both an important prerequisite for appreciating the material of this book and the main beneficiary of the methods it develops.

Straightforward manipulation and simple proofs, particularly when similar to previously displayed work, are often omitted, but the corresponding results are marked at the left-hand margin, by the symbol ►, as are illustrative exercises in the body of the text. The symbol ► indicates something which most definitely should be carried through by the student. The required workings are generally neither long nor difficult and notes on their solutions are given in many cases at the end of the book, but most students will benefit greatly in technical skill, understanding and confidence, by carrying out these parts of the arguments themselves.

Many equations (especially partial differential equations) can be written more compactly by using subscripts, e.g. u_{xy} for a second partial derivative instead of the more familiar $\partial^2 u/\partial y\, \partial x$, and this certainly saves typographical space. However, for many students, trying to put physical meaning to the equations rather than just manipulating them according to a set of allowed rules, the labour of mentally unpacking such equations, especially if u or x or y carry further subscripts, indices or primes, is sufficiently great that it is not possible to think of the physical interpretation at the same time. Consequently it has been decided to write out such expressions in their more obvious but longer form. For ordinary differential equations, where less confusion arises, a prime notation has generally been used, except where the independent variable has a clear physical connection with time when a dot notation is employed to make this connection immediate.

The summation convention is introduced in the course of the first chapter, and where appropriate is used in later ones, but with a brief reminder given as to its meaning after each substantial intermission.

It is a pleasure to record my sincere thanks to Sue Mackenzie and Jacky Walters for their patience and care in typing a difficult text in a short time, and to Helen Maczkiewicz for the quality of the drawings and the speed with which she produced them.

I am also greatly indebted to those of my colleagues and students who have read parts of the original manuscript, for pointing out errors and making constructive suggestions. In particular I would like to thank C. M. M. Nex and A. C. Steven of the Cavendish Laboratory, and several of the undergraduates of Clare College, Cambridge, notably R. J. Citron, D. Deutsch, P. N. Jones, H. J. Manning, P. D. A. Orr and O. B. R. Strimpel. Of course all errors and ambiguities remaining are entirely the responsibility of the author, and I would be most grateful to have them brought to my attention.

My thanks also go to the University of Cambridge for permission to use some past examination questions, and to the Cavendish teaching staff whose lecture handouts have collectively provided the basis for some of the examples included.

Finally I wish to place on record my appreciation of the care in setting and printing taken by the staff of William Clowes & Sons, Limited, and also my sincere thanks to the editorial staff of The Cambridge University Press for their much-appreciated advice on the preparation and presentation of the contents of this book.

Cambridge, 1973 K.F.R.

Mathematical symbols

$\equiv$	identically equal to; definition of
$\approx$	approximately equal to
$\simeq$	asymptotically equal to (see section 1.1)
$\triangleq$	corresponds to
$\propto$	proportional to
$\rightarrow$	approaches; tends to; becomes
$\pm$	plus or minus
$\mp$	minus or plus
$\| \ \|$	magnitude of; determinant of (according to context)
$\leqslant$	less than or equal to
$\geqslant$	greater than or equal to
$\ll$	much smaller than
$\gg$	much greater than
$O(\ )$	(see section 1.1)
$o(\ )$	(see section 1.1)
$\ldots$	similar terms included but not explicitly stated

$n!$ factorial $n = 1 \times 2 \times 3 \times \cdots \times (n - 1) \times n;\quad 0! = 1$

$\dbinom{n}{r}$ binomial coefficient $= \dfrac{n!}{r!(n - r)!}$

$\ln x$ natural logarithm of x

$\arcsin x$ the quantity whose sine is x (often $\sin^{-1} x$)

$\arccos x$ the quantity whose cosine is x (often $\cos^{-1} x$)

$\arctan x$ the quantity whose tangent is x (often $\tan^{-1} x$)

$\lim\limits_{x \to a} f(x)$ the limit to which $f(x)$ tends as x approaches a

$\displaystyle\sum_{i=1}^{n} a_i$ the sum $a_1 + a_2 + \cdots + a_n$

▶ exercises or working the reader should carry through

1
Preliminary calculus

Although the major part of this book is concerned with mathematics of direct value in describing situations arising in physical science and engineering, this opening chapter, although directed to the same end, is of a less obviously 'applied' nature. It is concerned with those techniques of mathematics, principally in the field of calculus, which are the nuts and bolts of the more particularly orientated methods presented in later chapters.

Two particular factors have to be taken into account in its presentation; firstly the various levels of previous knowledge which different readers will possess, and secondly the fact that the subjects to be treated in this chapter form a less coherent whole than do those in any other.

The first of these has been approached at the 'highest common factor' level, namely, knowledge has been presumed only of those topics which will normally be familiar to a student who, in his previous studies, has taken mathematics in conjunction with other science subjects, rather than as his main or only subject. As a result, although several parts of this chapter will almost certainly be unfamiliar to him, the reader with more than this presumed level of knowledge may in some sections find it sufficient to make sure he can solve the corresponding exercises, marked by the symbol ►, and then pass on to the next section.

As a result of the rather diverse nature of the topics considered, the degree of difficulty of the material does not vary 'monotonically' throughout the chapter. Rather, the order of presentation has been chosen so as to reduce, as far as possible, both abrupt changes in subject matter and forward references.

1.1   Notations and symbols

Throughout the text the notations and symbols adopted have, where possible, been made consistent with those recommended by international agreement.†

> † *Quantities, units and symbols*, The Symbols Committee of the Royal Society, 1971.

For ease of future location and reference, all but the commonest of the symbols used for mathematical operations are set out in a list of mathematical symbols facing page 1. In nearly all cases the brief explanations given there will be sufficient for the reader to understand the way in which the symbols are used.

However, the symbols O, o and $\simeq$ need some further explanation. These three symbols are used to compare the behaviour of two functions, as a variable upon which they both depend tends to a particular limit, usually infinity or zero, and obvious from the context. The variable may be a discrete integer variable n or a continuous one x. If the two functions are denoted by f and ϕ, and ϕ is positive, then the *definitions* of these symbols are as follows:

(i) If there exists a constant k such that $|f| \leqslant k\phi$ as the limit is approached, then $f = O(\phi)$. The statement $f = O(1)$ means that f is bounded.

(ii) If f/ϕ tends to 0 as the limit is approached, then $f = o(\phi)$. The statement $f = o(1)$ means that f tends to zero.

(iii) If f/ϕ tends to a limit l, where $l \neq 0$, as the limit of n or x is approached, then $f \simeq l\phi$. The statement $a \simeq b$ means that the ratio of the two sides tends to unity.

Although equations relating the various symbols, and a resulting 'algebra', may be established, these notations will not be used in this book for anything more than shortening the presentation of equations, and to reduce the repetitive use of wordy additions to equations. However, the reader is urged to verify for himself the following examples of statements involving use of the symbols.

If $f(n) = 3n^2 + 2n$ and $g(x) = ax^{1/2} + bx^{3/2}$, then:

(i) As $n \to \infty$: $f(n) = O(n^2)$, $f(n) = O(n^3)$, $f(n) = o(n^3)$, $f(n) \simeq 3n^2$.

(ii) As $x \to \infty$: $g(x) = O(x^{3/2})$, $g(x) = o(x^2)$, $g(x) \simeq bx^{3/2}$, $\cos \lambda x = O(1)$, $x^{-1} \cos \lambda x = o(1)$.

(iii) As $x \to 0$: $g(x) = O(x^{1/2})$, $g(x) = o(x^{1/4})$, $g(x) \simeq ax^{1/2}$, $\cos \lambda x \simeq 1$, $x^{-1} \cos \lambda x \simeq x^{-1}$.

1.2   Complex numbers

The notions of a complex number and of a complex function of a real number [e.g. exp (ix)] are used extensively later and so we begin by summarizing without proofs the basic properties of complex numbers.

We denote a general complex number z by

$$z = x + iy, \tag{1.1}$$

where x is the real part (Re z) and y the imaginary part (Im z) of z and i is usually described as 'the square root of -1'. The numbers x and y are themselves real. An alternative notation is

$$z = r \cos \theta + ir \sin \theta, \tag{1.2}$$

where $r = (x^2 + y^2)^{1/2}$ and $\theta = \arctan(y/x)$ [taking regard of the signs of x and y individually].

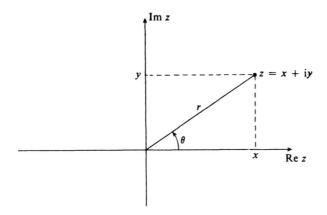

Fig. 1.1 An Argand diagram.

Complex numbers can be represented by points on a diagram (an **Argand diagram**), such as that shown in fig. 1.1, in which the relationships between equations (1.1) and (1.2) are self-evident. The quantity r is called the **modulus** of z, written $r = \text{mod } z$ or $r = |z|$, and θ is known as the **argument** of z, written $\theta = \arg z$. It is apparent that $\arg z$ is undetermined to the extent of $2\pi n$, where n is any integer; for this reason the **principal value** of $\arg z$ is also defined, and is given by that value which lies in $-\pi < \arg z \leqslant \pi$.

The defining laws of addition, subtraction, multiplication and division for two complex numbers z and z' are summarized in the following formulae. The commutative, associative and distributive laws hold to exactly the same extent as they do for real numbers.

Addition:

$$\begin{aligned} z + z' &= (x + iy) + (x' + iy') \\ &= (x + x') + i(y + y'). \end{aligned} \tag{1.3}$$

Subtraction:

$$\begin{aligned} z - z' &= (x + iy) - (x' + iy') \\ &= (x - x') + i(y - y'). \end{aligned} \tag{1.4}$$

Multiplication:

$$\begin{aligned} zz' &= (x + iy) \times (x' + iy') \\ &= (xx' - yy') + i(xy' + yx'). \end{aligned} \tag{1.5}$$

Division: $$\frac{z'}{z} = \frac{x' + iy'}{x + iy}$$

$$= \frac{xx' + yy'}{x^2 + y^2} + i\frac{xy' - yx'}{x^2 + y^2}. \tag{1.6}$$

Two particular points should be noted:

(i) If all imaginary parts are zero, i.e. $y = y' = 0$, then all of equations (1.3) to (1.6) reduce to those appropriate to real numbers.

(ii) A particular case of (1.5) occurs when $x = x' = 0$ and $y = y' = 1$; we then have the fundamental result

$$i^2 = i \times i = -1. \tag{1.7}$$

These results taken in total show that we may for practical purposes treat x, y and i as though they are ordinary real numbers, provided that whenever i^2 appears it is replaced by -1.

For multiplication and division, the representation of complex numbers in terms of modulus and argument is particularly convenient, since, as

$$(xx' - yy')^2 + (xy' + yx')^2 = (x^2 + y^2)(x'^2 + y'^2),$$

we have from (1.5) that

$$\text{mod } zz' = (\text{mod } z) \times (\text{mod } z'). \tag{1.8}$$

Further, since

$$\arctan \frac{xy' + yx'}{xx' - yy'} = \arctan \frac{y}{x} + \arctan \frac{y'}{x'},$$

[from the formula $\tan(A + B) = (\tan A + \tan B)/(1 - \tan A \tan B)$] we also conclude that

$$\arg(zz') = \arg(z) + \arg(z'). \tag{1.9}$$

Thus if $z = r(\cos\theta + i\sin\theta)$ and $z' = r'(\cos\theta' + i\sin\theta')$ then $zz' = rr'[\cos(\theta + \theta') + i\sin(\theta + \theta')]$.

Similarly from (1.6) it can be shown that:

►1. $\text{mod}(z'/z) = (\text{mod } z')/(\text{mod } z),$ (1.10)

and

►2. $\arg(z'/z) = (\arg z') - (\arg z).$ (1.11)

►3. Show that multiplying a complex number $a + ib$ by i corresponds to rotating its vector on the Argand diagram by $\pi/2$ in the positive (anti-clockwise) sense.

A more convenient and shorter expression for $\cos\theta + i\sin\theta$ may be obtained as follows. Recall that for a real number x the function exponential (x) or $\exp(x)$ or e^x is defined by the series

$$\exp(x) = 1 + x + \frac{x^2}{2!} + \frac{x^3}{3!} + \cdots = \sum_{r=0}^{\infty} \frac{x^r}{r!}. \tag{1.12}$$

If we replace x by $i\theta$ (with θ real) in this and reduce any power of i by writing i^2 as -1 (see (1.7)) we obtain

$$\exp(i\theta) = 1 + i\theta - \frac{\theta^2}{2!} - \frac{i\theta^3}{3!} + \cdots. \tag{1.13}$$

The real and imaginary parts of this expression are

$$\text{Re}\,[\exp(i\theta)] = 1 - \frac{\theta^2}{2!} + \frac{\theta^4}{4!} - \cdots \tag{1.14 a}$$

and

$$\text{Im}\,[\exp(i\theta)] = \theta - \frac{\theta^3}{3!} + \frac{\theta^5}{5!} - \cdots. \tag{1.14 b}$$

But these two latter series are just the power series expansions for the functions $\cos\theta$ and $\sin\theta$ respectively. Thus using this in (1.13) establishes that

$$\exp(i\theta) = \cos\theta + i\sin\theta. \tag{1.15 a}$$

Changing θ to $-\theta$ and recalling that $\cos(-\theta) = \cos\theta$ whilst $\sin(-\theta) = -\sin\theta$, shows that

$$\exp(-i\theta) = \cos\theta - i\sin\theta. \tag{1.15 b}$$

Solving (1.15 a) and (1.15 b) for $\cos\theta$ and $\sin\theta$ gives

$$\cos\theta = \tfrac{1}{2}[\exp(i\theta) + \exp(-i\theta)], \tag{1.16 a}$$

$$\sin\theta = \frac{1}{2i}[\exp(i\theta) - \exp(-i\theta)]. \tag{1.16 b}$$

In this notation, (1.2) becomes

$$z = r\exp(i\theta), \tag{1.17}$$

and, for example, (1.5) takes the form

$$r\exp(i\theta)r'\exp(i\theta') = rr'\exp[i(\theta + \theta')]. \tag{1.18}$$

It is apparent that, by repeated applications of (1.18), the following more general result may be obtained:

$$\begin{aligned} z_1 z_2 \ldots z_n &= r_1\exp(i\theta_1)r_2\exp(i\theta_2)\ldots r_n\exp(i\theta_n) \\ &= r_1 r_2 \ldots r_n \exp[i(\theta_1 + \theta_2 + \cdots + \theta_n)]. \end{aligned} \tag{1.19}$$

A particular case of this is that in which all the z_j are equal and of unit modulus, i.e. $z_j = \exp(i\theta)$ for all j; then

$$[\exp(i\theta)]^n = \exp(in\theta). \tag{1.20}$$

Written in terms of cosines and sines, this result is known as **de Moivre's theorem**

$$(\cos\theta + i\sin\theta)^n = \cos n\theta + i\sin n\theta. \tag{1.21}$$

The complex exponential form is a very useful way of expressing functions which undergo sinusoidal variation, either in time or with distance. Thus the function

$$f(t) = R\exp(i\omega t) = R\cos\omega t + iR\sin\omega t \tag{1.22}$$

is one whose real and imaginary parts both undergo sinusoidal variations of amplitude R and period $2\pi/\omega$, although they are out of phase with each other by a quarter of a cycle.

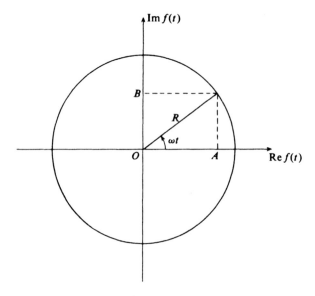

Fig. 1.2   The representation of $\exp(i\omega t)$ in an Argand diagram.

In the Argand diagram (fig. 1.2) the point which represents $f(t)$ at any particular time lies on a circle of radius R. As time increases the point moves in the positive sense (anticlockwise) with angular velocity ω. It is easily seen that the lengths OA and OB (reckoned positively or negatively) give the real and imaginary parts of (1.22).

The use of complex exponentials to describe travelling waves (in the form $\exp [i(kx - \omega t)]$) is discussed in chapter 9.

It may also be noted that (1.15 a) and (1.16 a, b) are closely related to the functions $\cosh (x)$ and $\sinh (x)$ which are defined as

$$\cosh (x) = \tfrac{1}{2}[\exp (x) + \exp (-x)], \tag{1.23 a}$$

$$\sinh (x) = \tfrac{1}{2}[\exp (x) - \exp (-x)]. \tag{1.23 b}$$

These two functions, also known as the *hyperbolic cosine* and *hyperbolic sine* respectively, satisfy the identity

▶4. $$\cosh^2 (x) - \sinh^2 (x) = 1, \tag{1.24}$$

as is easily verified.

If x is set equal to $i\theta$ in (1.23 a, b) and comparison is made between the resulting right-hand sides and the right-hand sides of (1.16 a, b) then the equalities

$$\cosh (i\theta) = \cos (\theta), \tag{1.25 a}$$

$$\sinh (i\theta) = i \sin (\theta), \tag{1.25 b}$$

are established. Conversely, putting θ equal to ix in (1.16 a, b) shows that

$$\cos (ix) = \cosh (x), \tag{1.26 a}$$

$$\sin (ix) = i \sinh (x). \tag{1.26 b}$$

Finally we define the **complex conjugate** (or simply **conjugate**) of $z(= x + iy)$ as the complex number

$$z^* \equiv x - iy. \tag{1.27}$$

The following properties of the complex conjugate are almost immediate, and others can be derived from them,

$$(z^*)^* = z, \tag{1.28 a}$$

$$z + z^* = 2 \operatorname{Re} z, \tag{1.28 b}$$

$$z - z^* = 2i \operatorname{Im} z, \tag{1.28 c}$$

$$zz^* = |z|^2. \tag{1.28 d}$$

In (r, θ) notation the conjugate of z is clearly given by

$$z^* = r \exp (-i\theta). \tag{1.29}$$

▶5. Verify that (1.29) is consistent with (1.28 a–d).

1.3   Convergence of series

In several places later in this book we will have occasion to consider the sum of a series of terms a_n which are all given by a single formula involving n, but each for its own value of n. For example in the two series defined by:

(i) $b_n = 1/2^n, \qquad n = 1, 2, \ldots$ (1.30 a)

(ii) $c_n = 1/n, \qquad n = 1, 2, \ldots$ (1.30 b)

b_7 has the value $1/128$ and $c_7 = 1/7$. Only series with *real* terms will be considered in this section.

Our concern will be with the quantity S_N which is the sum of the first N terms

$$S_N = \sum_{n=1}^{N} a_n,$$ (1.31)

and the particular question to be answered is whether or not, if we allow N to tend to infinity, S_N tends to a definite limit, or increases or decreases without limit, or oscillates 'finitely or infinitely'. We will correspondingly say that the series $\sum a_n$ **converges, diverges** to $+\infty$ or $-\infty$, or **oscillates** finitely or infinitely.

A formal analytic treatment of these matters involves careful study of the properties of bounds and limiting processes. However we will take a more heavy-handed approach and use expressions such as:

(*a*) 'tends to infinity' when we should say 'exceeds any given quantity M';
(*b*) 'a series converges to the sum S' when we should say 'given *any* $\epsilon > 0$ there exists an N_0, which may depend upon ϵ, such that $|S_N - S| < \epsilon$ for *all* N greater than N_0'.

Even if we can establish in a particular case that a series does converge, this may not automatically determine the value of its sum S. The evaluation of the sum is often a separate problem, and one which we will sometimes leave unsolved.

Suppose first that a series $\sum a_n$ is convergent, then it is apparent that, however small a quantity ϵ we may have chosen, we must be able to find a value N_0 of N such that

$$\left| \sum_{n=N_0}^{n = N_0 + N_1} a_n \right| < \epsilon,$$ (1.32)

whatever the value of N_1 (> 0). Put roughly in words, 'we can always find a value N_0 of n such that the sum of any number of terms from N_0 onwards is less than any preassigned positive quantity, however small'.

The smaller the preassigned quantity, the larger the value of N_0 needed, but one can always be found.

The converse of this is also true, namely that if, given any $\epsilon > 0$, we can find an N_0 such that the statement involving inequality (1.32) is true, then the series must be convergent. This follows since the sum of $N_0 + N_1$ terms must be within $\pm \epsilon$ of S_{N_0} for any positive N_1, and ϵ itself could have been chosen arbitrarily small.

It is not usual to actually assign values to ϵ, but for the sake of illustration we may do so and consider the series $\sum b_n$ given by (1.30 a). Suppose we choose ϵ as 10^{-3}. The eleventh term of the series $b_{11} = 2^{-11} = 1/2048$, the next term is one half of this, and the one after that (the thirteenth) one quarter of it, and so on. It is apparent that the addition of each further term only halves the gap between the sum starting at the eleventh term and the value $2/2048$. Hence

$$\left| \sum_{n=11}^{n=11+N_1} b_n \right| < \frac{2}{2048} < 10^{-3} = \epsilon,$$

for any N_1, thus showing that $\sum b_n$ is convergent. If instead of 10^{-3} for ϵ we had chosen 10^{-6}, then we would have had to take N_0 as 21 rather than 11. What has to be shown is that *whatever* value of ϵ is given, provided it is positive, we can find a corresponding N_0.

The actual sum S for this particular series is rather obvious if the first few terms are written out, and in any case as this series is a geometric one the formula for its sum, $\frac{1}{2}/(1 - \frac{1}{2})$, is probably well known to the reader.

It is almost as straightforward to show that the series $\sum c_n$ defined by (1.30 b) is *not* convergent. This can be done by grouping its terms in the following way

$$S_N = \sum_1^N c_n = 1 + (\tfrac{1}{2}) + (\tfrac{1}{3} + \tfrac{1}{4}) + (\tfrac{1}{5} + \tfrac{1}{6} + \tfrac{1}{7} + \tfrac{1}{8})$$

$$+ (\tfrac{1}{9} + \cdots + \tfrac{1}{16}) + \cdots. \quad (1.33)$$

The sum of the terms in each bracket is in every case $\geqslant \tfrac{1}{2}$, and since it is apparent that as many such brackets can be made up as we desire, the sum S_N can be made bigger than any finite value simply by making N large enough. This shows that S_N does not tend to a limit S and that the series $\sum c_n$ does not converge. Such a series is said to diverge.

To investigate whether or not a given series $\sum a_n$ converges, it is useful to have a number of tests and theorems of general applicability available. Some of these we will merely state, since once they have been stated they become almost self-evident – but are no less useful for that.

1. *Crucial consideration.* In all tests for, or discussions of, the convergence of a series, it is not what happens in the first ten, or the first thousand, or the first million terms (or any other finite number of terms) that matters, but what happens *ultimately*.

2. A *necessary* (but certainly not sufficient) condition for convergence is that $|a_n| \to 0$ as $n \to \infty$.

In the next four tests we will assume that all terms of all series mentioned are non-negative (or, more precisely, that the number of negative ones is finite).

3. If there exists a constant M such that $S_N = \sum {}^N a_n$ is $< M$ for all values of N, then $\sum a_n$ is convergent.† If no such constant exists $\sum a_n$ diverges.

4. *Comparison test.* (i) If for some N_0, $a_n \leqslant \lambda d_n$ for some fixed λ and *all* $n > N_0$, and $\sum d_n$ is convergent, then $\sum a_n$ is convergent.
(ii) If for some N_0, $a_n \geqslant \lambda d_n$ for some fixed λ and *all* $n > N_0$, and $\sum d_n$ diverges, then $\sum a_n$ diverges also.

5. *Ratio test* (*D'Alembert*). A series $\sum a_n$ converges or diverges according as

$$\lim_{n \to \infty} \frac{a_{n+1}}{a_n} \tag{1.34}$$

is <1 or >1 respectively. If the limit (1.34) is equal to 1, no conclusion can be drawn.
    To prove this we observe that if the limit (1.34) is λ where $\lambda < 1$, then we can find a value r in the range $\lambda < r < 1$ and a value N_0 such that

$$\frac{a_{n+1}}{a_n} < r,$$

for all $n > N_0$. Now the terms of the series a_n after a_{N_0} are

$$a_{N_0+1}, \qquad a_{N_0+2}, \qquad a_{N_0+3}, \dots, \tag{1.35 a}$$

and each of these is less than the corresponding term of

$$r a_{N_0}, \qquad r^2 a_{N_0}, \qquad r^3 a_{N_0}, \dots. \tag{1.35 b}$$

However, the terms of (1.35 b) are those of a geometric series with a common ratio r less than 1; the geometric series consequently converges

---

† The result is intuitively obvious, but a strict analytic proof requires a careful discussion of bounds.

and therefore, by the comparison test (result 4) so must the series (1.35 a). The observation in statement 1 is now enough to establish the validity of the ratio test.

The divergent case, where $\lambda > 1$, is proved by an analogous argument.

6. *Root test (Cauchy)*. A series $\sum a_n$ converges or diverges according as

$$\lim_{n \to \infty} (a_n)^{1/n} \tag{1.36}$$

is < 1 or > 1 respectively. If the limit (1.36) is equal to 1, no conclusion can be drawn.

The proof follows, almost exactly, the lines of that for the ratio test and so will not be given here.

▶6. Construct this proof.

Apart from obvious modifications, results 3, 4, 5 and 6 hold for series consisting entirely of negative terms, or [more liberally] with only a finite number of positive ones.

The one type of series for which we have as yet made no provision is one containing infinite numbers of both positive and negative terms. The two signs may appear in alternate terms or in a less symmetric way, e.g.

$$\tfrac{1}{1} - \tfrac{1}{2} + \tfrac{1}{3} - \tfrac{1}{4} + \cdots, \tag{1.37 a}$$

or $\qquad -\tfrac{1}{1} + \tfrac{1}{2} + \tfrac{1}{3} + \tfrac{1}{4} - \tfrac{1}{5} + \tfrac{1}{6} + \tfrac{1}{7} + \tfrac{1}{8} - \tfrac{1}{9} + \cdots. \tag{1.37 b}$

The characteristic of this type of series is not that there are equal numbers of positive and negative terms, but that the numbers of each should be infinite. Specific to this type of series we will give only one particular test.

7. *Alternating signs test*. A series $\sum (-1)^n a_n$, where the a_n are positive, converges if

(i) $a_n \to 0$ as $n \to \infty$, and (ii) an N_0 can be found such that $a_n > a_{n+1}$ for all $n \geq N_0$.

To prove this suppose for definiteness that N_0 is even and consider the series starting at a_{N_0}. First of all group the sum of its first $2m$ terms as

$$S_{2m-1} = (a_{N_0} - a_{N_0+1}) + (a_{N_0+2} - a_{N_0+3})$$
$$+ \cdots + (a_{N_0+2m-2} - a_{N_0+2m-1}).$$

By condition (ii) all the brackets are positive, and so S_{2m-1} increases as m increases. But, on the other hand

$$S_{2m-1} = a_{N_0} - (a_{N_0+1} - a_{N_0+2})$$
$$- \cdots - (a_{N_0+2m-3} - a_{N_0+2m-2}) - a_{N_0+2m-1},$$

and, since each bracket is positive, we must have S_{2m-1} is less than a_{N_0}. These two properties of S_{2m-1}, together with result 3, establish the validity of the test.

It is clear that the proof does not depend in any essential way on N_0 being even, and this is in no way a restriction on the general result.

If from the series of terms a_n, the series $|a_n|$ is formed and $\sum |a_n|$ is convergent, then the series $\sum a_n$ is said to be **absolutely convergent**. Clearly there is no distinction between convergence and absolute convergence for series all of whose terms are positive anyway. Series containing only a finite (but non-zero) number of terms of a particular sign will be absolutely convergent if they are convergent; the actual value of the sums $S = \sum a_n$ and $S' = \sum |a_n|$ will be different however.

It is apparent that a series can be convergent without being absolutely convergent. An example of such a series is (1.37 a), which is convergent by the alternating signs test of result 7, but is shown by the divergence of (1.30 b) not to be absolutely convergent.

On the other hand an absolutely convergent series is necessarily convergent.

▶7. Determine how the behaviour of $S_N = \sum_{n=0}^{N} ax^n$ as $N \to \infty$ depends upon the value of x.

▶8. Find the sum S_N of the first N terms of the following series and hence determine whether the series are convergent, divergent, or oscillatory.

$$(a) \sum_0^\infty (a + nd), \qquad (b) \sum_1^\infty \ln\left(\frac{n+1}{n}\right), \qquad (c) \sum_0^\infty (-2)^n,$$

$$(d) \sum_1^\infty \frac{1}{n(n+1)} \quad \text{[write as partial fractions]},$$

$$(e) \sum_1^\infty (-1)^{n+1} \frac{n}{3^n}.$$

▶9. Determine whether the following series are convergent. (Their sums, where they exist, are not required.)

$$(i) \sum_1^\infty \frac{1}{n^2}, \qquad (ii) \sum_1^\infty \frac{1}{n^{1/2}}, \qquad (iii) \sum_1^\infty \frac{\cos (n\theta)}{n(n+1)},$$

$$(iv) \sum_1^\infty \frac{n^{1/2}}{(1+n)^{1/2}}, \qquad (v) \sum_1^\infty \frac{n^2}{n!}, \qquad (vi) \sum_1^\infty \frac{(\ln n)^n}{n^{n/2}},$$

$$(vii) \sum_1^\infty \frac{n^n}{n!}.$$

▶10. Are the following series absolutely convergent, convergent or oscillatory?

$$(a) \sum_1^\infty \frac{(-1)^n}{n^{5/2}}, \qquad\qquad (b) \sum_1^\infty \frac{(-1)^n(2n+1)}{n},$$

$$(c) \sum_1^\infty \frac{(-1)^n(n^2+1)^{1/2}}{n \ln n}, \qquad (d) \sum_0^\infty \frac{(-1)^n|x|^n}{n!},$$

$$(e) \sum_0^\infty \frac{(-1)^n}{n^2+3n+2}, \qquad (f) \sum_1^\infty \frac{(-1)^n 2^n}{n^{1/2}}.$$

1.4   Differentiation

If $f(x)$ is a function of x only, then the first **derivative** (or **differential coefficient**) of $f(x)$, denoted by $f'(x)$ or df/dx is defined to be

$$f'(x) \equiv \frac{df(x)}{dx} \equiv \lim_{\Delta x \to 0} \frac{f(x+\Delta x) - f(x)}{\Delta x}, \tag{1.38}$$

provided that the limit exists. The value of the limit will in almost all cases depend upon that of x of course.

In graphical terms $f'(x)$ also gives the slope of $f(x)$ at the particular value of x as illustrated in fig. 1.3.

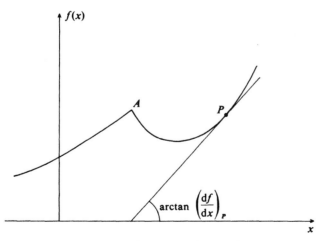

Fig. 1.3   The derivative $f'(x)$ as the slope of $f(x)$. At point A the derivative is discontinuous.

In the definition (1.38) we allow Δx to tend to zero from either positive or negative values and require the same limit to be obtained in both cases. We will not concern ourselves with cases in which neither or only one of the limits exists nor, except to notice that they correspond to kinks in the graph of $f(x)$, with cases in which the two limits exist but are not equal, such as at point A in fig. 1.3.

The second derivative of $f(x)$ can be obtained by using (1.38) again, but with $f(x)$ replaced by $f'(x)$, and so on. If we denote the nth (order) derivative of $f(x)$ by df^n/dx^n or $f^{(n)}(x)$, with $f^{(1)}(x) \equiv f'(x)$, [and $f^{(0)}(x)$ formally $\equiv f(x)$] then

$$f^{(n+1)}(x) = \lim_{\Delta x \to 0} \frac{f^{(n)}(x + \Delta x) - f^{(n)}(x)}{\Delta x}, \tag{1.39}$$

provided that the limit exists.

All this should be familiar to the reader, as should the derivatives of many standard functions and forms. Because of this, further discussion will not be given; rather, the student is encouraged to carry out the following exercises.

▶11. Obtain the following from first principles, i.e. using (1.38):

$$(a) \frac{d}{dx}(3x + 4), \qquad (b) \frac{d}{dx}(x^2 + x), \qquad (c) \frac{d^2}{dx^2}(x^2 + x),$$

$$(d) \frac{d^3}{dx^3}(x^2 + x), \qquad (e) \frac{d}{dx}(\sin x).$$

▶12. Write down or obtain the first derivatives of the following:

(a) x^8, (b) x^{-3}, (c) $(1 + x^2)^{1/2}$, (d) $\cos(2x)$, (e) $\tan(ax)$, (f) $\arctan(ax)$, (g) $\exp(x)$, (h) $\cosh(x)$, (i) $\operatorname{arsinh}(x)$, (j) $\ln(kx)$, (k) $\sin^2 3x$.

▶13. Show from first principles that

$$\frac{d}{dx}\left(\frac{1}{f(x)}\right) = -\frac{1}{f^2(x)}\frac{df}{dx}. \tag{1.40}$$

Hence obtain the (first) derivative of:

$$(a) \frac{1}{(2x + 3)^3}, \qquad (b) \frac{1}{\exp(2x)}, \qquad (c) \sec^2 x,$$

$$(d) \operatorname{cosech}^3(3x).$$

1.5 Differentiation of products and quotients

Let us consider a case in which the function $f(x)$ can be written as the product of two functions of x, namely $f(x) = g(x)h(x)$. For example if $f(x)$ were given by $x^3 \sin x$ then we might take $g(x) = x^3$ and $h(x) = \sin x$. Clearly in many cases the separation is not unique. (In the given example possible alternative break-ups would be $g(x) = x^2$, $h(x) = x \sin x$, or even, to stretch a point, $g(x) = x^4 \tan x$, $h(x) = x^{-1} \cos x$.) We may ask if there is a relationship between the derivative of f and those of g and h.

To answer this, we apply definition (1.38) to $f(x)$. In doing this we will have to consider $f(x + \Delta x) - f(x)$, which can be rearranged, by subtracting and adding $g(x + \Delta x)h(x)$, as follows

$$
\begin{aligned}
f(x + \Delta x) - f(x) &= g(x + \Delta x)h(x + \Delta x) - g(x)h(x) \\
&= g(x + \Delta x)[h(x + \Delta x) - h(x)] \\
&\quad + h(x)[g(x + \Delta x) - g(x)].
\end{aligned}
$$

Hence

$$
\begin{aligned}
\frac{df}{dx} &= \lim_{\Delta x \to 0} \frac{f(x + \Delta x) - f(x)}{\Delta x} \\
&= \lim_{\Delta x \to 0} \left\{ g(x + \Delta x)\left[\frac{h(x + \Delta x) - h(x)}{\Delta x} \right] \right. \\
&\qquad\qquad \left. + h(x)\left[\frac{g(x + \Delta x) - g(x)}{\Delta x} \right] \right\}.
\end{aligned}
$$

In the limit $\Delta x \to 0$, the factors in square brackets become dh/dx and dg/dx (by the definitions of these quantities) and $g(x + \Delta x)$ simply becomes $g(x)$. Consequently we obtain

$$
\frac{d}{dx}[g(x)h(x)] = \frac{df}{dx} = g(x)\frac{dh}{dx} + h(x)\frac{dg}{dx}, \tag{1.41}
$$

which is a general result obtained without making any assumptions about the specific forms f, g and h, other than $f(x) = g(x)h(x)$. In words, the result reads,

> The derivative of the product of two functions is equal to {the first function times the derivative of the second} plus {the second function times the derivative of the first}.

For the example $f(x) = x^3 \sin x$ given earlier, (1.41) gives

$$
\frac{d}{dx}(x^3 \sin x) = x^3 \frac{d}{dx}(\sin x) + \sin x \frac{d}{dx}(x^3)
$$

$$
= x^3 \cos x + 3x^2 \sin x. \tag{1.42}
$$

This result for the derivative of a product containing two factors can be used to obtain one for a product of three factors by writing one of the two factors, say $h(x)$, itself as a product of two factors, viz. $h(x) = j(x)k(x)$, so that $f(x)$ has the three-product form

$$f(x) = g(x)j(x)k(x). \tag{1.43}$$

Then result (1.41) shows that

$$\frac{df}{dx} = g(x)\frac{d}{dx}[j(x)k(x)] + j(x)k(x)\frac{dg}{dx},$$

and (1.41) can again be used to expand the first term on the right, giving the complete result as

$$\frac{d}{dx}[g(x)j(x)k(x)] = g(x)j(x)\frac{dk}{dx} + g(x)k(x)\frac{dj}{dx} + j(x)k(x)\frac{dg}{dx}. \tag{1.44 a}$$

In primed notation, without the argument x written explicitly each time, this is

$$(gjk)' = gjk' + gj'k + g'jk. \tag{1.44 b}$$

It is readily apparent that this can be extended to products containing any general number of factors n, and that the expression for the derivative will consist of n terms with a prime appearing in successive terms on each of the n factors in turn.

▶14. Verify these results formally for the previous function $f = x^3 \sin x$ by writing it variously as (a) $x^2 \cdot x \cdot \sin x$, (b) $x \cdot x \cdot x \cdot \sin x$, (c) $2x^3 \cdot \frac{1}{2} \sin x$, (d) $x^4 \cdot \tan x \cdot x^{-1} \cos x$ and obtaining result (1.42) each time.

▶15. Obtain the first derivatives of the following:

(a) $x^2 \exp(x)$, (b) $\sin x \cosh x$, (c) $x(\ln x - 1)$, (d) $x^3 \tan^3 x$, (e) $x \sin(ax) \cdot \exp(\lambda x)$, (f) $x^2 \cos(1 - x^2)$, (g) $(a^2 + x^2)^{-1}(b^2 + x^2)^{-2}(c^2 + x^2)^{-3}$.

By applying result (1.41) for the derivative of a product to a function $f(x)$ of the form $f(x) = g(x) \cdot [1/h(x)]$, we may obtain the derivative of the quotient of two factors. Thus

$$f' = \left(\frac{g}{h}\right)' = g\left(\frac{1}{h}\right)' + \left(\frac{1}{h}\right)g' = g\left(-\frac{1}{h^2}h'\right) + \frac{g'}{h},$$

where the result of ▶13 has been used to evaluate $(1/h)'$. This can now be rearranged in the convenient and memorizable form

$$f' = \left(\frac{g}{h}\right)' = \frac{hg' - gh'}{h^2}. \tag{1.45}$$

As usually expressed in words [hardly mathematically, but sufficiently clearly];

> The derivative of a quotient is equal to 'the bottom times the derivative of the top minus the top times the derivative of the bottom, all over the bottom squared'.

▶16. Find the (first) derivatives of the following:

(a) $\dfrac{x}{(a + x)^2}$, (b) $\dfrac{x}{(1 - x)^{1/2}}$,

(c) tan x in the form $\dfrac{\sin x}{\cos x}$, (d) $\dfrac{3x^2 + 2x + 1}{8x^2 - 4x + 2}$.

1.6   Leibniz theorem

Following on from the question posed in the first paragraph of the previous section about the existence of a relationship between the first derivative of a product and the derivatives of its constituent factors, and its answer contained in (1.41), we may further ask if corresponding results are obtainable for any arbitrary derivative, say the nth.

It will be found that the answer is 'yes' with the general result expressed by Leibniz theorem, but before proving this let us carry out some trials to get a feel for what is involved.

Again suppose f is of the form $f = gh$, where f, g and h are all functions of x, and the nth derivative of, for example, g is denoted by $g^{(n)}$. Formally we denote the undifferentiated g itself by $g^{(0)}$. Then from (1.41) we have

$$f^{(1)} = g^{(1)}h^{(0)} + g^{(0)}h^{(1)}. \tag{1.46 a}$$

If we differentiate this directly several times, as is done below, it will be noticed that some of the terms obtained at any particular stage are repeated and can be gathered together before proceeding to the next differentiation. Starting from (1.46 a) we obtain successively, by repeated applications of (1.41) in its general form

$$(ab)' = a'b + b'a,$$

that

$$\begin{aligned}
f^{(2)} &= [g^{(2)}h^{(0)} + g^{(1)}h^{(1)}] + [g^{(1)}h^{(1)} + g^{(0)}h^{(2)}] \\
&= g^{(2)}h^{(0)} + 2g^{(1)}h^{(1)} + g^{(0)}h^{(2)},
\end{aligned} \tag{1.46 b}$$

$$f^{(3)} = [g^{(3)}h^{(0)} + g^{(2)}h^{(1)}] + 2[g^{(2)}h^{(1)} + g^{(1)}h^{(2)}]$$
$$+ [g^{(1)}h^{(2)} + g^{(0)}h^{(3)}]$$
$$= g^{(3)}h^{(0)} + 3g^{(2)}h^{(1)} + 3g^{(1)}h^{(2)} + g^{(0)}h^{(3)}, \qquad (1.46\text{ c})$$

▶17. $f^{(4)} = g^{(4)}h^{(0)} + 4g^{(3)}h^{(1)} + 6g^{(2)}h^{(2)} + 4g^{(1)}h^{(3)} + g^{(0)}h^{(4)}.$ (1.46 d)

The forms (1.46 a–d) are certainly very suggestive, and, if it were not for the fact that $g^{(n)}$ means the nth derivative $d^n g/dx^n$, rather than the nth power of g, they would simply be statements of the binomial theorem expansions of f^n for an f given by $g + h$, rather than gh. Even so the forms make a very clear suggestion as to the general expression for the nth derivative of the product $g(x)h(x)$. This is embodied in the following theorem.

Leibniz theorem. If $f(x) = g(x)h(x)$, then

$$f^{(n)} = \sum_{r=0}^{n} \binom{n}{r} g^{(n-r)}h^{(r)}, \qquad (1.47)$$

where the symbols $f^{(r)}$, $g^{(n-r)}$ and $h^{(r)}$, have the meaning previously assigned for $r = 0, 1, 2, \ldots, n$, and $\binom{n}{r}$ is the binomial coefficient (see list of mathematical symbols).

We note that we have already proved (1.47) for values up to $n = 4$ by explicit construction. The general proof proceeds by induction as follows.

Assume (1.47) holds for some particular value N of n, and then construct $f^{(N+1)}$, which will be shown to be given by (1.47) also, but with $n = N + 1$. From the assumed form

$$f^{(N+1)} = \sum_{r=0}^{N} \binom{N}{r} [g^{(N-r)}h^{(r)}]'$$
$$= \sum_{r=0}^{N} \binom{N}{r} [g^{(N-r+1)}h^{(r)} + g^{(N-r)}h^{(r+1)}]. \qquad (1.48)$$

Every term of (1.48) is of the form $g^{(N+1-m)}h^{(m)}$ for some value of m, which is the same as one of the terms of (1.47) with $n = N + 1$. Hence the general form is correct and all that remains is to show that the corresponding multiplying constants are also correctly given by (1.47).

For any particular value of m there are two terms in (1.48) containing $g^{(N+1-m)}h^{(m)}$ [one arising from $r = m$, the other from $r = m - 1$]. The sum of the two corresponding binomial coefficients can be rewritten as one binomial coefficient as follows,†

† This argument formally fails when $m = 0$ since we have not defined $\binom{N}{-1}$ but the result (1.49) is obviously true for all N if $m = 0$.

$$\binom{N}{m} + \binom{N}{m-1} = \frac{N!}{m!(N-m)!} + \frac{N!}{(m-1)!(N-m+1)!}$$

$$= \frac{N!(N-m+1+m)}{m!(N-m+1)!}$$

$$= \frac{(N+1)!}{m!(N+1-m)!} = \binom{N+1}{m}. \tag{1.49}$$

Thus the term containing $g^{(N+1-m)}h^{(m)}$ in $f^{(N+1)}$ is

$$\binom{N+1}{m} g^{(N+1-m)}h^{(m)}, \tag{1.50}$$

which is exactly what is given by (1.47) if n is set equal to $N + 1$.

Thus the assumption that (1.47) is true for $n = N$ enables us to prove that it is true for $n = N + 1$, and hence, by repeated applications, for any n greater than N. But we have explicitly shown (1.47) is true for $n = 1$ [actually for $n = 2, 3, 4$ as well], and hence it must be true for all n. This general method of proceeding is called the *method of induction*.

As an example we may write down immediately the third derivative of our earlier function $f(x) = x^3 \sin x$ [to begin with, reference to the explicit formula (1.46 c) may be helpful],

$$\frac{d^3f}{dx^3} = 6 \cdot \sin x + 3 \cdot 6x \cdot \cos x + 3 \cdot 3x^2 \cdot (-\sin x) + x^3 \cdot (-\cos x). \tag{1.51}$$

In practice, a convenient way is first to write h and its derivatives [up to the third in this case] in the various terms from left to right, and then g and its derivatives working from right to left, and finally to fill in the binomial coefficients.

▶18. Obtain (1.51) by direct differentiation of $f(x) = x^3 \sin x$, and note the relative amounts of labour involved.

▶19. Find:

(a) the 2nd derivative of $\tan x \sin 2x$,
(b) the 3rd derivative of $\sin x \ln x$,
(c) the 4th derivative of $(2x^3 + 3x^2 + x + 2) \exp(2x)$.

▶20. If $y(x) = x^2 \exp(x)$, show that

$$xy^{(n+1)} + (n - x - 2)y^{(n)} - ny^{(n-1)} = 0.$$

If a function $f(x)$ has derivatives of all orders in some neighbourhood containing the point $x = a$, then, at any point in the neighbourhood, the function can be represented by a series in powers of $x - a$, as follows:

$$f(x) = f(a) + (x - a)f^{(1)}(a) + \cdots + \frac{(x - a)^r}{r!}f^{(r)}(a) + \cdots$$

$$+ \frac{(x - a)^{n-1}}{(n - 1)!}f^{(n-1)}(a) + R_n. \quad (1.52)$$

Here n can have any value and R_n can be shown to have the form

$$R_n = \frac{(x - a)^n}{n!}f^{(n)}(a + \theta_n(x - a)), \quad (1.53)$$

where θ_n is some value which lies in the range $0 < \theta_n < 1$. This expansion is called a **Taylor series** and is very useful for obtaining a simple polynomial representation of a function which is valid near $x = a$. The 'remainder' R_n can be written in a variety of forms, the one used in (1.53) being known as Lagrange's form.

For many functions $R_n \to 0$ as $n \to \infty$, and in these cases the Taylor series can be written as

$$f(x) = \sum_{r=0}^{\infty} \frac{(x - a)^r}{r!}f^{(r)}(a). \quad (1.54\ a)$$

When $a = 0$ this reduces to **Maclaurin's series**

$$f(x) = \sum_{r=0}^{\infty} \frac{x^r}{r!}f^{(r)}(0). \quad (1.54\ b)$$

If such a series expansion is required, the methods of finding derivatives discussed earlier in this section often prove useful.

▶21. Find series expansions for the following functions about the points indicated: (a) $\sin 2x$ about $x = 0$; (b) $\tan x$ about $x = 0$ up to the x^5 term; (c) $\ln x$ about $x = 1$; (d) $\sin (x) \exp (x)$ about $x = 0$ up to the x^5 term.

1.7   Maxima and minima

By reference to its interpretation as the slope of the graph of $f(x)$ against x, when df/dx has the value 0 at some particular value of x, then f has a **stationary value** there. Three such points are indicated in the graph of fig. 1.4, the points B, Q and S.

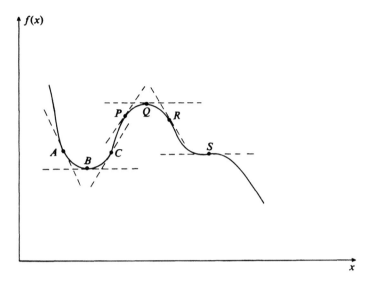

Fig. 1.4 Stationary points of a function of a single variable. A minimum occurs at B, a maximum at Q and a point of inflection at S.

The behaviour of $f(x)$ is different in kind at the points B and Q from that at S. For the former pair the value of f at $B(Q)$ is lower (higher) than that at any other point in the immediate neighbourhood of $B(Q)$, and the function f is said to have a (local) **minimum** at B and a (local) **maximum** at Q. Clearly, for either, a necessary condition is that

$$\frac{df}{dx} = 0. \tag{1.55}$$

At S, although $df/dx = 0$, f has neither a minimum nor a maximum since no neighbourhood about S can be defined in which the value of $f(x)$ is *everywhere* greater than, or *everywhere* less than, the value of f at S. Such a point is called a **point of inflection**.

To distinguish mathematically between minima and maxima we observe that as x increases and f passes through a minimum, the slope df/dx itself changes from a negative value (at A), through zero (at B), to a positive value (at C). Thus in this case df/dx is increasing, which means that its derivative d^2f/dx^2 must be positive. Likewise, at the maximum PQR in the figure we must have $d^2f dx^2$ is negative. It is less obvious, but intuitively reasonable, that at S, $d^2f/dx^2 = 0$.

Such tests will not be adequate for all cases – consider, for example, $f(x) = x^4$ which has a clear minimum at $x = 0$, but has both df/dx and

$d^2f/dx^2 = 0$ there† – but we may summarize the following *sufficient* conditions for maxima and minima of functions with first and second derivatives:

	$\dfrac{df}{dx}$	$\dfrac{d^2f}{dx^2}$	
maximum of $f(x)$	0	< 0	
minimum of $f(x)$	0	> 0	(1.56)

▶22. Find the positions and natures of the stationary points of the following functions: (a) $2x^3 - 3x^2 - 36x + 2$; (b) $\sin(ax)$ with $a > 0$; (c) $x^5 + x^3$; (d) $x^5 - x^3$.

▶23. Find the lowest value taken by the function $3x^4 + 4x^3 - 12x^2 + 6$.

1.8   Partial differentiation

In previous sections the function f depended upon only one variable x and was written $f(x)$. Certain constants and parameters may also have appeared in the definition of f, e.g. $f(x) = ax + 2$ contains the constant 2 and the parameter a, but only x was considered as a variable and only the derivatives $f^n(x) \equiv d^n f/dx^n$ defined.

However we may equally well consider functions which depend on more than one variable, for example, the function $f(x, y) \equiv x^2 + 3xy$ depends upon the two variables x and y [and the constant 3]. For any pair of values (x, y), $f(x, y)$ has a well defined value, e.g. $f(2, 3) = 22$. This notion can clearly be extended to functions dependent on more than two variables and if we wish to discuss one involving a fixed, but otherwise arbitrary, number n of them, we will simply write it as $f(x_i)$, meaning by this

$$f(x_1, x_2, \ldots, x_n). \tag{1.57}$$

When $n = 2$, x_1 and x_2 can be thought of as the x and y used above, but for many purposes a specification of n will not be needed.

Functions of one variable, like $f(x)$, can be represented by a graph on a plane sheet of paper, and it is apparent that functions of two variables can, with more effort, be represented by a surface in three-dimensional

---

† By studying Taylor's expansion about the stationary point, the student will easily verify that if the first derivative not to vanish at the stationary point is $f^{(n)}$, then n even gives a maximum or minimum, and n odd a point of inflection.

space. The analogy between this and physical models of mountains or other geographical features need hardly be mentioned. For functions of more than two variables such representations are not available; however, the mathematical properties and procedures are simply the natural extensions of those for the one- and two-variable cases. For the purposes of visualizing the physical interpretation of the mathematics to be developed, the reader is advised to think of the two-variable case even where an unspecified n-variable one is being discussed.

Following on from the work of earlier sections, derivatives of functions of several variables can be defined and studied. A lead as to how this may be done can be obtained by considering the role played by parameters in functions of a single variable x; although of unspecified value, they are treated as constants when derivatives with respect to x are found. From this it is only a small step to defining the derivative with respect to x of a function $f(x, y)$ of two variables, by saying that it is that for a one-variable function when y is held fixed and treated as a constant.

To signify that the derivative is with respect to x, but at the same time recognize that another derivative with respect to y (with x held constant) exists, it is denoted *not* by df/dx but by $\partial f/\partial x$, and called the **partial derivative** or **partial differential coefficient** of f with respect to x. To define it formally along the lines of (1.38) we have

$$\frac{\partial f(x, y)}{\partial x} \equiv \lim_{\Delta x \to 0} \frac{f(x + \Delta x, y) - f(x, y)}{\Delta x}, \tag{1.58 a}$$

provided that the limit exists.

The other derivative mentioned for the function $f(x, y)$ of the two variables x and y is $\partial f/\partial y$ defined as the limit (if it exists)

$$\frac{\partial f(x, y)}{\partial y} \equiv \lim_{\Delta y \to 0} \frac{f(x, y + \Delta y) - f(x, y)}{\Delta y}. \tag{1.58 b}$$

It is obvious that $\partial f/\partial x$ and $\partial f/\partial y$ may both depend on both x and y, and in general will do so. For the very simple example quoted earlier $f(x, y) = x^2 + 3xy$, we have directly $\partial f/\partial x = 2x + 3y$ and $\partial f/\partial y = 3x$.

It is a common practice in connection with partial derivatives of functions involving more than one variable, to indicate those which are held constant by writing them as subscripts to the derivative symbol. Thus the quantities defined by (1.58 a) and (1.58 b) would be written (respectively) as

$$\left(\frac{\partial f}{\partial x}\right)_y \quad \text{and} \quad \left(\frac{\partial f}{\partial y}\right)_x. \tag{1.59}$$

The extension of these definitions to the general n-variable case is straightforward and can be formally written

$$\frac{\partial f(x_i)}{\partial x_k} = \lim_{\Delta x_k \to 0} \frac{1}{\Delta x_k} \{ f(x_1, x_2, \ldots, x_k + \Delta x_k, \ldots, x_n)$$
$$- f(x_1, x_2, \ldots, x_k, \ldots, x_n)\}, \quad (1.60)$$

provided the limit exists. For the purposes of illustration in this introductory chapter, only two-variable functions will normally be considered, but use is made of the more general notation in some later chapters.

In terms of a three-dimensional model the partial derivatives $\partial f/\partial x$ and $\partial f/\partial y$ are easily visualized, representing as they do the 'slopes' or 'rates of change with distance' of the function, when moving parallel to the x- and y-axes respectively (in the positive senses).

Just as for one-variable functions, higher derivatives may be defined. For example, from $\partial f/\partial x$ a second derivative with respect to x, $\partial^2 f/\partial x^2$, may be found, or equally validly the derivative of $\partial f/\partial x$ with respect to y. This latter (which of course will not normally be equal to $\partial^2 f/\partial x^2$) is denoted by

$$\frac{\partial^2 f}{\partial y \, \partial x}, \quad \text{meaning} \quad \frac{\partial}{\partial y} \left(\frac{\partial f}{\partial x} \right). \quad (1.61)$$

For most functions with which we will have to deal,

$$\frac{\partial^2 f}{\partial y \, \partial x} = \frac{\partial^2 f}{\partial x \, \partial y}, \quad (1.62)$$

continuity of the derivatives being a sufficient condition to ensure this.

As simple but concrete examples using the same function $f(x, y) = x^2 + 3xy$ as previously, we have already noted that

$$\frac{\partial f}{\partial x} = 2x + 3y, \qquad \frac{\partial f}{\partial y} = 3x,$$

and so further obtain that

$$\frac{\partial^2 f}{\partial x^2} = 2, \qquad \frac{\partial^2 f}{\partial y \, \partial x} = 3, \qquad \frac{\partial^2 f}{\partial x \, \partial y} = 3, \qquad \frac{\partial^2 f}{\partial y^2} = 0.$$

▶24. (i) Find all the first partial derivatives of the following functions:
(a) $x^2 y$, (b) $x^2 + y^2 + 4$, (c) $\sin (x/y)$, (d) $\arctan (y/x)$, (e) $r(x, y, z) = (x^2 + y^2 + z^2)^{1/2}$.
(ii) For (a), (b) and (e), find $\partial^2 f/\partial x^2$, $\partial^2 f/\partial y^2$, $\partial^2 f/\partial x \, \partial y$.
(iii) For (d), say, verify that $\partial^2 f/\partial x \, \partial y = \partial^2 f/\partial y \, \partial x$.

Having defined partial derivatives for functions of more than one variable, we must next connect them with the changes which occur in the value of the function when all of the variables are changed at the same time. Again for discussion purposes we consider a function $f(x, y)$ of two variables x and y.

Suppose that finite changes Δx and Δy are made in x and y, and as a result f changes to $f + \Delta f$. Then we must have

$$
\begin{aligned}
\Delta f &= f(x + \Delta x, y + \Delta y) - f(x, y) \\
&= f(x + \Delta x, y + \Delta y) - f(x, y + \Delta y) + f(x, y + \Delta y) - f(x, y) \\
&= \left[\frac{f(x + \Delta x, y + \Delta y) - f(x, y + \Delta y)}{\Delta x} \right] \Delta x \\
&\quad + \left[\frac{f(x, y + \Delta y) - f(x, y)}{\Delta y} \right] \Delta y . \quad (1.63)
\end{aligned}
$$

In line (1.63) we note that the quantities in square brackets are very similar to those involved in the definitions of partial derivatives. For them to be strictly equal to the partial derivatives, Δx and Δy would need to be infinitesimally small. But even for finite (but not too large) Δx and Δy an approximate formula [useful in estimating errors in f knowing those in measured quantities x and y, for example]

$$
\Delta f \approx \frac{\partial f(x, y)}{\partial x} \Delta x + \frac{\partial f(x, y)}{\partial y} \Delta y \quad (1.64)
$$

can be obtained. It will be noticed that the first square bracket of (1.63) actually approximates to $\partial f(x, y + \Delta y)/\partial x$, but that this has been replaced by $\partial f(x, y)/\partial x$ in (1.64). This approximation clearly has the same degree of validity as that which replaces the square bracket by the partial derivative, has itself.

How valid an approximation (1.64) is to (1.63) depends not only on how small Δx and Δy are, but also upon the magnitudes of higher partial derivatives. This can be seen, for example, by treating x as fixed in the second square bracket of (1.63) and expanding the bracket by a Taylor series in Δy to give the exact equation,

$$
\left[\frac{f(x, y + \Delta y) - f(x, y)}{\Delta y} \right] = \left(\frac{\partial f}{\partial y} \right)_x + \frac{1}{2!} \left(\frac{\partial^2 f}{\partial y^2} \right)_x \Delta y + \cdots .
\quad (1.65)
$$

Approximation (1.64) corresponds to omitting all but the first term on the right-hand side.

It is perhaps appropriate to mention here the analogues of Taylor's and Maclaurin's series for functions of more than one variable. For our standard two-variable function, Taylor's expansion about the point (x_0, y_0) is

$$f(x, y) = f(x_0, y_0) + (x - x_0)\frac{\partial f}{\partial x} + (y - y_0)\frac{\partial f}{\partial y}$$

$$+ \frac{1}{2!}\left[(x - x_0)^2\frac{\partial^2 f}{\partial x^2} + 2(x - x_0)(y - y_0)\frac{\partial^2 f}{\partial x \, \partial y}\right.$$

$$\left. + (y - y_0)^2\frac{\partial^2 f}{\partial y^2}\right] + \cdots, \tag{1.66}$$

where all the derivatives are evaluated at (x_0, y_0). Maclaurin's series is obtained by putting $x_0 = y_0 = 0$. For functions of more than two variables the corresponding result is mentioned in the next section.

When we later come to discuss integrals and integration, we will need to consider expressions such as dx, dy and df which are known as **differentials** and represent arbitrarily small quantities, not just approximately but in principle. From what has been said previously it is apparent that for the function $f = f(x, y)$ they are exactly related by

$$df = \left(\frac{\partial f}{\partial x}\right)dx + \left(\frac{\partial f}{\partial y}\right)dy. \tag{1.67}$$

The left-hand side, df, of (1.67) is called the **total differential** of f.

If f were a function of n variables and denoted by $f(x_i)$ as in (1.57), then the corresponding expression for df would be

$$df = \left(\frac{\partial f}{\partial x_1}\right)dx_1 + \left(\frac{\partial f}{\partial x_2}\right)dx_2 + \cdots + \left(\frac{\partial f}{\partial x_n}\right)dx_n. \tag{1.68}$$

In some situations, despite the fact that several variables x_i appear to be involved, effectively only one of them is. This occurs if there are subsidiary relationships constraining all the x_i to have values dependent on the value of one of them, say x_1. These relationships are represented by equations, typically of the form

$$x_j = g_j(x_1), \qquad j = 2, 3, \ldots, n, \tag{1.69}$$

where the jth function g_j gives the dependence of x_j on x_1. In principle f can then be expressed as a function of x_1 alone by substituting from (1.69) for $x_2, x_3, \ldots, x_n$, and then the total derivative [or simply the derivative] of f with respect to x_1 obtained by ordinary differentiation.

Alternatively (1.68) can be used to give

$$\frac{df}{dx_1} = \left(\frac{\partial f}{\partial x_1}\right)\frac{dx_1}{dx_1} + \left(\frac{\partial f}{\partial x_2}\right)\frac{dx_2}{dx_1} + \cdots + \left(\frac{\partial f}{\partial x_n}\right)\frac{dx_n}{dx_1}, \tag{1.70}$$

or, including the functions g_j more explicitly,

$$\frac{df}{dx_1} = \frac{\partial f}{\partial x_1} + \frac{\partial f}{\partial x_2}\frac{dg_2}{dx_1} + \cdots + \frac{\partial f}{\partial x_n}\frac{dg_n}{dx_1}. \tag{1.71}$$

It should be noticed that the left-hand side of this equation is the total derivative df/dx_1, whilst the partial derivative $\partial f/\partial x_1$ forms only a part of the right-hand side. In evaluating this partial derivative only *explicit* appearances of x_1 in the function f must be taken account of, and *no* allowance must be made for the knowledge that 'as x_1 is changed, this necessarily changes x_2, and x_2 appears in f'. This latter contribution is taken care of by the remaining terms on the right-hand side of (1.71).

Naturally what has been shown using x_1 in the above, applies equally well to any other of the x_j, with the appropriate consequent changes.

Finally, to illustrate what has been said with a simple transparent example, suppose that for our earlier function $f(x, y) = x^2 + 3xy$, y is constrained to be $y = \arcsin x$ and we require to find df/dx. To evaluate (1.71) [here $n = 2$, $x_1 = x$, and $x_2 = y$] we need

$$\frac{\partial f}{\partial x} = 2x + 3y, \qquad \frac{\partial f}{\partial y} = 3x, \qquad \frac{dy}{dx} = \frac{1}{(1 - x^2)^{1/2}},$$

yielding

$$\frac{df}{dx} = 2x + 3y + 3x\frac{1}{(1 - x^2)^{1/2}}$$

$$= 2x + 3\arcsin x + \frac{3x}{(1 - x^2)^{1/2}}.$$

►25. Show that the same result is obtained if $y = \arcsin x$ is substituted in $f(x, y)$ before obtaining df/dx by 'one-variable differentiation'.

►26. Obtain df/dy using each of the two methods.

►27. Find df/dx if $f = (x^2 + y^2 + z^2)^{1/2}$, where x, y and z are given by $x = \sec^2 y$ and $z = xz^2 + x^2$.

1.9 The summation convention

In the previous section it was apparent that series of terms such as those in (1.68) and (1.70) are cumbersome to write out explicitly. They can be shortened by the use of the summation sign $\sum$; for example, (1.68) could be written

$$df = \sum_{i=1}^{n}\left(\frac{\partial f}{\partial x_i}\right)dx_i. \tag{1.72}$$

But by the use of the **summation convention**, such expressions can be compacted still further. The convention is:

> In any expression containing subscripted (or superscripted) variables, any lower-case alphabetic subscript (or superscript) which appears twice and only twice in any term of the expression is assumed to be summed over (unless the contrary is stated).

Thus a statement

$$df = \frac{\partial f}{\partial x_i} \, dx_i, \tag{1.73}$$

is completely equivalent to (1.68). Equally (1.70) can be replaced by

$$\frac{df}{dx_1} = \frac{\partial f}{\partial x_i} \frac{dx_i}{dx_1},$$

and the expression for the total derivative of f with respect to any arbitrary x_j written as

$$\frac{df}{dx_j} = \frac{\partial f}{\partial x_i} \frac{dx_i}{dx_j}. \tag{1.74}$$

In this last expression it should be noted that summation over values from 1 to n applies to i only. Since j appears only once on the right-hand side, no summation with respect to it is to be carried out. This is in line with the fact that j appears on the left-hand side but i does not.

The general form of Taylor series for a function of n variables x_i can be conveniently written using the summation convention. For an expansion about the point X_i

$$f(x_i) = f(X_i) + (x_j - X_j) \frac{\partial f}{\partial x_j}$$
$$+ \frac{(x_k - X_k)(x_l - X_l)}{2!} \frac{\partial^2 f}{\partial x_k \, \partial x_l} + \cdots. \tag{1.75}$$

We have here deliberately used different subscript letters $j, k, l, \ldots$ in each of the implied summations to emphasize which factors are multiplied together in each summation. Of course all the subscripts run over the range 1 to n inclusive and, for example, x_3 is the same quantity whether it arises from x_j with $j = 3$ or x_l with $l = 3$. All the derivatives are evaluated at the point X_i.

Although the convention is introduced here in the context of partial differentiation, it is adopted in far more general circumstances. In later chapters expressions such as:

(i) $a_1b_1 + a_2b_2 + \cdots + a_nb_n$,

(ii) $a_{i1}b_{1k} + a_{i2}b_{2k} + \cdots + a_{in}b_{nk}$,

(iii) $a_{11} + a_{22} + \cdots + a_{nn}$,

will appear, particularly in connection with vectors, tensors and matrices. All three can be expressed in substantially shorter forms by using the summation convention thus,

(i) a_ib_i,   (ii) $a_{ij}b_{jk}$,   (iii) a_{ii}.

This preliminary mention of the summation convention should prepare the reader for its more regular use in subsequent chapters, although on the first few occasions it appears, a brief reminder of its meaning is given.

▶28. Use the Taylor series (1.75) to find a polynomial expansion, up to quadratic terms in $x - \pi/4$, y and $z - 1$, of the function $\sin(x + yz)$ about the point $x = \pi/4$, $y = 0$, $z = 1$.

1.10   Change of variables

It is sometimes necessary or desirable to make a change of coordinate system during the course of an analysis, and consequently to have to change an equation expressed in one set of coordinates into an equation using another set. Effectively the same situation arises if a function f depends upon one set of variables x_i, so that $f = f(x_i)$, but the x_i are given in terms of a further set of variables y_j by equations of the form

$$x_i = x_i(y_j). \tag{1.76}$$

For each value of i, the function x_i on the right of this equation will be a different function of $y_1, y_2, \ldots$. The two subscripts i and j need not run over the same range, but if both the x's and y's are sets of independent variables they will do so.

In this section the behaviour of derivatives under changes of coordinates is considered. Differentials are treated later in connection with integration.

The simplest case occurs when there is only one y upon which each of the x_i depends. Perhaps the most common physical situation is that in which the one y represents time [then usually denoted by t of course] and the x_i are the spatial coordinates of a point moving along a prescribed track. With only one y the total derivative df/dy becomes directly meaningful and is given by

$$\frac{df}{dy} = \frac{\partial f}{\partial x_i}\frac{dx_i}{dy} \quad \text{(summation convention)}. \tag{1.77}$$

This can be obtained manipulatively by dividing (1.68) through by dy, and a proper formal proof follows closely the lines adopted in the next few paragraphs.

The next situation to consider is that in which more than one y is involved. For the function $f(x_i)$ where $x_i = x_i(y_j)$ as discussed above, we determine the partial derivative of f with respect to one of the y's, which we denote by y_j. It is clear that f varies as y_j is changed, but because the link is through all of the x_i (y_j effecting each x_i, and f dependent on all x_i) the connection is an indirect one.

Suppose y_j increases to $y_j + \Delta y$, then each of the x_i will change by an amount

$$\Delta x_i = \frac{\partial x_i}{\partial y_j} \Delta y + O\{(\Delta y)^2\}. \tag{1.78}$$

That is Δx_i is given only approximately by $(\partial x_i/\partial y_j)\,\Delta y$ for small but finite Δy, but as $\Delta y \to 0$, (1.78) gives the ratio of Δx_i to Δy exactly. As a result of these changes, in each of the x_i, f changes its value from $f(x_i)$ to $f(x_i + \Delta x_i)$, i.e. to $f(x_1 + \Delta x_1, x_2 + \Delta x_2, \ldots, x_n + \Delta x_n)$. We are thus interested in the limiting value of the ratio

$$\frac{f(x_i + \Delta x_i) - f(x_i)}{\Delta y}.$$

The numerator of this can be expanded by a Taylor expansion (as in (1.75)) as

$$f(x_i + \Delta x_i) - f(x_i) = \frac{\partial f}{\partial x_i} \Delta x_i + O\{(\Delta x_i)^2\}. \tag{1.79}$$

[In obtaining this from (1.75), X_i and x_i of (1.75) have been replaced by x_i and $x_i + \Delta x_i$ respectively. The summation convention applies to the repeated subscript in the first term on the right-hand side.]

Next, substitution from (1.78) for Δx_i and division all through by Δy gives

$$\frac{f(x_i + \Delta x_i) - f(x_i)}{\Delta y} = \frac{\partial f}{\partial x_i} \frac{\partial x_i}{\partial y_j} + O(\Delta y). \tag{1.80}$$

[Since, as shown by (1.78), Δx_i and Δy are of the same order of smallness, the $O\{(\Delta x_i)^2\}$ of (1.79) can be considered as being replaced by $O\{(\Delta y)^2\}$.] Finally, letting Δy tend to zero yields the required expression,

$$\frac{\partial f}{\partial y_j} = \lim_{\Delta y \to 0} \frac{f(x_i + \Delta x_i) - f(x_i)}{\Delta y} = \frac{\partial f}{\partial x_i} \frac{\partial x_i}{\partial y_j}. \tag{1.81}$$

This result is known as the **chain rule** for partial differentiation and is the analogue of that for ordinary (or total) derivatives, to which it reduces if there is only one x and only one y,

$$\frac{df}{dy} = \frac{df}{dx}\frac{dx}{dy}. \tag{1.82}$$

▶29. If the position (x, y, z) of a particle at time t is given by

$$x = 2t^2, \qquad y = \sin 2t, \qquad z = \exp(-t),$$

find the rate of change of radial distance r of the particle at time t.

In deriving result (1.81) no assumptions have been made about f and so the chain rule is really a property of the operation of differentiation. That is, it could equally well be written

$$\frac{\partial}{\partial y_j} = \frac{\partial x_i}{\partial y_j}\frac{\partial}{\partial x_i}, \tag{1.83}$$

which in turn can be put into words as 'to change from the independent variables y_j to the independent variables x_i, replace each partial derivative with respect to y_j by $(\partial x_i/\partial y_j)$ times the partial derivative with respect to x_i and add the contributions for all i together'.

As an example of these procedures in practice, consider the following.

Example 1.1. Plane cylindrical polar coordinates r and ϕ are given in terms of Cartesian coordinates x and y by

$$x = r\cos\phi, \qquad y = r\sin\phi,$$

as in fig. 1.5. For an arbitrary function $f(x, y)$, which when put into (r, ϕ) coordinates is $g(r, \phi)$, transform the expression

$$\frac{\partial^2 f}{\partial x^2} + \frac{\partial^2 f}{\partial y^2},$$

into one in r and ϕ.

To identify this problem in terms of the previous notation, we have

$$y_1 \triangleq x, \qquad y_2 \triangleq y, \qquad x_1 \triangleq r, \qquad x_2 \triangleq \phi,$$

and

$$r^2 = x^2 + y^2, \qquad \phi = \arctan(y/x).$$

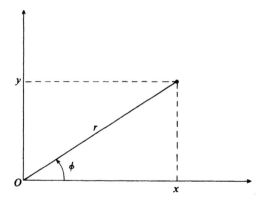

Fig. 1.5   The connection between Cartesian and polar coordinates in two dimensions.

The four partial derivatives needed are:

$$\frac{\partial r}{\partial x} = \frac{x}{(x^2 + y^2)^{1/2}} = \cos\phi, \qquad \frac{\partial\phi}{\partial x} = \frac{-(y/x^2)}{1 + (y/x)^2} = -\frac{\sin\phi}{r},$$

$$\frac{\partial r}{\partial y} = \frac{y}{(x^2 + y^2)^{1/2}} = \sin\phi, \qquad \frac{\partial\phi}{\partial y} = \frac{(1/x)}{1 + (y/x)^2} = \frac{\cos\phi}{r}.$$

Thus from (1.83) we may write

$$\frac{\partial}{\partial x} = \cos\phi\,\frac{\partial}{\partial r} - \frac{\sin\phi}{r}\frac{\partial}{\partial\phi}, \tag{1.84}$$

and

$$\frac{\partial}{\partial y} = \sin\phi\,\frac{\partial}{\partial r} + \frac{\cos\phi}{r}\frac{\partial}{\partial\phi}. \tag{1.85}$$

Now it is only a matter of writing

$$\frac{\partial^2 f}{\partial x^2} = \left(\cos\phi\,\frac{\partial}{\partial r} - \sin\phi\,\frac{\partial}{\partial\phi}\right)\left(\cos\phi\,\frac{\partial}{\partial r} - \frac{\sin\phi}{r}\frac{\partial}{\partial\phi}\right)g$$

$$= \cos\phi\,\frac{\partial}{\partial r}\left(\cos\phi\,\frac{\partial g}{\partial r} - \frac{\sin\phi}{r}\frac{\partial g}{\partial\phi}\right)$$

$$\qquad\qquad - \frac{\sin\phi}{r}\frac{\partial}{\partial\phi}\left(\cos\phi\,\frac{\partial g}{\partial r} - \frac{\sin\phi}{r}\frac{\partial g}{\partial\phi}\right),$$

and a similar expression for $\partial^2 f/\partial y^2$,

$$\frac{\partial^2 f}{\partial y^2} = \sin\phi\,\frac{\partial}{\partial r}\left(\sin\phi\,\frac{\partial g}{\partial r} + \frac{\cos\phi}{r}\frac{\partial g}{\partial\phi}\right)$$

$$\qquad\qquad + \frac{\cos\phi}{r}\frac{\partial}{\partial\phi}\left(\sin\phi\,\frac{\partial g}{\partial r} + \frac{\cos\phi}{r}\frac{\partial g}{\partial\phi}\right).$$

When these two expressions are added together and the slightly lengthy but straightforward differentiations carried out (an exercise left to the reader), the change of variables is completed.

▶30. $$\frac{\partial^2 f}{\partial x^2} + \frac{\partial^2 f}{\partial y^2} = \frac{\partial^2 g}{\partial r^2} + \frac{1}{r}\frac{\partial g}{\partial r} + \frac{1}{r^2}\frac{\partial^2 g}{\partial \phi^2}. \tag{1.86}$$

Although we have referred to f as $f(x, y)$ and to g as $g(r, \phi)$, both functions in fact represent the same quantity in the two-dimensional space, even though f and g will have quite different algebraic forms. For this reason, it is not uncommon to leave the name of the function unchanged and assume that it is expressed in an algebraic form suitable to the co-ordinate system currently being employed.

▶31. Express $\partial^2 f/\partial x\, \partial y$ in plane cylindrical polar coordinates.

▶32. New coordinates ξ and η are defined in terms of two-dimensional Cartesians x and y by

$$\xi = x + y, \qquad \eta = x - y.$$

Express the equation

$$\frac{\partial^2 f}{\partial x^2} - \frac{\partial^2 f}{\partial y^2} + \frac{\partial f}{\partial x} + \frac{\partial f}{\partial y} = 0$$

in the new coordinates.

1.11   Dependent functions

In obtaining (1.67) we assumed that f was a function of x and y, but equally we could have said that y was a function of f and x, $y = y(f, x)$, since if the latter two are given, y is determined. Equally $x = x(f, y)$. To emphasize this symmetry of relationship let us replace f by z, without in any way intending to imply that any or all of x, y and z are coordinate positions.

Writing (1.67) in terms of x, y, z we have

$$dz = \left(\frac{\partial z}{\partial x}\right)_y dx + \left(\frac{\partial z}{\partial y}\right)_x dy, \tag{1.87 a}$$

and equally

$$dx = \left(\frac{\partial x}{\partial y}\right)_z dy + \left(\frac{\partial x}{\partial z}\right)_y dz. \tag{1.87 b}$$

In equations (1.87) the subscript showing which variable is held fixed in each partial differentiation is explicitly attached to the corresponding partial derivative. Substitution for dx from (1.87 b) into (1.87 a) gives

$$dz = \left(\frac{\partial z}{\partial x}\right)_y \left(\frac{\partial x}{\partial y}\right)_z dy + \left(\frac{\partial z}{\partial x}\right)_y \left(\frac{\partial x}{\partial z}\right)_y dz + \left(\frac{\partial z}{\partial y}\right)_x dy. \qquad (1.88)$$

Now it is clear that if $(\partial z/\partial x)_y$ exists and is not equal to zero, then $(\partial x/\partial z)_y$ also exists and is equal to the reciprocal of $(\partial z/\partial x)_y$

$$\left(\frac{\partial x}{\partial z}\right)_y = 1 \bigg/ \left(\frac{\partial z}{\partial x}\right)_y . \qquad (1.89)$$

Of course for equations like (1.89) to be valid, the same quantities must be held constant in the two derivatives [here y in both].

Using this general result shows that the terms in (1.88) involving dz cancel and hence that

$$\left(\frac{\partial z}{\partial x}\right)_y \left(\frac{\partial x}{\partial y}\right)_z + \left(\frac{\partial z}{\partial y}\right)_x = 0. \qquad (1.90)$$

By using (1.89) again to write the second term of this as $(\partial y/\partial z)_x^{-1}$, the result may be written in the more symmetrical form

$$\left(\frac{\partial z}{\partial x}\right)_y \left(\frac{\partial x}{\partial y}\right)_z \left(\frac{\partial y}{\partial z}\right)_x = -1, \qquad (1.91)$$

provided none of the derivatives vanish.

▶33. Show that

$$\left(\frac{\partial z}{\partial y}\right)_x = - \left(\frac{\partial z}{\partial x}\right)_y \bigg/ \left(\frac{\partial y}{\partial x}\right)_z . \qquad (1.92)$$

▶34. Verify explicitly the general result (1.91) for the special cases (i) $z^2 = x^2 + y^2$, (ii) $z = \arctan(x/y)$.

1.12   Maxwell's relations and Jacobians

To illustrate some of the ideas of the previous section, **Maxwell's thermodynamic relations** will now be obtained. They express relationships between four thermodynamic quantities describing unit amount of a substance. The quantities are p the pressure, V the volume, T the thermodynamic temperature and S the entropy of the substance. These four quantities are not independent, only two of them being independently variable. For the moment we will not specify which, but merely denote them by x and y.

The first law of thermodynamics may be expressed as

$$dU = T\,dS - p\,dV, \tag{1.93}$$

where U is the internal energy of the substance. Comparing this with (1.68) and (1.70) we see that we may write

$$\left(\frac{\partial U}{\partial x}\right)_y = T\left(\frac{\partial S}{\partial x}\right)_y - p\left(\frac{\partial V}{\partial x}\right)_y \tag{1.94 a}$$

and

$$\left(\frac{\partial U}{\partial y}\right)_x = T\left(\frac{\partial S}{\partial y}\right)_x - p\left(\frac{\partial V}{\partial y}\right)_x. \tag{1.94 b}$$

Next (1.94 a) is differentiated with respect to y with x constant and (1.94 b) with respect to x with y constant, and the two results equated (since $\partial^2 U/\partial y\,\partial x = \partial^2 U/\partial x\,\partial y$). After cancelling $T\,\partial^2 S/\partial y\,\partial x - p\,\partial^2 V/\partial y\,\partial x$ from each side we obtain the relationship

$$\left(\frac{\partial T}{\partial y}\right)_x\left(\frac{\partial S}{\partial x}\right)_y - \left(\frac{\partial p}{\partial y}\right)_x\left(\frac{\partial V}{\partial x}\right)_y = \left(\frac{\partial T}{\partial x}\right)_y\left(\frac{\partial S}{\partial y}\right)_x - \left(\frac{\partial p}{\partial x}\right)_y\left(\frac{\partial V}{\partial y}\right)_x. \tag{1.95}$$

By substitution of various pairs of quantities chosen from among p, V, T and S, for x and y, Maxwell's equations may be obtained. For example, taking x as S and y as V yields

$$\left(\frac{\partial T}{\partial V}\right)_S \times 1 - \left(\frac{\partial p}{\partial V}\right)_S \times 0 = \left(\frac{\partial T}{\partial S}\right)_V \times 0 - \left(\frac{\partial p}{\partial S}\right)_V \times 1,$$

or

$$\left(\frac{\partial T}{\partial V}\right)_S = -\left(\frac{\partial p}{\partial S}\right)_V. \tag{1.96}$$

▶35. Prove the three other common Maxwell relations

(i) $\left(\dfrac{\partial T}{\partial p}\right)_S = \left(\dfrac{\partial V}{\partial S}\right)_p$,   (ii) $\left(\dfrac{\partial V}{\partial T}\right)_p = -\left(\dfrac{\partial S}{\partial p}\right)_T$,

(iii) $\left(\dfrac{\partial S}{\partial V}\right)_T = \left(\dfrac{\partial p}{\partial T}\right)_V$.

Equation (1.95) can be rearranged to read

$$\left(\frac{\partial T}{\partial y}\right)_x\left(\frac{\partial S}{\partial x}\right)_y - \left(\frac{\partial T}{\partial x}\right)_y\left(\frac{\partial S}{\partial y}\right)_x = \left(\frac{\partial p}{\partial y}\right)_x\left(\frac{\partial V}{\partial x}\right)_y - \left(\frac{\partial p}{\partial x}\right)_y\left(\frac{\partial V}{\partial y}\right)_x. \tag{1.97}$$

The reader who has an acquaintance with determinants will notice that both sides of (1.97) have the appearance of being the expansion of a 2×2 determinant. [Determinants and matrices are discussed at length in chapter 14.] The left- and right-hand sides have respectively the forms

$$
\begin{vmatrix} \left(\dfrac{\partial S}{\partial x}\right)_y & \left(\dfrac{\partial S}{\partial y}\right)_x \\[2mm] \left(\dfrac{\partial T}{\partial x}\right)_y & \left(\dfrac{\partial T}{\partial y}\right)_x \end{vmatrix} , \quad \begin{vmatrix} \left(\dfrac{\partial V}{\partial x}\right)_y & \left(\dfrac{\partial V}{\partial y}\right)_x \\[2mm] \left(\dfrac{\partial p}{\partial x}\right)_y & \left(\dfrac{\partial p}{\partial y}\right)_x \end{vmatrix} . \tag{1.98}
$$

Such forms occur sufficiently frequently in the theory of partial differentiation for them to be denoted by a special symbol defined by

$$
\frac{\partial(f, g, h, \ldots)}{\partial(x, y, z, \ldots)} \equiv \begin{vmatrix} \dfrac{\partial f}{\partial x} & \dfrac{\partial f}{\partial y} & \dfrac{\partial f}{\partial z} & \cdots \\[2mm] \dfrac{\partial g}{\partial x} & \dfrac{\partial g}{\partial y} & \dfrac{\partial g}{\partial z} & \cdots \\[2mm] \dfrac{\partial h}{\partial x} & \dfrac{\partial h}{\partial y} & \dfrac{\partial h}{\partial z} & \cdots \\[2mm] \vdots & \vdots & \vdots & \end{vmatrix} . \tag{1.99}
$$

The determinant in (1.99) is known as the **Jacobian** of $f, g, h, \ldots$ with respect to $x, y, z, \ldots$. With this notation the two forms in (1.98) can be written as Jacobians and (1.97) expressed by

$$
\frac{\partial(S, T)}{\partial(x, y)} = \frac{\partial(V, p)}{\partial(x, y)} . \tag{1.100}
$$

With a little more obscurity but compensating compactness, the Jacobian of a set of functions f_i $(i = 1, 2, \ldots, n)$ with respect to variables x_j $(j = 1, 2, \ldots, n)$ can be described as the determinant of the matrix whose (i, j)th element is $\partial f_i / \partial x_j$, and denoted by

$$
\frac{\partial(f_i)}{\partial(x_j)} \equiv \frac{\partial(f_1, f_2, \ldots, f_n)}{\partial(x_1, x_2, \ldots, x_n)} . \tag{1.101}
$$

If f and g are two functions of variables u and v, and u and v are in turn functions of x and y, then by a direct but slightly lengthy expansion it can be shown that the relevant Jacobians satisfy

$$
\frac{\partial(f, g)}{\partial(u, v)} \frac{\partial(u, v)}{\partial(x, y)} = \frac{\partial(f, g)}{\partial(x, y)} . \tag{1.102}
$$

This result has obvious extensions to sets of n functions and more than one intermediate set.

For readers who already have sufficient familiarity with matrices and their properties, a fairly compact proof for sets of n functions of the result corresponding to (1.102) can be given as follows. Other readers may turn straight to the results (1.106 b) and (1.107 b) and return to the proof at some later time.

Consider three sets of variables x_i, y_i and z_i, with i running from 1 to n for each set, and denote by $M(XY)$ the matrix whose (i, j)th element is $\partial x_i/\partial y_j$. From (1.73) applied twice

$$\mathrm{d}x_i = \frac{\partial x_i}{\partial y_j}\,\mathrm{d}y_j = \frac{\partial x_i}{\partial y_j}\frac{\partial y_j}{\partial z_k}\,\mathrm{d}z_k. \tag{1.103}$$

Thus

$$\frac{\partial x_i}{\partial z_k} = \frac{\partial x_i}{\partial y_j}\frac{\partial y_j}{\partial z_k}, \quad \text{[a chain rule]}$$

or

$$[M(XZ)]_{ik} = [M(XY)]_{ij}[M(YZ)]_{jk}, \tag{1.104}$$

which is the statement that the matrix $M(XZ)$ is the matrix product of matrices $M(XY)$ and $M(YZ)$, in that order.

We now use the general result for determinants of product matrices that $|AB| = |A| \times |B|$ and recall that the Jacobian

$$\frac{\partial(x_i)}{\partial(y_j)} = J_{xy} = |M(XY)|; \tag{1.105}$$

on taking the determinant of (1.104) we obtain

$$J_{xz} = J_{xy}J_{yz}. \tag{1.106 a}$$

As a special case, if set z_i is taken identical to set x_i and the obvious result $J_{xx} = 1$ used, we obtain

$$J_{xy}J_{yx} = 1. \tag{1.107 a}$$

Written in the usual notation, these two results are

$$\frac{\partial(x_1, x_2, \ldots, x_n)}{\partial(z_1, z_2, \ldots, z_n)} = \frac{\partial(x_1, x_2, \ldots, x_n)}{\partial(y_1, y_2, \ldots, y_n)} \cdot \frac{\partial(y_1, y_2, \cdots, y_n)}{\partial(z_1, z_2, \ldots, z_n)}, \tag{1.106 b}$$

and

$$\frac{\partial(x_1, x_2, \ldots, x_n)}{\partial(y_1, y_2, \ldots, y_n)} = \left[\frac{\partial(y_1, y_2, \ldots, y_n)}{\partial(x_1, x_2, \ldots, x_n)}\right]^{-1}. \tag{1.107 b}$$

The similarity between the properties of Jacobians and those of derivatives is apparent, and to some extent is suggested by the notation.

►36. Use (1.106 b) and (1.107 b) to show that result (1.100) can be written as

$$\frac{\partial(S, T)}{\partial(V, p)} = 1.$$

1.13   Stationary values for functions of several variables

Just as for functions of a single variable, we may seek values of the variables x_i upon which a function $f(x_i)$ depends, that give f a (local) maximum or minimum. However the general situation is more complex than that for a function of one variable as may be seen from fig. 1.6.

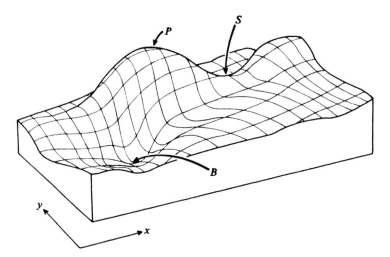

Fig. 1.6   Stationary points of a function of two variables. A minimum occurs at B, a maximum at P and a saddle point at S.

This figure gives a rough sketch of part of a three-dimensional model of a function $f(x, y)$. At positions P and B there are a peak and a bowl respectively – or, more mathematically, a local maximum and a local minimum. At position S the situation is more complicated, since a section parallel to the plane $x = 0$ would show a maximum, but one parallel to the plane $y = 0$ would show a minimum. A point such as S is known as a *saddle point*. The orientation of the 'saddle' in the xy-plane is irrelevant; it is as shown in the diagram solely for ease of discussion.

For a function of two variables, such as the one shown, it is geometrically obvious that a necessary condition for a stationary point (maximum, minimum or saddle point) to occur is that

$$\frac{\partial f}{\partial x} = 0 \quad \text{and} \quad \frac{\partial f}{\partial y} = 0. \tag{1.108}$$

The vanishing of the partial derivatives in directions parallel to the axes is enough to ensure that the partial derivative in any other direction does so too, since any such derivative can be resolved along the directions of the two axes.

A more difficult criterion to formulate mathematically is that which determines the nature – maximum, minimum or saddle point – of the stationary point. To do so we may employ the Taylor expansion (1.75) about the point (x_0, y_0) keeping only the first non-vanishing terms (in addition to $f(x_0, y_0)$), since they can always be made to dominate later ones by making $x - x_0$ and $y - y_0$ small enough. The second and third terms of (1.75) vanish on account of (1.108) and so we are left with

$$f(x, y) - f(x_0, y_0) = \frac{1}{2!}\left[(x - x_0)^2 \frac{\partial^2 f}{\partial x^2} \right.$$
$$\left. + 2(x - x_0)(y - y_0)\frac{\partial^2 f}{\partial x \, \partial y} + (y - y_0)^2 \frac{\partial^2 f}{\partial y^2}\right]. \tag{1.109}$$

Arrangement in this form shows that the variation of $f(x, y)$ from its value $f(x_0, y_0)$ at the stationary point is determined by the quantity in square brackets. For a maximum (such as P in fig. 1.6) this quantity must be negative whatever the values of $x - x_0$ and $y - y_0$, and similarly for a minimum (such as B) it must be positive for all values. In both cases we should more strictly say 'for all sufficiently small values of $x - x_0$ and $y - y_0$'.

To save space let us write the quantity in square brackets as

$$X^2 A + 2XYB + Y^2 C, \tag{1.110}$$

where A, B and C are real constants, and then rearrange it in the form of the sum of two squares thus,

$$A\left(X + \frac{BY}{A}\right)^2 + \left(C - \frac{B^2}{A}\right)Y^2. \tag{1.111}$$

For this to be >0 for *all* X and Y, we must have

(i) $A > 0$,   (ii) $C - (B^2/A) > 0$. \hfill (1.112 a)

In view of (i), (ii) can be written as $AC > B^2$.

If (for a maximum of f) expression (1.110) is to be < 0 for *all* X and Y, we must have

$$\text{(iii)}\ A < 0, \qquad \text{(iv)}\ C - (B^2/A) < 0. \qquad\qquad (1.112\ b)$$

In view of (iii), (iv) can, like (ii), be written as $AC > B^2$.

When expression (1.110) is positive for some values of X and Y and negative for others we have a situation corresponding to a saddle point. If it is zero for all X and Y, then all second derivatives of f are zero and our criterion fails.

Stated in terms of partial derivatives, results (1.112) become

$$\left.\begin{array}{l} \dfrac{\partial^2 f}{\partial x^2} > 0 \text{ for a minimum,} \\[2mm] \phantom{\dfrac{\partial^2 f}{\partial x^2}} < 0 \text{ for a maximum,} \end{array}\right\} \qquad\qquad (1.113\ a)$$

$$\frac{\partial^2 f}{\partial x^2}\frac{\partial^2 f}{\partial y^2} > \left(\frac{\partial^2 f}{\partial x\,\partial y}\right)^2 \text{ in either case.} \qquad (1.113\ b)$$

Failure of condition (1.113 b) implies (except where both sides are zero) that the stationary point is a saddle point. It should not be forgotten that the above criteria apply only at points where $\partial f/\partial x = \partial f/\partial y = 0$.

▶37. Show that the function $x^3 \exp(-x^2 - y^2)$ has a maximum at the point $((3/2)^{1/2}, 0)$, a minimum at $(-(3/2)^{1/2}, 0)$, and a stationary point, whose nature cannot be determined by the above procedures, at the origin.

What has been shown above for functions of two variables can be extended to those of an arbitrary number. We will only state the results; the general lines of the proofs, if not the details, are probably apparent from the statements.

For a function of n variables x_i, the condition that f should be stationary at the point $x_i = x_{i_0}$ is that $df = 0$ for *all* small values of $(x_k - x_{k_0})$. From (1.68) this implies that

$$\left[\frac{\partial f}{\partial x_i}\right]_{x_i = x_{i_0}} = 0, \qquad \text{for } i = 1, 2, \ldots, n. \qquad (1.114)$$

The condition that the stationary point should be a maximum is that the expression

$$Q = (x_i - x_{i_0})(x_j - x_{j_0})\frac{\partial^2 f}{\partial x_i\,\partial x_j} \text{ (summation convention)} \quad (1.115)$$

should be <0 for all sufficiently small values of $(x_k - x_{k_0})$. For a minimum Q must be >0 for all such values.

For the purely utilitarian purpose of practical application, we record here that the theory of matrices and quadratic forms shows that the question about Q can be answered as follows. Consider the series of determinants,

$$Q_1 = \frac{\partial^2 f}{\partial x_1^2}, \quad Q_2 = \begin{vmatrix} \dfrac{\partial^2 f}{\partial x_1^2} & \dfrac{\partial^2 f}{\partial x_2\,\partial x_1} \\[2ex] \dfrac{\partial^2 f}{\partial x_1\,\partial x_2} & \dfrac{\partial^2 f}{\partial x_2^2} \end{vmatrix},$$

$$Q_3 = \begin{vmatrix} \dfrac{\partial^2 f}{\partial x_1^2} & \dfrac{\partial^2 f}{\partial x_2\,\partial x_1} & \dfrac{\partial^2 f}{\partial x_3\,\partial x_1} \\[2ex] \dfrac{\partial^2 f}{\partial x_1\,\partial x_2} & \dfrac{\partial^2 f}{\partial x_2^2} & \dfrac{\partial^2 f}{\partial x_3\,\partial x_2} \\[2ex] \dfrac{\partial^2 f}{\partial x_1\,\partial x_3} & \dfrac{\partial^2 f}{\partial x_2\,\partial x_3} & \dfrac{\partial^2 f}{\partial x_3^2} \end{vmatrix}, \tag{1.116}$$

etc., up to the $n \times n$ determinant whose (i, j)th element is given by $\partial^2 f/\partial x_i\,\partial x_j$ for $1 \leqslant i \leqslant n$ and $1 \leqslant j \leqslant n$. Then $Q > 0$ for all values of $(x_k - x_{k_0})$ if all of $Q_1, Q_2, \ldots, Q_n$ are separately >0; also, $Q < 0$ for all values of $(x_k - x_{k_0})$ if $Q_1, Q_3, \ldots, Q_{2r+1}$ are each <0 whilst Q_2, $Q_4, \ldots, Q_{2r}$ are each >0.

The results obtained previously for a function of two variables are particular examples of these general ones, as may be checked by the reader.

▶38. Show that the function

$$10x^2 + 8y^2 + z^2 - 12xy - 4x - 24y - 2z + 8$$

has only one stationary point and that that is a minimum with a value of -33.

1.14   Stationary values under constraints

We have considered the problem of finding stationary values of a function of two or more variables when all the variables may be independently varied. However, it often happens in physical problems that not all the variables used to describe a situation are in fact independent; for example, the sum of the energies involving the different spatial variables may be required to be constant in time.

If f is a function of n variables x_i $(i = 1, 2, \ldots, n)$, but the x_i are not truly independent, being constrained by m $(\leqslant n - 1)$ relationships

$$g_j(x_i) = 0, \qquad j = 1, 2, \ldots, m, \tag{1.117}$$

then effectively there are $n - m$ independent variables available for variation. The form of equation given in (1.117) is no restriction on the types of constraints so long as they can be written in algebraic form. We assume that each of the m different functions g_j has partial derivatives with respect to each x_i appearing in it.

One can visualize the procedure for this situation as that of solving (1.117) for m of the x_i and then substituting for these m variables, in terms of the other $n - m$, into f. The function f would then contain $n - m$ independent variables and could be treated as in the previous section. In many cases, this procedure is feasible, but in others it is either cumbersome or impossible; for example, for some forms of the functions g_j it may not be possible to write an explicit solution for a particular variable. Consider, for example, an equation such as

$$x_1^2 + x_2^2 + \arcsin (x_1/x_2) = 0,$$

which cannot be solved explicitly for either x_1 or x_2 in terms of the other.

When such situations arise, use may be made of a method known as **Lagrange's undetermined multipliers**. At first sight the method appears somewhat artificial, but in later chapters it will be seen that the apparently arbitrarily introduced 'multipliers' in fact have close connections with the sought-for stationary value solutions.

At a stationary point x_{i_0} we must have $df = 0$ as before, but not now for all sufficiently small values of $x_k - x_{k_0}$, but rather only for those which are such that

$$g_j(x_k) = 0, \qquad j = 1, 2, \ldots, m. \tag{1.117 bis}$$

That is, only those variations of x_i are allowed which do not change the value of each of the g_j; as a formula,

$$dg_j = 0, \qquad j = 1, 2, \ldots, m. \tag{1.118}$$

Now consider a new function F of the x_i, defined by

$$F(x_i) = f(x_i) + \lambda_j g_j(x_i), \tag{1.119}$$

where the λ_j (m of them) are as yet unknown multiplying constants – hence the name of the method. The fact that they are taken as *constants* should be emphasized. In view of the requirements that at a stationary point df must be zero and from (1.118) that dg_j must also be zero, we must have that

$$dF = df + \lambda_j\, dg_j = 0. \tag{1.120}$$

From this we may write (see (1.68))

$$\frac{\partial F}{\partial x_k} = 0, \quad k = 1, 2, \ldots, n, \tag{1.121}$$

that is

$$\frac{\partial f}{\partial x_k} + \lambda_j \frac{\partial g_j}{\partial x_k} = 0, \quad k = 1, 2, \ldots, n. \tag{1.122}$$

Line (1.122) gives n equations which may be used together with the m equations (1.117) to provide $m + n$ equations for what are now $m + n$ unknowns, namely the n quantities x_i and the m values of λ_j.

Thus at the price of increasing the total number of unknowns to $m + n$, the problem has been made into one in which the changes in the x_i can be made independently.

To illustrate the undetermined-multiplier method described above, we will now (partially) derive a well known result from statistical mechanics – the *Boltzmann distribution*. We will not be concerned here with the physical arguments involved, but merely summarize these by stating the problem as follows.

A system contains a very large number N of particles, each of which can be in any of R energy levels with a corresponding energy E_i ($i = 1, 2, \ldots, R$). If the number of particles in the ith level is n_i, and the total energy of the system is E, then it is required to find the distribution n_i for $i = 1, 2, \ldots, R$.

Statistical arguments indicate that the distribution will be that which maximizes the expression

$$P = \frac{N!}{n_1!\, n_2! \ldots n_R!},$$

but of course subject to the constraints

$$g_1(n_i) \equiv N - \sum_{i=1}^{R} n_i = 0, \tag{1.123 a}$$

and

$$g_2(n_i) \equiv E - \sum_{i=1}^{R} n_i E_i = 0. \tag{1.123 b}$$

Maximizing P is equivalent to minimizing its denominator since $N!$ is fixed. This is most conveniently done by considering its logarithm and using an approximation due to Stirling† and valid for large n, that

$$\ln (n!) \simeq n[\ln (n) - 1].$$

The problem then becomes one of minimizing

$$\ln (n_1! \, n_2! \ldots n_R!) \simeq \sum_{i=1}^{R} n_i \ln (n_i) - \sum_{i=1}^{R} n_i \qquad (1.124)$$

subject to (1.123 a) and (1.123 b). The second term on the right of (1.124) necessarily has value $-N$, and so may be disregarded from the point of view of finding stationary values. The implicit assumption has been made here that for distributions near to the required one, all the n_i are reasonably large.

Thus we arrive at the situation described earlier, with the n_i playing the role of the x_i, $\sum n_i \ln (n_i)$ the part of f, and equations (1.123) providing two constraints on the R variables. The analogue of F in (1.119) is

$$F = \sum_i n_i \ln (n_i) + \lambda_1 \left(N - \sum_i n_i \right) + \lambda_2 \left(E - \sum_i n_i E_i \right). \qquad (1.125)$$

Taking its partial derivative with respect to a general n_k gives

$$\frac{\partial F}{\partial n_k} = n_k \frac{1}{n_k} + \ln (n_k) - \lambda_1 - \lambda_2 E_k, \qquad (1.126)$$

and setting this equal to zero at the value n_{k_0}, corresponding to the stationary value of P, yields

$$\ln (n_{k_0}) = \lambda_2 E_k + \lambda_1 - 1. \qquad (1.127)$$

Hence the distribution of the n_k follows the law

$$n_{k_0} = C \exp (\lambda_2 E_k), \qquad (1.128)$$

where C is independent of k, but of course C and λ_2 are such that the two equations

$$\sum_k C \exp (\lambda_2 E_k) = N,$$

and

$$\sum_k C E_k \exp (\lambda_2 E_k) = E,$$

† More accurately, a first-order Stirling formula gives

$$n! \simeq (2\pi)^{1/2} n^{n+1/2} \exp (-n)$$

for large n, and the form used in the text is an approximation to this in which all terms not increasing at least as quickly as n have been dropped.

are satisfied. [The reader is probably aware that λ_2 has the value $-1/kT$, where k is Boltzmann's constant and T the thermodynamic temperature.]

▶39. Find the maximum and minimum values taken by the expression $13x^2 + 8xy + 7y^2$ on the circle $x^2 + y^2 = 1$.

1.15 Integration

The notion of an integral as the area under a curve will be familiar to the reader. In fig. 1.7, in which the solid line is a plot of a function $f(x)$, the shaded area represents the quantity which is denoted by

$$I = \int_a^b f(x)\,dx. \tag{1.129}$$

This expression is known as the **definite integral** of $f(x)$ between the **lower limit** $x = a$ and the **upper limit** $x = b$, and $f(x)$ is called the **integrand**.

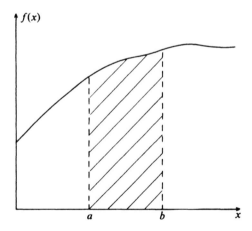

Fig. 1.7 An integral as the area under a curve.

This is not the formal definition of the integral, but a readily visualizable representation of it. The formal definition of I involves subdividing the interval $a \leqslant x \leqslant b$ into a large number of subintervals by defining intermediate points ξ_i such that $a = \xi_0 < \xi_1 < \xi_2 < \cdots < \xi_n = b$. After this the sum

$$S = \sum_{i=1}^{n} f(x_i)(\xi_i - \xi_{i-1}) \tag{1.130}$$

is formed, where x_i is an arbitrary point which lies in the range $\xi_{i-1} \leqslant x_i \leqslant \xi_i$. Then if n is allowed to tend to infinity in any way whatsoever, subject only to the restriction that the length of every subinterval ξ_{i-1} to ξ_i tends to zero, S may or may not tend to a unique limit I. If it does, then the definite integral of $f(x)$ between a and b is defined as having the value I and the fact expressed by (1.129). If no unique limit exists the integral is undefined.

This description in no way purports to be a proper analytical one, but merely a prosaic explanation of the steps involved. The reader interested in a rigorous development should consult a textbook on pure mathematics.

Some straightforward properties of definite integrals, for which general lines of proof are almost self-evident are:

$$\int_a^b 0 \, dx = 0, \tag{1.131 a}$$

$$\int_a^b f(x) \, dx = - \int_b^a f(x) \, dx, \tag{1.131 b}$$

$$\int_a^c f(x) \, dx = \int_a^b f(x) \, dx + \int_b^c f(x) \, dx; \tag{1.131 c}$$

if $m \leqslant f(x) \leqslant M$ for $a \leqslant x \leqslant b$, then

$$m(b - a) \leqslant \int_a^b f(x) \, dx \leqslant M(b - a), \tag{1.131 d}$$

$$\int_a^b [f(x) + g(x)] \, dx = \int_a^b f(x) \, dx + \int_a^b g(x) \, dx. \tag{1.131 e}$$

The alternative view of integration as the converse of differentiation can be connected to the graphical representation as follows. Instead of considering the definite integral of f between fixed limits a and b, form the integral from a lower limit x_0 to a variable upper limit y (as illustrated by the shading in fig. 1.8). This type of integral is formally defined just as previously, but because of its variable upper limit, it is known as an **indefinite integral**. Its value depends upon the value of y and so we will write it as $F(y)$ where

$$F(y) = \int_{x_0}^y f(x) \, dx. \tag{1.132}$$

It is apparent that $F(y)$ is a continuous function of y.

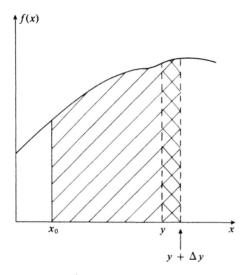

Fig. 1.8   The indefinite integral of $f(x)$ from x_0 to y.

Now consider what happens if y is increased to $y + \Delta y$; then

$$F(y + \Delta y) = \int_{x_0}^{y + \Delta y} f(x)\, dx, \tag{1.133}$$

which from (1.131 c) can be written as

$$F(y + \Delta y) = \int_{x_0}^{y} f(x)\, dx + \int_{y}^{y + \Delta y} f(x)\, dx$$

$$= F(y) + \int_{y}^{y + \Delta y} f(x)\, dx. \tag{1.134}$$

Rearranging (1.134) and dividing through by Δy yields

$$\frac{F(y + \Delta y) - F(y)}{\Delta y} = \frac{1}{\Delta y} \int_{y}^{y + \Delta y} f(x)\, dx. \tag{1.135}$$

Now if Δy is allowed to tend to zero, the left-hand side of (1.135) becomes, by definition, dF/dy, and the right-hand side becomes $f(y)$. This latter statement follows since all the x_i referred to in definition (1.130) have to come from an arbitrarily small region close to $x = y$ and so the $f(x_i)$ have to have the value $f(y)$ arbitrarily precisely; the sum on the right of (1.130) then becomes $f(y)$ times the sum of the subintervals, which is exactly $f(y) \times \Delta y$.

Equation (1.135) therefore, in the limit $\Delta y \to 0$, reduces to

$$\frac{dF(y)}{dy} = f(y), \tag{1.136}$$

that is, the indefinite integral $F(y)$ of a function $f(x)$, is that function which has f as its derivative. This is such an important result that it is worth repeating in a different form, namely

$$\frac{d}{dy}\left[\int_{x_0}^{y} f(x)\,dx\right] = f(y). \tag{1.137}$$

It is also apparent that the value I of the definite integral is given in terms of $F(y)$ by

$$I = \int_{a}^{b} f(x)\,dx = \int_{x_0}^{b} f(x)\,dx - \int_{x_0}^{a} f(x)\,dx$$
$$= F(b) - F(a). \tag{1.138}$$

In our presentation, different variables x and y have been used as the arguments of f and F, and a specific lower limit included. It is common practice to omit one or both of the limits in an indefinite integral, e.g. to write (1.132) as

$$F = \int f(x)\,dx, \tag{1.139}$$

and to insert them only when a specific evaluation has to be made. It is also not uncommon to use the same symbol for both variables

$$F(x) = \int^{x} f(x)\,dx,$$

but this practice can be confusing, and should be avoided for that reason.

1.16   Infinite and improper integrals

The prescription given earlier for the formulation of a definite integral does not accommodate the case in which the integration runs over an infinite range in x, that is a or b (or both) is infinite. Nor does it deal with the situation in which $f(x)$ is unbounded (becomes arbitrarily large in modulus) in some part of the range, e.g. $f(x) = (2 - x)^{-1/4}$ near the point $x = 2$.

The failure in the case of an infinite range for x comes about because any finite number of intermediate division points must leave at least one infinite interval. At some point in this interval $f(x)$ will be non-zero (except for trivial cases) and so the corresponding term in the sum S itself contributes an infinite amount. This difficulty is got round by taking the limit of, say, $b \to \infty$ *after* a finite integral with upper limit b has been defined. Thus, if for *finite* b

$$I = \int_a^b f(x) \, dx = F(b) - F(a), \tag{1.140}$$

and if, when $b \to \infty$, $F(b)$ approaches a definite finite limit L, i.e.

$$\lim_{b \to \infty} F(b) = L, \tag{1.141}$$

then we define the **infinite integral** of $f(x)$ from a to ∞ as having the value $L - F(a)$,

$$\int_a^\infty f(x) \, dx = L - F(a). \tag{1.142}$$

Other infinite integrals

$$\int_{-\infty}^b f(x) \, dx \quad \text{and} \quad \int_{-\infty}^\infty f(x) \, dx,$$

may be defined by similar limiting processes.

The second case of failure of the simple formulation, the case in which the integrand is unbounded, leads us to define an **improper integral** by the process of omitting from the range of integration a small piece of it containing the point at which the integrand is unbounded, evaluating the remaining integral, and then letting the size of the excluded range tend to zero. If with this procedure the integral tends to a finite limit, then the value of the improper integral is defined to be equal to that limit. The excluded piece may be entirely inside the range or may include one of the limits. An evaluation for the previously mentioned function will show what is meant. Suppose we have to find [or define]

$$J = \int_0^2 (2 - x)^{-1/4} \, dx. \tag{1.143}$$

The integrand tends to infinity near $x = 2$ and so we replace J by

$$J_1 = \int_0^{2-\epsilon} (2 - x)^{-1/4} \, dx, \tag{1.144}$$

where an interval of length ϵ has been removed between $x = 2 - \epsilon$ and $x = 2$. Now J_1 can be evaluated as

$$J_1 = [-\tfrac{4}{3}(2 - x)^{3/4}]_0^{2-\epsilon}$$
$$= -\tfrac{4}{3}\epsilon^{3/4} + \tfrac{4}{3} \times 2^{3/4}.$$

If ϵ is now allowed to tend to zero, J_1 tends to the limit $\tfrac{4}{3} \times 2^{3/4}$ which is thus the value of the improper integral (1.143).

▶40. Determine whether the following integrals exist, and where they do evaluate them:

(i) $\displaystyle\int_0^\infty \exp(-\lambda x)\,dx,$ (ii) $\displaystyle\int_{-\infty}^\infty \frac{x}{(x^2 + a^2)^2}\,dx,$

(iii) $\displaystyle\int_1^\infty \frac{1}{x+1}\,dx,$ (iv) $\displaystyle\int_0^1 \frac{1}{x^2}\,dx,$

(v) $\displaystyle\int_0^{\pi/2} \cot\theta\,d\theta,$ (vi) $\displaystyle\int_0^1 \frac{x}{(1-x^2)^{1/2}}\,dx.$

1.17   Integration by parts

Particular methods of evaluating definite and indefinite integrals will not be considered in general here, as the reader is presumed to have some familiarity with the integration of standard forms and the *method of substitutions* (changing to a new integration variable in order to reduce an integrand to a standard form).

However, one particular method known as **integration by parts** is of such general utility and will be employed so many times in the remaining chapters, that some discussion of it is amply justified.

The basis of the method is the result derived in (1.41) for the derivative of a product:

$$\frac{d}{dx}[g(x)h(x)] = g\frac{dh}{dx} + h\frac{dg}{dx}. \tag{1.41 bis}$$

If this equation is rearranged and each term then integrated between limits $x = a$ and $x = b$, we obtain

$$\int_a^b g\frac{dh}{dx}\,dx = \int_a^b \frac{d}{dx}[gh]\,dx - \int_a^b \frac{dg}{dx}h\,dx. \tag{1.145}$$

Now the first term on the right is the integral of the derivative of a function $[= g(x)h(x)]$ and is thus immediately evaluated as the difference between the two values of the function at the limits, namely

$$[g(x)h(x)]_a^b \qquad\qquad (1.146\,a)$$

or $\qquad g(b)h(b) - g(a)h(a).$ $\qquad\qquad (1.146\,b)$

Thus, (1.145) can be written as the equation giving the 'formula for integration by parts',

$$\int_a^b g \frac{dh}{dx} \, dx = [gh]_a^b - \int_a^b \frac{dg}{dx} h \, dx, \qquad\qquad (1.147)$$

or, expressed in terms of differentials of g and h,

$$\int_a^b g \, dh = [gh]_a^b - \int_a^b h \, dg. \qquad\qquad (1.148)$$

To see how this may be of value in evaluating an integral $\int f(x) \, dx$, where $f(x)$ is not a 'standard form' whose integral is known, suppose that f can be written as the product of two functions, one of which has a known integral and the other of which is differentiable. Then the first function may be taken as dh/dx in (1.147) and the second as g, and the integral replaced by the right-hand side of (1.147). The first term of the new expression can certainly be evaluated, and it may well be that the second term is now an integral which can be evaluated using known results. Even if this is not so, a second integration by parts [of course h must be integrated again and g further differentiated, or the work of the first integration by parts will be undone] may produce the desired situation and result in two definite expressions, each to be evaluated at the upper and lower limits, together with an integral which has a known value.

As with one or two earlier results in this chapter, integration by parts is often remembered for practical purposes in the form of a clumsy but recallable sentence.

> The integral of a product is equal to {the first times the integral of the second} minus the integral of {the derivative of the first times the integral of the second}.

Although the discussion here has been in terms of definite integrals, integration by parts applies equally well to indefinite integrals.

To demonstrate one or two ways in which integration by parts can be used, we will solve the following three examples.

Example 1.2. Evaluate

$$I = \int_a^b x \cos x \, dx. \qquad\qquad (1.149)$$

The function whose derivative is $x \cos x$ is not directly apparent, but if we identify the x and $\cos x$ of (1.149) with the g and dh/dx of (1.147) respectively, then integration by parts gives

$$I = [x \sin x]_a^b - \int_a^b 1 \times \sin x \, dx$$

$$= b \sin b - a \sin a - (-\cos b + \cos a)$$

$$= (b \sin b + \cos b) - (a \sin a + \cos a).$$

Example 1.3. Evaluate

$$\int_1^y \ln(x) \, dx.$$

Here we do not have an obvious product in the integrand, but one can be manufactured by taking $\ln x$ as one factor and unity as the other. Since we cannot integrate $\ln(x)$ [that is the whole problem] this is the factor which must be differentiated and unity is the factor to be integrated. Proceeding on these lines, we obtain

$$\int_1^y \ln(x) \, dx = [\ln(x) \, x]_1^y - \int_1^y \frac{1}{x} \cdot x \, dx.$$

The integral on the right is now trivial and the result

$$\int_1^y \ln(x) \, dx = y \ln(y) - 1 \ln(1) - [x]_1^y$$

$$= y \ln(y) - y + 1$$

follows.

Example 1.4. By integrating by parts twice, evaluate

$$I = \int_0^\infty e^{-\lambda x} \cos ax \, dx, \qquad \lambda > 0.$$

Integrating by parts once (using $\exp(-\lambda x)$ as the g of (1.147))

$$I = \left[e^{-\lambda x} \frac{\sin ax}{a} \right]_0^\infty - \int_0^\infty (-\lambda) e^{-\lambda x} \frac{\sin ax}{a} \, dx.$$

The infinite upper limit causes no difficulty and so

$$I = 0 - 0 + \frac{\lambda}{a} \int_0^\infty e^{-\lambda x} \sin ax \, dx.$$

Now integrating by parts a second time,

$$I = \frac{\lambda}{a} \left[e^{-\lambda x} \frac{(-\cos ax)}{a} \right]_0^\infty$$

$$- \frac{\lambda}{a} \int_0^\infty (-\lambda) e^{-\lambda x} \frac{(-\cos ax)}{a} \, dx.$$

But the integral on the right is just a simple multiple $[-\lambda^2/a^2]$ of the original integral I. Thus

$$I = \frac{\lambda}{a} \left[0 + \frac{1}{a} \right] - \frac{\lambda^2}{a^2} I,$$

and

$$I = \frac{\lambda/a^2}{1 + \lambda^2/a^2} = \frac{\lambda}{a^2 + \lambda^2}.$$

These examples, together with the exercises below, should be sufficient to give the reader the working knowledge necessary to follow the use of integration by parts made in subsequent chapters.

▶41. Use integration by parts to evaluate the following:

(i) $\int_0^y x^2 \sin x \, dx$,

(ii) $\int_1^y x \ln(x) \, dx$,

(iii) $\int^y \arcsin x \, dx$,

(iv) $\int^y \frac{\ln(a^2 + x^2)}{x^2} \, dx$,

(v) $\int_0^{\pi/2} \cos(2x) \sin(3x) \, dx$,

(vi) repeat (v) reversing the roles of the factors.

▶42. If

$$I_n = \int_0^\infty x^n e^{-x} \, dx,$$

show, by integrating by parts, that $I_n = nI_{n-1}$. Hence evaluate I_n.

▶43. If J_n is the integral

$$J_n = \int_0^\infty x^n \exp(-x^2) \, dx,$$

show that

(i) $J_{2r+1} = (r!)/2$,

(ii) $J_{2r} = \dfrac{(2r - 1) \cdot (2r - 3) \ldots 3 \cdot 1}{2^r} J_0.$

1.18   Multiple integrals

For a function of several variables, just as derivatives with respect to two or more of them may be considered, so may the integral of the function with respect to more than one variable be formed. The formal definitions of such **multiple integrals** follow very much the obvious extensions of the definition of the integral with respect to a single variable, discussed at the beginning of section 1.15. We will discuss almost exclusively integrals involving only two or three variables, partly because they are more readily visualizable, and partly because they illustrate all the essential features of those involving more. The number of variables is called the **dimension** of the integral; up to the present only one-dimensional integrals have been considered.

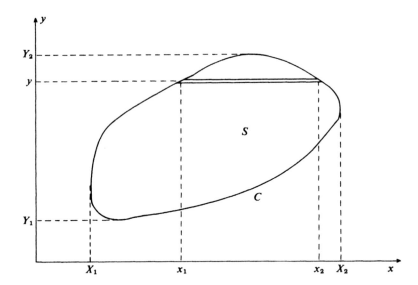

Fig. 1.9   The region of integration for the double integral given in equation (1.150). The elementary strip shown is appropriate to the integral in the form expressed by line (1.151 a).

For an integral involving two variables – a *double integral* – we have a function $f(x, y)$ to be integrated with respect to x and y between certain limits. The function f may also depend upon other variables but we will not write these explicitly. The limits may, for both x and y, be constants, or the limits for x may depend on the value of y or vice-versa. Whatever the relationship, the limits can usually be represented by a closed curve C in the xy-plane, bounding an area which we will denote by S (fig. 1.9).

The integral is written symbolically as

$$I = \iint_S f(x, y) \, dx \, dy, \qquad (1.150\ a)$$

or as

$$I = \int_S f(x, y) \, dA, \qquad (1.150\ b)$$

where dA (or dA_{xy}) stands for the element of area in the xy-plane.

Some authors use a single integration symbol whatever the dimension of the integral; others use as many symbols as the dimension. In different circumstances, both have their disadvantages, either of ambiguity or of clumsiness. We will adopt the convention, typified by (1.150 a, b), that as many integration symbols will be used as differentials *explicitly* written. This will allow the freedom to use the more compact form when no confusion is likely, but to be explicit when emphasis is needed or ambiguity is possible.

Form (1.150 b) makes no explicit statement about the limits of integration except by reference to a diagram such as fig. 1.9 or to an equation $c(x, y) = 0$ giving the boundary curve C. This same prescription for the limits may also be used in connection with form (1.150 a), but an explicit statement of them can be written in the form

$$I = \int_{Y_1}^{Y_2} \left\{ \int_{x_1(y)}^{x_2(y)} f(x, y) \, dx \right\} dy. \qquad (1.151\ a)$$

Expression (1.151 a) indicates (refer again to fig. 1.9) that $f(x, y)$ is to be integrated with respect to x between the two values $x = x_1$ and $x = x_2$, where x_1 and x_2 both depend on y, and then the result (considered as a function of y) is to be integrated between the limits $y = Y_1$ and $y = Y_2$.

The same result would be obtained if the order of the two integrations were reversed, and the area S 'cut up the other way'. A formula corresponding to (1.151 a) for this case would be

$$I = \int_{X_1}^{X_2} \left\{ \int_{y_1(x)}^{y_2(x)} f(x, y) \, dy \right\} dx. \qquad (1.151\ b)$$

As a simple example consider

$$I = \iint (x - y)^2 \, dx \, dy$$

over the triangular area (shown in fig. 1.10) whose sides are $x = 0$, $y = 0$, $x + y = 1$.

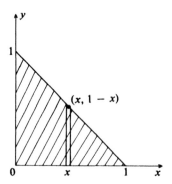

Fig. 1.10 The triangular area for the evaluation of I.

Suppose we choose to carry out the integration with respect to y first. With x fixed, the range of y is 0 to $1 - x$, as indicated in the figure. Thus with all its limits explicitly filled in, the expression for I becomes

$$I = \int_0^1 \left\{ \int_0^{1-x} (x - y)^2 \, dy \right\} dx,$$

which can be evaluated as

$$I = \int_0^1 \{ [-\tfrac{1}{3}(x - y)^3]_0^{1-x} \} \, dx$$

$$= \int_0^1 \{ -\tfrac{1}{3}(2x - 1)^3 + \tfrac{1}{3}x^3 \} \, dx$$

$$= [-\tfrac{1}{24}(2x - 1)^4 + \tfrac{1}{12}x^4]_0^1$$

$$= -\tfrac{1}{24} + \tfrac{1}{12} + \tfrac{1}{24} - 0 = \tfrac{1}{12}.$$

In this case the symmetry between x and y of both the integrand and integration region is sufficient by itself to show that the same result would be obtained if the order of integrations were reversed and the integral with respect to x evaluated first.

▶44. (i) Evaluate $\iint x^2 y \, dx \, dy$ over the same triangular region as in the above example (fig. 1.10).
(ii) Reverse the order of integrations and demonstrate that the same result is obtained.

▶45. Evaluate $\int_S (x^2 y + x) \, dA$, where S is the rectangle with corners $(0, 0)$, $(a, 0)$, (a, b), $(0, b)$.

Three-dimensional, or *triple* integrals, are described in an analogous way. Without going into details, we note that they will be written either as

$$\iiint_V f(x, y, z)\, dx\, dy\, dz, \qquad (1.152\,a)$$

or as

$$\int_V f(x, y, z)\, dV \quad \text{or} \quad \int_V f(x, y, z)\, dV_{xyz}. \qquad (1.152\,b)$$

The integration is over a volume whose surface determines the upper and lower limits for x, y and z. As with double integrals the particular values of the limits appropriate to any one of the variables of integration depends upon the order in which the individual integrations are carried out.

Multiple integrals over ranges which are infinite in extent or include points at which the integrand becomes infinite are treated in the same way as similar one-dimensional integrals.

▶46. Evaluate the integral $\int_V (x^2 + y^2 + z^2)\, dV$, where V is the rectangular volume whose six surfaces are the planes $x = \pm a$, $y = \pm b$, $z = \pm c$.

▶47. Evaluate $\int x^2 y\, dV$ over the volume bounded by the planes $x = 0$, $y = 0$, $z = 0$, $x + y + z = 1$. Use two different orders of carrying out the integrations, and show that the same result is obtained both times.

▶48. Evaluate, for $\lambda > 0$,

$$\iiint \exp(-\lambda r)\exp(ir\cos\theta)\, r\sin\theta\sin^2\phi\, dr\, d\theta\, d\phi$$

over the infinite volume $0 \leqslant r < \infty$, $0 \leqslant \theta \leqslant \pi$, and $0 \leqslant \phi \leqslant 2\pi$. [This is the integral of the function $r^{-1}\exp[r(-\lambda + i\cos\theta)]\sin^2\phi$ over all space, with everything expressed in spherical polar coordinates the element of volume being $r^2\sin\theta\, dr\, d\theta\, d\phi$.]

1.19   Change of variables in multiple integrals

It often happens that, either because of the form of the integrand involved, or because of the shape of the boundary of the region of integration, it is desirable to express a multiple integral in terms of a new set of variables.

As an illustration of what is meant by this, we may refer to the note in

▶48 and see that if the function to be integrated over all space were expressed in Cartesian coordinates it would be

$$f(x, y, z) = \frac{1}{(x^2 + y^2 + z^2)^{1/2}} \exp\left[-\lambda(x^2 + y^2 + z^2)^{1/2}\right]$$

$$\times \exp(iz) \frac{(x^2 + y^2)}{x^2 + y^2 + z^2}. \quad (1.153)$$

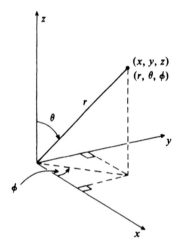

Fig. 1.11 The connection between Cartesian and spherical polar coordinates.

The connections used to link the two sets of coordinates are depicted in fig. 1.11 and given as formulae by

$$\left.\begin{array}{l} x = r \sin \theta \cos \phi, \\ y = r \sin \theta \sin \phi, \\ z = r \cos \theta, \end{array}\right\} \quad (1.154\ a)$$

and the inverse equations

$$\left.\begin{array}{l} r = +(x^2 + y^2 + z^2)^{1/2}, \\ \theta = \arccos\left[z/(x^2 + y^2 + z^2)^{1/2}\right], \\ \phi = \arctan(y/x). \end{array}\right\} \quad (1.154\ b)$$

Now suppose that in the course of a calculation it is required to integrate $f(x, y, z)$ of expression (1.153) over all space, i.e. to evaluate the integral,

$$I = \int_{-\infty}^{\infty} \int_{-\infty}^{\infty} \int_{-\infty}^{\infty} \frac{x^2 + y^2}{(x^2 + y^2 + z^2)^{3/2}}$$

$$\times \exp\left[-\lambda(x^2 + y^2 + z^2)^{1/2}\right] \exp(iz) \, dx \, dy \, dz. \quad (1.155)$$

Clearly the integrand has a simpler form when expressed in spherical polar coordinates [namely the previously noted $r^{-1} \exp [r(-\lambda + i \cos \theta)] \sin^2 \phi$] and changing this part of the integral to such new coordinates presents no difficulty. Neither, in this case, do the limits of integration which become $0 \leqslant r < \infty, 0 \leqslant \theta \leqslant \pi$ and $0 \leqslant \phi < 2\pi$. The part of the integral which requires more consideration is the element of volume $dx\,dy\,dz$, which, it is stated in ▶48, becomes $r^2 \sin \theta\,dr\,d\theta\,d\phi$.

The purpose of this section is to develop a general expression relating an element of volume (or area) in one coordinate system to the corresponding element in another. If the sets of coordinates in the two systems have corresponding physical dimensions (e.g. all are lengths, or two are lengths and one a time, in each set) then we may speak of the *local magnification* of the element of volume (or area) in going from one coordinate system to another.

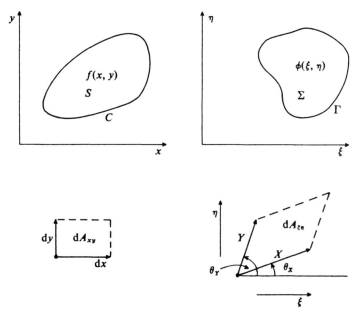

Fig. 1.12 The regions of integration and the elementary areas in the original xy-plane and the transformed $\xi\eta$-plane.

We will use a language which suggests that we are all the time dealing with lengths, areas and volumes, but the results obtained apply to all sets of variables so long as there is a one-to-one correspondence between the values taken by the variables in the two systems in describing the same situation. Our approach will be a simple geometrical one and we will start with the two-dimensional case.

Suppose that we require to change an integral

$$\int_S f(x, y)\, dA_{xy} \tag{1.156}$$

in terms of coordinates x, y into one expressed in new coordinates ξ and η which are given in terms of x and y by (differentiable) equations

$$\xi = \xi(x, y), \qquad \eta = \eta(x, y). \tag{1.157}$$

The area S and the curve C which bounds it will become a new area Σ and a new boundary Γ in the $\xi\eta$-plane, as shown in fig. 1.12. The function $f(x, y)$ becomes $\phi(\xi, \eta)$.

An element of area in the xy-plane is given by $dA_{xy} = dx\, dy$ as in the figure and the question is 'what is the size of the area which this corresponds to in the $\xi\eta$-plane?'. In general the corresponding element $dA_{\xi\eta}$ will not be the same as dA_{xy} in shape, but this does not matter since all elements are infinitesimally small and the value of the integrand is constant over the elementary area providing the latter is small.

Since dx and dy are infinitesimal, the lines of lengths X and Y corresponding to them will be straight and $dA_{\xi\eta}$ will have the shape of a parallelogram. We work out the area of the parallelogram as follows (see fig. 1.12 for the notation),

$$
\begin{aligned}
dA_{\xi\eta} &= XY \sin(\theta_Y - \theta_X) \\
&= Y \sin \theta_Y\, X \cos \theta_X - Y \cos \theta_Y\, X \sin \theta_X \\
&= \frac{\partial \eta}{\partial y}\, dy\, \frac{\partial \xi}{\partial x}\, dx - \frac{\partial \xi}{\partial y}\, dy\, \frac{\partial \eta}{\partial x}\, dx
\end{aligned}
\tag{1.158}
$$

$$
= \left(\frac{\partial \xi}{\partial x} \frac{\partial \eta}{\partial y} - \frac{\partial \xi}{\partial y} \frac{\partial \eta}{\partial x} \right) dA_{xy}. \tag{1.159}
$$

The substitutions made in line (1.158) follow from noticing that, for example, $Y \sin \theta_Y$ is the projection on the η-axis in the $\xi\eta$-plane of the line corresponding to the infinitesimal dy (parallel to the y-axis) in the xy-plane. Thus

$$Y \sin \theta_Y = \frac{\partial \eta}{\partial y}\, dy. \tag{1.160}$$

Similarly for the other three terms.

The quantity in brackets in (1.159) is just the Jacobian of ξ, η with respect to x, y, discussed in section 1.12. Thus we conclude that an element of area dA in the xy-plane corresponds to one

$$J = \frac{\partial(\xi, \eta)}{\partial(x, y)}$$

times as big in the $\xi\eta$-plane, i.e. the Jacobian at any particular point gives the local magnification of the size of an element of area in changing from the (x, y) set of coordinates to the (ξ, η) set.

If the 'physical' dimensions of x, y and ξ, η are different, as they are in example 1.5 below, then it seems inappropriate to speak of magnification. Nevertheless, the relationship between the size of the differential of area generated by dx, dy and the size of the corresponding differential of area generated by $d\xi, d\eta$ is

$$d\xi\, d\eta = \frac{\partial(\xi, \eta)}{\partial(x, y)}\, dx\, dy. \tag{1.161}$$

Of course the value of the Jacobian can, and in general will, vary over the region of integration.

The order in which a new set of variables is arranged (e.g. ξ, η rather than η, ξ) must agree with conventional use (i.e. a rotation from the ξ- to the η-axis of $\pi/2$ is in the positive (anticlockwise) direction) or spurious minus signs can be introduced into the double integral.

Rather than explain further in general terms the actual procedure, let us give an elementary example of a two-dimensional change of variable [to illustrate the method, but hardly to obtain a previously unknown result].

Example 1.5. Find the area of the circle $x^2 + y^2 = a^2$.

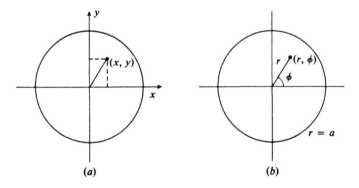

Fig. 1.13 The region of integration for example 1.5 in (a) the xy-plane and (b) the $r\phi$-plane.

In the given Cartesian coordinate system the required area α is

$$\alpha = \int_{x=-a}^{x=a} \int_{y=-(a^2-x^2)^{1/2}}^{y=(a^2-x^2)^{1/2}} dy\, dx. \tag{1.162}$$

[In this case the function $f(x, y)$ of (1.151) is unity.] Because of the circular boundary of the integration region a change of variables to plane polar coordinates r, ϕ is indicated (fig. 1.13). The limits of the integral are then simple, namely $0 \leqslant r \leqslant a$, $0 \leqslant \phi \leqslant 2\pi$. This leaves only the Jacobian of the transformation to be computed.

For this purpose we have

$$x = r \cos \phi, \qquad y = r \sin \phi. \tag{1.163}$$

It is just as satisfactory to calculate $\partial(x, y)/\partial(r, \phi)$ as to calculate $\partial(r, \phi)/\partial(x, y)$, since these two are, by (1.107), reciprocals of each other. In this case it is more convenient, since the value of the Jacobian is automatically expressed in r and ϕ rather than in x and y, and it is $dx\,dy$ we wish to replace. Carrying out the calculation, we obtain

$$\frac{\partial(x, y)}{\partial(r, \phi)} = \begin{vmatrix} \cos \phi & -r \sin \phi \\ \sin \phi & r \cos \phi \end{vmatrix} = r. \tag{1.164}$$

Thus we have, using (1.161) in the form

$$dx\,dy = \frac{\partial(x, y)}{\partial(\xi, \eta)} d\xi\,d\eta, \tag{1.165}$$

that

$$dx\,dy = \frac{\partial(x, y)}{\partial(r, \phi)} dr\,d\phi = r\,dr\,d\phi.$$

Putting this into (1.162) expressed in the new coordinates we obtain

$$\alpha = \int_{r=0}^{a} \int_{\phi=0}^{2\pi} r\,dr\,d\phi = \int_{r=0}^{a} 2\pi r\,dr = \pi a^2.$$

[Of course, since $f(x, y)$ equals 1 in (x, y) coordinates, it does so in (ξ, η) coordinates also.]

A change of variable in a three-dimensional integral follows the same general lines as that in a two-dimensional one. Without the same amount of subsidiary explanation as given above for the double integral, we will outline the main points of the derivation.

Suppose we wish to change from coordinates (x, y, z) to coordinates (ξ, η, χ). A rectangular volume of sides dx, dy and dz in the original system will become a parallelepiped in the new one, with edges X, Y and Z, corresponding to dx, dy and dz as in fig. 1.14 (a) and (b).

Consider the (straight) line element X, corresponding in the ξ, η, χ

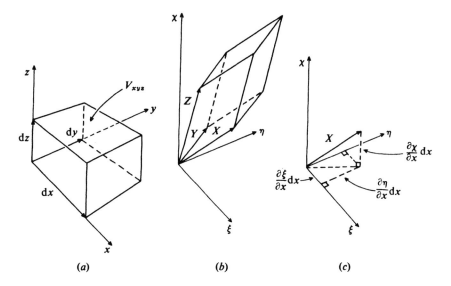

Fig. 1.14 (*a*) The rectangular volume in Cartesian coordinates. (*b*) In (ξ, η, χ) co-ordinates the volume is a parallelepiped. (*c*) The components of X in (ξ, η, χ) space.

system to dx. It has components in the directions of the ξ-, η- and χ-axes of

$$\frac{\partial \xi}{\partial x} dx, \qquad \frac{\partial \eta}{\partial x} dx, \qquad \frac{\partial \chi}{\partial x} dx \qquad (1.166)$$

as illustrated in fig. 1.14 (*c*). The lines Y and Z have components given by (1.166), only with x replaced by y and z respectively.

At this point we must use a result which is not proved until later (equations (2.23) and (2.25) of section 2.8) but which, it should be added, does not depend upon the present work for its proof of validity. The result gives the volume of the parallelepiped in terms of the components of its edges obtained above as

$$dV_{\xi\eta x} = \begin{vmatrix} \dfrac{\partial \xi}{\partial x} dx & \dfrac{\partial \xi}{\partial y} dy & \dfrac{\partial \xi}{\partial z} dz \\[2mm] \dfrac{\partial \eta}{\partial x} dx & \dfrac{\partial \eta}{\partial y} dy & \dfrac{\partial \eta}{\partial z} dz \\[2mm] \dfrac{\partial \chi}{\partial x} dx & \dfrac{\partial \chi}{\partial y} dy & \dfrac{\partial \chi}{\partial z} dz \end{vmatrix} . \qquad (1.167)$$

When the factor $dx \, dy \, dz$ has been taken out of the determinant, we again arrive at a Jacobian form of result, namely

$$dV_{\xi\eta x} = \frac{\partial(\xi, \eta, \chi)}{\partial(x, y, z)} \, dx \, dy \, dz = \frac{\partial(\xi, \eta, \chi)}{\partial(x, y, z)} \, dV_{xyz}. \tag{1.168}$$

This shows that the ratio of elementary volumes in the two systems in the neighbourhood of any point is given by the Jacobian at that point.

Although we will not prove it, the general result for a change of coordinates in an n-dimensional space from a set x_i to a set y_j (i, j both run from 1 to n) is

$$dx_1 \, dx_2 \ldots dx_n = \frac{\partial(x_1, x_2, \ldots, x_n)}{\partial(y_1, y_2, \ldots, y_n)} \, dy_1 \, dy_2 \ldots dy_n. \tag{1.169}$$

As a final example we use relationship (1.168) to obtain the result with which we started this section, namely that in a change from Cartesian to spherical polar coordinates, $dx \, dy \, dz$ becomes $r^2 \sin \theta \, dr \, d\theta \, d\phi$.

From equations (1.154 a) we calculate the required Jacobian as

$$J = \frac{\partial(x, y, z)}{\partial(r, \theta, \phi)} = \begin{vmatrix} \sin \theta \cos \phi & r \cos \theta \cos \phi & -r \sin \theta \sin \phi \\ \sin \theta \sin \phi & r \cos \theta \sin \phi & r \sin \theta \cos \phi \\ \cos \theta & -r \sin \theta & 0 \end{vmatrix}.$$

$$\tag{1.170}$$

The determinant can be evaluated most simply by expanding it with respect to the last row.

$$\begin{aligned} J &= \cos \theta [r^2 \sin \theta \cos \theta (\cos^2 \phi + \sin^2 \phi)] \\ &\quad + r \sin \theta [r \sin^2 \theta (\cos^2 \phi + \sin^2 \phi)] + 0 \\ &= r^2 \sin \theta (\cos^2 \theta + \sin^2 \theta) \\ &= r^2 \sin \theta. \end{aligned}$$

This establishes the stated result, that

$$dx \, dy \, dz = \frac{\partial(x, y, z)}{\partial(r, \theta, \phi)} \, dr \, d\theta \, d\phi = r^2 \sin \theta \, dr \, d\theta \, d\phi.$$

▶49. Verify formally that the general result (1.169) is trivially valid for a one-dimensional integral.

▶50. By defining new variables $\xi = x - y$ and $\eta = x + y$, evaluate the integral discussed in section 1.18 and fig. 1.10 (page 56)

$$\iint (x - y)^2 \, dx \, dy,$$

over the triangle with sides $x = 0$, $y = 0$, $x + y = 1$.

▶51. The moment of inertia I about the origin of a uniform solid body of density ρ is given by the integral

$$I = \int_V (x^2 + y^2 + z^2)\rho \, dV$$

over the volume of the body. Show that the moment of inertia of a right circular cylinder of radius a, length $2b$ and mass M, about its centre is

$$M\left(\frac{a^2}{2} + \frac{b^2}{3}\right).$$

[Transform to cylindrical polar coordinates r, ϕ, z given by $x = r \cos \phi$, $y = r \sin \phi$, $z = z$.]

1.20 Examples for solution

1. In the theory of special relativity the position coordinate x and time coordinate t in one frame of reference, are related to those in another by equations of the form:

$$x' = x \cosh \phi - ct \sinh \phi,$$
$$ct' = -x \sinh \phi + ct \cosh \phi.$$

Express x and ct in terms of x', ct' and ϕ and show that

$$x^2 - c^2 t^2 = (x')^2 - (ct')^2.$$

2. The $N + 1$ complex numbers ω_m are given by

$$\omega_m = \exp(2\pi im/N), \qquad m = 0, 1, 2, \ldots, N.$$

(a) Evaluate

(i) $\sum_{m=0}^{N} \omega_m$,   (ii) $\sum_{m=0}^{N} \omega_m^2$,   (iii) $\sum_{m=0}^{N} \omega_m x^m$.

(b) Use these results to evaluate

(i) $\sum_{m=0}^{N} \left(\cos \frac{2\pi m}{N} - \cos \frac{4\pi m}{N}\right)$,   (ii) $\sum_{m=0}^{3} 2^m \sin \frac{2\pi m}{3}$.

3. In a quantum theory of a system of oscillators the average energy $\bar{E}$ of the system is given by the expression

$$\bar{E} = \frac{\sum_{n=0}^{\infty} nh\nu \exp(-nx)}{\sum_{n=0}^{\infty} \exp(-nx)},$$

where $x = h\nu/kT$. Carry out the summation to find $\bar{E}$ and show that for small x, $\bar{E} \approx kT$, whilst for large x, $\bar{E} \approx h\nu \exp(-h\nu/kT)$.

4. How does the convergence of the infinite series $\sum\limits_{n=r}^{\infty} \dfrac{(n-r)!}{n!}$ depend on r (integral)?

5. For what positive values of x is $\sum\limits_{n=1}^{\infty} \dfrac{x^{n/2} e^{-n}}{n}$ convergent?

6. Prove that $\sum\limits_{n=2}^{\infty} \ln \left[\dfrac{n^r + (-1)^n}{n^r} \right]$ is absolutely convergent for $r = 2$, and convergent, but not absolutely convergent, for $r = 1$.

7. Find the values of x for which $f^{(4)}(x) = 0$ when $f(x) = x^2 \times \exp(-x/a)$. What are the values of $f^{(8)}(x)$ at these points? $(a > 0)$.

8. If $y(x) = \exp(-\tfrac{1}{2}x^2)$, show that $dy/dx + xy = 0$ and hence that

$$y^{(n+1)} + xy^{(n)} + ny^{(n-1)} = 0, \qquad n \geqslant 0.$$

If $g_n(x) = \exp(\tfrac{1}{2}x^2) y^{(n)}(x)$, show that for $n \geqslant 0$,

(i) g_n satisfies $g_n'' - xg_n' + ng_n = 0$,
(ii) $g_n' + ng_{n-1} = 0$,
(iii) $g_{n+1} + xg_n + ng_{n-1} = 0$.

Calculate $g_n(x)$ for $n = 0, 1, 2, 3$.

9. Find a polynomial expansion of $\operatorname{arsinh} x$ up to the term in x^5.

10. Use a Taylor expansion to show that

$$\lim_{x \to a} \frac{f(x)}{g(x)} = \lim_{x \to a} \frac{f'(x)}{g'(x)},$$

provided the second limit exists, given that $f(a) = g(a) = 0$. Use this result to evaluate

(a) $\lim\limits_{x \to 0} \dfrac{\sin(3x)}{\sinh(x)}$,

(b) $\lim\limits_{x \to 0} \dfrac{\sin x - x \cosh(x)}{\sinh(x) - x}$,

(c) $\lim\limits_{x \to 0} \dfrac{\tan(x) - x}{\cos(x) - 1}$.

11. A surface is defined in Cartesian coordinates by $z = x^2 + y^2$, where x and y are both functions of two other variables s and t. Calculate $\partial z/\partial t$ for the following cases,

(i) $x = s \cos t,\ y = s \sin t$,
(ii) $x = \tfrac{1}{2}(s^2 - t^2),\ y = st$,
(iii) $x = \cosh s \cos t,\ y = \sinh s \sin t$.

12. Find the locations and characters of the stationary points of the function $z = (x^2 - y^2) \exp(-x^2 - y^2)$.

13. Two particles each of unit mass move in a vertical plane; the
x-axis is horizontal and the y-axis vertical. The first particle is constrained to lie on the left branch of the hyperbola $x_1^2 - y_1^2 = a^2$
and the second on the vertical line $x_2 = 2a$. Both particles are
subject to gravity and to a mutually attractive force of magnitude
g/a times their relative displacement.

   Assuming that at equilibrium the potential energy of the whole
system is a minimum, show that the first particle lies at a point on
the hyperbola

$$(x_1 - 2a)(y_1 + 2a) + 4a^2 =$$

14. Evaluate the integrals

$$I = \int_0^\infty e^{-\alpha t} \cos \omega t \, dt, \qquad J = \int_0^\infty e^{-\alpha t} \sin \omega t \, dt,$$

by integrating $\exp(-\alpha t + i\omega t)$ between 0 and ∞. [Compare with
the methods used in example 1.4 (page 52).]

15. (i) Find positive constants a, b such that $ax \leqslant \sin x \leqslant bx$ for
$0 \leqslant x \leqslant \frac{1}{2}\pi$. Use this inequality to find upper and lower bounds
for the integral

$$I = \int_0^{\pi/2} (1 + \sin x)^{1/2} \, dx.$$

(Evaluate the bounds to 2 significant figures.)
(ii) Using the substitution $t = \tan(x/2)$, evaluate I exactly.

16. (More difficult.) Assume that the following equation is valid

$$\frac{d}{dy} \int_a^b f(x, y) \, dx = \int_a^b \frac{\partial f}{\partial y}(x, y) \, dx.$$

Use it to show that, for any real y,

$$I(y) \equiv \int_0^{\pi/2} \arcsin(\tanh y \sin x) \, dx = \int_0^y \frac{2t \, dt}{\sinh 2t},$$

and hence evaluate

$$K = \int_0^\infty \frac{t \, dt}{\sinh t}.$$

[Assume that $-\frac{1}{2}\pi \leqslant \arcsin x \leqslant \frac{1}{2}\pi$ if $-1 \leqslant x \leqslant 1$.]

17. Evaluate $\iint (x + 2y)^2 \, dx \, dy$ over the triangle bounded by the
lines $y = x$, $y = -x$, $x = 1$.

18. Evaluate $\int_S f(x, y) \, dA_{xy}$ over the rectangle $0 \leqslant x \leqslant \mu$, $0 \leqslant y \leqslant v$ for the functions

$$\text{(i)} \ f(x, y) = \frac{x}{x^2 + y^2}, \qquad \text{(ii)} \ f(x, y) = (v - y + x)^{-3/2}.$$

19. Sketch the two families of curves

$$y^2 = 4u(u - x), \qquad y^2 = 4v(v + x),$$

where u and v are parameters.

By transforming to the uv-plane evaluate the integral of $y/(x^2 + y^2)^{1/2}$ over that part of the quadrant $x > 0$, $y > 0$ bounded by the lines $x = 0$, $y = 0$ and the curve $y^2 = 4a(a - x)$.

2
Vector algebra

This chapter is of a preliminary nature and is designed to indicate the level of knowledge assumed in the development of the third and subsequent chapters. It deals with those elementary properties of vectors and their algebra which will be used later. The results and properties are usually stated without proof, but with illustrations, and a set of exercises is included at the end in section 2.10 to enable the student to decide whether or not further preliminary study is needed. It is suggested that the reader who already has some working familiarity with vector algebra might first attempt the exercises and return to this chapter only if he has difficulty with them.

2.1   Definitions

The simplest kind of physical quantity is one which can be completely specified by its magnitude, a single number together with the units in which it is measured. Such a quantity is called a **scalar** and examples include temperature, time, work, and [scalar] potential.

Quantities which require both a magnitude ($\geqslant 0$) and a direction in space† to specify them are known (with a few exceptions, such as finite rotations, discussed below) as vectors; familiar examples include position with respect to a fixed origin, force, linear momentum and electric field. Using an arbitrary but generally accepted convention, vectors can be used to represent angular velocities and momenta, the axis of rotation being taken as the direction of the vector and the sense being such that the rotation appears clockwise when viewed parallel [as opposed to anti-parallel] to the vector. The magnitude of the angular velocity or momentum gives the magnitude of the corresponding vector.

A further, less intuitively obvious use of vectors is to represent a plane element of area of a surface or to give the local orientation of the surface

---

† We will usually consider three-dimensional space, but most of the algebra of vectors has meaning in a space of another dimension.

[more precisely, of the local tangent plane to the surface]. In these cases the normal to the surface determines the direction of the vector, whilst its length (magnitude) is given by the size of the element in the first case or is unity in the second.

From what was said in the last paragraph but one, it is clear that finite rotations can be characterized by magnitudes and directions. However, they are not vectors, since they do not obey some of the 'working rules' given later in this chapter – for example, they do not add like vectors.

2.2   Notation

The fact that a vector has a sense of direction as well as a magnitude means that more than one number is needed to specify it. This specification can be done in many ways, but the simplest is in terms of **components** which give the amounts of each of a standard set of vectors [usually, but not always, mutually orthogonal] which must be added together [see section 2.5] to produce the given vector.

In common with most other textbooks, we will denote a vector by clarendon (bold) type, e.g. **a**, and its magnitude by italics a, except where confusion may arise, when the modulus sign $|\mathbf{a}|$ will be used explicitly. Both upper and lower case letters will be used to represent vectors.

The components of a vector with respect to a given standard set are written as $(a_1, a_2, \ldots, a_n)$, where the order of the standard vectors and of the components correspond. The magnitude, or modulus, of a vector **a** is given in terms of its components by $a^2 = a_1^2 + a_2^2 + \cdots + a_n^2$ if the standard set consists of mutually orthogonal unit vectors [see section 2.5]. Vectors for which $a = 1$ are called unit vectors and will be denoted by $\hat{\mathbf{a}}$. Clearly a unit vector $\hat{\mathbf{a}}$ in the same direction as **a** is obtained by dividing each of the components of **a** by a.

For the sake of definiteness and brevity, and since most physical problems occur in three-dimensional Euclidean space, we will continue our discussion of vectors as if they were all in three dimensions. Many later results and definitions in this chapter are expressed in terms of three-dimensional Cartesian coordinates (x, y, z), but with obvious modifications are equally valid for other coordinates and also in other dimensions. When a coordinate is denoted by x_i, where i is a general subscript running from 1 to n, it will be understood to refer to x, y or z for $i = 1, 2$ or 3 respectively when applied to a Cartesian coordinate system.

Two vectors **a** and **b** are equal, $\mathbf{a} = \mathbf{b}$, if they have equal moduli and the same direction in space. Equivalently their corresponding components, in terms of the same standard set, are individually equal, $a_i = b_i$ ($i = 1, 2, 3$).

2.3   Combining vectors

All students of elementary mechanics are familiar with the fact that the result of applying two separate forces **a** and **b** to a body is the same as applying a single force **c** to the body. This single force or resultant is written in vector form as

$$\mathbf{c} = \mathbf{a} + \mathbf{b}, \tag{2.1}$$

and can be determined graphically by the triangle law of addition (fig. 2.1). The components of **c** are given in terms of those of **a** and **b** by $c_i = a_i + b_i$ ($i = 1, 2, 3$), i.e. $c_1 = a_1 + b_1$, etc. Naturally both sets of components must be expressed in terms of the same set of standard vectors.

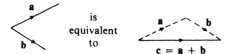

Fig. 2.1   Addition of vectors.

The **addition** of vectors in general is defined in accordance with this procedure, and it is clear from either way of constructing **c** that the addition of vectors is commutative,

$$\mathbf{a} + \mathbf{b} = \mathbf{b} + \mathbf{a}. \tag{2.2}$$

It can be generalized to the addition of any number of vectors and in this is associative, namely, in the case of four vectors for example,

$$(\mathbf{a} + \mathbf{b}) + \mathbf{c} + \mathbf{d} = \mathbf{a} + (\mathbf{b} + \mathbf{c}) + \mathbf{d} = \mathbf{a} + \mathbf{b} + (\mathbf{c} + \mathbf{d}). \tag{2.3}$$

Thus the order in which a number of vectors to be added appear is immaterial, but naturally they must all have the same dimension, and to make physical sense must represent quantities of like kind.

Fig. 2.2   Subtraction of vectors.

The extension to **subtraction** is obvious and may be summarized as: **a** − **b** is the vector given by **a** + (−**b**), where −**b** is the vector having the same magnitude and line of action as **b**, but in the opposite sense. The components of **a** − **b** are given by $(a_1 - b_1, a_2 - b_2, a_3 - b_3)$ and the cor-

responding graphical construction is illustrated in fig. 2.2 [note the directions of the arrows]. The subtraction of two equal vectors yields the
zero vector 0, which has zero modulus and no direction associated with it.

2.4   Multiplication of vectors

Apart from the multiplication together of two scalars to produce a further
scalar [e.g. volume × density = mass, all are scalars], the simplest
multiplication of physical quantities is that of a vector **a** by a scalar λ
to produce another vector **b** = λ**a**. The scalar λ may be positive, negative
or zero [or even complex for some applications]. If it is zero the resultant
b is the zero vector **0**, but apart from this case the direction of **b** is that
of **a** [or opposite to it if λ is negative] and its magnitude is $|\lambda|$ times that
of **a**.

Multiplication of a vector by a scalar is associative, commutative and
distributive, and these properties may be summarized for arbitrary vectors **a** and **b** and arbitrary scalars λ and μ by

$$(\lambda\mu)\mathbf{a} = \lambda(\mu\mathbf{a}) = \mu(\lambda\mathbf{a}), \tag{2.4}$$

and $$(\lambda + \mu)(\mathbf{a} + \mathbf{b}) = \lambda\mathbf{a} + \lambda\mathbf{b} + \mu\mathbf{a} + \mu\mathbf{b}. \tag{2.5}$$

As elementary illustrations of the addition and multiplication of vectors we consider the following examples. All vector positions are taken
from a common origin O.

Example 2.1. To find the vector equation of a line, one point **a** of which
is given and whose direction is parallel to another given vector **n**.

Let **r** be the position vector of an arbitrary point R on the line (fig. 2.3).

Fig. 2.3   The line in example 2.1.

Then **r** is the sum of **a** and a vector representing the displacement A to R.
But this displacement is parallel [or antiparallel] to **n** and hence is given
by λ**n** for some appropriate scalar λ. Thus the position of any arbitrary
point R on the line, and hence the equation of the line, is given by

$$\mathbf{r} = \mathbf{a} + \lambda\mathbf{n}, \tag{2.6}$$

where the parameter λ ranges over all real values.

Example 2.2. Any three non-collinear points in three dimensions determine a unique plane containing them. Find the (vector) equation of the plane containing the points (whose position vectors are) **a**, **b** and **c**.

Denote the points by A, B and C and an arbitrary point of the plane by R. Through R draw lines parallel to AC and AB as in fig. 2.4. Then,

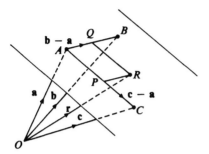

Fig. 2.4   The plane in example 2.2.

from the figure it is clear that the position **r** of R is given by the sum of **a**, a vector $[AQ]$ parallel to $\mathbf{b} - \mathbf{a}$, and a vector $[QR]$ parallel to $\mathbf{c} - \mathbf{a}$. Thus $\mathbf{r} = \mathbf{a} + h(\mathbf{b} - \mathbf{a}) + k(\mathbf{c} - \mathbf{a})$, or written more symmetrically with scalar parameters λ, μ, ν,

$$\mathbf{r} = \lambda\mathbf{a} + \mu\mathbf{b} + \nu\mathbf{c}, \text{ where } \lambda + \mu + \nu = 1. \tag{2.7}$$

2.5   Base vectors

We have mentioned already the standard set of vectors in terms of which the components of a vector generate the vector, namely that if the standard set are the vectors $\mathbf{e}_i$, where i runs from 1 to 3 [more generally 1 to n], then

$$\mathbf{a} = a_1\mathbf{e}_1 + a_2\mathbf{e}_2 + a_3\mathbf{e}_3, \tag{2.8}$$

or $\mathbf{a} = a_i\mathbf{e}_i$ (using the summation convention, described in section 1.9, that repeated subscripts are implicitly summed over, unless the contrary is stated). The vectors in this set are called the **base vectors**.

The choice of suitable base vectors is by no means unique and all that is required of a chosen set is that any vector in the space can be written as in (2.8) for some set of numbers a_1, a_2, a_3. This is so provided that,

(i) there are as many $\mathbf{e}_i$ as the dimension of the space, and
(ii) no one of them can be expressed as a linear sum [the form of (2.8)] of the others. [In three dimensions – they are not coplanar.]

In more formal language, the base vectors must (i) span the space, and (ii) be linearly independent. The linear independence can be expressed in the form that: if $\lambda_i e_i = \lambda_1 e_1 + \lambda_2 e_2 + \cdots = 0$, then all λ_i must be 0 if the vectors e_i are linearly independent. It is natural to use a set all of which are unit vectors and we will always assume that this is done.

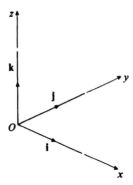

Fig. 2.5   The base vectors **i**, **j**, **k** of the Cartesian system.

When working in three-dimensional space the usual choice of base vectors is the mutually orthogonal set of three unit vectors in the directions of the Cartesian x-, y- and z-axes. They are denoted by **i**, **j** and **k** respectively [rather than the more strictly correct **î**, **ĵ** and **k̂**] and like the axes form a right-handed set when taken in their alphabetical order (see fig. 2.5). That is, **i**, **j** and **k** are in the same relative directions as the thumb, index finger and second finger respectively, of a right hand, when the fingers and thumb are held mutually at right angles. Components of a vector **a** with respect to this base are written (a_x, a_y, a_z) and so

$$\mathbf{a} = a_x \mathbf{i} + a_y \mathbf{j} + a_z \mathbf{k}. \tag{2.9}$$

Clearly the vector **i** itself is given by $(1, 0, 0)$, **j** by $(0, 1, 0)$, etc.

2.6   Scalar products

We turn now to the simplest form of multiplication of one vector **a** by another **b**. The result, called the **scalar product** of **a** and **b** and denoted by **a**·**b**, is a scalar quantity given by $ab \cos \theta$ where θ is the angle between the two vectors $(0 \leqslant \theta \leqslant \pi)$. If the two vectors are perpendicular $(\theta = \pi/2)$ their scalar product is zero. The converse of this is also true, unless at least one of the vectors is the zero vector.

From its definition, the scalar product is clearly equal to the length of **a** multiplied by the projection of **b** on the direction of **a** [or vice-versa], and so the process of taking a scalar product follows both the commutative and distributive laws.

▶1. By noting that

$$\mathbf{i} \cdot \mathbf{j} = \mathbf{j} \cdot \mathbf{k} = \mathbf{k} \cdot \mathbf{i} = 0, \tag{2.10}$$

and $$\mathbf{i} \cdot \mathbf{i} = \mathbf{j} \cdot \mathbf{j} = \mathbf{k} \cdot \mathbf{k} = 1, \tag{2.11}$$

show that for arbitrary vectors **a** and **b**, the scalar product is also given by

$$\mathbf{a} \cdot \mathbf{b} = a_x b_x + a_y b_y + a_z b_z, \tag{2.12}$$

and hence that

$$\cos \theta = l_a l_b + m_a m_b + n_a n_b, \tag{2.13}$$

where the direction cosines l_a, m_a, n_a are given respectively by a_x/a, a_y/a, a_z/a and similarly for vector **b**. Note that $l_a = \hat{\mathbf{a}} \cdot \mathbf{i}$, $m_a = \hat{\mathbf{a}} \cdot \mathbf{j}$, etc.

Examples of scalar products arise naturally throughout physics and in particular in connection with energy. Perhaps the simplest is the work done in moving the point of application of a constant force **F** through a displacement **r** when the work done is $\mathbf{F} \cdot \mathbf{r}$; notice that, as expected, if the displacement is perpendicular to the direction of the force then $\mathbf{F} \cdot \mathbf{r} = 0$ and no work is done. A second simple example is afforded by the potential energy $-\mathbf{m} \cdot \mathbf{B}$ of a magnetic dipole, represented in strength and orientation by a vector **m**, placed in an external magnetic field **B**.

Example 2.3. To find the equation of a plane that contains a given point **a** and is perpendicular to a given unit vector $\hat{\mathbf{n}}$.

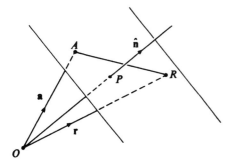

Fig. 2.6   The plane of example 2.3 containing the point **a** and with unit normal $\hat{\mathbf{n}}$.

If R is a general point of the plane with position **r** then the line AR, represented by vector $\mathbf{r} - \mathbf{a}$, lies in the plane and is therefore perpendicular to $\hat{\mathbf{n}}$. Thus we have for the equation of the plane

$$\hat{\mathbf{n}} \cdot (\mathbf{r} - \mathbf{a}) = 0, \quad \text{or} \quad \hat{\mathbf{n}} \cdot \mathbf{r} = \hat{\mathbf{n}} \cdot \mathbf{a}. \tag{2.14}$$

►2. Deduce that (2.14) can be written as $\hat{\mathbf{n}} \cdot \mathbf{r} = p$, where p is the perpendicular distance from O to the plane (the length of OP in fig. 2.6).

For some [less geometrical] applications of vectors to physical systems it is convenient to work with complex rather than purely real components. In these cases the scalar product definition has to be modified to

$$\mathbf{a}^* \cdot \mathbf{b} = a_i^* b_i = (a_i b_i^*)^* = (\mathbf{a} \cdot \mathbf{b}^*)^*, \tag{2.15}$$

where * indicates complex conjugation. With this wider definition, the property of commutation is lost in the sense that $\mathbf{a}^* \cdot \mathbf{b} \neq \mathbf{b}^* \cdot \mathbf{a}$ for general complex vectors $\mathbf{a}$ and $\mathbf{b}$. Of course $\mathbf{a}^* \cdot \mathbf{b}$ is still equal to $\mathbf{b} \cdot \mathbf{a}^*$. The determination of magnitude however from $a^2 = \mathbf{a} \cdot \mathbf{a}$ [see equation (2.12) and section 2.2] is essentially unaltered since $\mathbf{a}^* \cdot \mathbf{a} = \mathbf{a} \cdot \mathbf{a}^* = $ real quantity, even if $\mathbf{a}$ has complex components.

Finally it should be noted that the scalar product of two vectors is an **invariant,** that is its value does not depend upon the set of base vectors chosen [although its algebraic form may well do so]. This is to be compared with the vector components themselves, which are very much determined by the choice of base vectors.

2.7   Vector products

In addition to the scalar product, it is possible to obtain from two vectors a further quantity, which is itself a vector. This vector is called the **vector** (or **cross**) product of the two vectors and will be denoted by $\mathbf{a} \wedge \mathbf{b}$. It is defined as a vector $\mathbf{v}$ which is simultaneously perpendicular to both $\mathbf{a}$ and $\mathbf{b}$ [i.e. to the plane containing them], and has magnitude $ab \sin \theta$,

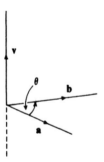

Fig. 2.7   The direction of the vector product $\mathbf{v} = \mathbf{a} \wedge \mathbf{b}$.

where again θ is the angle between $\mathbf{a}$ and $\mathbf{b}$. The sense of the vector $\mathbf{v}$ is that along which a rotation about $\mathbf{v}$ from the direction of $\mathbf{a}$ to the direction of **b** must be viewed in order to appear clockwise (see fig. 2.7). Stated in another way, **a**, **b** and **v** form a right-handed set. It is apparent from this definition that

$$\mathbf{b} \wedge \mathbf{a} = -(\mathbf{a} \wedge \mathbf{b}), \tag{2.16}$$

but that the distributive law holds for vector products

$$\mathbf{a} \wedge (\mathbf{b} + \mathbf{c}) = (\mathbf{a} \wedge \mathbf{b}) + (\mathbf{a} \wedge \mathbf{c}). \tag{2.17}$$

Other obvious relationships include,

$$\mathbf{a} \wedge \mathbf{a} = 0, \tag{2.18 a}$$

$$\mathbf{a} \cdot (\mathbf{a} \wedge \mathbf{b}) = 0 = \mathbf{b} \cdot (\mathbf{a} \wedge \mathbf{b}), \tag{2.18 b}$$

and $\quad (\mathbf{a} \cdot \mathbf{b})(\mathbf{a} \cdot \mathbf{b}) + (\mathbf{a} \wedge \mathbf{b}) \cdot (\mathbf{a} \wedge \mathbf{b}) = a^2 b^2. \tag{2.19}$

Two straightforward examples, taken from mechanics, of physical quantities which can be represented by the vector (cross) product of two other vectors, are the moment of a force about a point and the linear velocity of a point in a rotating body.

Consider first the force **F** (fig. 2.8) acting at the point **r**. Its moment about O is $F \times$ distance OP, which numerically is just $Fr \sin \theta$; in addition, the sense of the moment is clockwise about an axis through O perpendicularly into the plane of the paper. Thus the moment is completely represented by the vector **r** ∧ **F**, in both magnitude and spatial sense. [Imagine the lines of **r** and **F** extended beyond R when determining the sense of the vector product.]

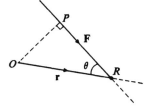

Fig. 2.8   The moment of the force **F** about O is **r** ∧ **F** (into the plane of the paper).

For the second example consider the point **r** of a body which is rotating with angular velocity ω about an axis OA through the origin. The angular rotation is represented, as discussed in section 2.1, by the vector **ω** in fig. 2.9. The instantaneous linear velocity of R is into the plane of the paper and its magnitude is $\omega \times (PR) = \omega r \sin \theta$. It is therefore completely represented by a vector **v** given by $\mathbf{v} = \boldsymbol{\omega} \wedge \mathbf{r}$.

Other examples of the vector product taken from the area of electromagnetism include the force on a current-carrying wire in a magnetic field, given in an obvious notation by $\mathbf{F} = \mathbf{I} \wedge \mathbf{B}$, and the energy flow or Poynting vector of an electromagnetic wave, $\mathbf{S} = \mathbf{E} \wedge \mathbf{H}$.

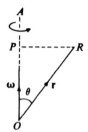

Fig. 2.9   The velocity of the point R in a rotating body is $\boldsymbol{\omega} \wedge \mathbf{r}$.

In Cartesian coordinates it is apparent that,

$$\mathbf{i} \wedge \mathbf{i} = \mathbf{j} \wedge \mathbf{j} = \mathbf{k} \wedge \mathbf{k} = 0, \tag{2.20}$$

$$\mathbf{i} \wedge \mathbf{j} = \mathbf{k} = -(\mathbf{j} \wedge \mathbf{i}), \qquad (\mathbf{j} \wedge \mathbf{k}) = \mathbf{i} = -(\mathbf{k} \wedge \mathbf{j}),$$
$$\mathbf{k} \wedge \mathbf{i} = \mathbf{j} = -(\mathbf{i} \wedge \mathbf{k}), \tag{2.21}$$

and from these that if $\mathbf{v} = \mathbf{a} \wedge \mathbf{b}$, then the components of $\mathbf{v}$ are given by

▶3. $(v_x, v_y, v_z) = (a_y b_z - a_z b_y, a_z b_x - a_x b_z, a_x b_y - a_y b_x).$ (2.22 a)

This last result is sometimes conveniently written in determinantal form,

$$\mathbf{a} \wedge \mathbf{b} = \begin{vmatrix} a_x & a_y & a_z \\ b_x & b_y & b_z \\ \mathbf{i} & \mathbf{j} & \mathbf{k} \end{vmatrix}. \tag{2.22 b}$$

2.8   Triple products

By combining the notions of scalar and vector products of two vectors, it is straightforward to construct quantities from three or more vectors. The two commonest of these, which merit some further discussion, are the triple products.

We first consider the **triple scalar product** formed in an obvious way from three vectors, namely $(\mathbf{a} \wedge \mathbf{b}) \cdot \mathbf{c}$ which is a simple number, and, using the form of $\mathbf{a} \wedge \mathbf{b}$ given in (2.22 b), can be written as

$$(\mathbf{a} \wedge \mathbf{b}) \cdot \mathbf{c} = \begin{vmatrix} a_x & a_y & a_z \\ b_x & b_y & b_z \\ c_x & c_y & c_z \end{vmatrix}. \tag{2.23}$$

From this form it follows immediately that

$$(\mathbf{a} \wedge \mathbf{b})\cdot\mathbf{c} = (\mathbf{c} \wedge \mathbf{a})\cdot\mathbf{b} = (\mathbf{b} \wedge \mathbf{c})\cdot\mathbf{a}$$
$$= -(\mathbf{b} \wedge \mathbf{a})\cdot\mathbf{c} = -(\mathbf{a} \wedge \mathbf{c})\cdot\mathbf{b} = -(\mathbf{c} \wedge \mathbf{b})\cdot\mathbf{a}. \quad (2.24)$$

The simplest geometrical interpretation of the triple scalar product is in terms of the volume of a parallelepiped whose edges are given by $\mathbf{a}$, $\mathbf{b}$ and $\mathbf{c}$ (see fig. 2.10). The vector $\mathbf{v} = \mathbf{a} \wedge \mathbf{b}$ is perpendicular to the base of the solid and has magnitude $v = ab \sin \theta$, i.e. the area of the base. Further $\mathbf{v}\cdot\mathbf{c} = vc \cos \phi$. Thus, since $c \cos \phi = OP = $ vertical height of the parallelepiped, it is clear that

$$(\mathbf{a} \wedge \mathbf{b})\cdot\mathbf{c} = (\text{area of the base}) \times (\text{perpendicular height})$$
$$= \text{volume}. \quad (2.25)$$

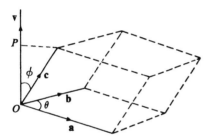

Fig. 2.10 The volume of the parallelepiped is the triple scalar product $(\mathbf{a} \wedge \mathbf{b})\cdot\mathbf{c}$.

The triple scalar product [and the volume of the solid] vanishes when $\mathbf{c}$ lies in the plane of $\mathbf{a}$ and $\mathbf{b}$, since then $\mathbf{v}$ is perpendicular to $\mathbf{c}$. This condition can also be expressed as $\mathbf{c} = \lambda\mathbf{a} + \mu\mathbf{b}$ for some λ and μ, a condition which is both necessary and sufficient for the vanishing of the triple scalar product.

Finally, from both its geometrical interpretation and the fact that it is a scalar product, we note that the triple scalar product is invariant with respect to the choice of base vectors.

We turn now to the vector form of triple product, which is given by $(\mathbf{a} \wedge \mathbf{b}) \wedge \mathbf{c}$ and can be written in terms of simpler products as

$$(\mathbf{a} \wedge \mathbf{b}) \wedge \mathbf{c} = (\mathbf{a}\cdot\mathbf{c})\mathbf{b} - (\mathbf{b}\cdot\mathbf{c})\mathbf{a}. \quad (2.26)$$

This alternative expression for the **triple vector product** can be understood in general form by reference to fig. 2.11. As remarked previously, $\mathbf{a} \wedge \mathbf{b}$ is perpendicular to the plane containing $\mathbf{a}$ and $\mathbf{b}$, and since $(\mathbf{a} \wedge \mathbf{b}) \wedge \mathbf{c}$ must be perpendicular to $\mathbf{a} \wedge \mathbf{b}$ it must lie in the plane containing $\mathbf{a}$ and $\mathbf{b}$. Hence it must be expressible as $\lambda\mathbf{a} + \mu\mathbf{b}$ for some λ and μ.

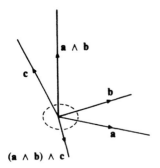

(a ∧ b) ∧ c

Fig. 2.11 The triple vector product (a ∧ b) ∧ c.

▶4. By applying (2.22 a) twice, show that $\lambda = -(\mathbf{b}\cdot\mathbf{c})$ and $\mu = (\mathbf{a}\cdot\mathbf{c})$.

▶5. By studying the form of (2.26) and/or by drawing diagrams similar to fig. 2.11, show that forming the triple vector product is in general non-associative, i.e.

$$(\mathbf{a} \wedge \mathbf{b}) \wedge \mathbf{c} \neq \mathbf{a} \wedge (\mathbf{b} \wedge \mathbf{c}).$$

2.9   Reciprocal vectors

We conclude this introductory chapter by defining the notion of **reciprocal vectors**. They have considerable use in solid state physics and crystallography but they will be only briefly discussed in this book.

Given a set of non-coplanar vectors **a**, **b**, **c**, their reciprocal vectors **a′**, **b′**, **c′** are defined by

$$\left.\begin{aligned} \mathbf{a'} &= \lambda^{-1}(\mathbf{b} \wedge \mathbf{c}), \\ \mathbf{b'} &= \lambda^{-1}(\mathbf{c} \wedge \mathbf{a}), \\ \mathbf{c'} &= \lambda^{-1}(\mathbf{a} \wedge \mathbf{b}), \end{aligned}\right\} \tag{2.27}$$

where λ is the triple scalar product of the three vectors. Clearly **a′** is perpendicular to **b** and **c**, but not to **a** since the three vectors are non-coplanar, and in fact **a′** is of such a length that its scalar product with **a** is unity.

▶6. Verify that

$$\begin{aligned} \mathbf{a'}\cdot\mathbf{a} &= \mathbf{b'}\cdot\mathbf{b} = \mathbf{c'}\cdot\mathbf{c} = 1, \\ \mathbf{a'}\cdot\mathbf{b} &= \mathbf{a'}\cdot\mathbf{c} = 0, \text{ etc.} \end{aligned} \tag{2.28}$$

►7. Show that the reciprocal vectors of $\mathbf{a}'$, $\mathbf{b}'$, $\mathbf{c}'$ are the original vectors.

It should be noted that $\mathbf{a}'$, $\mathbf{b}'$ and $\mathbf{c}'$ are not mutually perpendicular unless $\mathbf{a}$, $\mathbf{b}$ and $\mathbf{c}$ are.

2.10 Examples for solution

1. Which of the following statements are true? (The vectors $\mathbf{a}$, $\mathbf{b}$ and $\mathbf{c}$ are general vectors.)

(a) $\mathbf{c}\cdot(\mathbf{a} \wedge \mathbf{b}) = (\mathbf{b} \wedge \mathbf{a})\cdot\mathbf{c}$. ($b$) $\mathbf{a} \wedge (\mathbf{b} \wedge \mathbf{c}) = (\mathbf{a} \wedge \mathbf{b}) \wedge \mathbf{c}$. ($c$) $\mathbf{a} \wedge (\mathbf{b} \wedge \mathbf{c}) = (\mathbf{a}\cdot\mathbf{c})\mathbf{b} - (\mathbf{a}\cdot\mathbf{b})\mathbf{c}$. ($d$) $\mathbf{d} = \lambda\mathbf{a} + \mu\mathbf{b}$ implies $(\mathbf{a} \wedge \mathbf{b})\cdot \mathbf{d} = 0$. ($e$) $\mathbf{a} \wedge \mathbf{c} = \mathbf{b} \wedge \mathbf{c}$ implies $\mathbf{c}\cdot\mathbf{a} - \mathbf{c}\cdot\mathbf{b} = c|\mathbf{a} - \mathbf{b}|$. ($f$) $(\mathbf{a} \wedge \mathbf{b}) \wedge (\mathbf{c} \wedge \mathbf{b}) = \mathbf{b}(\mathbf{b}\cdot(\mathbf{c} \wedge \mathbf{a}))$.

2. Treating the earth as a sphere of radius R, find the distance measured on the earth's surface, between two points at co-latitude–longitude (θ_1, ϕ_1) and (θ_2, ϕ_2). In physics, θ is measured from the North Pole not from the Equator and is called co-latitude.

3. Prove Lagrange's identity

$$(\mathbf{a} \wedge \mathbf{b})\cdot(\mathbf{c} \wedge \mathbf{d}) = (\mathbf{a}\cdot\mathbf{c})(\mathbf{b}\cdot\mathbf{d}) - (\mathbf{a}\cdot\mathbf{d})(\mathbf{b}\cdot\mathbf{c}).$$

4. Find the length of the common perpendicular to the two non-parallel lines $\mathbf{r} = \mathbf{a} + \lambda\mathbf{b}$ and $\mathbf{r} = \mathbf{a}' + \mu\mathbf{b}'$ ($-\infty < \lambda, \mu < \infty$).

5. Three non-coplanar vectors $\mathbf{a}$, $\mathbf{b}$, $\mathbf{c}$, have as their respective reciprocal vectors the set $\mathbf{a}'$, $\mathbf{b}'$, $\mathbf{c}'$. Show that the normal to the plane containing the points $h^{-1}\mathbf{a}$, $k^{-1}\mathbf{b}$ and $l^{-1}\mathbf{c}$ is in the direction of the vector $h\mathbf{a}' + k\mathbf{b}' + l\mathbf{c}'$.

6. $ABCDEFGH$ is a cube of side $2a$ m and whose centre is O. J is the centre of the face $BCGF$ (see fig. 2.12).

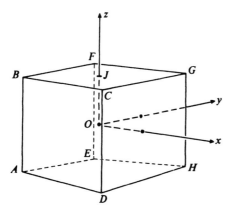

Fig. 2.12 The cube. All edges are $2a$ m in length.

(a) Express the area AEC as a vector.

(b) What is the angle between the face diagonal AF and the body diagonal AG?

(c) Find the equation of the plane through B, parallel to the plane CGE.

(d) Find (i) the perpendicular distance from J onto the plane OCG, and (ii) the volume of the tetrahedron $JOCG$.

(e) A force of 5 N acts at H in the direction $\overrightarrow{HJ}$. What is (i) its moment $\mathbf{M}$ about the point A, and (ii) its moment about the line $\overrightarrow{AF}$? (iii) If the force moves its point of application from H to G to F to J, what is the total work done?

(f) The cube is given two simultaneous angular velocities, one of 2 rad s^{-1} about the axis OJ and the other of $\sqrt{3}$ rad s^{-1} about OG. Find (i) the distance from J of the stationary point on the face $BCGF$, and (ii) the speed of corner C.

3
Calculus of vectors

This chapter is concerned with the differentiation and integration of vectors, both of vectors describing particular bodies, such as the velocity of a particle, and of vector fields in which a vector is defined as a function of the coordinates throughout some volume [one-, two- or three-dimensional].

Since our aim is to develop methods for handling problems arising from physical situations, we limit ourselves to functions which have sufficiently amenable mathematical properties. In particular we will assume that the functions with which we have to deal are continuous and differentiable, except where the opposite is physically obvious [e.g. a square-wave electrical input, or a sharp corner of an electrical conductor].

3.1   Derivatives of vectors and vector expressions

When a vector is a function of a scalar quantity it is possible to consider the derivatives of the vector with respect to that scalar. The most common such scalar is time and for the sake of definiteness we will call the scalar t and interpret it as time, although clearly all our results will be valid for any other analogous physical scalar.

Consider a vector $\mathbf{a}$, with value $\mathbf{a}(t)$ at time t, and value $\mathbf{a}(t + \Delta t)$ a short time Δt later. Then the small change in the vector during Δt is itself a vector $\Delta \mathbf{a} = \mathbf{a}(t + \Delta t) - \mathbf{a}(t)$, and we define the derivative (also a vector) as,

$$\frac{d\mathbf{a}}{dt} = \lim_{\Delta t \to 0} \frac{\mathbf{a}(t + \Delta t) - \mathbf{a}(t)}{\Delta t}, \tag{3.1}$$

assuming that this limit exists.

The most obvious example of a vector derived in this way is the velocity vector $\mathbf{v}$ of a particle which moves along a path given as a function of time, $\mathbf{r} = \mathbf{r}(t)$. Then the velocity vector is obtained as

$$\mathbf{v}(t) = \frac{d\mathbf{r}(t)}{dt},$$

and, being itself both a vector and a function of t, also gives rise to a further vector, the acceleration

$$\mathbf{f}(t) = \frac{d\mathbf{v}}{dt} = \frac{d^2\mathbf{r}}{dt^2}.$$

Clearly the magnitude of $\mathbf{v}$ at any time gives the speed of the particle, whilst its direction is along the tangent to the path at the instantaneous position of the particle.

In terms of coordinates [taken here as Cartesian] the derivative of a vector is given by

$$\frac{d\mathbf{a}}{dt} = \frac{da_x}{dt}\mathbf{i} + \frac{da_y}{dt}\mathbf{j} + \frac{da_z}{dt}\mathbf{k}. \tag{3.2}$$

[Note that $\mathbf{i}$, $\mathbf{j}$ and $\mathbf{k}$ are taken as fixed and time independent – this may not be true of base vectors in all cases.]

In the composite vector expressions discussed in chapter 2, each of the vectors and scalars may [in general] be functions of t, or perhaps functions of other scalars which are themselves functions of t. However, the procedures for obtaining their derivatives (with respect to t) are intuitively obvious from, and readily verifiable by, ordinary differential calculus.

They may be summarized by noting that differentiation is distributive and that

$$\frac{d\mathbf{a}(s)}{dt} = \frac{ds}{dt}\frac{d\mathbf{a}}{ds}, \text{ where scalar } s = s(t), \tag{3.3 a}$$

$$\frac{d(\lambda\mathbf{a})}{dt} = \lambda\frac{d\mathbf{a}}{dt} + \frac{d\lambda}{dt}\mathbf{a}, \text{ where } \lambda \text{ is a scalar}, \tag{3.3 b}$$

$$\frac{d}{dt}(\mathbf{a}\cdot\mathbf{b}) = \mathbf{a}\cdot\frac{d\mathbf{b}}{dt} + \frac{d\mathbf{a}}{dt}\cdot\mathbf{b}, \tag{3.3 c}$$

$$\frac{d}{dt}(\mathbf{a}\wedge\mathbf{b}) = \mathbf{a}\wedge\frac{d\mathbf{b}}{dt} + \frac{d\mathbf{a}}{dt}\wedge\mathbf{b}, \tag{3.3 d}$$

where of course in (3.3 d) the order of the factors in the products on the right is just as important as it was in the original vector product. The derivatives of more complicated vector expressions can be evaluated by the repeated use of these equations.

The integration of a vector (or an expression containing vectors) with respect to a scalar t can be regarded in a straightforward way as the inverse of differentiation. The only points to be noticed are,

(i) the integral has the same nature (vector or scalar) as the integrand, and
(ii) the constant of integration must be of the same nature as the integral.

Two simple illustrations will suffice.

► 1. Show $\int \left[\left(\mathbf{a} \wedge \dfrac{d^2\mathbf{b}}{dt^2} \right) + \left(\mathbf{b} \wedge \dfrac{d^2\mathbf{a}}{dt^2} \right) \right] dt$

$$= \left(\mathbf{a} \wedge \dfrac{d\mathbf{b}}{dt} \right) + \left(\mathbf{b} \wedge \dfrac{d\mathbf{a}}{dt} \right) + \mathbf{h}.$$

► 2. Show $\int \left[\dfrac{d\mathbf{a}}{dt} \cdot (\mathbf{b} \wedge \mathbf{c}) + \mathbf{a} \cdot \left(\dfrac{d\mathbf{b}}{dt} \wedge \mathbf{c} \right) + \mathbf{a} \cdot \left(\mathbf{b} \wedge \dfrac{d\mathbf{c}}{dt} \right) \right] dt$

$$= \mathbf{a} \cdot (\mathbf{b} \wedge \mathbf{c}) + k.$$

Here $\mathbf{h}$ is an arbitrary constant vector and k an arbitrary constant scalar.

3.2   Vectors with several arguments

The ideas of the last section are easily extended to cases where the vectors [or scalars] are functions of more than one independent scalar, $t_1, t_2, \ldots, t_n$, when the derivatives become partial derivatives $\partial \mathbf{a}/\partial t_i$ defined as in ordinary differential calculus. Equation (3.3 a), in particular, generalizes to the chain rule of partial differentiation. If $\mathbf{a} = \mathbf{a}(s_1, s_2, \ldots, s_n)$, where each s_j is a function $s_j(t_1, t_2, \ldots, t_n)$ of the t_i, then

$$\frac{\partial \mathbf{a}}{\partial t_i} = \frac{\partial \mathbf{a}}{\partial s_1}\frac{\partial s_1}{\partial t_i} + \frac{\partial \mathbf{a}}{\partial s_2}\frac{\partial s_2}{\partial t_i} + \cdots + \frac{\partial \mathbf{a}}{\partial s_n}\frac{\partial s_n}{\partial t_i} = \frac{\partial \mathbf{a}}{\partial s_j}\frac{\partial s_j}{\partial t_i}. \tag{3.4}$$

A special case of this rule arises when $\mathbf{a}$ is a function of t explicitly as well as of other scalars $(s_1, s_2, \ldots, s_n)$ which are themselves functions of t. Then we have,†

$$\frac{d\mathbf{a}}{dt} = \frac{\partial \mathbf{a}}{\partial t} + \frac{\partial \mathbf{a}}{\partial s_j}\frac{\partial s_j}{\partial t}. \tag{3.5}$$

In the derivation of (3.1), the notion of a small [and, in the limit $\Delta t \to 0$, infinitesimal] vector $\Delta \mathbf{a}$ was used. In sections 3.4 onwards, we will have repeated need for this idea and, as can be seen from the definition, the **differential** $d\mathbf{a}$ can be written as

$$d\mathbf{a} = \frac{d\mathbf{a}}{dt}\, dt, \tag{3.6}$$

or for vectors dependent on several variables,

$$d\mathbf{a} = \frac{\partial \mathbf{a}}{\partial t_i}\, dt_i. \tag{3.7}$$

† In the interests of compactness we will henceforth use the summation convention, wherever it does not lead to ambiguity.

As specific examples we may consider the infinitesimal change in position of a particle moving with velocity $\mathbf{v}$ in infinitesimal time dt

$$\mathbf{v}\,dt = d\mathbf{r} = v_x\,dt\,\mathbf{i} + v_y\,dt\,\mathbf{j} + v_z\,dt\,\mathbf{k}, \tag{3.8}$$

or the change in an electric field $\mathbf{E}$ in moving from one position $\mathbf{r}$ to a neighbouring one $\mathbf{r} + d\mathbf{r}$ [here x, y, z are the scalar arguments]

$$d\mathbf{E} = \frac{\partial \mathbf{E}}{\partial x}\,dx + \frac{\partial \mathbf{E}}{\partial y}\,dy + \frac{\partial \mathbf{E}}{\partial z}\,dz. \tag{3.9}$$

Written more explicitly, $d\mathbf{a}$ is a vector whose x-component is

$$\frac{\partial a_x}{\partial x}\,dx + \frac{\partial a_x}{\partial y}\,dy + \frac{\partial a_x}{\partial z}\,dz,$$

and similarly for the other components with a_x replaced by a_y and a_z in turn. [In chapter 4, expression (3.9) will be written more compactly using the vector operator ∇ as $(d\mathbf{r}\cdot\nabla)\mathbf{E}$.]

A further important differential expression is that used to represent an infinitesimal element of area dS on a surface [in the limit it can be accurately represented as plane]. In this case the vector is as for a finite area (see section 2.1) except for its magnitude, thus,

$$d\mathbf{S} = dS(n_x\mathbf{i} + n_y\mathbf{j} + n_z\mathbf{k}), \tag{3.10}$$

where (n_x, n_y, n_z) are the components of a unit vector in the direction of the *outward* normal to the surface [outward with respect to the volume whose surface is being considered].

In connection with integration in particular, it should always be borne in mind that the differentials $d\mathbf{r}$ and $d\mathbf{S}$ are vectors with a sense of direction, which in general will vary over the range of the integration. This is to be contrasted with, for example, integrals with respect to volume, where the infinitesimal element is a scalar e.g. $dV = r^2 \sin\theta\,dr\,d\theta\,d\phi$, and carries no directional sense.

3.3   Fields and line integrals

We now turn to the case when a particular scalar or vector quantity is defined not just at one point in space, but continuously throughout some volume of the space [often the whole space]. The space then has associated with it a **scalar** or **vector field,** the variation of which from point to point we will assume to be both continuous and differentiable.

Simple examples of scalar fields include the pressure at a point in a fluid and the electrostatic potential in the presence of an electric charge, whilst vector fields drawn from the same areas are the velocity vector in the fluid [giving the local strength and direction of flow] and the electric field. The familiar notion of a line of force in the electrostatic case is then simply a curve whose tangent anywhere is in the direction of the vector field at that point. Through each point of the space there is in general one and only one such line. In the hydrodynamic case the corresponding physical interpretation is that of a streamline. So long as the field does not vary with time these tangent lines are also the paths that would be followed by small free test charges in the field or particles in the fluid.

With the study of continuously varying vector and scalar fields there arises the need to consider the integration of the field quantities, along lines, over surfaces and throughout volumes in the space of the field. We start with **line integration** between two given points [or one fixed one and a variable upper limit] and along a prescribed path. The path may be given parametrically by $x = x(u)$, $y = y(u)$, $z = z(u)$ or by means of simultaneous equations relating x, y, z of the path.

Before considering the important line integrals discussed in section 3.4 where the 'variable of integration' is a vector, we deal briefly, by means of examples, with two cases in which the differential is a scalar. In the first of these the integrand is also a scalar [field], whilst in the second it is a vector.

Example 3.1. For the scalar function $\phi(x, y) = (x - y)^2$, evaluate $\int_L \phi \, dl$ where dl is an elemental length of the curve L which is the semicircle of radius a joining $(a, 0)$ and $(-a, 0)$ [in that sense] and for which $y \geq 0$.

For this path $dl = a \, d\theta$, where the parameter θ has an obvious geometrical meaning, and $(x - y)^2 = a^2(1 - \sin 2\theta)$. Thus,

$$\int_L \phi \, dl = \int_0^\pi a^3(1 - \sin 2\theta) \, d\theta = \pi a^3.$$

Example 3.2. Evaluate $\int \mathbf{F} \, dt$ along the path $x = ct$, $y = c/t$, $z = d$, $(1 \leq t \leq 2$, c and d are constants) and $\mathbf{F}$ is the vector (field) with Cartesian components $(xy^2, 2, x)$.

We observe that the integral will itself be a vector, with its three components each given by a [scalar] integral. On substituting we obtain

$$\int_1^2 \mathbf{F} \, dt = \mathbf{i} \int_1^2 \frac{c^3}{t} \, dt + \mathbf{j} \int_1^2 2 \, dt + \mathbf{k} \int_1^2 ct \, dt$$

$$= c^3 \ln 2 \, \mathbf{i} + 2 \, \mathbf{j} + \frac{3}{2} c \, \mathbf{k}.$$

▶3. Show, by writing $dy = (dy/dt) \, dt$, that $\int_c^{c/2} \mathbf{F} \, dy = -\frac{3}{8}c^4 \, \mathbf{i} - c \, \mathbf{j} - c^2 \ln 2 \, \mathbf{k}$. It will be seen that this $\neq \int \mathbf{F} \, dt$.

3.4   Line integrals with respect to vectors

We now obtain an expression for the total work done by a force **F** when it moves its point of application along a continuous curve C joining two points A and B, which may or may not coincide (see fig. 3.1). We allow the magnitude and direction of **F** to vary along the curve. Let the force

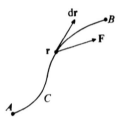

Fig. 3.1   Tangential line integral of **F** along curve C.

be acting at a point **r** and consider a small displacement d**r** along the curve, then the small amount of work done $dW = \mathbf{F} \cdot d\mathbf{r}$, as discussed in section 2.6. [Note that this can be either positive or negative.] It follows in the usual way that the total work done is

$$W_C = \int_C \mathbf{F} \cdot d\mathbf{r} = \int_A^B (F_x \, dx + F_y \, dy + F_z \, dz). \qquad (3.11)$$

Once the curve C has been specified, this is the sum of three straightforward integrals.

The scalar quantity formed by an integral of the form (3.11) is called the **tangential line integral** of **F** from A to B along C. Naturally its occurrence is not confined to cases involving forces explicitly although this is the commonest physical origin; for example, the electrostatic potential energy gained by moving a charge e along a path C in an electric field **E** is $-e \int_C \mathbf{E} \cdot d\mathbf{r}$. We may also note that Ampère's law concerning the magnetic field associated with a current-carrying wire can be written as a tangential line integral,

$$\int_C \mathbf{B} \cdot d\mathbf{r} = \mu_0 I, \qquad (3.12)$$

where I is the current enclosed by a closed path C traversed in a right-handed sense with respect to the current direction.

A very simple example of such a line integral, in which **F** is a force, independent of position, is provided by part (iii) of example 6(e) of section 2.10. There, the curve C consists of the straight line segments, HG,

GF and FJ, whilst F_x, F_y, F_z are $-5/\sqrt{6}$, $-5/\sqrt{6}$, $10/\sqrt{6}$ respectively. As a slightly more complicated evaluation we consider the following.

Example 3.3. Evaluate $\int \mathbf{F} \cdot d\mathbf{r}$ for the $\mathbf{F}$ and path of example 3.2.

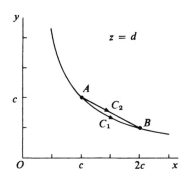

Fig. 3.2   The paths C_1 and C_2 of example 3.3 and example 3 of section 3.8.

The path joining A, (c, c, d), and B, $(2c, \frac{1}{2}c, d)$ and denoted by C_1 in fig. 3.2 is given by

$$x = ct, \qquad y = c/t, \qquad z = d.$$

Thus $dx = c\, dt$, $dy = -c/t^2\, dt$, $dz = 0$, and on substituting in (3.11) we obtain

$$\int_{C_1} \mathbf{F} \cdot d\mathbf{r} = \int_1^2 \left(\frac{c^4}{t} - \frac{2c}{t^2} + 0 \right) dt$$

$$= c^4 \ln 2 - c.$$

▶4. Evaluate the same integral between A and B, but this time along the curve C_2 (straight line) $2y = 3c - x$, $z = d$. Show that it has value $\frac{13}{16}c^4 - c$.

It will be seen from these two examples that the value of the integral depends not only on the vector $\mathbf{F}$ and the end points A and B, but also on the path C taken. [The two values are numerically close because the two curves C_1 and C_2 nearly coincide.] However, anticipating one of the results of chapter 4, we record here that for certain important kinds of vectors $\mathbf{F}$, the integral is in fact independent of the path taken. In these cases, since $\int_A^B = -\int_B^A$ for all integrals, the value of $\int \mathbf{F} \cdot d\mathbf{r}$ taken round any simply connected closed loop $APBQA$ is zero for arbitrary positions of P and Q.

We conclude this section by obtaining a line integral which is a vector. Consider a loop of wire L carrying a current I and placed in a magnetic field $\mathbf{B}$. Then the force $d\mathbf{F}$ on a small length $d\mathbf{r}$ of the wire is given by Fleming's left-hand rule as $d\mathbf{F} = I \, d\mathbf{r} \wedge \mathbf{B}$, and so the total force on the loop is

$$\mathbf{F} = I \int_L d\mathbf{r} \wedge \mathbf{B}. \tag{3.13}$$

▶5. Write the Biot–Savart law of electromagnetism in vector form and hence obtain as a double integral an expression for the total force between two current-carrying loops.

3.5   Surface integrals

As with line integrals, integrals taken over surfaces can involve vectors and scalars, and equally can result in either a vector or a scalar. The simplest case involves entirely scalars and is exemplified by the computation of the total electric charge on a surface, or of the mass of a shell, when the charge or mass density is known,

$$\int_S \rho(\mathbf{r}) \, dS. \tag{3.14}$$

This is a double integral, since dS is a general notation for an element of area [note that here we are only concerned with the magnitude of the area, hence dS not $d\mathbf{S}$], and in particular cases could be $dx \, dy$ or $r^2 \sin \theta \, d\theta \, d\phi$ or many others. However once a suitable coordinate system has been chosen and dS and ρ expressed or given in terms of it, such a scalar integral is straightforward to evaluate.

Next consider a vector field $\mathbf{F}$ defined throughout a region containing the surface S and take, as a particular case of the scalar density, the normal component of $\mathbf{F}$. If, as discussed in section 3.2, we define a unit vector $\hat{\mathbf{n}}$ as the local outward normal to the surface, then the normal component of $\mathbf{F}$ is the scalar density $\mathbf{F} \cdot \hat{\mathbf{n}}$. The resultant integral is called the (*outward*) *flux* of $\mathbf{F}$ through S,

$$\int_S \mathbf{F} \cdot \hat{\mathbf{n}} \, dS = \int_S \mathbf{F} \cdot d\mathbf{S}, \tag{3.15}$$

this last form being justified by (3.10). This is illustrated in fig. 3.3.

Ready examples of physically important flux integrals are (i) the total volume or mass of a fluid crossing a given area in unit time inside a fluid in motion, $\int_S \rho(\mathbf{r})\mathbf{v}(\mathbf{r}) \cdot d\mathbf{S}$ where $\mathbf{v}(\mathbf{r})$ is the velocity field, and (ii) the electromagnetic flux of energy out of a given volume bounded by a surface S, $\int_S (\mathbf{E} \wedge \mathbf{H}) \cdot d\mathbf{S}$.

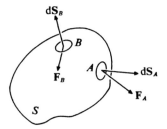

Fig. 3.3   The flux of **F** through surface S is $\int \mathbf{F} \cdot d\mathbf{S}$. The contribution to the integral is positive at A and negative at B.

As an additional example of a flux integral we note also that Gauss's theorem of electrostatics – over any closed surface, the integral of the normal component of the electric displacement is equal to the total charge enclosed – can be written in this form,

$$\int_S \mathbf{D} \cdot d\mathbf{S} = \sum_i q_i. \tag{3.16}$$

If no dielectric is present $\mathbf{D} = \epsilon_0 \mathbf{E}$, and for an isotropic dielectric $\mathbf{D} = \epsilon \epsilon_0 \mathbf{E}$.

The solid angle subtended at a point O by a surface [closed or otherwise] can also be represented by an integral of the form (3.15) although it is not strictly a flux integral [except for imaginary isotropic rays radiating from O]. The integral

$$\Omega = \int_S \frac{\mathbf{r} \cdot d\mathbf{S}}{r^3}, \tag{3.17}$$

gives the solid angle subtended at O by surface S if $\mathbf{r}$ is the position of an element of the surface measured from O. A little thought will show that the expression $\mathbf{r} \cdot d\mathbf{S}/r^3$ or $\hat{\mathbf{r}} \cdot d\mathbf{S}/r^2$ takes account of all three relevant factors, the size of the element of surface, its inclination to the line joining the element to O and the distance from O. Such a general expression is often useful for computing solid angles when the three-dimensional geometry is complicated.

It will be noted that (3.17) remains valid when the surface S is not convex and a single ray from O in certain directions would cut S in more than one place [but we exclude multiply-connected surfaces]. In particular, if the surface is closed, (3.17) yields zero if O is outside S, and 4π if it is an interior point.

Surface integrals resulting in vectors are also possible, although they occur less frequently in physical situations. However, by way of illustration, we may quote the example of the total resultant force experienced by a body immersed in a stationary fluid, in which the hydrostatic pressure p is given as $p(\mathbf{r})$. The pressure is everywhere inwardly directed and so the resultant force is the integral of $-p\,d\mathbf{S}$ taken over the whole surface. An exercise relating this to the Archimedean upthrust is given in the examples at the end of this chapter.

3.6   Volume integrals

Volume integrals are defined in an obvious way, and they too may result in scalars or vectors. Two closely related examples, one of each kind, are provided by the total mass of a fluid contained in a volume V and given by $\int_V \rho(\mathbf{r})\,dV$, and the total linear momentum of that same fluid $\int_V \rho(\mathbf{r})\mathbf{v}(\mathbf{r})\,dV$, where $\mathbf{v}(\mathbf{r})$ is the velocity vector field in the fluid.

As a slightly more complicated example of a volume integral we may consider the following example.

Example 3.4. Find an expression for the angular momentum of a solid body rotating with angular velocity $\boldsymbol{\omega}$ about an axis through the origin.

Consider a small volume element dV situated at position $\mathbf{r}$; its linear momentum is $\rho\,dV\dot{\mathbf{r}}$, where $\rho = \rho(\mathbf{r})$ is the density distribution, and its angular momentum about O is $\mathbf{r}\wedge\rho\dot{\mathbf{r}}\,dV$. Thus for the whole body the angular momentum $\mathbf{H}$ is

$$\mathbf{H} = \int_V (\mathbf{r}\wedge\dot{\mathbf{r}})\rho\,dV.$$

Putting $\dot{\mathbf{r}} = \boldsymbol{\omega}\wedge\mathbf{r}$ yields

$$\mathbf{H} = \int_V (\mathbf{r}\wedge(\boldsymbol{\omega}\wedge\mathbf{r}))\rho\,dV$$

$$= \int_V \boldsymbol{\omega}r^2\rho\,dV - \int_V (\mathbf{r}\cdot\boldsymbol{\omega})\mathbf{r}\rho\,dV.$$

▶6. Show that in a Cartesian system this result can be written in terms of coordinates as

$$H_i = T_{ij}\omega_j \qquad (i,j = x, y, z),$$

where T_{ij} is a 3×3 array in which

$$T_{xx} = \int_V (r^2 - x^2)\rho\,dV, \text{ similarly } T_{yy}, T_{zz},$$

$$T_{xy} = -\int_V xy\rho \, dV = T_{yx}, \text{ similarly } T_{yz}, T_{zx}.$$

[The T_{xx}, etc. are called the moments of inertia and $-T_{xy}$ the products of inertia. T_{ij} is known as the inertia tensor and is discussed further in chapter 15.]

3.7 Non-Cartesian systems

Because of the nature or symmetry of some physical situations it is useful to be able to describe a system in other than Cartesian coordinates, and for the work described in this chapter to be able to express, in particular, lengths, areas and volumes in these other systems.

In three dimensions we will need three coordinates, call them u_i ($i = 1, 2, 3$), in order to describe uniquely the position of a point in space by a set of particular values (u_1, u_2, u_3) [in Cartesians they are x, y, z]. In this brief treatment, we will confine ourselves to orthogonal systems of coordinates, that is ones in which the three surfaces on which the u_i have constant values $u_1 = \alpha$, $u_2 = \beta$, $u_3 = \gamma$ and which pass through the point (α, β, γ), meet at that point at right angles. For example, in spherical polar coordinates, $u_1 = r$, $u_2 = \theta$, $u_3 = \phi$ and the three surfaces through the point (R, Θ, Φ) are the sphere $r = R$, the circular cone $\theta = \Theta$, and the plane $\phi = \Phi$.

We remark that u_1, u_2, u_3 need not themselves have the dimensions of length. For example, of the polar coordinates r, θ, ϕ only r has dimensions of length. However a small change du_i in one of them, causes the point originally at (α, β, γ) to move through a small length dl given by $dl = h_i \, du_i$ (no summation), where h_i may be a function of the values α, β, γ, but is independent of the size of du_i so long as the latter is small. To take a concrete example, if again u_1, u_2, u_3 are the polar coordinates r, θ, ϕ, then for a change $d\phi$, $dl = r \sin \theta \, d\phi$, i.e. $h_3 = r \sin \theta$.

Since we are considering only orthogonal systems,† and hence the changes of position in space corresponding to changes du_1, du_2 and du_3 separately are at right angles to each other, the total change is given by

$$|d\mathbf{r}|^2 = h_i^2 (du_i)^2 \qquad \text{(summation convention)}. \qquad (3.18)$$

This is to be considered as the defining equation for the h_i.

We will be concerned almost exclusively with three common cases, Cartesians, spherical polars and cylindrical polars. For these the elements

---

† The extension to non-orthogonal systems can be seen by writing $d\mathbf{r} = (\partial \mathbf{r}/\partial u_i) \times du_i$ and then forming $d\mathbf{r} \cdot d\mathbf{r}$. For non-orthogonal systems some terms of the form $du_i \, du_j$ ($i \neq j$) will remain.

are set out in table 3.1 below and are sufficiently familiar for no further explanation to be needed.

[Whilst r is the internationally recommended symbol in both spherical and cylindrical coordinates, the two r's are not the same, and the use of ρ for the distance from the z-axis in cylindrical polar systems has something to recommend it.]

Table 3.1

	Cartesian	Spherical polar	Cylindrical polar
$u_1\, u_2\, u_3$	$x\ y\ z$	$r\ \theta\ \phi$	$r\ \phi\ z$
$h_1\, h_2\, h_3$	$1\ 1\ 1$	$1\ r\ r\sin\theta$	$1\ r\ 1$
$(\mathrm{d}r)^2$	$\mathrm{d}x^2 + \mathrm{d}y^2 + \mathrm{d}z^2$	$\mathrm{d}r^2 + r^2\,\mathrm{d}\theta^2 + r^2\sin^2\theta\,\mathrm{d}\phi^2$	$\mathrm{d}r^2 + r^2\,\mathrm{d}\phi^2 + \mathrm{d}z^2$
$\mathrm{d}S_1$	$\mathrm{d}y\,\mathrm{d}z$	$r^2\sin\theta\,\mathrm{d}\theta\,\mathrm{d}\phi$	$r\,\mathrm{d}\phi\,\mathrm{d}z$
$\mathrm{d}S_2$	$\mathrm{d}x\,\mathrm{d}z$	$r\sin\theta\,\mathrm{d}r\,\mathrm{d}\phi$	$\mathrm{d}r\,\mathrm{d}z$
$\mathrm{d}S_3$	$\mathrm{d}x\,\mathrm{d}y$	$r\,\mathrm{d}r\,\mathrm{d}\theta$	$r\,\mathrm{d}r\,\mathrm{d}\phi$
$\mathrm{d}V$	$\mathrm{d}x\,\mathrm{d}y\,\mathrm{d}z$	$r^2\sin\theta\,\mathrm{d}r\,\mathrm{d}\theta\,\mathrm{d}\phi$	$r\,\mathrm{d}r\,\mathrm{d}\phi\,\mathrm{d}z$

3.8   Examples for solution

1. Write the following physical laws or descriptions in vector notation:

(a) Newton's second law for a variable mass particle;

(b) the equation of motion of a simple harmonic oscillator;

(c) Faraday's law for the electric field in a conductor cutting a magnetic field;

(d) the motion of a charged particle in an electric and magnetic field;

(e) the moment of all external forces acting on a system of particles equals the rate of change of the total angular momentum.

2. Evaluate $\int [\mathbf{a}(\dot{\mathbf{b}}\cdot\mathbf{a} + \mathbf{b}\cdot\dot{\mathbf{a}}) + \dot{\mathbf{a}}(\mathbf{b}\cdot\mathbf{a}) - 2(\dot{\mathbf{a}}\cdot\mathbf{a})\mathbf{b} - \dot{\mathbf{b}}a^2]\,\mathrm{d}t$, where $\dot{x}$ stands for $\mathrm{d}x/\mathrm{d}t$.

3. For the vector with components $(xy^2 + z, x^2y + 2, x)$ evaluate the tangential line integrals along the curves C_1 and C_2 of fig. 3.2 (p. 89). Show that they are the same, each equal to $c(d - 1)$. [The equations of C_1 and C_2 are given in example 3.3 and ▶4.]

4. A single-turn coil C of arbitrary shape is placed in a magnetic field $\mathbf{B}$. Show that the couple on the coil can be written as

$$\mathbf{M} = I\int_C (\mathbf{B}\cdot\mathbf{r})\,\mathrm{d}\mathbf{r} - I\int_C \mathbf{B}(\mathbf{r}\cdot\mathrm{d}\mathbf{r}),$$

where I is the current carried by the coil.

For a planar rectangular coil of sides $2a$ and $2b$ placed with its plane vertical, at an angle ϕ to a uniform horizontal field $\mathbf{B}$, evaluate $\mathbf{M}$, showing that it gives the expected answer $(0, 0, 4abBI \cos \phi)$.

5. Show that the solid angle subtended by a rectangular aperture of sides $2a$ and $2b$ at a point a distance c away from the plane of the aperture, and on that plane normal which passes through the centre of the rectangle, is

$$\Omega = 4 \int_0^b \frac{ac \, dy}{(y^2 + c^2)(y^2 + c^2 + a^2)^{1/2}}.$$

Verify that if $c \gg a$ and b, then the expected approximate value $4ab/c^2$ is obtained. [The integral can be evaluated exactly to give $\Omega = 4 \arctan [ab/c(a^2 + b^2 + c^2)^{1/2}]$.]

6. An axially symmetric body with its axis AB vertical is immersed in an incompressible fluid of density ρ. By evaluating the vertical (z direction) component of the resultant force $(-\int p \, d\mathbf{S})$ on the body show that the Archimedian upthrust is as expected $\rho g V$, with an obvious notation. [Take the radius as $r = r(z)$, with $r(z_A) = r(z_B) = 0$.]

7. In a Cartesian system A and B are the points $(0, 0, -1)$ and $(0, 0, 1)$ respectively. In a new coordinate system a general point P is given by (u_1, u_2, u_3) with $u_1 = \frac{1}{2}(r_1 + r_2)$, $u_2 = \frac{1}{2}(r_1 - r_2)$, $u_3 = \phi$, where r_1, r_2 are the distances AP and BP respectively and ϕ is the angle between the planes ABP and $y = 0$.

(a) Express z and the distance ρ from P to the z-axis in terms of u_1, u_2, u_3.
(b) Evaluate $\partial x/\partial u_i$, $\partial y/\partial u_i$, $\partial z/\partial u_i$ ($i = 1, 2, 3$).
(c) Show that the Cartesian components of $\hat{\mathbf{u}}_j$ are $(\partial x/\partial u_j, \partial y/\partial u_j, \partial z/\partial u_j)$. Hence show that the new coordinates are mutually orthogonal, and evaluate the quantities h_1, h_2, h_3 and the element of volume in the new coordinate system.
(d) Determine and sketch the forms of the surfaces $u_i = $ constant ($i = 1, 2, 3$).

4
Vector operators

The next step in the theory of vectors is the introduction of vector operators and in particular the differential operators denoted by div, grad and curl. It is usual to define these operators from a strictly mathematical point of view as follows. For an arbitrary scalar ϕ and an arbitrary vector $\mathbf{a}$,

(i) grad ϕ is a vector with components given by

$$\text{grad } \phi \equiv \left(\frac{\partial \phi}{\partial x}, \frac{\partial \phi}{\partial y}, \frac{\partial \phi}{\partial z} \right), \tag{4.1}$$

(ii) div $\mathbf{a}$ is a scalar given by

$$\text{div } \mathbf{a} \equiv \frac{\partial a_x}{\partial x} + \frac{\partial a_y}{\partial y} + \frac{\partial a_z}{\partial z}, \tag{4.2}$$

(iii) curl $\mathbf{a}$ is a vector with components given by

$$\text{curl } \mathbf{a} \equiv \left(\frac{\partial a_z}{\partial y} - \frac{\partial a_y}{\partial z}, \frac{\partial a_x}{\partial z} - \frac{\partial a_z}{\partial x}, \frac{\partial a_y}{\partial x} - \frac{\partial a_x}{\partial y} \right). \tag{4.3}$$

These mathematical definitions are then related to particular physical situations.

However, in this book it is our aim to introduce ideas from as physical a basis as possible, and we will use as our definitions more cumbersome but more physically transparent expressions. We will then show that they are equivalent to the above if certain (mathematical) conditions are met, which in fact they are although we will not prove this.

4.1 Gradient of a scalar

Let us first consider a scalar field $\phi(x, y)$ defined throughout a two-dimensional space. For example, ϕ might represent the height above sea-level of the ground at map reference (x, y) and would be represented on a map by a series of contours which joined together all points with the same value of ϕ. [On any realistic map of course, only representative contours would be shown and each of these may consist of several disjoint parts.] Through any point $P = (x, y)$ in the space there will be one such contour and we can ask about the gradient (of the height) at that point. Clearly the gradient will have a direction which is perpendicular to the contour through P [since the direction of the contour is the direction of no change in height by definition] and a magnitude which depends upon the 'steepness'. On moving parallel to the coordinate axes, the rate of change of height with distance is $\partial\phi/\partial x$ or $\partial\phi/\partial y$, and the total gradient is represented completely by a two-dimensional vector with components $(\partial\phi/\partial x, \partial\phi/\partial y)$. Its direction gives the direction of the local gradient and its modulus the steepness.

This can all be recast for three dimensions and for a general scalar field ϕ, defined throughout some volume. In this case, instead of lines or contours of constant ϕ, we have surfaces of constant ϕ, one through each point (x, y, z). The normal to the surface through any point is in the direction of a vector with components $(\partial\phi/\partial x, \partial\phi/\partial y, \partial\phi/\partial z)$, and the magnitude of this vector is precisely equal to the rate of change of ϕ with distance in the direction of the normal. We therefore define the **gradient** of ϕ, or grad ϕ, to be the vector with these components, just as in (4.1),

$$\text{grad } \phi \equiv \left(\frac{\partial\phi}{\partial x}, \frac{\partial\phi}{\partial y}, \frac{\partial\phi}{\partial z} \right). \tag{4.1 bis}$$

It will be noticed that this can be written in the form of a **vector operator** (associated with the operation of taking a gradient) acting upon a scalar (giving the properties of the physical system), namely

$$\text{grad } \phi = \left(\frac{\partial\phi}{\partial x}, \frac{\partial\phi}{\partial y}, \frac{\partial\phi}{\partial z} \right) = \left(\frac{\partial}{\partial x}, \frac{\partial}{\partial y}, \frac{\partial}{\partial z} \right) \phi$$

$$= (\nabla_x, \nabla_y, \nabla_z)\phi = \nabla\phi, \tag{4.4}$$

where ∇ stands for the linear vector operator

$$\nabla \equiv \mathbf{i}\frac{\partial}{\partial x} + \mathbf{j}\frac{\partial}{\partial y} + \mathbf{k}\frac{\partial}{\partial z}, \tag{4.5}$$

and is called **del** (or sometimes **nabla**). Equation (4.4) has the form 'vector operator acting upon a scalar producing a vector' and should be distinguished from $\phi\nabla$ which has components $(\phi \,\partial/\partial x, \phi \,\partial/\partial y, \phi \,\partial/\partial z)$ and is still a vector operator.

From its physical definition, grad ϕ is clearly a vector field associated with the space and does not change its magnitude or direction in space, even if the coordinates used to describe the space are changed – although of course the values of its components will change from one coordinate system to another.

For a particular direction given by $\hat{n}$, the rate of change of ϕ is simply obtained from grad ϕ by forming the scalar product $\hat{n} \cdot \nabla \phi$. Other immediate extensions include

▶1. $\nabla(uv) = u\nabla v + v\nabla u,$ (4.6)

▶2. $\nabla(\phi(u)) = (\partial\phi/\partial u)\nabla u.$ (4.7)

4.2   Divergence of a vector

An important property of physical systems is the way in which the vector fields that are associated with them vary with position. If the fields are ones with which field lines are readily associated (e.g. electrostatic fields or fluids in motion) then these variations are easily visualized as the converging or diverging of the field lines.

In order to describe these variations in a quantitative way at the different points in space, we consider the net flux (per unit volume) of the vector out of a small volume surrounding any particular point. Clearly if field lines start or stop in the volume [sources or sinks present] then this net flux will not be zero. Obvious examples include a small volume surrounding an isolated charge, where the vector electric field is outwards in all directions and must produce a non-zero net flux, and a medium which is being raised in temperature as heat flows through it, more entering any particular volume than leaving it.

In a space in which a vector field $\mathbf{a} = (a_x, a_y, a_z)$ is defined, consider a small volume V with surface S positioned in the neighbourhood of a point $P = (x, y, z)$. We will take the volume as a rectangular box with P at one corner (fig. 4.1). The flux *out* of any surface element dS is $\mathbf{a} \cdot d\mathbf{S}$ and the flux over the whole surface is $\int_S \mathbf{a} \cdot d\mathbf{S}$, where S stands for the six faces of the box, denoted by S_x, S_x' (marked in the figure), S_y, S_y', S_z and S_z'. The volume V of the box is just $dx\, dy\, dz$. We are thus led to consider the scalar quantity

$$\text{div } \mathbf{a} \equiv \lim_{V \to 0} \frac{1}{V} \int_S \mathbf{a} \cdot d\mathbf{S}. \qquad (4.8)$$

It is not apparent from our construction that a limit as defined in (4.8) exists which is independent of the shape of the volume assumed, and from this point of view a definition using (4.2) is to be preferred. However the limit can be shown to exist and, as it is our approach to adopt the most

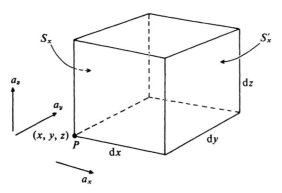

Fig. 4.1 The elementary rectangular volume for the calculation of the divergence of **a**.

physical starting point, (4.8) is our definition of the **divergence** of the vector **a** at the point P.

We now show the equivalence of this to the usual definition. On the plane surface S_x the flux is given by

$$\mathbf{a} \cdot d\mathbf{S} = -a_x \, dy \, dz,$$

the minus sign appearing because the outward normal to S_x is in the negative x-direction. For the parallel plane S_x', a distance dx away, we have by a Taylor expansion

$$\mathbf{a} \cdot d\mathbf{S} = a_x \, dy \, dz + \frac{\partial}{\partial x} (a_x \, dy \, dz) \, dx + \text{higher orders.}$$

Adding these two contributions together and retaining only terms which are less than fourth order in small quantities, we have

$$\frac{\partial a_x}{\partial x} \, dx \, dy \, dz.$$

Similar contributions come from the other pairs of surfaces to give for the evaluation of (4.8),

$$\text{div } \mathbf{a} = \lim_{dx, dy, dz \to 0} \left(\frac{\partial a_x}{\partial x} + \frac{\partial a_y}{\partial y} + \frac{\partial a_z}{\partial z} \right) dx \, dy \, dz \Big/ dx \, dy \, dz$$

$$= \frac{\partial a_x}{\partial x} + \frac{\partial a_y}{\partial y} + \frac{\partial a_z}{\partial z}, \qquad (4.2 \text{ bis})$$

which is just the definition more usually adopted.

Like grad, the divergence can also be written in terms of ∇. Equation (4.2) is just $\nabla_x a_x + \nabla_y a_y + \nabla_z a_z$ which can be written

$$\text{div } \mathbf{a} = \left(\mathbf{i} \frac{\partial}{\partial x} + \mathbf{j} \frac{\partial}{\partial y} + \mathbf{k} \frac{\partial}{\partial z} \right) \cdot (\mathbf{i} a_x + \mathbf{j} a_y + \mathbf{k} a_z) = \nabla \cdot \mathbf{a}. \quad (4.9)$$

The **divergence theorem**, which relates the total flux of a vector field out of a surface S surrounding a finite volume V to the properties of the field inside the volume, follows almost immediately from our definition of the divergence. The only step needed is that of going from infinitesimal volumes to finite ones. To do this V, in which $\mathbf{a}$ is continuous and differentiable, is divided into a large number of smaller volumes V_i, each of which will ultimately be allowed to tend to zero. We then apply the definition to each infinitesimal volume and sum the results. The flux contributions over all the corresponding surfaces S_i, except those in common with S, cancel in pairs to give the required result.

Carrying this out formally gives, using (4.8) for each small volume V_i, that

$$V_i(\text{div } \mathbf{a})_{\text{in } V_i} = \int_{S_i} \mathbf{a} \cdot d\mathbf{S}.$$

Both sides of this equation are now summed over i and each V_i allowed to tend to zero size. The left-hand side is then exactly the definition of $\int_V \text{div } \mathbf{a} \, dV$, whilst the right-hand side $\sum_i \int_{S_i} \mathbf{a} \cdot d\mathbf{S}$ can be divided into two parts, one integral over those parts of the S_i which are also a part of S, the other over those parts which are internal to S. For these latter, each surface element appears in two terms with opposite signs, since the outward normals in the two terms are equal and opposite, and so this part of $\sum_i \int_{S_i} \mathbf{a} \cdot d\mathbf{S}$ gives zero contribution.

Writing out the non-vanishing parts gives the divergence theorem

$$\int_V \text{div } \mathbf{a} \, dV = \int_S \mathbf{a} \cdot d\mathbf{S}. \quad (4.10)$$

Interpreted physically, it states that the sum of all net losses of flux for all parts of a body or region equals the total loss from its surface.

The theorem finds most use as a tool in formal manipulations, but sometimes it is of value in evaluating surface integrals of the type $\int \mathbf{F} \cdot d\mathbf{S}$ over a complicated surface S if a simple surface S' can be found so that S and S' enclose a volume V in which div $\mathbf{F}$ is easily obtained. (See example 3 of section 4.8.)

The reader will easily verify by direct calculation that the divergence operation satisfies the following.

▶3. (i) $\nabla \cdot (\lambda \mathbf{a}) = \lambda \nabla \cdot \mathbf{a}$ (λ is constant, independent of position).
▶4. (ii) $\nabla \cdot (\mathbf{a} + \mathbf{b}) = \nabla \cdot \mathbf{a} + \nabla \cdot \mathbf{b}$.
▶5. (iii) $\nabla \cdot \mathbf{r} = 3$.
▶6. (iv) $\nabla \cdot [\mathbf{r}f(r)] = 3f(r) + rf'(r)$, where $r^2 = x^2 + y^2 + z^2$.

If the vector $\mathbf{a}$ is itself derived from a scalar as $\mathbf{a} = \text{grad } \phi$ then $\nabla \cdot \mathbf{a}$ has the form $\nabla \cdot \nabla \phi$ or, as it is usually written, $\nabla^2 \phi$, where ∇^2 (del squared) is the differential operator

$$\nabla^2 \equiv \frac{\partial^2}{\partial x^2} + \frac{\partial^2}{\partial y^2} + \frac{\partial^2}{\partial z^2}. \tag{4.11}$$

As an example of the use of the divergence theorem, involving ∇^2 as well, we will now prove Green's theorems.

Consider two scalar functions ϕ and ψ satisfying our usual differentiability conditions, and apply the divergence theorem to the vector $\phi \nabla \psi$, giving

$$\int_S \phi \nabla \psi \cdot d\mathbf{S} = \int_V \nabla \cdot (\phi \nabla \psi) \, dV$$

$$= \int_V [\phi \nabla^2 \psi + (\nabla \phi) \cdot (\nabla \psi)] \, dV. \tag{4.12}$$

Reversing the roles of ϕ and ψ in (4.12) and subtracting the two equations gives

$$\int_S (\phi \nabla \psi - \psi \nabla \phi) \cdot d\mathbf{S} = \int_V (\phi \nabla^2 \psi - \psi \nabla^2 \phi) \, dV. \tag{4.13}$$

Equation (4.12) is usually known as Green's first theorem and (4.13) as his second. In both it is common to replace $\nabla \psi \cdot d\mathbf{S}$ by $(\partial \psi / \partial n) \, dS$. The expression $(\partial \psi / \partial n)$ stands for $\nabla \psi \cdot \hat{\mathbf{n}}$, the rate of change of ψ in the direction of the outward normal to S. [Despite its formal appearance, $\partial \psi / \partial n$ is not a partial derivative with respect to the magnitude of $\hat{\mathbf{n}}$, which is unity and therefore constant anyway.]

4.3   Curl (or rotation) of a vector

First consider a body which is rotating with angular velocity ω about the z-axis (fig. 4.2).
At the point (x, y, z) the velocity components are $v_x = -\omega r \sin \theta = -\omega y$ and $v_y = \omega r \cos \theta = \omega x$ and so the velocity vector field $\mathbf{a}$ has the form

$$\mathbf{a} = -\omega y \mathbf{i} + \omega x \mathbf{j} + 0 \mathbf{k}. \tag{4.14}$$

To make the connection with our previous usage, we would like to be able to derive from this field the angular velocity vector $(0, 0, \omega)$ which characterizes the rotation.

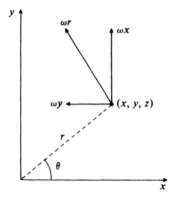

Fig. 4.2   The velocity field of a rigid body rotating with angular velocity ω about the z-axis.

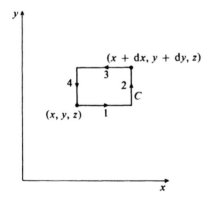

Fig. 4.3   A rectangular closed path in the xy-plane.

If, instead of a solid body, it were a liquid which had this velocity distribution, the rotation would be immediately recognized if it were required to swim round a simple closed path in the liquid. Clearly it would be much easier to swim one way round than the other, and this would be so even if the path did not enclose the axis. This help or hindrance, depending on which way round the path is traversed, is an indication of the rotation or curl of the liquid. Some quantitative measure of this is the product of the velocity (counted positive or negative) and the distance travelled in the direction of the velocity i.e. for an infinitesimal path length the product $\mathbf{a} \cdot \mathbf{dl}$ or for the complete course C, $\int_C \mathbf{a} \cdot \mathbf{dl}$. In general the bigger the area A of C, the bigger this integral, and as we are interested in a local property, we consider the quantity

$$\lim_{A \to 0} \left[\frac{1}{A} \int_C \mathbf{a} \cdot \mathbf{dl} \right], \tag{4.15}$$

around a simple closed curve C lying in one plane. The analogy with (4.8) should be noticed, surface integrals have become line integrals and volumes become areas.

We will take as C a rectangle in the xy-plane as in fig. 4.3. Then, taking the sides in the order they are numbered, (4.15) is evaluated using (4.14) for $\mathbf{a}$ as

$$\lim_{A \to 0} \frac{1}{A} \int_C \mathbf{a} \cdot \mathbf{dl}$$

$$= \lim_{dx, dy \to 0} \frac{-\omega y \, dx + \omega(x + dx) dy - \omega(y + dy)(-dx) + \omega x(-dy)}{dx \, dy}$$

$$= \lim_{dx, dy \to 0} \frac{\omega \, dy \, dx + \omega \, dx \, dy}{dx \, dy} = 2\omega.$$

Now a path in the xy-plane described in the sense used, is associated with a vector perpendicular to the plane, i.e. in the positive z-direction. Thus we can by this procedure associate with the z-direction a value 2ω.

If we now draw another rectangle C' in the yz-plane (with normal in the x-direction) and recalculate the line integral we obtain,

$$\lim_{A \to 0} \frac{1}{A} \int_{C'} \mathbf{a} \cdot \mathbf{dl} = \lim_{dy, dz \to 0} \frac{\omega x \, dy + 0 + \omega x(-dy) + 0}{dy \, dz} = 0.$$

Thus the value associated with the x-direction is 0, and similarly with the y-direction. Hence we can define by this procedure a new vector with components $(0, 0, 2\omega)$, which, within a constant, is just the angular velocity we sought. It should be noticed that this example gives the same vector wherever the path is chosen; in a more general case than rigid body rotation, the vector would change with position.

The general definition of the **curl of a vector** is based upon the same notion. Namely that the component of curl $\mathbf{a}$ in a direction $\hat{\mathbf{n}}$ at a point P is given by (4.15) when $\hat{\mathbf{n}}$ is a normal to the plane contour C of area A enclosing (or in the neighbourhood of) the point P. The vector field $\mathbf{a}$ is here of course a general one. Similar mathematical questions to the ones mentioned in connection with the definition of divergence arise here, but it can be shown that any shaped plane contour produces the same limit.

We conclude this section by showing the equivalence of this physics based definition to the mathematical one of (4.3). The rectangular contour of fig. 4.3 will again be used, but now **a** is a general vector field with components (a_x, a_y, a_z) which in general will all be functions of position. Again taking the numbered sides in order,

$$\mathbf{a} \cdot \mathbf{dl} = a_x(x, y)\, dx + a_y(x + dx, y)\, dy$$
$$+ a_x(x + dx, y + dy)(-dx)$$
$$+ a_y(x, y + dy)(-dy) + \text{second order.}$$

Grouping the first and third terms together and letting $dx \to 0$, and the second and fourth with $dy \to 0$, we have on substituting in (4.15)

▶7. $(\text{curl } \mathbf{a})_z = \lim\limits_{dy \to 0} \left\{ \dfrac{a_x(x, y) - a_x(x, y + dy)}{dy} \right\}$

$$+ \lim\limits_{dx \to 0} \left\{ \dfrac{a_y(x + dx, y) - a_y(x, y)}{dx} \right\} + \lim (\text{first order})$$

$$= -\dfrac{\partial a_x}{\partial y} + \dfrac{\partial a_y}{\partial x} + 0.$$

The other components of curl **a** are obtained by cyclic permutation of x, y, z to give complete agreement with (4.3)

$$\text{curl } \mathbf{a} = \mathbf{i}\left(\dfrac{\partial a_z}{\partial y} - \dfrac{\partial a_y}{\partial z}\right) + \mathbf{j}\left(\dfrac{\partial a_x}{\partial z} - \dfrac{\partial a_z}{\partial x}\right) + \mathbf{k}\left(\dfrac{\partial a_y}{\partial x} - \dfrac{\partial a_x}{\partial y}\right).$$

$$\text{(4.3 bis)}$$

An alternative form of (4.3), easy to remember, is a determinantal one,

$$\text{curl } \mathbf{a} = \begin{vmatrix} \mathbf{i} & \mathbf{j} & \mathbf{k} \\ \dfrac{\partial}{\partial x} & \dfrac{\partial}{\partial y} & \dfrac{\partial}{\partial z} \\ a_x & a_y & a_z \end{vmatrix}. \qquad\qquad (4.3\text{ b})$$

[If determinants are unfamiliar, a discussion of them may be found in the early part of chapter 14.]

▶8. Using the relations (2.20) and (2.21) show that curl **a** can be written as

$$\text{curl } \mathbf{a} = \nabla \wedge \mathbf{a}. \qquad\qquad (4.16)$$

▶9. Rework the case of a uniformly rotating body (about the z-axis) using cylindrical polars (r, ϕ, z) and planar paths, the sides of which lie in coordinate planes. Show that the vector obtained is $(0, 0, 2\omega)$ thus supporting (but not proving) the stated independence of the limit in (4.15) of the path shape used.

▶10. Verify by direct calculation that

$$\nabla \cdot (\mathbf{a} \wedge \mathbf{b}) = \mathbf{b} \cdot (\nabla \wedge \mathbf{a}) - \mathbf{a} \cdot (\nabla \wedge \mathbf{b}) \text{ and that } \nabla \wedge \mathbf{r} = 0.$$

4.4 Stokes' theorem

Stokes' theorem is the 'curl analogue' of the divergence theorem and is proved in an analogous manner. It states that if a simply connected smooth, but not necessarily plane, surface S is bounded by a line L (closed) then

$$\int_S \text{curl } \mathbf{a} \cdot d\mathbf{S} = \int_L \mathbf{a} \cdot d\mathbf{l}, \tag{4.17}$$

where $\mathbf{a}$ is a vector field defined in a region containing S.

Following the same line as previously, this time we divide the surface S into many small areas S_i with boundaries L_i. Then, for one such area $\int_{S_i} \text{curl } \mathbf{a} \cdot d\mathbf{S}$ is equal to {the component of curl $\mathbf{a}$ parallel to the normal to $d\mathbf{S}$} multiplied by dS. But from the definition (4.15) of curl $\mathbf{a}$ this is just $\int_{L_i} \mathbf{a} \cdot d\mathbf{l}$. Adding together all such results we have

$$\sum_i \int_{S_i} (\nabla \wedge \mathbf{a}) \cdot d\mathbf{S} = \sum_i \int_{L_i} \mathbf{a} \cdot d\mathbf{l}.$$

The left-hand side of this is just the left-hand side of (4.17), whilst on the right-hand side, all parts of all boundaries which are not part of L as well, are included twice, being traversed in opposite directions on the two occasions and thus contributing nothing. Those parts which are common with L add up to produce exactly the right-hand side of (4.17) thus proving the theorem. In all parts of the proof, boundaries are considered as traversed in the positive sense with respect to the outward normal of the surface they bound.

As a simple example of Stokes' theorem consider the following.

Example 4.1. From Ampère's rule derive one of Maxwell's equations in the case when the currents are steady.

Ampère's rule for distributed current with current density $\mathbf{J}$ is

$$\int_C \mathbf{B} \cdot d\mathbf{l} = \mu_0 \int_S \mathbf{J} \cdot d\mathbf{S}$$

for any circuit C bounding a surface S. Using Stokes' theorem, the left-hand side can be transformed into $\int_S (\nabla \wedge \mathbf{B}) \cdot d\mathbf{S}$, and hence

$$\int_S (\nabla \wedge \mathbf{B} - \mu_0 \mathbf{J}) \cdot d\mathbf{S} = 0,$$

for *any* surface S. This can only be so if

$$\nabla \wedge \mathbf{B} = \mu_0 \mathbf{J},$$

which is the required relation.

▶11. From Faraday's law of electromagnetic induction, derive Maxwell's equation curl $\mathbf{E} = -\partial \mathbf{B}/\partial t$.

4.5   Vector operator identities

As in chapter 2 for ordinary vectors, certain identities exist for vector operators. Some of these we will merely record, leaving it for the reader to verify them for himself. One or two will be discussed in a little more detail.

A vector field $\mathbf{a}$ for which div $\mathbf{a} = \nabla \cdot \mathbf{a} = 0$ is said to be *solenoidal*, and one for which curl $\mathbf{a} = 0$ to be *irrotational*. If the vector $\mathbf{a}$ is itself derived as the gradient of a scalar then it is necessarily irrotational. To see this we simply write out the expression for curl $\mathbf{a}$

$$\nabla \wedge \mathbf{a} = \left(\frac{\partial a_z}{\partial y} - \frac{\partial a_y}{\partial z}, \ldots, \ldots \right)$$

$$= \left(\frac{\partial}{\partial y} \left(\frac{\partial \phi}{\partial z} \right) - \frac{\partial}{\partial z} \left(\frac{\partial \phi}{\partial y} \right), \ldots, \ldots \right),$$

i.e.   curl $\mathbf{a} = (0, 0, 0)$ if $\mathbf{a} = \mathrm{grad}\ \phi$.   (4.18)

[This result might also be expected since curl $\mathbf{a}$ can be written as $\nabla \wedge \nabla \phi$ and looks like the vector product of a vector with itself, but with vector operators such formal results cannot be safely presumed.]

▶12. If $\mathbf{a}$ is derived as curl $\mathbf{b}$ for some vector $\mathbf{b}$, then $\mathbf{a}$ is solenoidal, i.e.

$$\nabla \cdot \mathbf{a} = 0.$$ (4.19)

The analogue of the triple vector product is worth a little further study and in particular the vector given by curl (curl $\mathbf{a}$) and obtained when both $\mathbf{b}$ and $\mathbf{c}$ are replaced by ∇ in $\mathbf{c} \wedge (\mathbf{b} \wedge \mathbf{a})$. The relationship is

$$\nabla \wedge (\nabla \wedge \mathbf{a}) = \nabla(\nabla \cdot \mathbf{a}) - \nabla^2 \mathbf{a},$$ (4.20 a)

or in words,

$$\text{curl (curl } \mathbf{a}) = \text{grad (div } \mathbf{a}) - \text{del squared } \mathbf{a}. \qquad \text{(4.20 b)}$$

It should be remarked that the right-hand side of (4.20) is not identically zero, as can be easily verified by writing it out in Cartesian components. For example,

$$[\mathbf{\nabla}(\mathbf{\nabla} \cdot \mathbf{a})]_x = \frac{\partial}{\partial x}\left(\frac{\partial a_x}{\partial x} + \frac{\partial a_y}{\partial y} + \frac{\partial a_z}{\partial z}\right)$$

$$= \frac{\partial^2 a_x}{\partial x^2} + \frac{\partial^2 a_y}{\partial x\,\partial y} + \frac{\partial^2 a_z}{\partial x\,\partial z},$$

whilst $\quad [\nabla^2 \mathbf{a}]_x = \dfrac{\partial^2 a_x}{\partial x^2} + \dfrac{\partial^2 a_x}{\partial y^2} + \dfrac{\partial^2 a_x}{\partial z^2}.$

In the expression $\nabla^2 \mathbf{a}$, the quantity ∇^2 (called the **Laplacian**) is a linear differential operator acting upon a vector [as opposed to the scalar discussed towards the end of section 4.2] with the vector itself consisting of a sum of unit vectors multiplied by components. Two cases arise:

(i) If the unit vectors are constants [independent of the values of the coordinates] the differential operator gives a non-zero contribution only when acting upon the coordinates [with the unit vectors merely as multipliers].

(ii) If the unit vectors vary as the values of the coordinates change [i.e. are not constant in direction and magnitude throughout the whole space] then the derivatives of these vectors appear as contributions to $\nabla^2 \mathbf{a}$.

Cartesians are an example of the first case in which $(\nabla^2 \mathbf{a})_i = \nabla^2 a_i$. In this case (4.20) can be applied to each component separately

$$[\text{curl (curl } \mathbf{a})]_i = [\text{grad (div } \mathbf{a})]_i - \nabla^2 a_i. \qquad \text{(4.21)}$$

On the other hand spherical and cylindrical polars come in the second class. For them (4.20) is still true but the further step to (4.21) cannot be made.

Equation (4.21) is proved for Cartesians in chapter 15, section 15.5, whilst a counter example for cylindrical polar coordinates is indicated in example 7 of section 4.8 at the end of this chapter.

4.6   Conservative fields and potentials

In this section we consider further the important result (4.18). This, together with some other previous results and the ones proved below, can be used to express four equivalent statements.

A vector field **a** is **conservative** if and only if any one (and hence all) of the following is true:

> There exists a scalar function ϕ whose value at the point (x, y, z) depends only on the point (x, y, z) and is such that $\mathbf{a} = \nabla\phi$. (4.22 a)

> The integral $\int_A^B \mathbf{a} \cdot d\mathbf{l}$ is independent of the path from A to B. (4.22 b)

> The integral $\int \mathbf{a} \cdot d\mathbf{l}$ around any closed loop is zero. (4.22 c)

> $\nabla \wedge \mathbf{a} = \mathbf{0}$. (4.22 d)

The validity or otherwise of any one of these statements implies the same for the other three.

The full equivalence of (4.22 b) and (4.22 c) is almost self-evident and is briefly discussed in section 3.4. Equation (4.18) shows that (4.22 a) implies (4.22 d), and Stokes' theorem (4.17) shows that (4.22 d) implies (4.22 c) and hence (4.22 b). It thus only remains to show the equivalence of (4.22 a) and (4.22 b), to imply the equivalence of all four.

If (4.22 a) is true, then

$$\int_A^B \mathbf{a} \cdot d\mathbf{r} = \int_A^B \left\{ \left(\frac{\partial\phi}{\partial x} \right) dx + \left(\frac{\partial\phi}{\partial y} \right) dy + \left(\frac{\partial\phi}{\partial z} \right) dz \right\}$$

$$= \int_A^B d\phi = \phi_B - \phi_A.$$

Now $\phi_B - \phi_A$ depends only on the values of ϕ at A and B, and is independent of the path taken between them, which establishes that (4.22 a) implies (4.22 b).

Now suppose (4.22 b) is true, and consider a quantity $\phi(P)$ defined by

$$\phi(P) = \int_A^P \mathbf{a} \cdot d\mathbf{r},$$

where A is a fixed point and P variable. Because of (4.22 b), $\phi(P)$ is single valued since the right-hand side is independent of the path from A to P. Let Q be another point d**r** away from P, then

$$\phi(Q) = \phi(P) + \mathbf{a} \cdot d\mathbf{r},$$

since the path APQ is one way of reaching Q from A and **a** is substantially constant along the infinitesimal path. Thus $d\phi = \phi(Q) - \phi(P) = \mathbf{a} \cdot d\mathbf{r}$ and this means that **a** is just grad ϕ, completing the establishment of (4.22 a) and the equivalence of the four forms of (4.22).

Example 4.2. In example 3 of section 3.8 it was shown that, at least for two different paths, the tangential line integral of the vector field $\mathbf{a} = (xy^2 + z, x^2y + 2, x)$ was independent of the path between the end points. This illustrates but does not prove condition (4.22 b); show that the field is in fact conservative (a) by establishing condition (4.22 d), and (b) by establishing condition (4.22 a).

(a) Applying (4.3) to $\mathbf{a}$ gives

$$\text{curl } \mathbf{a} = (0 - 0, 1 - 1, 2yx - 2xy) = (0, 0, 0),$$

thus immediately establishing condition (4.22 d).

(b) Suppose $\phi(x, y, z)$ is to be the required function. Then firstly we must have

$$\partial\phi/\partial x = xy^2 + z.$$

Thus, $\phi = \frac{1}{2}x^2y^2 + zx + f(y, z)$ for some function f. Secondly $\partial\phi/\partial y$, $= x^2y + \partial f/\partial y$, must also $= x^2y + 2$. Hence $f = 2y + g(z)$ and $\phi = \frac{1}{2}x^2y^2 + zx + 2y + g(z)$ for some function g. Finally $\partial\phi/\partial z$, $= x + \partial g/\partial z$, must also $= x$. Hence $g = \text{constant} = k$. So we have explicitly constructed a function

$$\phi(x, y, z) = \frac{1}{2}x^2y^2 + zx + 2y + k,$$

whose gradient is the vector field $\mathbf{a}$, and thus established condition (4.22 a).

▶13. For the vector field of example 3.2 (section 3.3), namely $\mathbf{a} = (xy^2, 2, x)$, show that condition (4.22 d) is not satisfied and attempt to construct a suitable function ϕ, as above, showing where the procedure breaks down. [Example 3.2 itself has already shown that condition (4.22 b) and hence (4.22 c) are not valid.]

The quantity ϕ which figures so prominently in this section [and in many subsequent chapters] is called the **potential function** of the vector field $\mathbf{a}$. More precisely it is the scalar potential function, since it is both possible and useful at times to define a vector potential function. Its uses in physics and engineering are so wide-spread that there is little point here in giving particular examples, and so we merely record its great importance in such representative areas as electrostatics, gravitation, fluid dynamics, magnetostatics, and atomic and nuclear physics. But before leaving it we summarize the main properties of the potential:

(i) It is a scalar.
(ii) Its value at the point $P = (x, y, z)$ depends only on the quantities x, y, z (and other fixed constants).

(iii) Its value is equal to the tangential line integral of the vector field from an arbitrary fixed point A [at which the potential is thus defined to be zero] to P, the value of the integral being independent of the path from A to P.

(iv) Its gradient $\nabla\phi$ is the vector field.

Scalar potentials which are multivalued functions of position (but in simple ways) are also of value in describing some physical situations, the most obvious example being the magnetic potential associated with a current-carrying wire.

4.7   Non-Cartesian systems

The operators which we have discussed in this chapter, grad, div, curl and ∇^2, were all defined in terms of Cartesian coordinates, but for many physical situations other coordinate systems are more natural. For example, many systems, such as a single isolated charge in space, have spherical symmetry and spherical polar coordinates would be the obvious choice. For axisymmetric systems, spherical or cylindrical polars are the natural system of coordinates. The physical laws governing the behaviour of the systems are often expressed in terms of the operators we have been considering and so it is necessary to be able to express these operators in these other, non-Cartesian, coordinates.

We will use the notation of section 3.7 and concentrate mostly on spherical and cylindrical polars.

Gradient. The element of length associated with a change du_i in one of the coordinates is $h_i\,du_i$ (no summation) and, since the gradient of a function is the rate of change of that function with distance, the vector grad Φ is given by†

$$\text{grad } \Phi = \nabla\Phi = \frac{1}{h_1}\frac{\partial\Phi}{\partial u_1}\hat{\mathbf{e}}_1 + \frac{1}{h_2}\frac{\partial\Phi}{\partial u_2}\hat{\mathbf{e}}_2 + \frac{1}{h_3}\frac{\partial\Phi}{\partial u_3}\hat{\mathbf{e}}_3. \tag{4.23}$$

It should be borne in mind that the unit vectors $\hat{\mathbf{e}}_i$ are not constant throughout the space in general, but are defined locally to give the local direction of increase of the coordinates u_i. This is apparent from fig. 4.4 which shows the unit vectors for two different positions (r, θ, ϕ) and (r', θ', ϕ') in spherical polars.

† Here we use Φ for the potential to avoid confusion with the azimuthal angle ϕ.

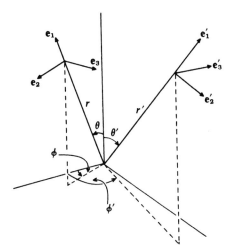

Fig. 4.4   The unit vectors e_i and e_i' in spherical polar coordinates at two different positions **r** and **r**'.

Using table 3.1 and equation (4.23) we have in particular that the components of grad Φ are

$$\left(\frac{\partial\Phi}{\partial r}, \frac{1}{r}\frac{\partial\Phi}{\partial\theta}, \frac{1}{r\sin\theta}\frac{\partial\Phi}{\partial\phi}\right) \text{ in spherical polars,} \tag{4.24}$$

and

$$\left(\frac{\partial\Phi}{\partial r}, \frac{1}{r}\frac{\partial\Phi}{\partial\phi}, \frac{\partial\Phi}{\partial z}\right) \text{ in cylindrical polars.} \tag{4.25}$$

Divergence. Consider for definiteness the case of spherical polars illustrated in fig. 4.5, and let us apply the definition of divergence in (4.8) to the small near-rectangular volume generated by making small increments in each of the coordinates. As in the argument of section 4.2 the contribution to $\mathbf{a}\cdot d\mathbf{S}$ from faces S_r and S_r' together is

$$\frac{\partial}{\partial r}(a_r\, r\, d\theta\, r\sin\theta\, d\phi)\, dr.$$

Similarly the contributions from $\{S_\theta$ and $S_\theta'\}$ and $\{S_\phi$ and $S_\phi'\}$ are

$$\frac{\partial}{\partial\theta}(a_\theta\, dr\, r\sin\theta\, d\phi)\, d\theta \quad \text{and} \quad \frac{\partial}{\partial\phi}(a_\phi\, dr\, r\, d\theta)\, d\phi$$

respectively, giving for the total surface integral

$$\int_s \mathbf{a}\cdot d\mathbf{S} = \left[\frac{\partial}{\partial r}(r^2\sin\theta\, a_r) + \frac{\partial}{\partial\theta}(r\sin\theta\, a_\theta) + \frac{\partial}{\partial\phi}(ra_\phi)\right]dr\, d\theta\, d\phi.$$

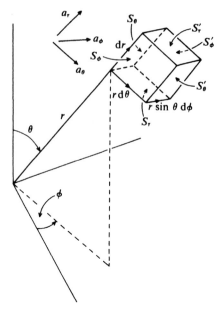

Fig. 4.5 The elementary volume in spherical polar coordinates for the calculation of the divergence of **a**.

Now the volume of the solid is $V = dr \times r\,d\theta \times r \sin \theta\,d\phi$, so that from (4.8) we obtain for the divergence of **a** in polar coordinates

$$\nabla \cdot \mathbf{a} = \frac{1}{r^2 \sin \theta} \left[\frac{\partial}{\partial r}(r^2 \sin \theta\, a_r) + \frac{\partial}{\partial \theta}(r \sin \theta\, a_\theta) + \frac{\partial}{\partial \phi}(r a_\phi) \right]$$

$$= \frac{1}{r^2} \frac{\partial}{\partial r}(r^2 a_r) + \frac{1}{r \sin \theta} \frac{\partial}{\partial \theta}(\sin \theta\, a_\theta) + \frac{1}{r \sin \theta} \frac{\partial a_\phi}{\partial \phi}. \qquad (4.26)$$

Following this same method it is straightforward to obtain the result for a general orthogonal system.

▶14. $$\nabla \cdot \mathbf{a} = \frac{1}{h_1 h_2 h_3} \left[\frac{\partial}{\partial u_1}(h_2 h_3 a_1) + \frac{\partial}{\partial u_2}(h_3 h_1 a_2) + \frac{\partial}{\partial u_3}(h_1 h_2 a_3) \right].$$

$$(4.27)$$

In particular, for cylindrical polars,

▶15. $$\nabla \cdot \mathbf{a} = \frac{1}{r} \frac{\partial}{\partial r}(r a_r) + \frac{1}{r} \frac{\partial a_\phi}{\partial \phi} + \frac{\partial a_z}{\partial z}. \qquad (4.28)$$

The operator ∇^2. If we take the particular case of $\mathbf{a} = \nabla\Phi$ then (4.27) yields an expression for ∇^2 for a general orthogonal system. Substituting for a_1, a_2, a_3 in (4.27) from (4.23) gives

$$\nabla^2\Phi = \frac{1}{h_1 h_2 h_3}\left[\frac{\partial}{\partial u_1}\left(\frac{h_2 h_3}{h_1}\frac{\partial\Phi}{\partial u_1}\right) + \frac{\partial}{\partial u_2}\left(\frac{h_3 h_1}{h_2}\frac{\partial\Phi}{\partial u_2}\right)\right.$$
$$\left. + \frac{\partial}{\partial u_3}\left(\frac{h_1 h_2}{h_3}\frac{\partial\Phi}{\partial u_3}\right)\right]. \quad (4.29)$$

For the two particular cases, we have in spherical polars;

▶16. $\quad \nabla^2\Phi = \dfrac{1}{r^2}\dfrac{\partial}{\partial r}\left(r^2\dfrac{\partial\Phi}{\partial r}\right) + \dfrac{1}{r^2\sin\theta}\dfrac{\partial}{\partial\theta}\left(\sin\theta\dfrac{\partial\Phi}{\partial\theta}\right) + \dfrac{1}{r^2\sin^2\theta}\dfrac{\partial^2\Phi}{\partial\phi^2},$

$$(4.30)$$

and in cylindrical polars;

▶17. $\quad \nabla^2\Phi = \dfrac{1}{r}\dfrac{\partial}{\partial r}\left(r\dfrac{\partial\Phi}{\partial r}\right) + \dfrac{1}{r^2}\dfrac{\partial^2\Phi}{\partial\phi^2} + \dfrac{\partial^2\Phi}{\partial z^2}. \quad (4.31)$

It will be noticed that in the expressions for $\nabla^2\Phi$, the last term in (4.30) and the last two terms in (4.31) have the formal appearance of being the result of repeating the operations which produced the corresponding terms in the expressions for $\nabla\Phi$ (equations (4.24) and (4.25) respectively). This they have in common with all three terms of $\nabla^2\Phi$ when it is expressed in Cartesians. However, the first two terms of (4.30) and the first of (4.31) do not possess this property.

The reason for the differences is to be found in the fact that for the coordinates r and θ, the volume swept out in space by a given increment in that coordinate (keeping the others constant) is dependent on the value of that coordinate at the time, whereas for the others this is not so. For example, the volume between the coordinate surfaces $r = r_1$ and $r = r_1 + dr$ in spherical polars is $4\pi r_1^2\,dr$, but that between the coordinate planes $\phi = \phi_1$ and $\phi = \phi_1 + d\phi$ is $(d\phi/2\pi)$ of the whole space and independent of ϕ_1. So for r and θ the geometry of the space itself has a 'built-in' divergence which adds to that occurring in the scalar Φ. To illustrate this specifically, the r term of (4.30) is

$$\frac{1}{r^2}\frac{\partial}{\partial r}\left(r^2\frac{\partial\Phi}{\partial r}\right) = \frac{\partial^2\Phi}{\partial r^2} + \frac{2}{r}\frac{\partial\Phi}{\partial r},$$

the first term on the right being the 'expected' one and the second that deriving from the 'geometry of the space'. Of course both must be included in any use of the formulae.

The reader is reminded also of the discussion at the end of section 4.5 in connection with the application of ∇^2 to vectors, and of the complications which may arise when working with a general orthogonal system. It is usually simplest in these circumstances to express $\nabla^2\mathbf{a}$, by means of (4.20 a), in terms of other vector operators.

Curl. Following from the defining equation (4.15) and the discussion of section 4.3, we can construct in a straightforward manner the expression for curl $\mathbf{a}$ in a general orthogonal system. By considering a small contour lying in a plane $u_3 = $ constant [and traversed in the right-handed sense with respect to a normal in the positive u_3 direction] we obtain

$$(\text{curl }\mathbf{a})_3 = \lim_{du_1, du_2 \to 0} \left\{ \frac{1}{h_1\,du_1\,h_2\,du_2} \left[\frac{\partial}{\partial u_1}(a_2 h_2\,du_2)\,du_1 \right.\right.$$
$$\left.\left. - \frac{\partial}{\partial u_2}(a_1 h_1\,du_1)\,du_2 \right]\right\}$$
$$= \frac{1}{h_1 h_2}\left[\frac{\partial}{\partial u_1}(h_2 a_2) - \frac{\partial}{\partial u_2}(h_1 a_1) \right]. \tag{4.32}$$

The other two components are found by cyclically permuting subscripts 1, 2, 3.

▶18. Show that in spherical and cylindrical polars curl $\mathbf{a}$ has components:

spherical polars   cylindrical polars

$$r;\ \frac{1}{r}\frac{\partial a_\phi}{\partial \theta} + \frac{\cot\theta}{r}a_\phi - \frac{1}{r\sin\theta}\frac{\partial a_\theta}{\partial \phi}, \qquad r;\ \frac{1}{r}\frac{\partial a_z}{\partial \phi} - \frac{\partial a_\phi}{\partial z},$$

$$\theta;\ \frac{1}{r\sin\theta}\frac{\partial a_r}{\partial \phi} - \frac{a_\phi}{r} - \frac{\partial a_\phi}{\partial r}, \qquad \phi;\ \frac{\partial a_r}{\partial z} - \frac{\partial a_z}{\partial r},$$

$$\phi;\ \frac{a_\theta}{r} + \frac{\partial a_\theta}{\partial r} - \frac{1}{r}\frac{\partial a_r}{\partial \theta}, \qquad z;\ \frac{a_\phi}{r} + \frac{\partial a_\phi}{\partial r} - \frac{1}{r}\frac{\partial a_r}{\partial \phi}.$$

4.8   Examples for solution

1. Distinguish between the following expressions involving vector $\mathbf{a}$ and scalar ϕ, showing in particular: $(a) = (d) \neq (c)$; $(d) \neq (e)$; $(c) = (a) + (b)$; $(b) = -(e)$.

(a) $\phi\nabla \wedge \mathbf{a}$, (b) $\nabla\phi \wedge \mathbf{a}$, (c) $\nabla \wedge \phi\mathbf{a}$, (d) $(\nabla \wedge \mathbf{a})\phi$, (e) $(\mathbf{a} \wedge \nabla)\phi$.

2. The field of a magnetic dipole of moment $\mathbf{M}$, placed in vacuum

at the origin, is given by $H = -\nabla\Omega$, where Ω is the magnetostatic potential $\Omega = -(1/4\pi)M\cdot\nabla(1/r)$. Show that

$$H = (1/4\pi)[3\hat{r}(M\cdot r)r^{-4} - Mr^{-3}]$$

and deduce the fields in the Gauss A and B positions, of a short bar magnet. (A position is on the axis of the magnet; B position on the 'equatorial' line.)

3. Evaluate the surface integral $\int_S F\cdot dS$ over the open hemispherical surface $x^2 + y^2 + z^2 = a^2$, $z \geq 0$, when $F = (y - x)i + x^2zj + (z + x^2)k$, (a) directly, and (b) by the use of the divergence theorem.

4. Show that the vector $r^n r$ is (i) irrotational for any value of n, and (ii) solenoidal only if $n = -3$. Deduce the physically obvious result that 'any spherically symmetric vector field which is everywhere directed away from the origin is irrotational'.

5. A vector field a has components $(-zxr^{-3}, -zyr^{-3}, (x^2 + y^2)r^{-3})$, where $r^2 = x^2 + y^2 + z^2$. Show that the field is conservative (a) by showing curl $a = 0$, and (b) by constructing its potential function ϕ.

6. A force $F(r)$ acts on a particle at r. In which of the following cases can F be represented in terms of a potential? When it can, find the potential.

(a) $F = [i - j - 2r(x - y)]\exp(-r^2)$.
(b) $F = [zk + r(x^2 + y^2 - 1)]\exp(-r^2)$.
(c) $F = k + (r \wedge k)r^{-2}$.

7. Using cylindrical polar coordinates (r, ϕ, z) and a vector field a with components $(u, v, 0)$, where for simplicity u and v are taken as functions of r and ϕ only, show that the r component of $\nabla(\nabla\cdot a) - \nabla \wedge (\nabla \wedge a)$ is

$$\frac{\partial^2 u}{\partial r^2} - \frac{u}{r^2} + \frac{1}{r}\frac{\partial u}{\partial r} - \frac{2}{r^2}\frac{\partial v}{\partial \phi} + \frac{1}{r^2}\frac{\partial^2 u}{\partial \phi^2},$$

but that $\nabla^2 u$ contains only the first, third and last of these terms. This shows that (4.21) is not valid in a general orthogonal system. As a very specific example, take $u = $ constant $ = c \neq 0$ and $v = 0$ and evaluate $\nabla^2 a$ comparing it with $\nabla^2 a_i$.

8. For the coordinate system of example 7 in section 3.8, find the most general function f of u_1 only that satisfies $\nabla^2 f = 0$. (The required expressions for h_1, h_2, h_3 are given in the solutions to chapter 3.)

5
Ordinary differential equations

In the mathematical description of a physical system the basic laws are usually expressed as equations connecting local quantities. The equations consist of products and sums of the quantities and their derivatives, the actual combination of forms depending upon which variables are chosen to describe the situation.

The effect of external factors on the system is represented by the tying of the values of some of the quantities to particular values of others. The latter quantities are usually called the independent variables, and the former the dependent ones. A complete description (solution) of the physical system is obtained when the values of the dependent variables are known for all values of (i.e. as a function of) the independent ones.

In cases where derivatives appear in the expression of the physical laws it is thus necessary to convert a differential equation into a standard, tabulated, or computable function. This may take the form of a series or an integral which can be evaluated numerically. Methods by which the conversion can be effected form the content of this and the next two chapters and also (for equations containing more than one independent variable) of chapters 9 and 10.

Equations of the types considered in the first three of these chapters – those containing only one independent variable – are called **ordinary differential equations.** The two other chapters deal with some methods of solving partial differential equations, i.e. ones containing derivatives of dependent variables with respect to more than one independent variable.

In discussing ordinary differential equations in a general way we will take x as the independent variable and $y = y(x)$ as the dependent one. The derivatives of y with respect to x, dy/dx, $d^2y/dx^2, \ldots, d^ny/dx^n$, will, where no confusion arises, be denoted for brevity by y', $y'', \ldots, y^{(n)}$, respectively.

Two further definitions need to be made at this point. The first of these is the **order** of a differential equation, which is the order of the highest derivative appearing in the equation. The second is the **degree** of the equation, this being the exponent of the highest derivative, after the equation has been rationalized and all non-integral exponents of y and its derivatives eliminated (if necessary).† Thus, for example,

$$[1 + (y')^2]^{1/2} = y''' + y$$

is of third order and second degree, since y''' appears and after rationalization it is raised to the second power.

We will be concerned almost exclusively with equations of the first degree and will generally be dealing only with equations of first and second order. Much of our attention will be on **linear** equations in which *all* derivatives present and y itself appear only to the first power and at most are multiplied by a function of x (not of y or its derivatives).

5.1   General form of solution

It is helpful when considering the general form of the solutions of an ordinary differential equation (d.e.) to consider the inverse process, namely that of how a differential equation can be obtained from a given group‡ of functions, such that any one of them is a solution of the d.e. Suppose the members of the group are given by a particular form

$$y = f(x, a_1, a_2, \ldots, a_n), \tag{5.1}$$

where the a_i are a set of n parameters, with each member given by a different set of parameters. For example, if the group were all functions of the [familiar] form

$$y = a_1 \cos x + a_2 \sin x, \tag{5.2}$$

then n would be 2.

Since a d.e. is required for which *any* of the group is a solution, the equation clearly must not contain any of the a_i. As there are n of the a_i in expression (5.1), we must obtain $n + 1$ equations involving them in order that, by elimination, we can obtain one final equation without them. Initially we have only (5.1), but if this is differentiated n times, a total of $n + 1$ equations is obtained from which (in principle) all the a_i can be eliminated to give one d.e. satisfied by all the group. The n differentiations will result in $y^{(n)}$ being present in one of the $n + 1$ equations and hence in the final equation, which will therefore be of the nth order.

In the example of equation (5.2),

$$y' = -a_1 \sin x + a_2 \cos x, \tag{5.3 a}$$

---

† Some authors define degree as the highest exponent which appears of any derivative (including y itself), after the equation has been so treated. In this case 'of first degree' is synonymous with 'linear'.

‡ The word 'group' is not used here in any technical algebraic sense.

and $y'' = -a_1 \cos x - a_2 \sin x.$ (5.3 b)

Here the elimination of a_1 and a_2 is trivial (because of the similarity of form of y'' and y), resulting in

$$y'' + y = 0,\tag{5.4}$$

a second-order equation.

Thus to summarize – a group of functions with n parameters satisfy an nth-order differential equation in general.† The intuitive converse of this – that the general solution of an nth-order differential equation contains n arbitrary parameters (constants) – will, for our purposes of physical applications, be assumed to be valid, although a totally general proof is difficult.

▶1. Suppose that a solution of (5.4) can be found of the form $y(x) = \sum_0^\infty a_i x^i$. Show that exactly two of the a_i are arbitrary.

▶2. If (5.4) is multiplied through by y', it can then be arranged in the form $[(y')^2]' + [y^2]' = 0$.

Show, by carrying it out, that to obtain from this an expression for y in terms of x requires 2 integrations, resulting in 2 arbitrary constants.

These two exercises illustrate, but of course do not prove, the validity of the assumed result.

As mentioned earlier, external factors affect a system described by a d.e., by fixing the values of dependent variables at particular values of the independent ones. These externally imposed (or **boundary**) conditions on the solution are thus the means of determining the parameters which specify precisely which function is the required solution. It is apparent that the number of boundary conditions should match the number of parameters and hence the order of the equation, if a unique solution is to be obtained. Fewer independent boundary conditions than this will lead to a number of undetermined parameters in the solution, whilst an excess will usually mean that no acceptable solution is possible.

For an nth-order equation the required n boundary conditions can take many forms, for example, the value of y at n different values of x, or the value of any $n - 1$ of the n derivatives $y', y'', \ldots, y^{(n)}$ together with that of y, all for the same value of x, or any of many intermediate combinations.

† In some degenerate cases a d.e. of less than nth order is obtained.

▶3. Solve, where possible, equation (5.4) with the boundary conditions,

(a) $y(0) = 4$, $y(\pi/2) = -1$;

(b) $y(0) = 4$, $y'(0) = 2$;

(c) $y(\pi/4) = 2$;

(d) $y(0) = 1$, $y''(0) = -1$;

(e) $y(\pi/6) = \frac{1}{2} + 3^{1/2}$, $y(\pi/4) = 3(2)^{-1/2}$, $y'(\pi/4) = -(2)^{-1/2}$;

(f) $y(\pi/6) = \frac{1}{2}$, $y(\pi/4) = 3(2)^{-1/2}$, $y'(\pi/4) = -(2)^{-1/2}$.

5.2   First-order equations

We begin our study of more specific forms of differential equations with those of first order (and first degree). Our main interest from the physical point of view will be in linear equations but to start with we consider a slightly wider class.

Let us take as the general form of first-order equation

$$Qy' + P = 0, \tag{5.5}$$

where Q and P are, in general, both functions of x and y [but of course not of y', y'', etc.].

If, by a rearrangement or by multiplying the equation through by some function of x and y, it can be brought to a form where the quantity multiplying y' is a function of only one of x and y and the second term a function only of the other, then the variables are said to be **separated** and (5.5) can be written as

$$p(x)\,dx + q(y)\,dy = 0,$$

with solution

$$\int^x p(x_1)\,dx_1 + \int^y q(y_1)\,dy_1 = c.$$

Since the general equation (5.5) is of first order, its solution will contain one arbitrary constant. Suppose that it is possible to express the solution so as to make this constant c the subject of the equation expressing the solution, that is

$$c = f(x, y). \tag{5.6}$$

Then by differentiating (5.6) with respect to x we should recover the original equation

$$0 = \frac{\partial f}{\partial x} + \frac{\partial f}{\partial y}\frac{dy}{dx}, \tag{5.7}$$

where $\partial f/\partial x$ is the partial derivative† of f with respect to x with y held fixed, and similarly for $\partial f/\partial y$.

Comparing (5.7) with the original equation, we see that for a solution of the form (5.6) we must have

$$\frac{\partial f}{\partial x} \bigg/ \frac{\partial f}{\partial y} = P/Q. \tag{5.8}$$

Alternatively we may say that to obtain a solution of the form $f(x, y) = c$, the original equation (5.5) after possible multiplication through by a suitable function $g(x, y)$ should take the form (5.7). Since $\partial^2 f/\partial x\,\partial y = \partial^2 f/\partial y\,\partial x$, this implies as a necessary condition that

$$\frac{\partial}{\partial y}(gP) = \frac{\partial}{\partial x}(gQ). \tag{5.9}$$

It can also be shown that (5.9) is a sufficient condition for a solution of (5.5) to exist in the form (5.6).

Equations like (5.5) for which

$$\frac{\partial P}{\partial y} = \frac{\partial Q}{\partial x}, \tag{5.10}$$

i.e. with $g(x, y) = $ constant, are called **exact** equations and can be solved immediately:

$$c = f(x, y) = \int^x P(x_1, y)\,dx_1 + h(y),$$

where $h(y)$ is chosen to make

$$\frac{\partial}{\partial y}\left[\int^x P(x_1, y)\,dx_1\right] + h'(y) = Q. \tag{5.11 a}$$

Clearly, depending on the actual forms of P and Q, it may be computationally easier to evaluate

$$c = f(x, y) = \int^y Q(x, y_1)\,dy_1 + k(x),$$

with

$$\frac{\partial}{\partial x}\left[\int^y Q(x, y_1)\,dy_1\right] + k'(x) = P. \tag{5.11 b}$$

---

† This section assumes a little familiarity with partial differentiation, section 1.8 may be useful preliminary reading for the student without such familiarity.

Example 5.1. Solve $(x^3 + 2y)y' + 3x^2y = 0$.

The solution is obvious, but for illustrative purposes we will go pedantically through the above procedures.

Here, comparison with (5.5) gives $Q(x, y) = x^3 + 2y$ and $P(x, y) = 3x^2y$. Hence (5.10) is satisfied, since each side of it has values $3x^2$, and the equation is therefore exact. Using (5.11 a), its solution is

$$c = \int^x 3x_1^2 y\, dx_1 + h(y) = x^3y + h(y),$$

with

$$\frac{\partial}{\partial y}(x^3y) + h'(y) = x^3 + 2y.$$

Thus $h(y) = y^2$ and the solution is

$$x^3y + y^2 = c.$$

[All constants of integration may be taken as zero, i.e. formally absorbed into c.]

Alternatively [carrying right through with our pedantic illustration], using (5.11 b) the solution is

$$c = \int^y (x^3 + 2y_1)\, dy_1 + k(x) = x^3y + y^2 + k(x),$$

with $\dfrac{\partial}{\partial x}(x^3y + y^2) + k'(x) = 3x^2y.$

Thus $k(x) = 0$ and the solution is as before.

We now return to (5.9) and study the quantity $g(x, y)$, which because of the purpose it serves is known as an **integrating factor** (IF). Expanding (5.9) we obtain after rearrangement

$$g\left(\frac{\partial P}{\partial y} - \frac{\partial Q}{\partial x}\right) = Q\frac{\partial g}{\partial x} - P\frac{\partial g}{\partial y}. \tag{5.12}$$

If we now substitute for P from the original equation (5.5) into the last term of (5.12) and recall that $dg/dx = \partial g/\partial x + y'\, \partial g/\partial y$, then we obtain

$$\frac{1}{Q}\left(\frac{\partial P}{\partial y} - \frac{\partial Q}{\partial x}\right) = \frac{1}{g}\frac{dg}{dx}, \tag{5.13}$$

as the equation satisfied by the integrating factor g.

An explicit formula for g can be given for those cases in which the left-hand side of (5.13) is a function of x only (and not of y), for then simple integration gives

$$g(x) = \exp\left[\int^x \frac{1}{Q(x_1, y)} \left(\frac{\partial P(x_1, y)}{\partial y} - \frac{\partial Q(x_1, y)}{\partial x_1}\right) dx_1\right] \quad (5.14)$$

as the required integrating factor.

Often simple integrating factors can be determined by inspection, but in more complicated cases, provided its integrand is a function of x only,† (5.14) will provide the necessary IF and hence solution of the equation. Notice that if the integrand is zero, the original equation is already exact.

▶4. Solve the following by separation of variables to obtain $y = y(x)$:

(a) $y' - xy^3 = 0$;
(b) $y' \arctan x - y(1 + x^2)^{-1} = 0$;
(c) $x^2 y' + xy^2 = 4y^2$.

▶5. Show that the following equations are either exact or can be made so, and solve them:

(a) $y(2x^2 y^2 + 1)y' + x(y^4 + 1) = 0$;
(b) $2xy' + 3x + y = 0$;
(c) $(\cos^2 x + y \sin 2x)y' + y^2 = 0$.

If in (5.5), $P(x, y)/Q(x, y)$ can be expressed entirely as a function of (y/x), then the equation is said to be *dimensionally-homogeneous* and a general form of solution can be found. This method is treated in example 1 of section 5.12.

5.3   Linear first-order equations

Turning to linear equations of the first order produces some simplification of our general results.

For linear equations we have, using the notation of (5.5), that (i) $Q = Q(x)$ (and not $Q(x, y)$), and (ii) $P(x, y) = yR(x) + S(x)$. This means that the important criterion quantity $\partial P/\partial y - \partial Q/\partial x$ has the form

$$\partial P/\partial y - \partial Q/\partial x = R(x) - Q'(x)$$

† This is not intended to imply that an IF cannot be found if the integrand does depend on y – only that it must be sought by other means.

and is therefore necessarily a function of x only. Thus an integrating factor

$$g(x) = \exp\left[\int^x \frac{1}{Q(x_1)} [R(x_1) - Q'(x_1)] \, dx_1\right]$$

$$= \frac{1}{Q(x)} \exp\left[\int^x \frac{R(x_1)}{Q(x_1)} \, dx_1\right] \tag{5.15}$$

can always be found.

It is usually convenient in practice to divide the linear equation

$$Q(x)y' + R(x)y + S(x) = 0 \tag{5.16}$$

through by $Q(x)$ before starting to find the IF; the equation then reduces to

$$y' + r(x)y + s(x) = 0, \tag{5.17}$$

and the integrating factor to

$$\exp\left[\int^x r(x_1) \, dx_1\right]. \tag{5.18}$$

Example 5.2. A particle of mass m starts from rest at time $t = 0$ and is acted upon by an accelerating force $m \exp(-\beta t)$. Its motion is opposed by a viscous force of magnitude $m\eta$ times its velocity. Find its velocity at subsequent times and the total distance it travels.

Denoting the distance by y and its velocity dy/dt by v, we have as the equation of motion

$$m\dot{v} = m \exp(-\beta t) - m\eta v,$$

or $\dot{v} + \eta v = \exp(-\beta t).$

From (5.18) the integrating factor is

$$\exp\left(\int^t \eta \, dt_1\right) = \exp(\eta t),$$

giving $\dot{v} \exp(\eta t) + \eta v \exp(\eta t) = \exp(\eta t - \beta t).$

Then, either by inspection or by applying (5.11 b) we obtain

$$v \exp(\eta t) = (\eta - \beta)^{-1} \exp[(\eta - \beta)t] + c.$$

The initial condition, $v(0) = 0$, gives

$$\frac{dy}{dt} = v(t) = \frac{1}{\eta - \beta} [\exp(-\beta t) - \exp(-\eta t)].$$

The variables here are trivially separable, and integration gives the total distance travelled as

$$y(\infty) - y(0) = \frac{1}{(\eta - \beta)}\left[-\frac{1}{\beta}\exp(-\beta t) + \frac{1}{\eta}\exp(-\eta t)\right]_0^\infty = \frac{1}{\eta\beta}.$$

▶6. Solve, by finding a suitable IF:

(a) $(1 - x^2)y' + 2xy = (1 - x^2)^{3/2}$;
(b) $y' - y\cot x + \csc x = 0$;
(c) $(x + y^4)y' = y$ [treat y as the independent variable].

5.4   Higher degree first-order equations

First-order equations of higher degree than the first do not occur often in the description of physical systems,† since squared and higher powers of first derivatives usually arise from resistive or driving mechanisms and then an 'acceleration' or other higher derivative is present, and the equation is not first-order. They do however sometimes appear in connection with geometrical equations. If the equation can be explicitly solved for one of x, y or y', then either an explicit or a parametric solution can sometimes be obtained.

If the nth degree equation for y' can be solved for y', we obtain instead, n equations of the first degree, which in simple cases can be treated by previous methods. Somewhat more interesting from a technique point of view, are the cases where attempted solution along these lines produces an insoluble first-order equation.

Consider

$$x(y')^2 + y' - y = 0. \tag{5.19}$$

Solving directly for y' gives two equations

$$y' = -(2x)^{-1} \pm (2x)^{-1}(1 + 4xy)^{1/2},$$

which cannot be solved by simple methods. However a parametric solution can be obtained as follows.

Denote y' by z, and write (5.19) as

$$y = z^2 x + z. \tag{5.20}$$

---

† With the alternative definition of degree given in the footnote at the beginning of this chapter, this statement is no longer true, since Ricatti-type equations involving y' and powers of y higher than the first, have significant physical importance.

Differentiating with respect to x gives

$$z = y' = z^2 + (2zx + 1)z',$$

which can be made into a linear equation for x, using z as the independent variable

$$(z - z^2)\frac{dx}{dz} - 2zx - 1 = 0.$$

The integrating factor for this is

▶7. $(1 - z)z^{-1}$

yielding,

$$\frac{d}{dz}[x(1 - z)^2] = \frac{1 - z}{z}.$$

This on solution gives

$$(1 - z)^2 x = \ln z - z + c. \tag{5.21}$$

Together with (5.20), (5.21) gives a parametric representation of the solution of (5.19) with z as the parameter.

▶8. Obtain essentially the same result by the following procedure in which the roles of x and y are reversed.

(i) Solve (5.19) for x.
(ii) Differentiate with respect to y, putting $dx/dy = z^{-1}$.
(iii) Arrange as a linear equation in y.
(iv) Find the integrating factor.
(v) Solve for y in terms of the parameter z.
(vi) Compare with y as obtained from (5.20) and (5.21).

5.5   Second-order equations

It is an empirical fact that many natural processes and laws, when put into mathematical form, appear as second-order equations of the first degree. Some, such as the wave equation

$$\frac{\partial^2 y}{\partial x^2} = \frac{1}{c^2}\frac{\partial^2 y}{\partial t^2},$$

contain two second (partial) derivatives; others, for example the equation of an undriven damped oscillator

$$m\ddot{y} + 2b\dot{y} + ky = 0, \tag{5.22}$$

involve only a single second derivative.

A further feature common to these two equations and to many others of physical science is that they are linear. We will in consequence concern ourselves almost exclusively with linear second-order equations in the remainder of this chapter.

As we have shown previously, to obtain a unique solution for a second-order equation, two independent boundary conditions are required, in order to determine the values of two otherwise arbitrary parameters. We next demonstrate the general method whereby the solutions of such linear second-order equations may be sought and the two parameters determined.

We take the general linear equation for solution as

either $f_2(x)y''(x) + f_1(x)y'(x) + f_0(x)y(x) = f(x),$ (5.23 a)

or $f_2(x)y''(x) + f_1(x)y'(x) + f_0(x)y(x) = 0,$ (5.23 b)

in which for the present we will assume that, in the range of x under consideration, the $f_i(x)$ are continuous and, where necessary, differentiable.

It should be noted that (5.23 a) is not really linear, in that, if $y_1(x)$ and $y_2(x)$ are both solutions of it then $y(x) = \alpha_1 y_1(x) + \alpha_2 y_2(x)$ is also a solution only if constants α_1 and α_2 add up to unity. However, if $y_1(x)$ is a solution of (5.23 a) but $y_2(x)$ is a solution of (5.23 b) in which the right-hand side is zero, then $y_1(x) + y_2(x)$ is a solution of (5.23 a). Taking this one step further, and recalling that the general solutions of both (5.23 a) and (5.23 b) will each contain two arbitrary constants, we obtain the general result as follows.

If $y_1(x)$ is *any* solution of (5.23 a) and $y_2(x)$ is the general solution of (5.23 b), containing two arbitrary constants, then the general solution of (5.23 a) is given by $y(x) = y_1(x) + y_2(x)$.

The solution $y_1(x)$ for (5.23 a) with the full right-hand side is called the **particular integral**, whilst that of (5.23 b) is known as the **complementary function**. The result can be stated as 'the general solution is the sum of a particular integral and a complementary function'.†

It should be emphasized that for the practical purpose of obtaining an explicit solution, *any* particular integral (PI), however simple, will suffice; roughly speaking, the difference between any two possible PI will always be compensated by the consequent need to assign a different pair of values to the parameters in the complementary function (CF), when matching the boundary conditions.

To illustrate these general results we consider a very simple example in which all the $f_i(x)$ are constants.

---

† This form of solution generalizes to any nth-order linear equation, the only difference being that the complementary function then contains n arbitrary parameters.

Example 5.3. Find the solution of

$$y'' - 2y' - 3y = 6 \qquad (5.24)$$

which satisfies $y(0) = y'(0) = 0$.

We first seek the CF which will be the general solution of

$$y'' - 2y' - 3y = 0. \qquad (5.25)$$

For reasons which are discussed more fully later, we try a solution of the form $y = A \exp(mx)$ and find that this is indeed a solution if

▶9. $m = 3$ or $m = -1$.

Now (5.25) is linear and its general solution containing two arbitrary constants is therefore the complementary function

$$y_2(x) = A \exp(3x) + B \exp(-x).$$

All that remains now is to find one [the simplest possible] solution of (5.24). The most trivial is $y_1(x) = -2$, but this is a perfectly adequate PI, and gives as the general solution of (5.24)

$$y(x) = A \exp(3x) + B \exp(-x) - 2. \qquad (5.26)$$

The two boundary conditions require that

$$A + B - 2 = 0,$$
$$3A - B = 0,$$

thus determining A and B and yielding as the final solution,

$$y(x) = \tfrac{1}{2} \exp(3x) + \tfrac{3}{2} \exp(-x) - 2.$$

5.6 Linear equations with constant coefficients

In this section we consider specifically the class of linear second-order equations typified by the damped oscillator equation (5.22), and given generally by (5.23) when all the $f_i(x)$ are restricted to be constants. Thus we take our equation as

$$a_2 y'' + a_1 y' + a_0 y = f(x), \qquad a_2 \neq 0, \qquad (5.27)$$

with the a_i as constants.

Here we treat only cases in which the right-hand side $f(x)$ is either zero or a given manageable form. Methods available when $f(x)$ is unspecified or complicated are developed in chapter 7.

Our first task is to find the complementary function $y_2(x)$ for (5.27) since this will be applicable whatever the form of $f(x)$. We need

$$a_2 y_2'' + a_1 y_2' + a_0 y_2 = 0. \qquad (5.28)$$

It is clear that the best chance of finding a function which satisfies this is to look for one which on repeated differentiation essentially reproduces its own form, which can then be cancelled out of (5.28) as a factor. On this reasoning, an exponential function naturally suggests itself, and so as a test solution we try

$$y_2(x) = A \exp(\lambda x), \qquad\qquad (5.29)$$

where λ is at present undetermined.

Trying (5.29) in (5.28) leads to

$$(a_2\lambda^2 + a_1\lambda + a_0)A \exp(\lambda x) = 0. \qquad\qquad (5.30)$$

From this it is apparent that (5.29) is satisfactory provided that the **auxiliary equation** (an ordinary quadratic in λ)

$$a_2\lambda^2 + a_1\lambda + a_0 = 0, \qquad\qquad (5.31)$$

is satisfied. There will in general be two values λ_1, λ_2 of λ which do this,

$$\lambda_{\genfrac{}{}{0pt}{}{1}{2}} = \frac{-a_1 \pm (a_1^2 - 4a_2a_0)^{1/2}}{2a_2}. \qquad\qquad (5.32)$$

Recalling that (5.28) is linear and that (5.30) is satisfied for any A provided λ satisfies (5.31), we conclude that any solution of the form

$$y_2(x) = A \exp(\lambda_1 x) + B \exp(\lambda_2 x) \qquad\qquad (5.33)$$

will satisfy (5.28). This contains two arbitrary constants and thus is the most general solution of (5.28) and hence the required CF. The general form remains valid even if λ is complex, the only exception being when $\lambda_1 = \lambda_2$, i.e. $a_1^2 = 4a_2a_0$. In this case the general solution is

$$(A + Bx) \exp(\lambda_1 x), \qquad\qquad (5.34)$$

as is shown in part (iii) of the following example.

Example 5.4. Find the complementary functions for the equations:

(i) $y'' - 6y' + 8y = 16x + 12$;
(ii) $y'' - 6y' + 13y = 6 \cos 2x + 33 \sin 2x$;
(iii) $y'' - 6y' + 9y = \exp(2x)$.

For all three cases the auxiliary equation is

$$\lambda^2 - 6\lambda + \mu = 0,$$

with $\mu = 8, 13, 9$, respectively. The roots of these quadratics are:

(i) 2, 4; (ii) $3 + 2i$, $3 - 2i$; (iii) 3 twice.

Thus the complementary functions for (i) and (ii) follow straightforwardly from (5.33);

(i) $y_2(x) = A \exp(2x) + B \exp(4x)$,
(ii) $y_2(x) = \exp(3x)[A \exp(2ix) + B \exp(-2ix)]$, or
$\quad\quad = \exp(3x)[A' \cos(2x) + B' \sin(2x)]$.

For case (iii), so far we have only a one-parameter solution $y_2(x) = (A + B)\exp(3x) = C \exp(3x)$. To obtain a second, but still trying to maintain a 'self-reproducing' form like (5.29), let us try a solution

$$y_2(x) = g(x) \exp(\lambda x).$$

Substituting this into (5.28) we get, after a little rearrangement,

$$[(a_2\lambda^2 + a_1\lambda + a_0)g + (2a_2\lambda + a_1)g' + a_2g''] \exp(\lambda x) = 0.$$

But λ is a solution of the auxiliary equation and therefore the coefficient of g vanishes. Also, since we are considering the case of equal roots of the auxiliary equation, $\lambda = -a_1/2a_2$ and therefore the coefficient of g' vanishes as well. Thus, since $a_2 \neq 0$ we must have $g'' = 0$. This is trivially satisfied if $g(x) = D + Ex$ and the second part of the CF is $Ex \exp(\lambda x)$. The $D \exp(\lambda x)$ term merely repeats, or is part of, the first solution found. The completed complementary function is thus of the form given in (5.34). For case (iii) of the example we obtain $y_2(x) = (C + Ex) \exp(3x)$.

Although we will not pursue it in this chapter, the trial function method can be applied in an obvious way to higher-order equations with constant coefficients. Some exercises are set in the examples at the end of the chapter.

The second part of the task of solving (5.27) consists of finding any particular integral – to be added to the complementary function already found.

Often a PI of a simple form can be obtained by inspection or by assuming a parameterized form similar to $f(x)$, but there is in general no straightforward way of doing this. There are however a number of rather *ad hoc* ways of finding particular integrals, and since we have seen that the important thing is to find one such integral by some means or other [the complementary function taking care of the rest of the generality], we describe in the following sections some of these methods.

▶10. Find particular integrals for the equations of example 5.4.

5.7   The D operator

Instead of writing the original equation as it appears in (5.27) let us present it in a more symbolic way as

$$L(D)y \equiv (a_2D^2 + a_1D + a_0)y = f(x). \tag{5.35}$$

Here D is shorthand for the operator d/dx which acts upon the quantities [y only in this case] which appear to the right of it. The second derivative is

$$\frac{d^2}{dx^2} = \frac{d}{dx}\left(\frac{d}{dx}\right) = D\left(\frac{d}{dx}\right) = D(D),$$

and is therefore written as D^2; similarly d^n/dx^n would be denoted by D^n. The expression $L(D)$ is an abbreviated way of naming the *linear* operator $a_2D^2 + a_1D + a_0$, but may be considered to stand generically for an nth-order linear operator with constant coefficients for any n.

This appears a highly artificial and complicating procedure, but it will be seen that D, although it is an operator, can often be manipulated as if it were a purely algebraic quantity. This is so because differentiation obeys the distributive law and index laws $D^mD^n = D^{m+n} = D^nD^m$ [at least for $m, n \geqslant 0$], and commutes with constants [but not with functions of x].†

Before using the D operator to find particular integrals we need to establish certain of its properties. The proofs of these properties are given in bare mathematical outline but the student should go through them carefully to be sure that each step is followed.

In the proofs D is manipulated as an algebraic quantity except where it acts upon a function of x, when either nothing further can be done [e.g. when acting upon the arbitrary function $u(x)$] or the specific result is used [e.g. when acting upon $\exp(\lambda x)$ or $\cos(px)$]. It should be borne in mind that $L(D)$ is always a sum of terms of the kind a_nD^n.

(i) $D^n \exp(\lambda x) = \lambda^n \exp(\lambda x)$,

therefore $L(D) \exp(\lambda x) = L(\lambda) \exp(\lambda x)$. $\tag{5.36}$

(ii) $D^n[u \exp(\lambda x)] = \sum_{r=0}^{n} \binom{n}{r} [D^r \exp(\lambda x)][D^{n-r}u]$ (Leibniz),

---

† The methods of this section are, from a purist point of view, very dubious. But because of their great practical importance, particularly in the field of electrical engineering, the ends justify the means and the fact that derived solutions can be resubstituted for checking provides the necessary insurance.

$$= \exp{(\lambda x)} \sum_{r=0}^{n} \binom{n}{r} \lambda^r [D^{n-r} u], \quad \text{by (i)},$$

$$= \exp{(\lambda x)}(D + \lambda)^n u. \tag{5.37}$$

(iii) $D^2 \cos{(px)} = -p^2 \cos{(px)}$,

therefore $L(D^2) \cos{(px)} = L(-p^2) \cos{(px)}$. $\tag{5.38}$

Similarly, $L(D^2) \sin{(px)} = L(-p^2) \sin{(px)}$.

Result (i) is in fact that used in obtaining the auxiliary equation (5.31).

We now use D in a heuristic way to try to find means of producing particular integrals.

If it were permissible to treat (5.35) in a purely algebraic way then we would have as the solution

$$y(x) = \frac{1}{L(D)} f(x). \tag{5.39}$$

But since $L(D)$ is an operator, what meaning are we to attach to $[L(D)]^{-1} f(x)$?

Consideration of the simplest case provides a part of the answer. If $L(D)$ were simply D then the equation would be

$$y' = Dy = f(x),$$

and the formal and actual solutions would be

$$y = \frac{1}{D} f(x) = \int^{x} f(x_1) \, dx_1. \tag{5.40}$$

This clearly indicates that D^{-1} is to be interpreted as an integration symbol, and, by extension, that D^{-m} represents integration m times.

This still leaves unresolved the general meaning to be attached to (5.39), but the results (5.36)–(5.38) suggest that if $f(x)$ has particular forms such as $\exp{(\lambda x)}$ or $\cos{(px)}$, then it might mean, for example in the former case, $[L(\lambda)]^{-1} \exp{(\lambda x)}$. This certainly seems to be pushing pure symbolism rather far, but we know that any wrong solution will be shown up by resubstitution, and any acceptable one is all that is needed.

In this spirit, let us try as a solution of

$$L(D)y = A \exp{(\lambda x)} \tag{5.41}$$

the form

$$y = \frac{A}{L(\lambda)} \exp{(\lambda x)}. \tag{5.42}$$

Using (5.36), the verification is immediate,

$$L(D)\left[\frac{A}{L(\lambda)}\exp(\lambda x)\right] = \frac{A}{L(\lambda)}L(D)\exp(\lambda x)$$

$$= \frac{A}{L(\lambda)}L(\lambda)\exp(\lambda x) = A\exp(\lambda x).$$

This result assumes that $L(\lambda) \neq 0$. If it is equal to zero, then $(D - \lambda)$ must be a factor of $L(D)$ which can therefore be written as $L(D) = \Lambda(D)(D - \lambda)^m$ with $\Lambda(\lambda) \neq 0$. In the same way that (5.34) was derived for a special case, we can in this special case try an extension to solution (5.42) of the form

$$y = \frac{g(x)\exp(\lambda x)}{\Lambda(\lambda)}.$$

Then evaluating $L(D)y$,

$$L(D)y = \Lambda(D)(D - \lambda)^m \frac{[\exp(\lambda x)g(x)]}{\Lambda(\lambda)}$$

$$= \Lambda(D)\frac{\exp(\lambda x)}{\Lambda(\lambda)}D^m[g(x)],$$

where (5.37) has been used with λ replaced by $-\lambda$ and $u(x)$ by $\exp(\lambda x)g(x)$. Using (5.36), the right-hand side becomes $A\exp(\lambda x)$ provided that the last factor $D^m[g(x)]$ is equal to the constant A. This requires that $g(x) = Ax^m/m!$ and hence gives the solution

$$y = \frac{Ax^m}{m!}\frac{\exp(\lambda x)}{\Lambda(\lambda)}. \tag{5.43}$$

We will not continue any further with our generalized heuristic treatment, but rather solve a number of examples illustrating how some forms of $f(x)$ on the right-hand side of the original equation can be treated. We will here be concerned only with finding particular integrals. The method of example (iv) for dealing with odd powers of D when $f(x)$ is a sine or cosine function should be noted.

The examples are based on the same operators as those of examples 5.4 and ▶10, so as to illustrate the variety of forms $f(x)$ which can be dealt with, and so as to enable similar problems to be seen tackled by different methods.

Example 5.5. Use D operator methods to find particular integrals for the following equations:

(i) $y'' - 6y' + 9y = \exp(2x)$ [as in ▶10 (iii)].

$$y = \frac{1}{D^2 - 6D + 9} \exp(2x) = \frac{1}{4 - 12 + 9} \exp(2x)$$
$$= \exp(2x) \qquad \text{(using (5.42))}.$$

(ii) $y'' - 6y' + 8y = \exp(4x)$.
  Here $L(D) = (D - 4)(D - 2)$ and $L(4) = 0$, and so we must use (5.43) with $\Lambda(D) = (D - 2)$ and $m = 1$.

$$y = \frac{x \exp(4x)}{1! \; 4 - 2} = \tfrac{1}{2}x \exp(4x).$$

(iii) $y'' - 6y' + 9y = 2 \exp(3x)$.
  Again use (5.43) with $\Lambda(D) = 1$ and $m = 2$.

$$y = \frac{2x^2 \exp(3x)}{2! \quad 1} = x^2 \exp(3x).$$

(iv) $y'' - 6y' + 9y = \cos 2x$.

$$y = \frac{1}{D^2 - 6D + 9} \cos 2x = \frac{1}{-4 - 6D + 9} \cos 2x, \quad \text{using (5.38)}$$

$$= \frac{1}{5 - 6D} \frac{5 + 6D}{5 + 6D} \cos 2x = \frac{5 + 6D}{25 - 36D^2} \cos 2x$$

$$= \frac{5 + 6D}{25 - 36(-4)} \cos 2x = \frac{1}{169} (5 \cos 2x - 12 \sin 2x).$$

(v) $y'' - 6y' + 9y = x^2$.

$$y = \frac{1}{(D - 3)^2} x^2 = \frac{1}{9}\left(1 - \frac{D}{3}\right)^{-2} x^2$$

$$= \tfrac{1}{9}(1 + \tfrac{2}{3}D + \tfrac{1}{3}D^2 + \cdots)x^2$$

$$= \tfrac{1}{9}(x^2 + \tfrac{4}{3}x + \tfrac{2}{3}).$$

(vi) $y'' - 6y' + 8y = x^2$.

$$y = \frac{1}{(D-4)(D-2)} x^2 = \frac{1}{2}\left(\frac{1}{D-4} - \frac{1}{D-2}\right) x^2$$

$$= \left[-\frac{1}{8}\left(1 + \frac{D}{4} + \frac{D^2}{16} + \cdots\right) + \frac{1}{4}\left(1 + \frac{D}{2} + \frac{D^2}{4} + \cdots\right)\right] x^2$$

$$= -\tfrac{1}{8}(x^2 + \tfrac{1}{2}x + \tfrac{1}{8}) + \tfrac{1}{4}(x^2 + x + \tfrac{1}{2})$$

$$= \tfrac{1}{64}(8x^2 + 12x + 7).$$

▶11. Verify by direct differentiation and substitution that the solutions of example 5.5 do indeed satisfy the equations given.

The algebra-like use of the D operator can be extended to cases involving more than one dependent variable, which, for a solution, must be connected by more than one differential equation. Without solving the problem in full detail we show in the following example the general method [almost a complete parallel of elementary algebraic methods].

Example 5.6. Two electrical circuits, both of negligible resistance, each consist of a self inductance L and a capacitance $C = G^{-1}$. The mutual inductance of the two circuits is M. Find the current in the first coil after the transients have died away† if a generator in the first coil produces an e.m.f. $E \sin \omega t$.

If the charges on the two capacitances are q_1 and q_2, then the two equations are

$$L\ddot{q}_1 + M\ddot{q}_2 + Gq_1 = E \sin \omega t,$$
$$M\ddot{q}_1 + L\ddot{q}_2 + Gq_2 = 0,$$

or, in terms of the D operator [here D stands for d/dt],

(i) $(LD^2 + G)q_1 + MD^2 q_2 = E \sin \omega t,$
(ii) $MD^2 q_1 + (LD^2 + G)q_2 = 0.$

Multiplying (i) by $(LD^2 + G)$ and (ii) by MD^2 and subtracting

$$[(L^2 - M^2)D^4 + 2LGD^2 + G^2]q_1 = (LD^2 + G)E \sin \omega t$$
$$= (-\omega^2 L + G)E \sin \omega t.$$

$$(5.44)$$

† The actual resistance must be non-zero for this to happen.

The complementary function gives the behaviour of the transients in terms of the roots of

$$(L^2 - M^2)\lambda^4 + 2LG\lambda^2 + G^2 = 0.$$

For the stated problem we need a particular integral of (5.44). Using (5.38) this is immediate as

$$q_1 = \frac{(G - \omega^2 L)E \sin \omega t}{(L^2 - M^2)\,\omega^4 - 2LG\omega^2 + G^2},$$

with the required current as

$$i_1 = \dot{q}_1 = \frac{\omega(G - \omega^2 L)E \cos \omega t}{(L^2 - M^2)\,\omega^4 - 2LG\omega^2 + G^2}.$$

5.8   Variation of parameters

Having dealt at some length with the D operator techniques of finding particular integrals, we now turn to a second somewhat artificial [but successful] method of determining them. Again we will restrict our attention to second-order equations, but the generalization to higher orders proceeds without difficulty.

It should also be noted that the method to be described is applicable when the factors f_i multiplying the derivatives $y^{(i)}$ are functions of x, and not just constants. [In these circumstances D operator methods have to be used with care, since D does not commute with such functions; but see example 9 of section 5.12.]

The equation for which a particular integral is required is

$$L(y) \equiv f_2(x)y''(x) + f_1(x)y'(x) + f_0(x)y(x) = f(x). \qquad \text{(5.23 a bis)}$$

Let the CF be [following the notation of section 5.5]

$$y_2(x) = au_1(x) + bu_2(x), \qquad\qquad\qquad\qquad (5.45)$$

implying that $L(u_1) = L(u_2) = 0$. With a and b as constants, this also means that $L(y_2) = 0$. However if we replace a and b by two functions of x, say $A(x)$ and $B(x)$, then $L(y_2)$ will no longer be equal to zero, but might with suitable choices of A and B be made equal to $f(x)$ thus producing, not a complementary function, but a particular integral.

The reasons why such a procedure does not simply lead to a further second-order differential equation are two-fold. Firstly, repeated differentiation of products such as $A(x)u_1(x)$ always produces terms in which one of the factors is unchanged, and so the expression $L(A(x)u_1(x))$ includes a term which contains $A(x)$ only in the simple product form

$A(x)L(u_1(x))$, and this term will be zero. Secondly, with two functions $A(x)$ and $B(x)$ available and only one equation [(5.23 a)] to satisfy, an additional convenient constraint may be placed upon them. In these circumstances we will see that we can obtain, not a differential equation, but two simultaneous algebraic equations [for A' and B']. We now carry through the procedure which is known as the method of **variation of parameters**.

Assume that

$$y_1(x) = A(x)u_1(x) + B(x)u_2(x) \tag{5.46}$$

is a particular integral of (5.23 a). Differentiating (and rearranging),

$$y_1' = Au_1' + Bu_2' + (A'u_1 + B'u_2). \tag{5.47}$$

At this point we use the freedom to impose an extra constraint and require that the term in the brackets be zero.

$$A'u_1 + B'u_2 = 0. \tag{5.48}$$

This removes the derivatives of A and B from y_1' and results in (5.47) taking the form

$$y_1' = Au_1' + Bu_2'. \tag{5.49}$$

A further differentiation yields

$$y_1'' = Au_1'' + Bu_2'' + A'u_1' + B'u_2', \tag{5.50}$$

and substitution into (5.23 a) from (5.46), (5.49) and (5.50) gives

$$AL(u_1) + BL(u_2) + f_2A'u_1' + f_2B'u_2' = f(x). \tag{5.51}$$

As we have noted, $L(u_1) = L(u_2) = 0$, and so from (5.48) and (5.51) we obtain two simultaneous equations for A' and B', namely

$$A'u_1 + B'u_2 = 0, \tag{5.48 bis}$$

$$A'u_1' + B'u_2' = f(x)/f_2(x). \tag{5.52}$$

Finally provided that $u_1u_2' \neq u_2u_1'$ [u_2 is not a simple multiple of u_1], these can be solved to give

$$A' = \frac{u_2f(x)}{f_2(u_2u_1' - u_1u_2')} \quad \text{and} \quad B' = \frac{-u_1f(x)}{f_2(u_2u_1' - u_1u_2')}. \tag{5.53}$$

For specific forms of u_1, u_2 and $f(x)$, these may be integrated to yield closed expressions for $A(x)$ and $B(x)$, but even if this is not possible, (5.53) provides a formal solution, enabling the general solution of (5.23 a) to be written

$$y(x) = au_1(x) + bu_2(x) + A(x)u_1(x) + B(x)u_2(x). \tag{5.54}$$

Of course a and b must still be determined by boundary conditions. This generalized approach is used later in section 7.5 in connection with Green's functions.

Even though what we have just shown enables us to write down the solution from (5.53) and (5.54), we will, for illustration, go through the working again for a specific case in the following example.

Example 5.7. Solve $y'' + y = \operatorname{cosec} x$ subject to $y(0) = y(\pi/2) = 0$. The complementary function is trivially

$$y_2(x) = a \cos x + b \sin x.$$

Assume a particular integral

$$y_1(x) = A(x) \cos x + B(x) \sin x.$$

If we require, as in (5.48), that

(i) $\quad A' \cos x + B' \sin x = 0,$

then

$$y_1' = -A \sin x + B \cos x.$$

Differentiating again

$$y_1'' = (-A + B') \cos x - (A' + B) \sin x.$$

Substitution in the original equation gives

$$A(-\cos x + \cos x) + B(-\sin x + \sin x)$$
$$- A' \sin x + B' \cos x = \operatorname{cosec} x,$$

or .   (ii) $-A' \sin x + B' \cos x = \operatorname{cosec} x.$

Solving (i) and (ii) gives

$$A' = -\sin x \operatorname{cosec} x = -1, \qquad B' = \cos x \operatorname{cosec} x = \cot x.$$

Hence a solution is

$$A(x) = -x, \qquad B(x) = \ln(\sin x),$$

giving as the full solution of the original equation

$$y(x) = (a - x) \cos x + (b + \ln(\sin x)) \sin x.$$

The given boundary conditions require both a and b to be zero† thus making the final solution

$$y(x) = \sin(x) \ln(\sin x) - x \cos x.$$

† The limit of $\mu \ln \mu$, as $\mu \to 0$, is equal to 0.

Although our attention is focussed on second-order equations, it is worth remarking again that the method of variation of parameters can be applied to higher-order differential equations, additional conditions analogous to (5.48) being imposed on each derivative of the trial function except that of highest order.

5.9   Partially-known complementary function

An alternative and, in principle, easier method of using the complementary function to obtain a particular integral is described in this section. Like the variation of parameters method it can be used when the f_i in

$$L(y) \equiv f_2 y'' + f_1 y' + f_0 y = f(x) \tag{5.55}$$

are functions of x and not just constants. The method is easier in principle since only one of the integrals in the CF need be known. It is clear that in most cases the other has to be found sooner or later for a complete solution, but, as we shall see, it effectively comes from a first-order and not a second-order equation.

Suppose one solution $y_2(x)$ of the complementary function equation $L(y) = 0$ is known and try as a solution of (5.55)

$$y_1(x) = z y_2(x), \tag{5.56}$$

where z is also a function of x. We now rely on a similar property to the first of the two discussed in connection with the variation of parameters (section 5.8); namely that when y_1' and y_1'' are computed and substituted in (5.55) we obtain an equation of the form

▶12. $$f_2 y_2 z'' + (2 f_2 y_2' + f_1 y_2) z' + z L(y_2) = f(x). \tag{5.57}$$

Since $L(y_2) = 0$, this is a second-order equation in z, but with z absent. Thus it is in fact a linear first-order equation for z', which can be solved for z' and then integrated to give z and hence the particular integral $y_1(x)$.

To illustrate the method let us re-solve the equation of example 5.7, $y'' + y = \operatorname{cosec} x$. Suppose that only the part of the CF, $y_2 = \cos x$, has been noticed (or chosen to be noticed). To try to obtain a particular integral we use

$$y_1(x) = z \cos x,$$

giving

$$y_1'' + y_1 = z'' \cos x - 2 z' \sin x = \operatorname{cosec} x.$$

This is a linear first-order equation for z' with

▶13.   integrating factor $= \cos x$,

i.e. $(z' \cos^2 x)' = \cot x,$

$z' \cos^2 x = \ln (\sin x) + c.$

$$z(x) = \int^x \sec^2 x_1 \ln (\sin x_1) \, dx_1 + \int^x c \sec^2 x_1 \, dx_1$$

$$= \tan x \ln (\sin x) - x + k + c \tan x.$$

Therefore the PI is [as in example 5.7]

$$y_1(x) = \sin x \ln (\sin x) - x \cos x + k \cos x + c \sin x. \qquad (5.58)$$

It will be noticed that this includes *both* terms of the complete CF and thus the labour of finding the second part has been done.

▶14. Obtain the same solution (5.58), starting with $y_2(x) = \sin x$ as the partial CF.

5.10   General second-order equations

As mentioned before, the method of the previous section and the variation of parameters can be applied to linear differential equations even when the coefficients of the derivatives are functions of x. Examples using them appear in the exercises of section 5.12.

A further class of equations for which the methods of the previous three sections can be applied is the class of 'dimensionally-homogeneous' equations,† typified [for a second-order equation] by

$$a_2 x^2 y'' + a_1 x y' + a_0 y = f(x). \qquad (5.59)$$

By making the substitution $x = \exp(t)$, the equation can be reduced to one with constant coefficients. The general theory will not be worked out here, but an illustrative example is provided by exercise 14 of section 5.12.

The useful application of this substitution is actually somewhat wider than just equations of the form (5.59), and it usually results in a simplification when made in any equation where the 'dimensions' of x and y *individually* are the same in every term on the left-hand side. In this context $y^{(n)}$ has dimension 1 in y and dimension $-n$ in x and, for example, $2x^2 y' y$ has the same dimensions as $x^5 (y'')^2$. Equations involving expressions in which the dimensions of x and y *taken together* are the same in every term (e.g. $x^2 + 3xy + y^2 y'$) can be reduced to this amenable form by the substitution $y = zx$, z then having dimension 0.

---

† Equations of the form of (5.59) are called *homogeneous* without qualification in mathematics, but we will reserve this term for equations which are linear and in which each term contains y or one of its derivatives.

For the general second-order equation (linear or not) the occurrence of certain forms will permit the use of methods previously discussed, in particular those for first-order equations. The two most obvious cases are:

(i) If y is absent from the equation. Then a new variable $z = y'$ can be used and a first-order equation containing x, z and z' obtained. This is solved for z and then $y' = z$ solved for y.

(ii) If x is absent from the equation. Again use a new variable $z = y'$ and write $y'' = z' = (dz/dy) \times (dy/dx) = z(dz/dy)$. This gives a first-order equation in y, z and (dz/dy) which is processed as in (i).

5.11   Laplace transform methods

Before leaving the methods of solution of ordinary differential equations which yield solutions in the form of explicit or tabulated functions or in the form of an integral (quadrature), we must consider the method of **Laplace transforms.** Our treatment will be superficial since the area is a large one and has filled many books by itself. We will not concern ourselves at all with existence and uniqueness questions, but assume that all 'plausible' assumptions and operations are in fact valid.

The philosophy of the method is that a change in the variables appearing in a differential equation is made, with the object of removing from the equation all the differential and integral signs which appear so that the resulting equation is purely algebraic. This gain is not made without some penalty however, since, even if the resulting algebraic equations are soluble, the problem of transforming back to the original variables still remains.

Let us be more specific. Suppose that we have a function $f(x)$ defined for all positive x, then its Laplace transform $F(s)$ (or $L(f)$) is defined by

$$L(f) \equiv F(s) \equiv \int_0^\infty \exp{(-sx)} f(x)\,dx, \qquad (5.60)$$

provided this integral exists [which it will do for a wide range of functions $f(x)$ provided $s > 0$]. We here assume s is real, but complex values have to be considered in a more detailed study.

In this way (5.60) defines a *linear* transformation which converts from the variable x to a new variable s. As the simplest example, take $f(x) = 1$, then

$$L(1) = F(s) = \int_0^\infty \exp{(-sx)}\,dx = s^{-1}, \qquad \text{for } s > 0. \quad (5.61)$$

Further simple examples are

▶15. $L(x^n) = \displaystyle\int_0^\infty x^n \exp(-sx)\,dx = n!s^{-(n+1)}, \quad n = 0, 1, 2,\ldots,$

(5.62)

▶16. $L(\sin bx) = \displaystyle\int_0^\infty \sin(bx) \exp(-sx)\,dx = b(s^2 + b^2)^{-1}.$ (5.63)

Clearly the inversion of this transformation is not going to be an easy operation to perform, since an explicit formula for $f(x)$, given the form of $F(s)$, is not straightforwardly obtained from (5.60). However a fair amount of progress can be made without having to find such an explicit inverse, since we can prepare from (5.60) a 'dictionary' of Laplace transforms of common functions, and, when faced with an inversion to carry out, hope to find the given transform (together with its parent function) in the listing.

Before proceeding further we should show that this procedure will in fact do what is intended – namely transform derivatives and integrals to algebraic functions.

Consider first differentiation. We have to investigate the Laplace transform, $L(f')$, of the derivative of $f(x)$. Using the definition of $L(f')$

$$L(f') \equiv \int_0^\infty f'(x) \exp(-sx)\,dx,$$ (5.64)

we integrate by parts to obtain

$$L(f') = [\exp(-sx)f(x)]_0^\infty - \int_0^\infty (-s)\exp(-sx)f(x)\,dx$$
$$= -f(0) + sL(f).$$

Thus it is possible to relate the transform of the derivative f' to that of f,

$$L(f') = sL(f) - f(0).$$ (5.65)

This could be repeated explicitly for $f'', f''', \ldots, f^{(n)}$, or (5.65) can be used repeatedly to give

$$L(f^{(n)}) = s\{s[\ldots(sL(f) - f(0)) - \cdots$$
$$- f^{(n-3)}(0)] - f^{(n-2)}(0)\} - f^{(n-1)}(0)$$

▶17. $= s^n L(f) - \displaystyle\sum_{r=0}^{n-1} s^{n-r-1}f^{(r)}(0).$ (5.66)

This shows that the Laplace transform of any derivative of f can be written in terms of $L(f)$ together with the values of all lower derivatives evaluated at $x = 0$.

We have here assumed that f and its derivatives are continuous. If they are not, additional terms appear on the right-hand side; these can be found by breaking up the integral into ranges in which the integrand is continuous and repeating the above procedure. We will not consider such cases further except to note that if a discontinuity in f occurs at $x = x_0$ then (5.65) is modified to

$$L(f') = sL(f) - f(0) - [f(x_0+) - f(x_0-)] \exp(-sx_0),$$
$$(5.67)$$

where $f(x_0+)$ means the limit of $f(x)$ as $x \to x_0$ from above and similarly for $f(x_0-)$. It will be apparent that if $f(x)$ has a discontinuity at $x = 0$ then $f(0)$ is to be interpreted as $f(0+)$.

We turn now to integration; this is more straightforward. From the definition

$$L\left(\int_0^x f(x_1)\, dx_1\right) = \int_0^\infty dx \exp(-sx) \int_0^x f(x_1)\, dx_1$$

$$= \left[(-s)^{-1} \exp(-sx) \int_0^x f(x_1)\, dx_1\right]_0^\infty$$

$$- \int_0^\infty (-s^{-1}) \exp(-sx) f(x)\, dx.$$

The first term on the right vanishes at both limits to give

$$L\left(\int_0^x f(x_1)\, dx_1\right) = s^{-1} L(f). \qquad (5.68)$$

▶18. Deduce (5.68) directly from (5.65).

Of these two results, we will in this chapter be particularly concerned with that showing that differential coefficients can be replaced, in the transformed description, by polynomials. It will be apparent that the coefficients in these polynomials are closely connected with the boundary values of the derivatives.

Before using Laplace transform techniques to solve ordinary differential equations (their use in partial differential equations appears in section 9.11 of chapter 9), we must compile some standard 'dictionary' entries so that we will be in a position to complete the final stage of the operation – converting the solution back into the given variables. Three have already been covered in (5.61)–(5.63). These and several other common cases are given in table 5.1. In the last column is given the value s_0 for which s must satisfy $s > s_0$ in order that the transform exists. (This is not of importance for the present work so long as s_0 is finite.)

Table 5.1

	$f(x)$	$L(f)$	s_0
1	1	s^{-1}	0
2	x^n	$n!s^{-(n+1)}$	0
3	$\sin(bx)$	$b(s^2 + b^2)^{-1}$	0
4	$\cos(bx)$	$s(s^2 + b^2)^{-1}$	0
5	$\exp(ax)$	$(s - a)^{-1}$	a
6	$x^n \exp(ax)$	$n!(s - a)^{-(n+1)}$	a
7	$\sinh(ax)$	$a(s^2 - a^2)^{-1}$	$\lvert a \rvert$
8	$\cosh(ax)$	$s(s^2 - a^2)^{-1}$	$\lvert a \rvert$
9	$\exp(ax)\sin(bx)$	$b[(s - a)^2 + b^2]^{-1}$	a
10	$\exp(ax)\cos(bx)$	$(s - a)[(s - a)^2 + b^2]^{-1}$	a
11	$x^{1/2}$	$\frac{1}{2}(\pi s^{-3})^{1/2}$	0
12	$x^{-1/2}$	$(\pi/s)^{1/2}$	0
13	$U(x - x_0) = 1, x \geqslant x_0$ $ = 0, x < x_0$	$s^{-1}\exp(-sx_0)$	0

▶19. Verify from the definition some of the entries in the table, say results 5, 6, 7, 10 and 13.

▶20. Prove result 12 by using $\int_0^\infty \exp(-t^2)\,dt = \frac{1}{2}\pi^{1/2}$ and putting $t^2 = xs$. Deduce result 11 by integrating by parts.

In carrying out these exercises and from the results in table 5.1, it will be apparent that multiplying a function by $\exp(ax)$ has the effect on its transform that s is replaced by $s - a$. [Compare results 1 and 5, or 2 and 6, or 3 and 9]. This is easily proved generally

$$L(\exp(ax)f(x)) = \int_0^\infty f(x) \exp(ax) \exp(-sx)\,dx$$

$$= \int_0^\infty f(x) \exp[-(s - a)x]\,dx$$

$$= F(s - a), \text{ if } L(f(x)) = F(s). \tag{5.69}$$

As it were, multiplying $f(x)$ by $\exp(ax)$ moves the origin of s by an amount a.

If we multiply the transform by $\exp(-bs)$, $b > 0$, what does this correspond to in the original function? We take the negative exponent since $0 < s < \infty$. From the definition (5.60)

$$\exp(-bs)F(s) = \int_0^\infty \exp[-s(x + b)]f(x)\,dx$$

$$= \int_b^\infty \exp(-sz)f(z - b)\,dz, \qquad (5.70)$$

on putting $x + b = z$. Thus $\exp(-bs)F(s)$ is the Laplace transform of a function $g(x)$ defined to be

$$g(x) = 0 \quad \text{for} \quad 0 < x < b,$$
$$= f(x - b) \quad \text{for} \quad x > b. \qquad (5.71)$$

In words, the function has been translated to 'later' x by an amount b.

The results (5.69) and (5.70 and 5.71) are generally known as the **substitution** and **translation** properties respectively of Laplace transforms.

We now turn to the solution of differential equations using these methods. This will be treated by working through a number of examples of various kinds, and, so that Laplace transform methods can be seen in relation to other approaches, some of the same equations we have already considered will be re-solved here.

Consider first the solution of (5.4)

$$y'' + y = 0, \qquad (5.4 \text{ bis})$$

with the boundary conditions, say, $y(0) = A$, $y'(0) = B$. As on previous occasions, although the solution is immediate, we work pedantically through the steps of the method purely to illustrate them.

We first multiply the equation through by $\exp(-sx)$ and integrate with respect to x from 0 to ∞

$$\int_0^\infty y''(x)\exp(-sx)\,dx + \int_0^\infty y(x)\exp(-sx)\,dx = 0, \qquad (5.72\,\text{a})$$

or symbolically,

$$L(y'') + L(y) = L(0) = 0. \qquad (5.72\,\text{b})$$

If functions for which the Laplace transform is not known were to appear in the original equation, the next step would be to evaluate the corresponding integral along the lines of (5.72 a). There is of course always one function which falls into this 'transform-unknown' class, namely $y(x)$ itself. Let us denote its transform by $Y(s)$. It is an algebraic equation for $Y(s)$ which we must construct and solve in order to find the unknown function $y(x)$.

For the present example both terms are in convenient forms, and so, using (5.66) with $n = 2$ to deal with the first term, we have

$$[s^2 Y(s) - sy(0) - y'(0)] + Y(s) = 0.$$

Substituting the boundary values for $y(0)$ and $y'(0)$ and rearranging we have

$$Y(s) = \frac{As + B}{1 + s^2} = \frac{As}{1 + s^2} + \frac{B}{1 + s^2}.$$

Now if we consult table 5.1, results 3 and 4 [with $b = 1$] show that the right-hand side is the Laplace transform of

$$y(x) = A \cos x + B \sin x,$$

which is therefore the required solution [with the boundary conditions correctly included].

As a second illustration, consider [yet again] example 5.4 (i) (p. 128)

$$y'' - 6y' + 8y = 16x + 12.$$

This time we omit explicit intermediate steps and go, with the help of (5.66) and results 1 and 2 of table 5.1, straight to the transformed equation

$$[s^2 Y - sy(0) - y'(0)] - 6[s Y - y(0)] + 8 Y = 16s^{-2} + 12s^{-1}.$$

Rearranging,

$$Y(s) = \frac{s^2 B + s^2 A(s - 6) + 16 + 12s}{s^2(s^2 - 6s + 8)},$$

where $y(0)$ and $y'(0)$ have been written as A and B respectively. The right-hand side can be arranged in partial fractions as

$$Y(s) = \frac{3}{s} + \frac{2}{s^2} + \frac{4 - 2A + B}{2(s - 4)} - \frac{10 - 4A + B}{2(s - 2)}.$$

Again using table 5.1 to transform back the right-hand side term by term, this means that

$$y(x) = 3 + 2x + (2 - A + \tfrac{1}{2}B) \exp(4x) \\ - (5 - 2A + \tfrac{1}{2}B) \exp(2x),$$

which is of course the same form as has been obtained previously [example 5.4 and ▶10].

We lastly illustrate the use of the transform to solve simultaneous differential equations by basically re-solving another of our previous examples, example 5.6, but with slightly changed conditions. Suppose there is no external e.m.f. impressed, but that initially the second capacitance is given a charge CV_0 with the first one uncharged, and that at time $t = 0$, a switch in the second circuit is closed. Find the subsequent current in the first circuit.

Subject to the initial conditions $q_1(0) = \dot{q}_1(0) = \dot{q}_2(0) = 0$ and $q_2(0) = CV_0 = V_0/G$ we have to solve

$$L\ddot{q}_1 + M\ddot{q}_2 + Gq_1 = 0,$$
$$M\ddot{q}_1 + L\ddot{q}_2 + Gq_2 = 0. \tag{5.73}$$

Here L is the self-inductance of the circuits and is not to be confused with the symbolic Laplace transform operator. Writing the transform of $q_i(t)$ as $Q_i(s)$, we have on taking the transform

$$(Ls^2 + G)Q_1 + Ms^2Q_2 = sMV_0C, \qquad C = 1/G,$$
▶21. $\quad Ms^2Q_1 + (Ls^2 + G)Q_2 = sLV_0C.$

Eliminating Q_2 and rewriting as an equation for Q_1

▶22. $\quad Q_1(s) = \dfrac{MV_0 s}{[(L + M)s^2 + G][(L - M)s^2 + G]}$

$$= \frac{V_0}{2G}\left[\frac{(L + M)s}{(L + M)s^2 + G} - \frac{(L - M)s}{(L - M)s^2 + G}\right].$$

Thus using table 5.1

$$q_1(t) = \tfrac{1}{2}V_0C[\cos(\omega_1 t) - \cos(\omega_2 t)],$$

where $\quad \omega_1^2(L + M) = G \quad$ and $\quad \omega_2^2(L - M) = G. \tag{5.74}$

This gives finally that

$$i_1(t) = \tfrac{1}{2}V_0C[\omega_2 \sin(\omega_2 t) - \omega_1 \sin(\omega_1 t)].$$

With these three examples we finish our very superficial treatment of Laplace transforms for this chapter. As indicated earlier, the general subject is a large one and the interested reader can only be referred to one of the many books devoted to the subject for other aspects, e.g. the properties of the transform under differentiation and integration with respect to s, and all existence and uniqueness results. We will briefly return to the technique in connection with partial differential equations in chapter 9, and an outline of the method for finding an untabulated inverse by contour integration of a complex variable is given in chapter 16.

5.12   Examples for solution

1. *Dimensionally-homogeneous equations.* Suppose $P(x, y)/Q(x, y)$ of (5.5) can be written as $-f(y/x)$. Then by putting $y = zx$ show that the general solution is

$$\ln x = \int \frac{dz}{f(z) - z} + c.$$

Solve (a) $x^2y' - (y + \tfrac{1}{2}x)^2 = 0$; (b) $y' = (y/x) + \sin(y/x)$.

2. Find the general solution of $y' + 2(x + 1)^{-1}y = x + 1$.

3. Obtain a parametric solution of the equation $y'^2 - y' = \exp(x)$ with $y = 0$ when $x = 0$.

4. Using the auxiliary equation find the general solutions of the following:

(a) $y'' - 12y' + 16y = 32x - 8$;
(b) $y'''' - 6y''' + 11y'' - 6y' + 10y = 1$ [i is one root of the equation].

5. Use the D operator to solve $y'' - y' - 2y = 10\cos x$ with $y(0) = y'(0) = 0$.

6. Solve $y'' + 2y' + 3y = \exp(-x)\cos(\sqrt{2}\,x)$. Use D operator methods and consider the real part of $\exp[(-1 + \sqrt{2}\,i)x]$.

7. Continue the analysis of example 5.6 (p. 134). Show that if the resistance of the circuits is truly zero, then the transients do not die away but oscillate with a form

$$A\exp(i\omega_+ t) + B\exp(i\omega_- t),$$

where $\omega_\pm^2 = G(L \pm M)^{-1}$. [See line (5.74).] Find also the ultimate value of the current in the second circuit if the resistance is small but finite.

8. A particle of charge e and mass m moves in magnetic and electric fields. The electric field E is parallel to the x-axis and the magnetic field B to the z-axis. Its equation of motion is

$$m\ddot{\mathbf{r}} = e\dot{\mathbf{r}} \wedge \mathbf{B} + e\mathbf{E}.$$

Find the two simultaneous equations describing the motion in the xy-plane and solve them using D operator methods (to obtain uncoupled equations) for the case when the particle starts from rest at the origin.

9. *Factorization of the operator.* Suppose we need to solve $xy'' - (3 + x)y' + 3y = f(x)$. Using D operator notation this is

$$(x\mathbf{D}^2 - (3 + x)\mathbf{D} + 3)y = f,$$

in which the operator can be written as the *product* of two factors

$$(xD - 3)(D - 1)y = f.$$

[It should be noticed that this is *not* the same as $(D - 1) \times (xD - 3)y = f$ since D and x do not commute.] If such an arrangement is possible then the problem can be reduced to two consecutive first-order equations; solving first

$$(xD - 3)z = f,$$

and then using z from this to solve

$$(D - 1)y = z.$$

(a) If $f(x) = x^4 \exp(x)$, use this method to find the general solution.

(b) Apply the method to, say, equation (i) of example 5.4 (page 128) to show that the solution for constant coefficients is as found by more direct methods.

10. Use the method of variation of parameters to solve $y'' - 6y' + 8y = \exp(x)\cos(x)$.

11. Solve $x^2 y'' + 2xy' - n(n + 1)y = x^m$, (a) by trial function, (b) by variation of parameters. Assume $m \neq n$ and $m \neq -(n + 1)$.

12. Using the methods of section 5.9 solve:

(i) $x^2 y'' + (x^2 \tan x - 2x)y' + (2 - x \tan x)y = x^3 \sin 2x$,

(ii) $(1 - x^2)y'' + 2(x^2 + x - 1)y' - (x^2 + 2x - 1)y = 0$, [Try $y = \exp(x)$.],

(iii) $y'' - (1 - x^2)y' - x^2 y = 0$.

13. Solve: (i) $y'' + 2x(y')^2 = 0$; (ii) $yy'' + (y')^2 = yy'$.

14. By putting $x = \exp(t)$ reduce $x^2 y'' + 2xy' + \frac{1}{4}y = x^{-1/2}$ with $y(1) = y(e) = 0$, to a simpler equation and hence solve it.

15. Solve the non-linear equation

$$x^3 y'' - (x^2 + xy)y' + (y^2 + xy) = 0.$$

16. Use Laplace transforms to solve example 5.4 (iii) (page 128) with the conditions $y(0) = 3$, $y'(0) = 9$.

17. An electrical circuit consists of an uncharged capacitance G^{-1}, an inductance L and a resistance R, all in series with a generator of e.m.f. $V \sin \omega t$. A switch in the circuit is closed at time $t = 0$. If the charge on the condenser is $q(t)$, identify the transient and persistent terms in $Q(s)$, the Laplace transform of $q(t)$.

(a) Show that the transients die away with a time constant $2L/R$.

(b) Show that the character of the transient response depends on whether $4GL$ is greater than, less than, or equal to R^2. What general form is it in each case?

(c) Show that in the persistent response, the current is a phase angle ϕ ahead of the applied voltage where $\tan \phi = (G - L\omega^2)/R\omega$.

18. Two unstable isotopes A and B and a stable one C have the following decay rates per atom present: $A \to B$, 3 s^{-1}; $A \to C$, 1 s^{-1}; $B \to C$, 2 s^{-1}. Initially a quantity x_0 of A is present and none of the other two types. Find the amount of C present at any subsequent time. [Use Laplace transforms.]

19. The quantities $x(t)$, $y(t)$ satisfy the simultaneous equations

$$\ddot{x} + 2n\dot{x} + n^2 x = 0,$$
$$\ddot{y} + 2n\dot{y} + n^2 y = \mu \dot{x},$$

where $x(0) = y(0) = \dot{y}(0) = 0$, and $\dot{x}(0) = \lambda$. Show that

$$y(t) = \tfrac{1}{2}\mu\lambda t^2 \exp(-nt)(1 - \tfrac{1}{3}nt).$$

6

Series solutions of differential equations

In the preceding chapter the solution of ordinary differential equations in terms of standard functions or numerical integrals was discussed, and methods for obtaining such solutions explained and illustrated. The present chapter is concerned with a further method of obtaining solutions of ordinary differential equations, but this time in the form of a convergent series which can be evaluated numerically [and if sufficiently commonly occurring, named and tabulated]. As previously, we will be principally concerned with second-order linear equations.

There is no distinct borderline between this and the previous chapter; for consider the equation already solved many times in that chapter

$$y'' + y = 0. \tag{6.1}$$

The solution in terms of standard functions is of course

$$y(x) = a \cos x + b \sin x, \tag{6.2}$$

but an equally valid solution can be obtained as a series. Exactly as in ▶1 of chapter 5 we could try a solution

$$y(x) = \sum_{0}^{\infty} a_n x^n \tag{6.3}$$

and arrive at the conclusion that two of the a_n are arbitrary [a_0 and a_1] and that the others are given in terms of them by

$$a_0 = (-1)^n (2n)! a_{2n},$$
$$a_1 = (-1)^n (2n + 1)! a_{2n+1}. \tag{6.4}$$

Hence the solution is

$$y(x) = a_0 \left(1 - \frac{x^2}{2!} + \frac{x^4}{4!} - \cdots \right) + a_1 \left(x - \frac{x^3}{3!} + \frac{x^5}{5!} - \cdots \right). \tag{6.5}$$

It hardly needs pointing out that the series in the brackets are exactly those known as $\cos x$ and $\sin x$ and that the solution is precisely that of (6.2); it is simply that the cosine and sine functions are so familiar that they have a special name which is adequate to identify the corresponding series without further explanation.

It will also be true of most of our examples that they have a name (although their properties will be slightly less well known), but the methods we will develop can be applied to a variety of equations, both named and un-named.

Our principal 'demonstration models' will be Legendre's and Bessel's equations and since these two equations and their solutions play such a large role in the analysis of physics and engineering problems, we will also, at the same time, consider in detail some of their properties not strictly connected with their series representation. These properties will be used in later chapters (particularly in chapter 10).

In the course of establishing the properties of the Legendre and Bessel functions, the idea of generating functions will be introduced and illustrated as a convenient and compact way of representing and manipulating a series of parameter-dependent functions.

6.1   Ordinary and singular points

We take as the general form of second-order linear equation

$$y'' + P(x)y' + Q(x)y = 0. \tag{6.6}$$

[The general equation (5.23 b) of the previous chapter has been merely divided through by $f_2(x)$.] Our reason for making the coefficient of y'' unity is that with this form we can say in a fairly general way when a solution in the form of a series is possible.

Expressed in this form the two model equations are

Legendre's equation,

$$y'' - \frac{2x}{1 - x^2} y' + \frac{l(l + 1)}{1 - x^2} y = 0, \tag{6.7}$$

Bessel's equation,

$$y'' + \frac{1}{x} y' + \left(1 - \frac{m^2}{x^2}\right) y = 0. \tag{6.8}$$

In normal usage the x of Legendre's equation is the cosine of the polar angle in spherical polar coordinates and thus $-1 \leqslant x \leqslant 1$. In Bessel's equation, x is usually a multiple of a radial distance and therefore ranges from 0 to ∞.

Suppose now that we wish to find solutions of (6.6) expressed as power series in $(x - x_0)$ for any x_0 in a certain domain with the possible exception of a (finite) number of singular points. We are then led to consider whether both $P(x)$ and $Q(x)$ can themselves be expressed as series of (ascending) non-negative integer powers of $x - x_0$ (i.e. are analytic). If they can, x_0 is called an **ordinary point** of the equation. If either or both cannot be so expressed, then x_0 is a **singular point**. By the expression non-negative integer powers, we allow the possibility of constant terms in the expansions of P and Q.

Even though an equation may be singular at the point x_0, it does not follow that all its integrals are. [An 'integral' of a differential equation is the name often given to a function which is a solution of the equation.] In fact the necessary and sufficient condition† that a non-singular (finite) integral of (6.6) exists at the point $x = x_0$ is that $(x - x_0)P(x)$ and $(x - x_0)^2 Q(x)$ can be expressed as non-negative power series in $(x - x_0)$. [In complex variable language $(x - x_0)P(x)$ and $(x - x_0)^2 Q(x)$ are analytic at $x = x_0$.]

Points at which the equation is singular but a non-singular solution exists are called **regular singular points** of the equation. If both the equation and all its integrals are singular, x_0 is known as an **irregular singular point**.

In our subsequent discussion we will consider only ordinary and regular singular points. In addition we will take x_0, the point about which the expansion is to be made, as the origin, i.e. $x_0 = 0$. If this is not already the case, the substitution $X = x - x_0$ will make it so. With these limitations included we can multiply our original equation through by x^2 so that it becomes

$$x^2 y'' + xp(x)y' + q(x)y = 0, \tag{6.9}$$

where $p(x)$ and $q(x)$ are analytic, i.e.

$$p(x) = \sum_{i=0}^{\infty} p_i x^i, \qquad q(x) = \sum_{j=0}^{\infty} q_j x^j. \tag{6.10}$$

These last two expansions are valid for $0 \leqslant |x| < R$ for some non-zero R.

It will be readily noted that if the origin is an ordinary point of the original equation (6.6), then $p_0 = q_0 = q_1 = 0$.

▶1. Verify that regular singular points occur at $x = 0$ for Bessel's equation and at $x = \pm 1$ for Legendre's equation.

---

† See, for example, Jeffreys and Jeffreys, *Methods of mathematical physics*, 3rd ed. (Cambridge University Press, 1966) p. 479.

6.2   General method and indicial equation

We are seeking a solution of (6.9) of the form

$$y(x) = x^{\sigma} \sum_{0}^{\infty} a_n x^n = x^{\sigma}(a_0 + a_1 x + \cdots + a_n x^n + \cdots), \quad (6.11)$$

in which $a_0 \neq 0$ and the value of σ has yet to be determined. We expect this series to be convergent for $|x| <$ some R' except possibly (depending on the value of σ) at $x = 0$.

Differentiation of (6.11) and substitution into (6.9) leads to the following equation

$$\sum_{n} (\sigma + n)(\sigma + n - 1)a_n x^{\sigma + n} + \sum_{n, m} (\sigma + n)a_n p_m x^{\sigma + n + m}$$
$$+ \sum_{n, m} a_n q_m x^{\sigma + n + m} = 0, \quad (6.12)$$

where all summations are from 0 to ∞.

Each power of x must separately have its coefficient on the left-hand side equal to zero. The lowest power present is x^{σ} and so

$$[\sigma(\sigma - 1) + \sigma p_0 + q_0]a_0 = 0. \quad (6.13)$$

But $a_0 \neq 0$ and so [defining at the same time the function $\theta(\sigma)$]

$$\theta(\sigma) \equiv \sigma(\sigma - 1) + \sigma p_0 + q_0 = 0. \quad (6.14)$$

This equation is called the **indicial equation**. It is a quadratic in σ and in general has two roots, the nature of which determines the forms of possible series solutions, as will be seen in the next sections.

▶2. Show that for an equation for which the origin is an ordinary point, the indicial equation has roots 0 and 1.

Equating to zero the coefficients of higher powers of x, we get for $x^{\sigma + n}$ $(n \geqslant 1)$,

$$\theta(\sigma + n)a_n + \sum_{r=0}^{n-1} (\sigma + r)p_{n-r}a_r + \sum_{r=0}^{n-1} q_{n-r}a_r = 0, \quad (6.15)$$

where, as defined in (6.14),

$$\theta(\sigma + n) \equiv (\sigma + n)(\sigma + n - 1) + (\sigma + n)p_0 + q_0. \quad (6.16)$$

Thus (6.15) gives a recurrence relation for the a_n, expressing each a_n as a function of previous a_r $(0 \leqslant r \leqslant n - 1)$ and the coefficients of the original equation.

If it were not for special circumstances, the procedure from this point on would be straightforward, namely:

(i) Solve the indicial equation to obtain two roots, σ_1 and σ_2.

(ii) Select one of the roots, say σ_1, and also choose a_0 arbitrarily.

(iii) Compute $a_1, a_2, \ldots$ successively from (6.15) to generate one series $x^{\sigma_1} \sum_0^\infty a_t x^t$.

(iv) Repeat the last two steps using the other root σ_2.

(v) Take as the general solution

$$y(x) = x^{\sigma_1}(a_0 + a_1 x + a_2 x^2 + \cdots)$$
$$+ x^{\sigma_2}(b_0 + b_1 x + b_2 x^2 + \cdots). \quad (6.17)$$

As an example carry through the following exercise.

▶3. Consider an expansion about $x = 0$ of the solution of

$$x^2 y'' - \tfrac{3}{2} x y' + (1 + x) y = 0.$$

(i) Show the roots of the indicial equation are 2, $\tfrac{1}{2}$.

(ii) Find the recurrence relation corresponding to $\sigma = 2$.

(iii) Write down the general term of the series.

(iv) Similarly find the general term in the series corresponding to $\sigma = \tfrac{1}{2}$ and hence show that the general solution is

$$y(x) = 3a_0 x^2 \sum_0^\infty \frac{(-2x)^n}{n!(2n + 3)!!}$$

$$+ b_0 \left[x^{1/2} + 2x^{3/2} - x^{1/2} \sum_2^\infty \frac{(-2x)^n}{n!(2n - 3)!!} \right].$$

[The double factorial symbol $(2n + 1)!! \equiv 1 \cdot 3 \cdot 5 \cdot \ldots \cdot 2n + 1$, although not included in the list of internationally recognized symbols, is a convenient notation for the more cumbersome equivalent in terms of ordinary factorials, $(2n + 1)!/2^n n!$. However, it must not be mistaken for the factorial of a factorial.]

(v) Show that the series converges for all finite x.

The difficulties with the described procedure, apart from its cumbersomeness if $p(x)$ and $q(x)$ contain more than one or two terms, occur when either the roots of the indicial equation are not distinct or $\theta(\sigma + n) = 0$ for some positive integer $n (\neq 0)$. The latter case arises when σ_1 and σ_2 differ by an integer. [Since $\theta(\sigma) = 0$ is a quadratic and has only two roots and say $\theta(\sigma_1) = 0$, then $\theta(\sigma_1 + n) = 0$ implies that $\sigma_1 + n = \sigma_2$.] These special (but common) cases are considered in the following sections.

6.3   Equal roots of the indicial equation

We have just noted that if the indicial equation has only one distinct root σ_1 then the procedure of the previous section, whilst giving one valid series solution will not yield a second. The method of obtaining the second solution seems slightly *ad hoc* but proceeds as follows.

The calculation of a first series $x^{\sigma_1} \sum a_n x^n$ goes exactly as in the previous section, but rather than putting $\sigma = \sigma_1$ from the beginning of the calculation of the a_n, it is kept as a variable parameter. This means that the a_n computed are functions of σ and that the computed solution is now a function of σ and x.

$$y(\sigma, x) = x^\sigma \sum a_n(\sigma) x^n. \tag{6.18}$$

If we put $\sigma = \sigma_1$ in this we immediately obtain the first series, but for the moment we leave it as a parameter.

For brevity let us denote the differential operator on the left-hand side of (6.9) by L, i.e.

$$L \equiv x^2 \frac{d^2}{dx^2} + xp(x) \frac{d}{dx} + q(x), \tag{6.19}$$

and examine the effect of L on the y of (6.18).

Along the lines of the argument in section 6.2 we obtain

$$L[y(\sigma, x)] = a_0 x^\sigma \theta(\sigma) + \sum_{n=1}^{\infty} x^{\sigma+n} \left\{ \theta(\sigma + n) a_n(\sigma) \right.$$
$$\left. + \sum_{r=0}^{n-1} [(\sigma + r)p_{n-r} + q_{n-r}] a_r(\sigma) \right\}. \tag{6.20}$$

The factors $a_n(\sigma)$ for $n \geq 1$ have been written explicitly as functions of σ because they are calculated specifically to make each of the terms in the summation vanish individually for any given value of σ, and therefore depend upon it.

Since we are considering the case of equal roots of the indicial equation, $\theta(\sigma)$ has the form

$$\theta(\sigma) = (\sigma - \sigma_1)^2. \tag{6.21}$$

We therefore have the situation that the first term of $L[y(\sigma, x)]$ varies as $(\sigma - \sigma_1)^2$ as σ varies near σ_1 [actually for all σ] and equals zero at $\sigma = \sigma_1$, whilst the second term of $L[y(\sigma, x)]$ is always zero whatever value σ has [because of the way the a_n are calculated for $n \geq 1$]. Thus, not only do both terms vanish at $\sigma = \sigma_1$, but so too does their derivative with respect to σ, i.e.

$$\frac{\partial}{\partial \sigma} [L(y(\sigma, x))] = 0 \quad \text{at} \quad \sigma = \sigma_1. \tag{6.22}$$

But $\partial/\partial\sigma$ and L are operators which differentiate with respect to different variables [σ and x respectively] and so we can reverse their order, thus

$$L\left(\frac{\partial y}{\partial\sigma}(\sigma, x)\right) = 0 \quad \text{at} \quad \sigma = \sigma_1. \tag{6.23}$$

But $L(z) = 0$ is exactly the condition that z is a solution of the original equation (6.9) and so we have found that for the equal roots case $\partial y(\sigma, x)/\partial\sigma$ evaluated at $\sigma = \sigma_1$, is a second solution of that equation [the first being $y(\sigma_1, x)$].

If we carry out the differentiation explicitly, treating y as a product, we obtain

$$\frac{\partial y}{\partial\sigma} = \frac{\partial(x^\sigma)}{\partial\sigma}\left[\sum_0^\infty a_n(\sigma)x^n\right] + x^\sigma\left[\sum_1^\infty \frac{da_n(\sigma)}{d\sigma} x^n\right] \quad \text{at} \quad \sigma = \sigma_1$$

$$= (\ln x)x^{\sigma_1}\left[\sum_0^\infty a_n(\sigma_1)x^n\right] + x^{\sigma_1}\left[\sum_1^\infty \frac{da_n(\sigma_1)}{d\sigma} x^n\right]. \tag{6.24}$$

It will be seen from this that the general form of the second solution for the repeated-root case is always 'the first solution multiplied by $\ln x$ plus an additional series obtained from the first solution by differentiation with respect to the root at the value of the repeated root'.

We illustrate this with a worked example – Bessel's equation for $m = 0$.

Example 6.1. Find series solutions of the equation

$$y'' + x^{-1}y' + y = 0.$$

We first note that $x = 0$ is a regular singular point of the equation and if we write it as

$$x^2y'' + xy' + x^2y = 0,$$

then $p_0 = q_2 = 1$ and all other p_i and q_i are zero. The indicial equation is thus

$$\theta(\sigma) = \sigma(\sigma - 1) + \sigma = \sigma^2 = 0,$$

with repeated root $\sigma = 0$.

First the series for a general σ is calculated, taking a_0 as arbitrary. Applying (6.15), and recalling that $q_2 = 1$ is the only relevant non-zero quantity we have that

$$a_n = -\frac{a_{n-2}}{(\sigma + n)^2}$$

with $a_1 = 0$. Thus

$$y(\sigma, x) = a_0 x^\sigma \left\{ 1 - \frac{x^2}{(\sigma + 2)^2} + \frac{x^4}{(\sigma + 2)^2(\sigma + 4)^2} + \cdots \right.$$

$$\left. + \frac{(-1)^n x^{2n}}{[(\sigma + 2)(\sigma + 4)\ldots(\sigma + 2n)]^2} + \cdots \right\}. \quad (6.25)$$

We can now obtain both series. First putting $\sigma = 0$ in (6.25) yields (if we choose $a_0 = 1$)

$$J_0(x) \equiv y(0, x) = \sum_0^\infty \frac{(-1)^n}{(n!)^2} \left(\frac{x}{2}\right)^{2n}, \quad (6.26)$$

the zeroth-order Bessel function.

The second series solution is obtained by differentiating (6.25) with respect to σ at $\sigma = 0$;

$$\frac{\partial y}{\partial \sigma} = \ln x \, J_0(x) + \sum_1^\infty \alpha_n x^{2n},$$

where

$$\alpha_n = \frac{da_n(\sigma)}{d\sigma} \bigg|_{\sigma = 0} = \frac{d}{d\sigma} \left\{ \frac{(-1)^n}{[(\sigma + 2)(\sigma + 4)\ldots(\sigma + 2n)]^2} \right\}_{\sigma = 0}$$

▶ 4.

$$= -2a_n(\sigma) \sum_{r=1}^n \frac{1}{\sigma + 2r}, \quad \text{at } \sigma = 0$$

$$= \frac{-2(-1)^n}{2^{2n}(n!)^2} \sum_{r=1}^n \frac{1}{2r}.$$

Thus the second series solution is

$$y(x) = \ln x \, J_0(x) - \sum_{n=1}^\infty \frac{(-1)^n}{(n!)^2}$$

$$\times \left(1 + \tfrac{1}{2} + \tfrac{1}{3} + \cdots + \frac{1}{n} \right) \left(\frac{x}{2}\right)^{2n}, \quad (6.27)$$

and the general solution any linear combination of (6.26) and (6.27).

6.4   Roots differing by an integer

The second reason why the straightforward method of section 6.2 may fail to produce two series solutions for the second-order equation is that the two roots of the indicial equation may differ by an integer. Clearly no problem arises for the series determined by the larger of the roots σ_1 but it is easily seen that if $\sigma_1 = \sigma_2 + N$ where N is an integer, then the formula (6.15) for finding the a_n runs into trouble when calculating the series for $\sigma = \sigma_2$ at the stage when $n = N$. At this point $\theta(\sigma + n) = \theta(\sigma_2 + N) = \theta(\sigma_1)$, and this necessarily equals zero.

Two cases now arise – either the summation terms in (6.15) add up to zero or they don't. If they do, i.e.

$$\sum_{r=0}^{N-1} [(\sigma_2 + r)p_{N-r} + q_{N-r}]a_r = 0, \tag{6.28}$$

then (6.15) is automatically satisfied for any value of a_N which therefore becomes arbitrary and can be chosen at will. It is best (but not essential) to choose it equal to zero, since to do otherwise merely adds in a multiple of the σ_1 solution. The rest of the computation of the a_n then proceeds normally.

The more difficult case is when (6.28) is not satisfied and then a_N is formally required to be infinite. We can get around this difficulty by a device similar to that of the preceding section. In the present case the indicial equation has the form

$$\theta(\sigma) = (\sigma - \sigma_1)(\sigma - \sigma_1 + N). \tag{6.29}$$

Thus if in the series solution $y(\sigma, x)$ corresponding to a general value of σ, we replace the constant a_0 by $k(\sigma - \sigma_2)$ – where now we require that not a_0, but k, does not vanish – the extra factor will ensure that the new form of (6.28) is satisfied at $\sigma = \sigma_2$.

Denoting the new function by $y_1(\sigma, x)$, just as in (6.20)–(6.23) we will find that

$$L[y_1(\sigma, x)] = k(\sigma - \sigma_1)(\sigma - \sigma_2)^2 x^\sigma, \tag{6.30}$$

showing that both $y_1(\sigma_2, x)$ and $\partial y_1(\sigma_2, x)/\partial\sigma$ are solutions of $L(y) = 0$, i.e. equation (6.9).

These series are in addition to the solution already found in which $\sigma = \sigma_1$. However, only two of the three solutions are independent, since the series obtained from $y_1(\sigma_2, x)$ always has just enough of its early coefficients vanishing for it to be in fact only a simple multiple of $y(\sigma_1, x)$; this is shown in the next paragraph. Thus the two independent series which go to make up the general solution are $\partial y_1(\sigma_2, x)/\partial\sigma$ and either $y(\sigma_1, x)$ or $y_1(\sigma_2, x)$.

The quoted result in the previous paragraph is shown as follows.†
We first need to show that the coefficients $a_{n+N}(\sigma_2)$ satisfy the same defining relationships as $a_n(\sigma_1)$ and hence that they are related by a common factor for all n. [As it were, the coefficients have all been shifted by N

† This and the next paragraph could be omitted on a first reading.

between the two series.] This is easily seen since from (6.15), the $a_{n+N}(\sigma_2)$ satisfy

$$a_{n+N}(\sigma_2) = - \frac{\sum_{r=0}^{n+N-1} [(\sigma_2 + r)p_{n+N-r} + q_{n+N-r}] a_r(\sigma_2)}{\theta(\sigma_2 + n + N)}. \qquad (6.31)$$

Now if we write $r = N + s$ and $\sigma_2 = \sigma_1 - N$ where appropriate, this reads

$$a_{n+N}(\sigma_2) = - \frac{\sum_{s=-N}^{n-1} [(\sigma_1 + s)p_{n-s} + q_{n-s}] a_{N+s}(\sigma_2)}{\theta(\sigma_1 + n)}. \qquad (6.32)$$

But (6.32) is exactly the set of equations satisfied by $a_n(\sigma_1)$, thus showing that apart from a possible common multiplier [because of the arbitrariness of one coefficient in each series] the coefficients of $x^{\sigma_2}x^{n+N}$ in the one series and $x^{\sigma_1}x^n$ in the other are identical. But of course $\sigma_2 + n + N = \sigma_1 + n$ and this shows that the series are essentially the same for all powers of x greater than σ_1.

Further, with the additional factor $(\sigma - \sigma_2)$ in the coefficients of the σ_2 series, all the early values of $a_n(\sigma)$ will vanish for that series at $\sigma = \sigma_2$ until an n is reached such that $\theta(\sigma + n)$ in the denominator contains a factor $(\sigma + n - \sigma_1)$ which also vanishes at the same value of σ. This, of course, occurs when $n = N$ and so the first non-vanishing coefficient in this series is a_N, making $\sigma_2 + N$ the leading power of x, i.e. the same as in the σ_1 series. This and the previous paragraph together prove the result previously quoted.

Example 6.2. Solve Bessel's equation (a) when $m = \frac{1}{2}$, and (b) when $m = 1$.

The equation is as before

$$x^2 y'' + xy' + (x^2 - m^2)y = 0. \qquad (6.33)$$

The indicial equation is $\theta(\sigma) = \sigma(\sigma - 1) + \sigma - m^2 = 0$ and yields

$$\sigma = \pm m,$$

and hence for both cases the two roots differ by an integer.

(a) When $m = \frac{1}{2}$:

With only p_0, q_0 and q_2 non-zero the recurrence relation is simply

$$a_n = \frac{-a_{n-2}}{(\sigma + n)^2 - \frac{1}{4}}.$$

For $\sigma_1 = \frac{1}{2}$ and arbitrary a_0 this leads easily to

▶ 5.
$$a_{2n+1} = 0,$$
$$a_{2n} = \frac{(-1)^n a_0}{(2n+1)!},$$

and a series solution

▶ 6. $$y(x) = a_0 x^{1/2} \sum_{n=0}^{\infty} \frac{(-1)^n x^{2n}}{(2n+1)!} = a_0 x^{-1/2} \sin x. \qquad (6.34)$$

For $\sigma_2 = -\frac{1}{2}$ we again take an arbitrary a_0 and applying (6.15) require that

$$\theta(-\tfrac{1}{2} + 1)a_1 = (\tfrac{1}{2}p_1 - q_1)a_0.$$

The fact that $\theta(\tfrac{1}{2}) = 0$ does not cause difficulty here since, as it happens, both p_1 and q_1 are zero; so a_1 is arbitrary and we take the simplest choice, $a_1 = 0$. Proceeding as before, we obtain the second series solution as

▶ 7. $$y(x) = a_0 x^{-1/2} \sum_{n=0}^{\infty} \frac{(-1)^n x^{2n}}{(2n)!} = a_0 x^{-1/2} \cos x. \qquad (6.35)$$

If a_0 is chosen as $(2/\pi)^{1/2}$ in (6.34) and (6.35) the resulting functions are denoted by $J_{1/2}(x)$ and $J_{-1/2}(x)$ respectively, and the general solution of (6.33) for $m = \frac{1}{2}$ is

$$y(x) = aJ_{1/2}(x) + bJ_{-1/2}(x).$$

(b) When $m = 1$:
   The two roots here are $\sigma = \pm 1$. The larger root $\sigma_1 = 1$ will not give trouble and can be shown to yield

▶ 8. $$y(x) = a_0 x \sum_{n=0}^{\infty} \frac{(-1)^n}{n!(n+1)!} \left(\frac{x}{2}\right)^{2n}. \qquad (6.36)$$

However for $\sigma_2 = -1$ we have as the recurrence relation

$$a_n = \frac{-a_{n-2}}{n(n-2)}$$

and since a_0 is deliberately chosen non-zero, this implies an infinite a_2. We must therefore find the second solution by the method developed at the beginning of this section. The first job is to calculate $y(\sigma, x)$ for an arbitrary σ. The recurrence relation is

$$a_n = \frac{-a_{n-2}}{(\sigma + n)^2 - 1},$$

and thus $a_{2n+1} = 0$, whilst

$$a_{2n} = a_0(-1)^n \prod_{r=1}^{n} \frac{1}{(\sigma + 2r)^2 - 1} \qquad (n \geq 1).$$

The solution $y(\sigma, x)$ is therefore given by

$$y(\sigma, x) = a_0 x^\sigma \left[1 + \sum_{n=1}^{\infty} (-1)^n x^{2n} \prod_{r=1}^{n} \frac{1}{(\sigma + 2r)^2 - 1} \right]. \qquad (6.37)$$

From this we must construct $y_1(\sigma, x)$

$$= k(\sigma - \sigma_2) y(\sigma, x)/a_0$$

$$= kx^\sigma \left\{ (\sigma + 1) \right.$$

$$\left. + \sum_{n=1}^{\infty} \frac{(-1)^n x^{2n}}{[(\sigma + 3)(\sigma + 5) \ldots (\sigma + 2n - 1)]^2 (\sigma + 2n + 1)} \right\}, \qquad (6.38)$$

and its derivative $\partial y_1(\sigma, x)/\partial\sigma$

$$= \ln x \, y_1(\sigma, x) + kx^\sigma$$

$$\times \left\{ 1 + \sum_{n=1}^{\infty} \frac{(-1)^n x^{2n} f(\sigma, n)}{[(\sigma + 3)(\sigma + 5) \ldots (\sigma + 2n - 1)]^2 (\sigma + 2n + 1)} \right\}, \qquad (6.39)$$

where

▶9. $$f(\sigma, n) = -\frac{1}{\sigma + 2n + 1} - \sum_{s=1}^{n-1} \frac{2}{(\sigma + 2s + 1)}. \qquad (6.40)$$

These expressions are long, but evaluation of them is straightforward. It only remains to put $\sigma = -1$ in (6.38), (6.39) and (6.40) to obtain two series solutions of (6.33) with $m = 1$. The first few terms of the second [and independent] series solution are

$$\ln x \, y_1(-1, x) + kx^{-1} \left[1 - \frac{x^2}{2}(-\tfrac{1}{2}) + \frac{x^4}{2^2 \cdot 4}(-\tfrac{2}{2} - \tfrac{1}{4}) \right.$$

$$\left. - \frac{x^6}{2^2 \cdot 4^2 \cdot 6}(-\tfrac{2}{2} - \tfrac{2}{4} - \tfrac{1}{6}) + \cdots \right].$$

▶10. Put $\sigma = -1$ in (6.38) and evaluate the first few terms showing that it gives as a solution

$$y_1(-1, x) = kx^{-1} \left(-\frac{x^2}{2} + \frac{x^4}{2^2 \cdot 4} - \frac{x^6}{2^2 \cdot 4^2 \cdot 6} + \cdots \right).$$

Evaluate the corresponding terms of (6.36) in which $\sigma = 1$ and verify that the two series are related by a constant multiplier. [From the way it was obtained, the factor enclosed in the square brackets in the denominator of (6.38) takes the value 1, not 0, when $n = 1$ and $\sigma = -1$.]

▶11. Find series solutions of the equation $y'' - 2xy' - 2y = 0$. Identify one of the series as $y = \exp(x^2)$ and verify that this is a solution by direct substitution.

6.5   Convergence of series

Although we have discussed almost exclusively the solutions of Bessel's equation for various values of the parameter m^2, we have during the course of the discussion demonstrated each of the cases which can arise in the determination of two independent series solutions. Other equations for solution in ranges containing only ordinary and regular singular points can be treated by parallel methods. Some examples are provided in the exercises at the end of the chapter.

We have not mentioned, in any general sense, the range of x for which the solutions are valid. As indicated in connection with (6.11) we expect convergence for $|x| <$ some R' (except possibly at $x = 0$). Clearly we cannot expect the solution to be valid for any $|x|$ larger than that for which the representation (6.10) of $p(x)$ and $q(x)$ is valid, so $R' \leqslant R$. It can be shown as a general result that $R' = R$, so that (6.10) and the solutions have the same domain of convergence. Clearly, if, as is nearly always the case for any tractable recurrence relation, p and q are finite polynomials, then $R = \infty$ and the solution converges for all x.

▶12. Verify that all solutions of examples 6.1 and 6.2 converge for all x, except in some cases at $x = 0$.

6.6   Finite polynomials

As has been seen, the evaluation of successive terms of a series solution to a differential equation is carried out by means of a recurrence relation. The form of the relation for a_n depends upon n, the previous values of a_r ($r < n$), and the parameters of the equation. It may happen as a result of this that for some value of $n = N + 1$ the computed value a_{N+1} is zero, and [particularly if only one or two p_i and q_i are non-zero] that all higher a_r ($r \geqslant N + 1$) also vanish. If so, we are then left with a finite polynomial of degree N as a solution of the original equation

$$y(x) = \sum_{0}^{N} a_n x^n, \quad a_N \neq 0. \tag{6.41}$$

In many applications of theoretical physics (particularly in quantum mechanics) the termination of a potentially infinite series after a finite number of terms is of crucial importance in establishing physically acceptable descriptions and properties of systems. The condition that such a termination occurs is therefore of considerable significance.

For the discussion of finite polynomials, Legendre's equation

$$y'' - \frac{2x}{1 - x^2} y' + \frac{l(l + 1)}{1 - x^2} y = 0, \tag{6.7 bis}$$

will be our 'demonstration model'. Writing it in the standard form we have

$$x^2 y'' - \frac{2x^2}{1 - x^2} xy' + \frac{l(l + 1)x^2}{1 - x^2} y = 0, \tag{6.42}$$

in which, for an expansion about the origin,

$$p(x) = -2(x^2 + x^4 + x^6 + \cdots),$$
$$q(x) = l(l + 1)(x^2 + x^4 + x^6 + \cdots). \tag{6.43}$$

Recurrence relation (6.15) can now be used repetitively to find the conditions under which a_N vanishes for each N and the two values $\sigma = 0$, $\sigma = 1$. However, as is almost always the case when $p(x)$ and $q(x)$ contain more than one or two terms, the general condition for a general N is complicated to deduce.

▶13. Show by explicit use of (6.15) that all $a_{2n+1} = 0$ and

(i) for $\sigma = 0$, $a_2 = 0$ if $l = 0$ or -1, $a_4 = 0$ if $l = 2$ or -3;
(ii) for $\sigma = 1$, $a_2 = 0$ if $l = 1$ or -2, $a_4 = 0$ if $l = 3$ or -4.

A simpler method of obtaining finite polynomial solutions is to *assume* a solution of form (6.41) exists for some N and then substitute it in (6.7 bis) rearranged as

$$(1 - x^2)y'' - 2xy' + l(l + 1)y = 0. \tag{6.44}$$

To avoid any confusion with our previous use of a_n we will take (6.41) as

$$y(x) = \sum_0^N b_n x^n, \quad b_N \neq 0. \tag{6.41 bis}$$

This gives

$$\sum_{n=0}^N [(1 - x^2)n(n - 1)b_n x^{n-2} - 2xnb_n x^{n-1} + l(l + 1)b_n x^n] = 0. \tag{6.45}$$

Instead of starting with the lowest power of x, we this time start with the highest; such a one now exists because of our assumed form of solution. Thus the coefficients of x^N yield

$$[-N(N - 1) - 2N + l(l + 1)]b_N = 0,$$

i.e. $(l - N)(l + N + 1)b_N = 0.$ (6.46)

Since $b_N \neq 0$ we must therefore have either $l = N$ or $l = -(N + 1)$ as a condition for obtaining a finite polynomial solution.† With either choice $l(l + 1) = N(N + 1)$.

Moving to the more general coefficient of x^m, we have that

$$(m + 2)(m + 1)b_{m+2} - m(m - 1)b_m - 2mb_m + l(l + 1)b_m = 0,$$

which can be rearranged as the recurrence relationship

$$b_{m+2} = -\frac{l(l + 1) - m(m + 1)}{(m + 1)(m + 2)} b_m.$$ (6.47)

Since $b_{N+1} = 0$ this immediately shows that $b_{N-1} = b_{N-3} = \cdots = 0$, the final equation being $b_0 = 0$ if N is odd and $b_1 = 0$ if N is even.

Thus to summarize: A polynomial solution of degree N can be found for Legendre's equation provided that $l = N$ or $-(N + 1)$ with the properties:

(i) All terms in the polynomial have the same oddness or evenness in x (parity) as $x^N = x^l$ has.

(ii) The coefficients of the powers present in the polynomial are related by (6.47), one coefficient being arbitrary.

6.7   Legendre polynomials

Apart from the arbitrary constant multiplier, the polynomials developed in the preceding section are known as **Legendre polynomials** of degree l and denoted by $P_l(x)$. For most physical purposes $-1 \leqslant x \leqslant 1$, since x is the cosine of a polar angle, but some applications involve $|x| > 1$ or x complex. In addition the second solution of (6.44) [which is not analytic at $x = \pm 1$] is sometimes relevant.

Our purpose in the remainder of this section will be to study in some detail the particular properties of the finite polynomials $P_l(x)$.

---

† A result of great importance in the quantum mechanical theory of angular momentum.

It is conventional to normalize $P_l(x)$ (that is to choose the value of the constant multiplier) so that

$$P_l(1) = 1, \tag{6.48}$$

and thus, as a consequence of our previous observations,

$$P_l(-1) = (-1)^l. \tag{6.49}$$

We begin by explicitly constructing the first few Legendre polynomials.

(i) $l = 0$. Only a constant term is present, which must therefore be unity,

$$P_0(x) = 1. \tag{6.50}$$

(ii) $l = 1$. $b_1 \neq 0$, and thus from (6.47), $b_0 = 0$. Together with the normalization condition (6.48) this determines that

$$P_1(x) = x. \tag{6.51}$$

(iii) $l = 2$. $b_2 \neq 0$. From (6.47), $b_1 = 0$ and

$$b_2 = -\left[\frac{2(2+1) - 0(0+1)}{(0+1)(0+2)}\right] b_0 = -3b_0.$$

Requiring that $P_2(1) = 1$ establishes b_0 as $-\frac{1}{2}$, to give

$$P_2(x) = (-3x^2 + 1)b_0 = \tfrac{1}{2}(3x^2 - 1). \tag{6.52}$$

▶14. (a) Show that $P_3(x) = \tfrac{1}{2}(5x^3 - 3x)$, $P_4(x) = \tfrac{1}{8}(35x^4 - 30x^2 + 3)$.
(b) Verify (6.49) for $l = 0, 1, \ldots, 4$.
(c) Sketch as a function of x the first few $P_l(x)$ for (say) $l = 0$ to 3.

Mutual orthogonality of Legendre polynomials. The first additional property of the $P_l(x)$ that will be established is their mutual orthogonality, i.e. that

$$\int_{-1}^{1} P_l(x)P_k(x) \, dx = 0 \quad \text{if} \quad l \neq k. \tag{6.53}$$

This result will be established on more general grounds in chapter 7, but we will prove it here specifically, as well.

First write the defining equation (6.44) in the form

$$[(1 - x^2)P_l']' + l(l + 1)P_l = 0, \tag{6.54}$$

where P_l' stands for $dP_l(x)/dx$. Next multiply this through by P_k and integrate from $x = -1$ to $x = 1$,

$$\int_{-1}^{1} P_k[(1 - x^2)P_l']' \, dx + \int_{-1}^{1} P_k l(l + 1)P_l \, dx = 0.$$

Integrating the first term by parts and noting that the boundary contribution vanishes at both limits because of the $(1 - x^2)$ factor, we obtain

$$-\int_{-1}^{1} P_k'(1 - x^2)P_l' \, dx + \int_{-1}^{1} P_k l(l + 1)P_l \, dx = 0.$$

Now if we reverse the roles of l and k, and start again at the stage (6.54) with the defining equation for $P_k(x)$ we obtain

$$-\int_{-1}^{1} P_l'(1 - x^2)P_k' \, dx + \int_{-1}^{1} P_l k(k + 1)P_k \, dx = 0.$$

Subtracting the last two equations, we conclude that

$$[k(k + 1) - l(l + 1)]\int_{-1}^{1} P_k P_l \, dx = 0,$$

and therefore since $k \neq l$ we must have the result (6.53).

As a particular case we note that if we put $k = 0$,

$$\int_{-1}^{1} P_l(x) \, dx = 0 \quad \text{for} \quad l \neq 0. \tag{6.55}$$

▶15. Verify by inspection of the sketches of ▶14 (c), that they are not inconsistent with this orthogonality result.

Rodrigue's formula. As an aid to establishing further properties of the Legendre polynomials we next develop Rodrigue's representation of the polynomials. This is

$$P_l(x) = \frac{1}{2^l l!} \frac{d^l}{dx^l} (x^2 - 1)^l. \tag{6.56}$$

To prove this representation we denote $(x^2 - 1)^l$ by z, then $z' = 2lx(x^2 - 1)^{l-1}$ and

$$z'(x^2 - 1) - 2lxz = 0. \tag{6.57}$$

If we differentiate (6.57) $l + 1$ times using Leibniz' theorem

$$[z^{(l+2)}(x^2 - 1) + (l + 1)z^{(l+1)}2x + \tfrac{1}{2}(l + 1)lz^{(l)}2]$$
$$- 2l[xz^{(l+1)} + (l + 1)z^{(l)}] = 0,$$
$$(x^2 - 1)z^{(l+2)} + 2xz^{(l+1)} - l(l + 1)z^{(l)} = 0. \tag{6.58}$$

Changing the sign all through (6.58) and comparing with (6.44) we see that $z^{(l)}$ satisfies the same equation as $P_l(x)$. Thus

$$z^{(l)}(x) = \beta_l P_l(x), \tag{6.59}$$

and to establish β_l we note that the only term in the expression for the lth derivative of $(x^2 - 1)^l$ which does not contain a factor $(x^2 - 1)$ and therefore does not vanish at $x = 1$, is $(2x)^l l!(x^2 - 1)^0$. Putting $x = 1$ in (6.59) therefore shows that β_l must be $2^l l!$, thus completing the proof of (6.56).

As an immediate use of Rodrigue's formula let us show

$$I_l = \int_{-1}^{1} P_l(x) P_l(x)\, dx = \frac{2}{2l + 1}. \tag{6.60}$$

The result is trivially obvious for $l = 0$ and so assume $l \geq 1$, then by Rodrigue's formula

$$I_l = \frac{1}{2^{2l}(l!)^2} \int_{-1}^{1} \left[\frac{d^l(x^2 - 1)^l}{dx^l} \right] \left[\frac{d^l(x^2 - 1)^l}{dx^l} \right] dx.$$

Repeated integration by parts, with all boundary terms vanishing, reduces this to

▶16.
$$I_l = \frac{(-1)^l}{2^{2l}(l!)^2} \int_{-1}^{1} (x^2 - 1)^l \frac{d^{2l}}{dx^{2l}} [(x^2 - 1)^l]\, dx$$

$$= \frac{(2l)!}{2^{2l}(l!)^2} \int_{-1}^{1} (1 - x^2)^l\, dx. \tag{6.61}$$

If we write

$$J_l = \int_{-1}^{1} (1 - x^2)^l\, dx,$$

then integration by parts gives the recurrence relation

▶17. $(2l + 1)J_l = 2l J_{l-1}$,

and

$$J_l = \frac{2l}{2l + 1} \frac{2l - 2}{2l - 1} \cdots \frac{2}{3} J_0 = 2^l l! \frac{2^l l!}{(2l + 1)!} 2 = \frac{2^{2l + 1}(l!)^2}{(2l + 1)!}. \tag{6.62}$$

Finally putting (6.62) into (6.61) establishes (6.60).

6.8   Generating functions

A useful device for manipulating and studying sequences of functions or quantities which are labelled by an integer variable [here the functions $P_l(x)$ labelled by l] is the **generating function**. The generating function has perhaps its greatest utility in the areas of probability theory and statistics. However, we will also find it a great convenience in our present study.

The generating function for (say) a series of functions $f_n(x)$ for $n = 0, 1, 2, \ldots$ is a function $G(x, h)$ containing, as well as x, a dummy variable h, and such that

$$G(x, h) = \sum_{n=0}^{\infty} f_n(x)h^n, \tag{6.63}$$

i.e. $f_n(x)$ is the coefficient of h^n in the expansion of G in powers of h. The utility of the device lies in the fact that sometimes it is possible to get a closed form for $G(x, h)$.

For our study of Legendre polynomials let us consider the functions $P_n(x)$ defined by the equation

$$G(x, h) = (1 - 2xh + h^2)^{-1/2} = \sum_{n=0}^{\infty} P_n(x)h^n. \tag{6.64}$$

We will prove as a result of the considerations that the $\{P_n(x)\}$ so defined are identical with the Legendre polynomials and that the function $(1 - 2xh + h^2)^{-1/2}$ is in fact the generating function for them. In the process we will deduce several useful relationships between the various polynomials and their derivatives; these will be displayed for future reference in their final form, in anticipation of the ultimate identification of the functions, in (6.70).

In the following derivation $dP_n(x)/dx$ will be denoted as usual by P'_n. First differentiate (6.64) with respect to (wrt) x

$$h(1 - 2xh + h^2)^{-3/2} = \sum P'_n h^n. \tag{6.65}$$

Also differentiate it wrt h

$$(x - h)(1 - 2xh + h^2)^{-3/2} = \sum nP_n h^{n-1}. \tag{6.66}$$

Equation (6.65) can be written

$$h \sum P_n h^n = (1 - 2xh + h^2) \sum P'_n h^n,$$

and thus equating coefficients of h^{n+1}

$$P_n = P'_{n+1} - 2xP'_n + P'_{n-1}. \tag{6.67}$$

Equations (6.65) and (6.66) can be combined as

$$(x - h) \sum P'_n h^n = h \sum nP_n h^{n-1},$$

from which the coefficient of h^n yields

$$xP'_n - P'_{n-1} = nP_n. \tag{6.68}$$

Eliminating P'_{n-1} between (6.67) and (6.68) gives the further result

$$(n + 1)P_n = P'_{n+1} - xP'_n. \tag{6.69}$$

If we now take result (6.69) with n replaced by $n - 1$ and add x times result (6.68) to it, we obtain

▶18. $(1 - x^2)P'_n = n(P_{n-1} - xP_n)$.

Finally differentiating both sides wrt x and using (6.68) again we find

$$\begin{aligned}
(1 - x^2)P''_n - 2xP'_n &= n[(P'_{n-1} - xP'_n) - P_n] \\
&= n(-nP_n - P_n) \\
&= -n(n + 1)P_n, \tag{6.70}
\end{aligned}$$

i.e. the $P_n(x)$ defined by (6.64) satisfy Legendre's equation. It only remains to verify the normalization. This is easily done at $x = 1$ when G becomes $[(1 - h)^2]^{-1/2} = 1 + h + h^2 + \ldots$, and thus all the $P_n(x)$ so defined have $P_n(1) = 1$ as required.

This completes the proof that $G(x, h) = (1 - 2xh + h^2)^{-1/2}$ is a generating function for the Legendre polynomials and validates all the formulae (6.64) to (6.69) for the properties of the Legendre polynomials.

▶19. (a) Substitute from (6.64) into (6.66) and hence prove the recurrence relation relating Legendre polynomials of different l at the same value of x,

$$(l + 1)P_{l+1}(x) - (2l + 1)xP_l(x) + lP_{l-1}(x) = 0. \tag{6.71}$$

(b) Start with $P_0 = 1$ and use (6.71) to generate P_1, P_2, P_3, P_4. Check your results against previous derivations.

To summarize the position concerning Legendre polynomials, we now have three possible starting points which have been shown to be equivalent, the defining equations (6.44) and (6.48), Rodrigue's formula (6.56), and the generating function (6.64). In addition we have proved a variety of relationships and recurrence formulae [not easily rememberable, but collectively useful] and, as will be apparent later and particularly from the work of chapter 10, developed a powerful tool for use in axially symmetric situations in which the ∇^2 operator is involved and spherical polar coordinates are employed.

6.9   General remarks

As was our intention, we have concentrated to a very marked degree on Bessel's equation in connection with infinite series and Legendre's equation for finite polynomials. The techniques used are, however, applicable to many more functions than these, but since they are in all essentials the

same, we will not deal with them explicitly. Some further examples, e.g.
Hermite polynomials, will be found in the exercises of the next section.

An extended form of Legendre polynomials, the spherical harmonics
$Y_l^m(\theta, \phi)$ in which $x = \cos \theta$ and ϕ is an azimuthal variable, are used extensively in chapter 10 in connection with physical situations involving
Laplace's, Poisson's or Schrodinger's equation. Our present motivation
has been the solution of differential equations to a large extent in a
purely mathematical context, but this has at the same time laid the foundation for the future treatment of more directly physical equations.

6.10   Examples for solution

1. Find solutions, as power series in x, of the equation

$$4xy'' + 2(1 - x)y' - y = 0.$$

Identify one of the solutions and verify it by direct substitution.

2. Find the solution $J_m(x)$ of Bessel's equation which is finite at the
origin, for a general integer m. The conventional definition of $J_m(x)$
takes $a_0 = 2^{-m}(m!)^{-1}$.

3. For what range of values of x is a power series solution in x
valid for the equation $(x^2 + 5x + 4)y'' + 2y = 0$?

4. Find the general power series solution about $x = 0$ of

$$xy'' + (2x - 3)y' + 4x^{-1}y = 0.$$

5. *Continuation of* ▶11 (p. 162). Using the methods of chapter 5
find the second solution of the original equation and hence show

$$\int_0^x \exp(-u^2) \, du = \exp(-x^2) \sum_{n=0}^{\infty} \frac{2^n x^{2n+1}}{(2n+1)!!}.$$

[Establish the constants involved by comparing coefficients for small
x.]

6. For the equation $y'' + x^{-3}y = 0$, show that the origin is a regular
singular point if the independent variable is changed to $\xi = 1/x$.
Hence find the solution of the form

$$y(x) = \sum_0^{\infty} \frac{a_n}{x^n}.$$

Do not use the general theory, but seek a second solution of the form
$f(x)$ times the first one, showing that as $x \to \infty$ the second solution
becomes $c(x + \ln x) + O(1)$ where c is an arbitrary constant.

7. Show that, if n is a positive integer, the differential equation

$y'' - 2xy' + 2ny = 0$ has a polynomial solution (Hermite polynomial) of degree n. Find the general term of the polynomial.

8. The Hermite polynomials† $H_n(x)$ are defined by

$$\Phi(x, h) = \exp(2xh - h^2) = \sum_{n=0}^{\infty} \frac{1}{n!} H_n(x)h^n.$$

Show that

$$\frac{\partial^2 \Phi}{\partial x^2} - 2x \frac{\partial \Phi}{\partial x} + 2h \frac{\partial \Phi}{\partial h} = 0,$$

and hence that $H_n(x)$ satisfy the equation of example 7. Use Φ to prove

(i) $H_n'(x) = 2nH_{n-1}(x)$,
(ii) $H_{n+1}(x) - 2xH_n(x) + 2nH_{n-1}(x) = 0$.

9. By writing $\Phi(x, h)$ of example 8, as a function of $(h - x)$ rather than h, and by noting that for a function of $(h - x)$

$$\frac{\partial f}{\partial h} = -\frac{\partial f}{\partial x},$$

show that an alternative representation of the nth Hermite polynomial is

$$H_n(x) = (-1)^n \exp(x^2) \frac{d^n}{dx^n} [\exp(-x^2)].$$

[Note that $H_n(x) = \partial^n \Phi / \partial h^n$ at $h = 0$.]

10. Carry through the following procedure as an alternative proof that

$$\int_{-1}^{1} [P_l(x)]^2 \, dx = \frac{2}{2l + 1}.$$

(i) Use the generating function equation for the Legendre polynomials and square both sides.
(ii) Express the right-hand side as a sum of powers of h, obtaining expressions for the coefficients.
(iii) Integrate the right-hand side from -1 to 1 and use the orthogonality results (6.53).
(iv) Similarly integrate the left-hand side and expand the result in powers of h.
(v) Compare coefficients.

11. A charge $+2e$ is situated at the origin and charges of $-e$ at distances $\pm a$ from it along the polar axis. By relating it to the gen-

---

† Of importance in the quantum mechanical harmonic oscillator problem.

erating function for the Legendre polynomials, show that the electrostatic potential Φ at the point (r, θ, ϕ), with $r > a$, is

$$\Phi(r, \theta, \phi) = \frac{2e}{r} \sum_{s=1}^{\infty} P_{2s}(\cos \theta) \left(\frac{a}{r}\right)^{2s}.$$

7
Superposition methods

In the previous two chapters we have dealt with the solution of a differential equation of order n by two methods. In one, by finding n independent solutions of the equation and then combining them, weighted with coefficients determined by the boundary conditions; in the other by finding a solution in terms of a series whose coefficients are related by (in general) an n term recurrence relation, and thence fixed by the boundary conditions. For both approaches the linearity of the equation was an important (essential) factor in the utility of the method, and in this chapter our aim will be to exploit the superposition properties of linear equations even further.

This present chapter is more formal than most in the book and is aimed at a general introduction to the more specific methods of chapters 8, 9 and 10. It is intended to be sufficiently concrete for general ideas to be grasped from it alone, but the reader may find some benefit from re-reading it after assimilating the material of these three later chapters, since they illustrate the ideas presented here.

We will be concerned with the solution of equations of the homogeneous form

$$L(y) = 0, \tag{7.1}$$

but with the ultimate object of treating the inhomogeneous variety

$$L(y) = f(x), \tag{7.2}$$

where $f(x)$ is a prescribed or general function, and the boundary conditions to be satisfied by the solution $y = y(x)$ at the limits $x = a$ and $x = b$ are given. In these two equations the expression $L(y)$ stands for a linear differential operator acting upon the function $y(x)$; for example, in the case of the damped harmonic oscillator in chapter 5, L has the form

$$L \equiv m\frac{d^2}{dx^2} + 2b\frac{d}{dx} + k,$$

x having the physical interpretation of time and y that of the displacement of the oscillator. For this system (7.1) then gives the equation of free oscillations whilst (7.2) corresponds to forced oscillations. For a more general differential operator the coefficients of d^n/dx^n may be functions of x [but clearly not of y if L is to be linear].

In general, unless $f(x)$ is both known and simple, it will not be possible to find particular integrals of (7.2), even if complementary functions can be obtained from (7.1). The idea is therefore to exploit the linearity of L by building up the required solution as a **superposition**, generally containing an infinite number of terms, of some set of functions which each individually satisfy the boundary conditions. This clearly brings in a quite considerable complication, but since, within reason, we may select the set of functions to suit ourselves, we can obtain sizeable compensation for this complication. Indeed, if the set chosen is one containing functions which, when acted upon by L, produce particularly simple results, we can 'show a profit' on the operation. In particular, if the set consists of those functions $y_i(x)$ for which the resultant function $L(y_i)$ is simply $-\lambda_i y_i$ where λ_i is purely a constant,† then a distinct advantage may be obtained from the manoeuvre because all the differentiation will have disappeared from (7.2).

Let us carry this out in a slightly more formal manner. Suppose that we can find a set of functions $\{y_i(x)\}$ where i is a label running from 0 to ∞, such that

$$L(y_i) = -\lambda_i y_i. \tag{7.3}$$

As a possible solution of (7.2) try

$$y(x) = \sum_i a_i y_i(x), \tag{7.4}$$

which automatically satisfies the boundary conditions, since each y_i does so. Now, making full use of the linearity of L we have

$$f(x) = L(y) = L(\sum a_i y_i) = \sum a_i L(y_i) = -\sum a_i \lambda_i y_i. \tag{7.5}$$

Thus we are left with an equation which contains no differential operators – but at the price of having introduced the unknown a_i. This, however, can also be put right if, in addition, the set $\{y_i(x)\}$ is in some sense mutually orthogonal, e.g.,

$$\int_a^b y_j^*(z) y_i(z) \, dz = 0, \quad \text{if} \quad i \neq j. \tag{7.6}$$

---

† The minus sign is conventional and is included here only to give the Sturm–Liouville equation (7.9) its usual form.

Here $y_j^*(x)$ is the complex conjugate of $y_j(x)$. The orthogonality expressed by this equation is more strictly called Hermitian orthogonality, although the word 'Hermitian' is usually omitted. Straightforward orthogonality of functions y_i and y_j would not involve taking the complex conjugate of one of them in the integrand. This latter type of orthogonality has some uses in physics, but we will take the word 'orthogonal' in the Hermitian sense expressed by (7.6).

If (7.6) is satisfied, both sides of (7.5) can be multiplied through by y_j^*, the integration carried out, and an explicit formula for a_j obtained

$$a_j = -\frac{1}{\lambda_j} \frac{\int_a^b y_j^*(z) f(z)\, dz}{\int_a^b y_j^*(z) y_j(z)\, dz}. \tag{7.7}$$

This would complete the solution of the equation since each a_j is given in terms of the original function $f(x)$ and the required solution is given by $y(x) = \sum a_i y_i(x)$. All that remains is to establish whether it is indeed possible to find a set of functions $\{y_i(x)\}$ with the property (7.3), which satisfy the boundary conditions, and have the necessary mutual orthogonality. Functions which satisfy (7.3) are called **eigenfunctions** of the operator L, and λ_i are the corresponding **eigenvalues**.

In the next section we will show that at least for linear operators L of a particular form, such suitable sets of functions can be found, in the sense that if they satisfy an equation like (7.3) then they will have the necessary orthogonality properties and that certain (fairly broad) types of boundary conditions can be accommodated.

7.1   Sturm–Liouville theory

Second-order linear differential equations in which $L(y)$ has the form,

$$L(y) \equiv p(x)y'' + r(x)y' - q(x)y \quad \text{with } r(x) = p'(x), \tag{7.8}$$

and p, q and r are real functions of x, were first intensively studied by the French mathematicians Sturm and Liouville in the 1830s. Here $y = y(x)$, and a prime denotes differentiation with respect to x.

As will be seen in the subsequent discussion the class of differential equations of the form

$$L(y) = -\lambda \rho(x) y, \tag{7.9}$$

with $L(y)$ of the form (7.8), includes very many of those occurring naturally in physics and engineering.

Writing (7.8) and (7.9) together we obtain the equation

$$(py')' - qy + \lambda \rho y = 0. \tag{7.10}$$

This is known as the Sturm–Liouville equation, and its properties and uses will form the main content of the remainder of this chapter. Second-order linear differential operators L, for which $L(y)$ can be written in the form

$$L(y) = (py')' - qy,$$

where p and q are functions of x, are known as **self-adjoint** operators.†

Equation (7.9) is a slight extension of (7.3) to include on the right-hand side a *weight function* $\rho(x)$ and so allow a somewhat wider class of equations to be treated. In particular a linear operator in an equation which is not already self-adjoint, can be made so by multiplying the equation through by a suitable function, as is explained more fully in section 7.2. In many applications $\rho(x)$ is unity for all x and then (7.3) is recovered; in general it is a function determined by the choice of coordinate system used in describing a particular physical situation. The only requirement on it is that it is real and does not change sign in the range $a \leqslant x \leqslant b$, and can therefore, without loss of generality, be taken to be non-negative throughout.

We now turn to demonstrating that the solutions of the Sturm–Liouville (S–L) equation have the required properties to enable us to carry through the solution of differential equations by superposition methods.

Let us introduce the assumed boundary conditions immediately, so as not to interrupt the general argument later. Naturally they will appear *ad hoc* by being introduced at this stage, but the reasons for the choice will soon be apparent. We assume they are such as to satisfy

$$[y_j^* p y_i']_{x=a} = [y_j^* p y_i']_{x=b}, \quad \text{for all } i, j, \tag{7.11}$$

where $y_i(x)$ and $y_j(x)$ are any two solutions of the S–L equation (perhaps corresponding to two different values λ_i, λ_j of the constant λ, if $i \neq j$). This is in fact a fairly mild assumption about the boundary conditions and is met by many commonly occurring cases, e.g. $y(a) = y(b) = 0$, $y(a) = y'(b) = 0$, $p(a) = p(b) = 0$, and many more. The important point to

---

† For our present purposes this will suffice. More generally, the (Hermitian) adjoint of L is found by integrating the expression $\int v(x)L[u(x)]\,dx$ by parts until it has the form $\int u(x)L^*[v(x)]\,dx$ together with boundary contributions. This then defines the adjoint L^*, and L is self-adjoint if L and L^* have the same form. In particular if, in terms of the D operator of chapter 5, $L(u) = \sum_{|n| \leqslant m} a_n(x)\, D^n u(x)$ then $L^*(u) = \sum_{|n| \leqslant m} (-1)^{|n|}\, D^n[a_n(x)u(x)]$. This procedure generalizes to equations involving more than one independent variable, when dx represents $dx_1\,dx_2 \ldots$. It should be emphasized that the * in L^* does not simply mean complex conjugation of the various terms in L.

notice is that to satisfy (7.11), one boundary condition must be specified at *each* end of the range.

Reality of the eigenvalues. Let y_i be a solution of (7.10) corresponding to a particular value of λ_i, and consider the complex conjugate of (7.10)

$$(py_i^{*\prime})' - qy_i^* + \lambda_i^* \rho y_i^* = 0. \tag{7.12}$$

Multiplying this through by y_j and integrating from a to b (the first term by parts) we obtain

$$[y_j(py_i^{*\prime})]_a^b - \int_a^b y_j'(py_i^{*\prime})\,dx - \int_a^b y_j(q - \lambda_i^* \rho)y_i^*\,dx = 0. \tag{7.13}$$

The first term vanishes by virtue of the complex conjugate of (7.11) and a second integration by parts gives

$$[-(py_j')y_i^*]_a^b + \int_a^b (py_j')'y_i^*\,dx - \int_a^b y_j(q - \lambda_i^* \rho)y_i^*\,dx = 0. \tag{7.14}$$

Again the first term is zero because of (7.11), yielding

$$\int_a^b [(py_j')' - qy_j + \lambda_i^* \rho y_j]y_i^*\,dx = 0. \tag{7.15}$$

But, from the original S–L equation for y_j, we obtain by multiplying through by y_i^* and integrating, that

$$\int_a^b [(py_j')' - qy_j + \lambda_j \rho y_j]y_i^*\,dx = 0. \tag{7.16}$$

Subtracting (7.16) from (7.15) gives

$$(\lambda_i^* - \lambda_j)\int_a^b y_i^* \rho y_j\,dx = 0. \tag{7.17}$$

However, $\rho(x)$ is non-negative and so, if we take $i = j$, $\int y_i^* \rho y_i\,dx$ cannot be zero, and so we conclude that $\lambda_i^* = \lambda_i$, i.e. that the eigenvalues of the S–L equation are real.

Further since $\lambda^* = \lambda$ comparison of (7.10) with (7.12) shows that y^* and y are eigenfunctions corresponding to the same eigenvalue and hence, because of the linearity of the S–L equation, that at least one of $(y^* + y)$ and $i(y^* - y)$ is a non-zero *real* eigenfunction corresponding to that eigenvalue. Henceforth we will therefore assume that the eigenfunctions are real or have been made so by taking suitable linear combinations (the necessity for this occurs non-trivially only if a particular λ is degenerate, i.e. corresponds to more than one linearly independent eigenfunction).

Orthogonality and normalization of the eigenfunctions. Two further results follow from (7.17). The first shows that two different eigenfunctions are orthogonal in the sense

$$\int_a^b y_i \rho y_j \, dx = 0,$$

where the weight function has been included. This is trivially obvious if $\lambda_i \neq \lambda_j$. The small extension of this proof needed for the case when $\lambda_i = \lambda_j$ in order to establish mutual orthogonality for a suitable set of independent eigenfunctions is considered in example 1 of section 7.7.

The second point concerns the normalization of the $y_i(x)$, which because of the linearity of the S–L equation is arbitrary. We will assume for definiteness that they are normalized so that $\int y_i \rho y_i \, dx = 1$ or, combining this with the other result, that†

$$\int_a^b y_i \rho y_j \, dx = \delta_{ij}. \tag{7.18}$$

A quadratic form. It is convenient to note here, as a by-product, a result which will be useful in the variational methods discussed in chapter 13. Referring to (7.13) and recalling that λ_i and y_i can be taken as real and that the first term is zero, we have

$$\int_a^b (y_j' p y_i' + y_j q y_i) \, dx = \lambda_i \int_a^b y_i \rho y_j \, dx$$

$$= \lambda_i \delta_{ij} \quad \text{(no summation)}, \tag{7.19}$$

the last step applying if the functions are normalized. Hence this integral of a particular quadratic form in the y_i and y_i' has a value related to the corresponding eigenvalue.

7.2   Examples of Sturm–Liouville equations

In order to illustrate the wide application of the S–L theory we will in this section make a short catalogue of some common equations of physics which have the Sturm–Liouville form. The sceptic will no doubt remark that such equations are common precisely because they have a manageable form, but it should be noted that any second-order linear differential equation

$$p(x)y'' + r(x)y' + q(x)y + \lambda \rho(x)y = 0, \tag{7.20}$$

---

† δ_{ij}, which has the value 1 when $i = j$ and the value 0 when $i \neq j$, is known as the *Kronecker delta*.

can be converted to the required type by multiplying through by the factor

$$F(x) = \exp\left[\int^x \frac{r(z) - p'(z)}{p(z)} \, dz\right],$$ (7.21)

provided that the indefinite integral is defined. It then takes on the S–L form

▶1. $[F(x)p(x)y']' - [-F(x)q(x)]y + \lambda F(x)\rho(x)y = 0,$

but clearly with a different [but still non-negative] weight function. Note that this procedure is analogous to the use of an integrating factor in section 5.2.

One example of an S–L equation, which has already been discussed and solved by the series method in chapter 6, is **Legendre's equation** (6.7) which has the form,

$$(1 - x^2)y'' - 2xy' + l(l + 1)y = 0,$$ (7.22 a)

or $[(1 - x^2)y']' + l(l + 1)y = 0.$ (7.22 b)

Clearly this is an S–L equation with $p = 1 - x^2$, $q = 0$, unit weight function, and eigenvalue $l(l + 1)$. The mutual orthogonality property of its solutions $y_l = P_l(x)$ has been previously demonstrated (equation (6.53)).

▶2. How should α_l be chosen so that the set $y_l(x) = \alpha_l P_l(x)$ satisfy (7.18)?

▶3. Use (7.19) to evaluate $\int_{-1}^{1} [P_l'(\mu)]^2(1 - \mu^2) \, d\mu.$

Legendre's equation and its solutions (the Legendre polynomials) appear most readily in the analysis of physical situations involving the operator ∇^2 and axial symmetry, since the linear differential operator involved has the form of the polar angle part of ∇^2, when the latter is expressed in spherical polars. Immediate examples include the solution of Laplace's equation in axially symmetric situations and Schrodinger's equation for a quantum mechanical system involving a central potential.

Very closely related to these situations are those involving the **associated Legendre equation**

$$[(1 - x^2)y']' + \left[l(l + 1) - \frac{m^2}{1 - x^2}\right]y = 0,$$ (7.23)

and its polynomial solutions. These arise in physical situations in which there is dependence on the azimuthal angle ϕ of the form $\exp(im\phi)$ or $\cos(m\phi)$. From the point of view of the S–L equation, they are the same as for the Legendre equation itself, except that $q(x)$ is now $m^2(1 - x^2)^{-1}$.

Both of these equations will appear again in connection with the method of separation of variables considered in chapter 10.

Bessel's equation which arises from similar physical situations, but expressed in a cylindrical polar coordinate system, is also treated in chapter 10. It has the form

$$x^2y'' + xy' + (x^2 - n^2)y = 0, \tag{7.24}$$

but on changing variables to $\xi = x/a$, it takes on the S–L form with

▶4. $p = \xi, \quad q = n^2/\xi, \quad \rho = \xi$ and $\lambda = a^2$.

The most trivial of Sturm–Liouville equations, in which $p = 1$, $q = 0$ and $\rho = 1$, is the simple harmonic motion equation

$$y'' + \omega^2 y = 0. \tag{7.25}$$

The whole of chapter 8 is concerned with solutions of this equation and their properties and we will not consider it further here.

Two further examples of solutions of linear second-order differential equations taken from the quantum mechanical study of simple physical systems are:

(i) the **Hermite polynomials** involved in the description of the wave function of a harmonic oscillator and satisfying

$$y'' - 2xy' + 2\alpha y = 0, \tag{7.26}$$

and (ii) the **Laguerre polynomial** solutions for the hydrogen atom, satisfying

$$xy'' + (1 - x)y' + \alpha y = 0. \tag{7.27}$$

Both of these can be converted to Sturm–Liouville form, although there are some formal difficulties associated with a singularity of the Laguerre equation at the origin.

▶5. Find the 'integrating factors' for (7.26) and (7.27) and arrange them in the form (7.10).

▶6. Do the same for the Chebyshev equation $(1 - x^2)y'' - xy' + n^2y = 0$.

7.3   Application of superposition

Now that we have developed in sections 7.1 and 7.2 a class of functions with the necessary properties to make an eigenfunction expansion solution of an equation of the type (7.2) a possibility, we will, in this section, work out in detail an application to a specific problem. We will use one containing only a very simple operator (d^2/dx^2) so that the corresponding eigenfunctions are familiar.

Example 7.1. Solve the equation $y'' + \tfrac{1}{4}y = f(x)$ with $y(0) = y(\pi) = 0$.

The operator on the left-hand side of this equation is already self-adjoint and so we seek its eigenfunctions satisfying the S–L equation

$$y'' + \tfrac{1}{4}y + \lambda y = 0,$$

with unit weight function. These are obviously

$$y_n(x) = A_n \sin nx + B_n \cos nx,$$

corresponding to eigenvalues λ_n given by

$$n^2 = \lambda_n + \tfrac{1}{4}.$$

The boundary conditions, which clearly satisfy (7.11), require that n is a positive integer and that $B_n = 0$. Thus the appropriate functions are given by

$$y_n(x) = A_n \sin nx$$

and the normalization condition (7.18) requires

$$\int_0^\pi A_n^2 \sin^2 nx \, dx = 1, \quad \text{hence } A_n = \left(\frac{2}{\pi}\right)^{1/2}.$$

Thus if we write as the solution of the original problem $y(x) = \sum_n a_n y_n(x)$, we obtain, as in (7.5) and (7.7),

$$a_n = -(\lambda_n)^{-1} \int_0^\pi y_n(z) f(z) \, dz$$

$$= -(n^2 - \tfrac{1}{4})^{-1} \int_0^\pi \left(\frac{2}{\pi}\right)^{1/2} \sin(nz) f(z) \, dz,$$

and finally that the solution in terms of the given function $f(x)$ is†

$$y(x) = -\frac{2}{\pi} \sum_{n=1}^\infty \frac{\sin(nx)}{n^2 - \tfrac{1}{4}} \int_0^\pi f(z) \sin(nz) \, dz. \tag{7.28}$$

Example 7.2. A particle moves in a potential such that if it is displaced from the origin it executes simple harmonic motion of angular frequency $\tfrac{1}{2}$. As it moves through the origin it is suddenly subjected to an additional

---

† This result is also the Fourier series form because of the particular form of linear operator involved. However the above method is a general model for all equations involving S–L-like operators.

acceleration of $+1$ for a period $\pi/2$ and then to one of -1 for the next $\pi/2$ period. Use the results of example 7.1 to find the particle's initial velocity, if it is to be at the origin at the end of the second period.

The equation of motion is clearly

$$\ddot{y} + \tfrac{1}{4}y = 1, \quad 0 \le t < \pi/2,$$
$$= -1, \quad \pi/2 \le t < \pi,$$

with $y(0) = y(\pi) = 0$. Thus writing t for x in (7.28) we have an explicit expression for $y(t)$ for all $0 \le t \le \pi$, once we have evaluated

▶ 7. $\displaystyle \int_0^\pi f(z) \sin(nz)\, dz = n^{-1}[1 - 2\cos(n\pi/2) + (-1)^n]$

$$= 4/n \text{ if } n = 4m + 2,$$

$$= 0 \text{ otherwise.}$$

Thus we obtain as the displacement at time t,

$$y(t) = -\frac{8}{\pi} \sum_{m=0}^{\infty} \frac{\sin[(4m+2)t]}{(4m+2)[(4m+2)^2 - \tfrac{1}{4}]},$$

and, by differentiation at $t = 0$,

$$\dot{y}(0) = -\frac{8}{\pi} \sum_{m=0}^{\infty} \frac{1}{(4m+2)^2 - \tfrac{1}{4}} \approx -0.81,$$

as the required initial velocity.

7.4   Green's functions

The Green's function method described in this section relies heavily on the properties of the δ-function, which is treated in the next chapter (section 8.8). The reader who does not have prior knowledge of these properties is advised to omit this section at present, and, as mentioned elsewhere, to return to it later.

In a preceding section we saw that

$$y = -\sum_j \lambda_j^{-1} y_j(x) \int_a^b y_j^*(z) f(z)\, dz \qquad (7.29)$$

gave a solution of

$$L(y) = f(x), \qquad (7.30)$$

where the set $\{y_j(x)\}$ depended upon L. Now let us assume that we may interchange the order of summation and integration and write (7.29) as

$$y = \int_a^b \left\{ \sum_j [-\lambda_j^{-1} y_j(x) y_j^*(z)] \right\} f(z)\, dz. \qquad (7.31)$$

In this form the solution has clearer properties. We first observe that the expression in the curly brackets, being summed over j, is a function of x and z only, and could therefore be written as $G(x, z)$. Equation (7.31) is then

$$y(x) = \int_a^b G(x, z) f(z)\, dz. \qquad (7.32)$$

Now the structure of the solution of (7.30) is even more apparent, being the integral of the product of two factors, of which:

(i) the first, $G(x, z)$, is determined entirely by the boundary conditions and the eigenfunctions y_j, and hence by L itself, and
(ii) the second, $f(z)$, depends purely on the right-hand side of (7.30).

It is apparent from this that we have the possibility of finding, once and for all, for any given L, a function $G(x, z)$ which will enable us to solve (7.30) for any right-hand side. The solution will be in the form of an integral which, at worst, can be evaluated numerically. This function, $G(x, z)$, is called the **Green's function** for the operator L.

One expression for the Green's function has already been given, namely

$$G(x, z) = -\sum_j \lambda_j^{-1} y_j(x) y_j^*(z). \qquad (7.33)$$

But for an alternative way of finding the form of $G(x, z)$, we also note that expression (7.32) is by construction a solution of (7.30). Hence,

$$L(y) = \int_a^b L[G(x, z)] f(z)\, dz = f(x). \qquad (7.34)$$

But, as in equation (8.48 iii) of the next chapter,

$$f(x) = \int_a^b f(z)\delta(z - x)\, dz, \quad a \leqslant x \leqslant b,$$

where $\delta(z - x)$ is the Dirac δ-function, and so

$$\int_a^b \{L[G(x, z)] - \delta(z - x)\} f(z)\, dz = 0. \qquad (7.35)$$

However this is to hold for any f, and so we must have

$$L[G(x, z)] = \delta(z - x). \qquad (7.36)$$

In words, the Green's function G is the solution of the differential equation obtained by replacing the right-hand side of (7.30) by a δ-function. Note that in (7.36) z is only a parameter and all the differential operations implicit in L act upon the variable x.

Looked at directly from the superposition point of view, our result is that the solution of (7.30) is the superposition of the effects of isolated 'impulses' of size $f(z)\,dz$ occurring at positions $x = z$. Each 'impulse' of course has effects (propagated by (7.36)) at positions other than that at which it occurs and so the total result at any particular x has to be obtained by integrating over all z.

▶8. Use the preceding results to show the *closure* property of the eigenfunctions of L, namely

$$\sum_j y_j(x)y_j^*(z) = \delta(z - x). \tag{7.37}$$

[If the spectrum of eigenvalues of L is anywhere continuous, then the eigenfunction $y_j(x)$ must be treated as $y(j, x)$ and an integration carried out over j.]

7.5   Forms of Green's functions

To illustrate the form that Green's functions may take, we will find expressions for them in two cases. In the first of these, boundary conditions will be prescribed at two different positions, whereas in the second we will consider a case in which time is the independent variable and the system starts from rest.

For the first example we will re-solve our earlier illustration of example 7.1, but by a different method, giving an alternative expression from which a Green's function can be extracted by inspection. This is done purely to illustrate the form of the function, since clearly the Green's function method would be pointless if it were always necessary to solve the problem another way first in order to extract the Green's function.

Example 7.3. Let us apply the method of variation of parameters (section 5.8) to

$$y'' + \tfrac{1}{4}y = f(x), \quad y(0) = y(\pi) = 0.$$

Putting $y = A \sin(x/2) + B \cos(x/2)$, we require $A(\pi) = 0$ and $B(0) = 0$. If we make A and B satisfy $A' \sin(x/2) + B' \cos(x/2) = 0$, we obtain in the usual way,

▶9. $\tfrac{1}{2}A' \cos(x/2) - \tfrac{1}{2}B' \sin(x/2) = f(x),$

and hence the results

▶10. $A' = 2f(x) \cos(x/2), \qquad B' = -2f(x) \sin(x/2).$

On integrating, and taking into account the boundary conditions on A and B, we obtain for $y(x)$ the solution,

▶11. $y(x) = -\cos (x/2) \int_0^x 2f(z) \sin (z/2) \, dz$

$$- \sin (x/2) \int_x^\pi 2f(z) \cos (z/2) \, dz.$$

From this it is apparent that the Green's function $G(x, z)$ has the form

$$\begin{aligned} G(x, z) &= -2 \cos (x/2) \sin (z/2), \quad 0 \leqslant z \leqslant x, \\ &= -2 \sin (x/2) \cos (z/2), \quad x \leqslant z \leqslant \pi. \end{aligned} \tag{7.38}$$

This is an [more readily computable] alternative to our previous expression (from (7.28)),

$$G(x, z) = -\frac{2}{\pi} \sum_1^\infty \frac{\sin (nx) \sin (nz)}{n^2 - \frac{1}{4}}. \tag{7.39}$$

It will be noticed in (7.38) that $G(x, z)$ changes its form as z passes through the value $z = x$. This is to be expected since (7.36) in the present case is

$$y'' + \tfrac{1}{4}y = \delta(x - z),$$

and if we formally integrate this with respect to x between $x = z_-$ and $x = z_+$, two values one each side of $x = z$ we obtain

$$[y']_{z_-}^{z_+} + \frac{1}{4} \int_{z_-}^{z_+} y \, dx = \int_{z_-}^{z_+} \delta(x - z) \, dx = 1.$$

As we let z_+ and z_- tend to z, the second term on the left tends to zero, showing that

$$y'(z_+) - y'(z_-) = 1, \tag{7.40}$$

i.e. the derivative of y has a discontinuity of unit magnitude at the point $x = z$ [and also that y itself is therefore continuous there].

In our next example we can expect that not only will the Green's function change its form at $x = z$, but that it will be identically zero for $z > x$.

Example 7.4. A damped harmonic oscillator of mass m initially at rest is driven by a time dependent external force $mf(t)$, for $t > 0$. Find the subsequent motion.

The equation of motion is of the form

$$\ddot{x} + 2\beta \dot{x} + \omega^2 x = f(t), \quad \text{with } x(0) = \dot{x}(0) = 0. \tag{7.41}$$

Formally we will have as the solution

$$x(t) = \int_0^\infty f(z)G(t, z) \, dz, \tag{7.42}$$

where $G(t, z)$ satisfies

$$\ddot{G} + 2\beta\dot{G} + \omega^2 G = \delta(z - t). \tag{7.43}$$

Our qualitative discussion of section 7.4 in terms of 'impulses' is even more physically appropriate here, since the x of that discussion now has the role of time and $mf(z)\,dz$ is an impulse in the mechanics sense. Since $G(t, z)$ effectively represents the effect at time t of a unit impulse occurring at time z, we would expect on physical grounds ['causality'] that

$$G(t, z) = 0 \quad \text{for } z \geqslant t. \tag{7.44}$$

We now have to solve (7.43), treating z as a fixed parameter, for the region $0 \leqslant z \leqslant t$. This we will do using the complementary functions for $z < t$ and the step function condition (7.40) on $\dot{G}$ at $t = z$.

Putting $G(t, z) = A \exp(pt)$ into (7.43) with the right-hand side set equal to zero, we obtain

$$p_\pm = -\beta \pm (\beta^2 - \omega^2)^{1/2}. \tag{7.45}$$

Thus $G = A \exp(p_+ t) + B \exp(p_- t)$,   for $t > z$.

Continuity at $t = z$, together with (7.44), gives

$$0 = A \exp(p_+ z) + B \exp(p_- z), \tag{7.46}$$

whilst the condition on $\dot{G}$ yields

$$1 = p_+ A \exp(p_+ z) + p_- B \exp(p_- z). \tag{7.47}$$

Solving (7.46) and (7.47) for A and B gives finally that

$$G(t, z) = (p_+ - p_-)^{-1}\{\exp[p_+(t - z)] - \exp[p_-(t - z)]\}$$

▶12.
$$= (\beta^2 - \omega^2)^{-1/2} \exp[-\beta(t - z)]$$
$$\times \sinh[(\beta^2 - \omega^2)^{1/2}(t - z)], \tag{7.48}$$

for $0 \leqslant z \leqslant t$. [Not surprisingly a function of $(t - z)$ only.] Combining (7.48) and (7.44) with (7.42) thus produces an integral expression for the motion of the oscillator

$$x(t) = \int_0^t \frac{f(z)}{(\beta^2 - \omega^2)^{1/2}} \exp[-\beta(t - z)]$$
$$\times \sinh[(\beta^2 - \omega^2)^{1/2}(t - z)]\,dz. \tag{7.49}$$

Figure 7.1 illustrates the form of the Green's function for a few sample values of β and ω. It should be remembered that the curves give the responses of the system to a unit impulse at the time $t = z$, under the various conditions.

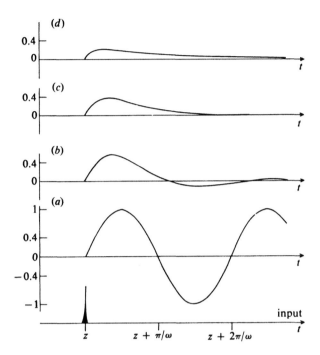

Fig. 7.1 Some Green's functions described by equation (7.48). All vertical scales are in units of ω^{-1}. The bottom curve shows the input to the system. (a) For $\beta = 0$ the response is an undamped sine wave. (b) For $\beta = \frac{1}{2}\omega$ the response is a damped sine wave. (c) For $\beta = \omega$ the response is critically damped of the form $(t - z) \exp[-\omega(t - z)]$. (d) For $\beta = 2\omega$ no oscillations occur but the effect of the impulse is long-lived.

7.6 Generalization

Finally, we formally collect together the elements of the Green's function method in a slightly more generalized way. If we seek solutions, for a self-adjoint [Hermitian] operator L, of the equation

$$L(y) + \mu\rho(x)y = f(x), \tag{7.50}$$

then, in terms of the functions $y_i(x)$ which satisfy [for generality taking i as continuous]

$$L(y_i) = -\lambda_i \rho y_i, \tag{7.51 a}$$

$$\int y_i^*(x)\rho(x)y_j(x)\,dx = \delta(i - j), \tag{7.51 b}$$

$$\int y_i^*(z)\rho(x)y_i(x)\,di = \delta(x - z), \tag{7.51 c}$$

y is given by

$$y(x) = \int G(x, z)f(z)\,dz, \tag{7.52}$$

with

$$G(x, z) = \int \frac{y_i(x)y_i^*(z)}{\mu - \lambda_i}\,di. \tag{7.53}$$

▶13. Verify by substitution that (7.52) is in fact a solution, when the original equation (7.50) contains a term involving a weight function $\rho(x)$.

7.7   Examples for solution

1. Suppose that a particular eigenvalue of (7.9) is N-fold degenerate i.e. $Ly_i + \lambda\rho y_i = 0$ for $i = 1, N$, and that in general $\int y_i^*\rho y_j\,dz \neq 0$ even if $i \neq j$. Then show that the new set of normalized functions $u_i(x)$ defined by

$$u_1 = \mathcal{N}\{y_1\},$$

$$u_i = \mathcal{N}\left\{y_i - \sum_{j=1}^{i-1} u_j\left(\int u_j^*\rho y_i\,dz\right)\right\}, \qquad i = 2, 3, \ldots, N,$$

are (a) all eigenfunctions of L with eigenvalue λ, and (b) mutually orthogonal with respect to the weight function ρ. Here $\mathcal{N}\{v(x)\}$ stands for $v(x)/\int v^*(z)\rho(z)v(z)\,dz$. [Use a process of induction.]

2. Express the Gauss equation

$$(x^2 - x)y'' + [(1 + \alpha + \beta)x - \gamma]y' + \alpha\beta y = 0,$$

where α, β and γ are parameters, in Sturm–Liouville form.

3. A particular associated Legendre function y_n corresponding to eigenvalue n is a solution of

$$xy'' + 2y' + \left(n - \frac{1}{2} - \frac{x}{4} - \frac{3}{4x}\right)y = 0,$$

with $y_n(0) = y_n(\infty) = 0$. Show that this implies

$$\int_0^\infty xy_n(x)y_m(x)\,dx = 0, \qquad \text{if } m \neq n.$$

4. Use the solutions of the equation $y'' + \lambda y = 0$ with boundary conditions $y(0) = 0$, $y'(\pi) = 0$ to find an expansion of the function $f(x) = x^2 - 2\pi x$. Hence solve $y'' + y = x^2 - 2\pi x$, with these same boundary conditions.

5. (a) Find the solution of $(1 - x^2)y'' - 2xy' + by = f(x)$, valid in the range $[-1, 1]$ and finite at $x = 0$, in terms of Legendre polynomials.

(b) If $b = 14$ and $f(x) = 5x^3$, find the explicit solution and verify it by direct substitution.

6. By substituting $x = \exp(t)$ find the normalized eigenfunctions $y_n(x)$ and eigenvalues λ_n of the operator L defined by

$$L(y) = x^2y'' + 2xy' + \tfrac{1}{4}y, \quad 1 \leqslant x \leqslant e,$$

with $y(1) = y(e) = 0$. Find, as a series $\sum a_n y_n(x)$, the solution of $L(y) = x^{-1/2}$.

7. Express the solution of Poisson's equation in electrostatics $\nabla^2 \phi(\mathbf{r}) = -\rho(\mathbf{r})/\epsilon_0$, where ρ is the non-zero charge density over a finite part of space, in the form of an integral and hence identify the Green's function for the ∇^2 operator.

8. Extension to example 7.4 (page 185). Consider the particular case when $f(t) = F \sin(pt)$ and $\beta^2 - \omega^2 < 0 = -n^2$ (say). Evaluate the integral (7.49) and show that it leads to decaying transients and steady state terms, the same as would be obtained by the complementary function and particular integral methods of chapter 5. [The algebra is rather lengthy but straightforward.]

9. In the quantum mechanical study of the scattering of a particle by a potential, a (Born approximation) solution can be obtained in terms of a function $y(\mathbf{r})$ which satisfies an equation of the form

$$(-\nabla^2 - K^2)y(\mathbf{r}) = F(\mathbf{r}).$$

Assuming that $y_k(\mathbf{r}) = (2\pi)^{-3/2} \exp(i\mathbf{k} \cdot \mathbf{r})$ is a suitably normalized eigenfunction of $-\nabla^2$ corresponding to an eigenvalue $-k^2$, find a suitable Green's function $G_K(\mathbf{r}, \mathbf{r}')$. By taking the direction of the vector $\mathbf{r} - \mathbf{r}'$ as polar axis for a k-space integration, show $G_K(\mathbf{r}, \mathbf{r}')$ can be reduced to

$$\frac{1}{4\pi^2|\mathbf{r} - \mathbf{r}'|} \int_{-\infty}^{\infty} \frac{w \sin w}{w^2 - w_0^2} \, dw,$$

where $w_0 = K|\mathbf{r} - \mathbf{r}'|$. [This integral can be evaluated by contour integration (chapter 16) to give $(4\pi|\mathbf{r} - \mathbf{r}'|)^{-1} \exp(iK|\mathbf{r} - \mathbf{r}'|)$.]

8
Fourier methods

It will undoubtedly have been observed by the reader who has only a moderate familiarity with the mathematical methods used for physical problems, that harmonic waves (of the form $\exp(i\omega t)$ or $\cos \omega t$) are very convenient functions to deal with. It is straightforward to differentiate, integrate and multiply them; their moduli are easily taken, and each contains only one frequency [or wavenumber, for forms like $\exp(ikx)$, using an obvious notation]. This last point is important since the response of many physical systems, such as an electronic circuit or a prism, depends most directly on the frequency content of the input the system receives.

Even if we were not familiar with the results of the Sturm–Liouville theory discussed in the previous chapter, these properties by themselves would indicate that it may be advantageous in some cases to express all the functions involved in a problem as superpositions of harmonic wave functions (**Fourier series** or **transforms**). The otherwise difficult parts of the problem might then be carried through more simply, and finally, if necessary, the output functions reconstituted from the 'processed' waves.

In fact, we recognize the harmonic wave $y(x) = \exp(ikx)$ as an eigenfunction of the simplest non-trivial Sturm–Liouville equation, with $p = 1$, $q = 0$, $\rho = 1$ and $\lambda = k^2$, and thus, provided that $[yy']_a^b = 0$, we may apply the general results of chapter 7. This boundary condition is clearly going to be satisfied if we consider periodic problems of period $b - a$, and so we are led to Fourier series, or if $a \to -\infty$ and $b \to \infty$, to Fourier transforms.

There are thus at least three different but connected ways in which the harmonic waves of Fourier methods may be considered to arise; as sets of mathematically simple and manageable functions, as the set of eigenfunctions of a particularly simple S–L equation, and, most physically, as the set of functions arising naturally in the description of a wide range of physical situations, such as the vibrations of a finite string, the scattering of light by a diffraction grating, and the transmission of an input signal by an electronic circuit.

8.1 Sets of functions

All that is required to be known about the sets of functions needed can be obtained directly from the Sturm–Liouville theory, namely their completeness [any 'reasonable' function can be expressed as a linear sum of them] and mutual orthogonality over an appropriate range, but it is both straightforward and instructive to obtain these properties in an empirical way as well.

Let us suppose that we are given a function which is periodic, e.g. that describing a circuit input voltage consisting of a regularly repeating wave train [periodic in time], or the potential experienced by an electron in a regular crystal [periodic in space], or a single-valued gravitational potential [periodic in azimuthal angle ϕ]. This function is to be represented as a (generally infinite) sum of Fourier terms.

We will work with t as the variable in which the periodicity occurs and with T as the period, but of course the physical interpretation of t will not necessarily be that of time. In periodic situations the origin of t is arbitrary but we will choose it so that the basic period is $-T/2$ to $T/2$.

Consider first the set of functions

$$h_n(t) = \cos(2\pi nt/T), \quad 0 \leqslant n < \infty. \tag{8.1}$$

Could this be a possible set for the expansion, using all integral values of n? We test first their mutual orthogonality over $(-T/2, T/2)$. Using ω to denote the quantity $(2\pi/T)$, we have

$$\int_{-T/2}^{T/2} \cos\left(\frac{2\pi nt}{T}\right) \cos\left(\frac{2\pi mt}{T}\right) dt$$

$$= \int_{-T/2}^{T/2} \tfrac{1}{2}[\cos(n+m)\omega t + \cos(n-m)\omega t] \, dt$$

$$= 0 \text{ unless } n = m, \tag{8.2}$$

since the integral of a cosine (or sine) function over a complete number of cycles [here $n+m$ and $n-m$ cycles] is zero. If $n = m$ the integral has the value $T/2$, as is easily verified. Thus the functions $h_n(t)$ are mutually orthogonal as required.

We now have to ask whether any periodic function $f(t)$ can be expressed in terms of them, i.e. $f(t) = \sum_n a_n h_n(t)$, with a_n constants. This is clearly not so, since $\cos(n\omega t)$ is an even function of t and so therefore is any function of the form $\sum_n a_n h_n(t)$; hence any odd periodic function of t such as $\tan(\omega t)$ cannot be represented by such a sum. We conclude therefore that the set of functions given by (8.1) is not suitable.

In a similar way, the set

$$g_n(t) = \sin(2\pi nt/T), \quad 0 < n(\text{integral}) < \infty, \tag{8.3}$$

is inappropriate.

▶1. Show that the $g_n(t)$ are mutually orthogonal, but do *not* enable all periodic functions to be expanded in terms of them.

The set $g_n(t)$ fail because even functions of t cannot be expressed in terms of them. However, any function $f(t)$ can be written as the sum of an odd and an even part,

$$f(t) = \tfrac{1}{2}[f(t) + f(-t)] + \tfrac{1}{2}[f(t) - f(-t)]$$
$$= f_{\text{even}}(t) + f_{\text{odd}}(t), \tag{8.4}$$

and so by combining the sets $h_n(t)$ and $g_n(t)$ together, $f(t)$ could be expressed in terms of the larger set so formed [f_{even} in terms of the h_n, and f_{odd} by the g_n]. All that remains is to determine whether the enlarged set is still a mutually orthogonal one. This is easily done since

$$\int_{-T/2}^{T/2} h_n(t)g_m(t)\,dt = \int_{-T/2}^{T/2} \cos(n\omega t)\sin(m\omega t)\,dt$$

$$= \int_{-T/2}^{T/2} \tfrac{1}{2}[\sin(m+n)\omega t + \sin(m-n)\omega t]\,dt$$

$$= 0 \text{ for all } m \text{ and } n \text{ [even } m = n]. \tag{8.5}$$

We thus arrive at the set of functions $\cos(2\pi nt/T)$ and $\sin(2\pi mt/T)$, with n and m running over all integral values. This is the same set as could have been obtained from the simple case of the Sturm–Liouville equation discussed above, but here arrived at in a more heuristic way.

As noted in chapter 7, the set of functions is not always unique if the eigenvalues are degenerate and in the present case an alternative, and in some cases preferable, set of functions can be obtained by taking linear combinations of the previous set,

$$\exp\left(\frac{i2\pi nt}{T}\right) = \cos\left(\frac{2\pi nt}{T}\right) + i\sin\left(\frac{2\pi nt}{T}\right),$$
$$-\infty < n \text{ (integral)} < \infty. \tag{8.6}$$

This alternative set allows some economy of expression, useful in formal manipulations, and is the natural set from which to proceed to the Fourier transform (section 8.6).

8.2   The expansion and coefficients

Returning to the expansion of a periodic function of period T (see fig. 8.1), we assume that $f(t)$ can be written as a Fourier series (FS), namely

$$f(t) = \tfrac{1}{2}A_0 + \sum_{n=1}^{\infty} [A_n \cos (n\omega t) + B_n \sin (n\omega t)], \tag{8.7}$$

or

$$f(t) = \sum_{n=-\infty}^{\infty} C_n \exp (in\omega t), \tag{8.8}$$

where, as before, $2\pi/T$ has been written as ω, the fundamental frequency.

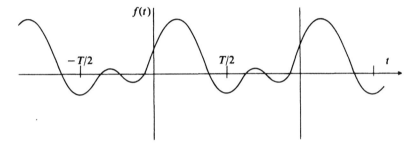

Fig. 8.1   A periodic function of period T.

Physically both formulae contain a superposition of amplitudes having the fundamental frequency and its harmonics, together with a possible constant [$n = 0$ term]. The factor $\tfrac{1}{2}$ in (8.7) is conventional and, as will be seen later, is included in order to make the calculation of A_0 the same as that of other A_n. Clearly $C_0 = \tfrac{1}{2}A_0$.

By writing cosine and sine in terms of exponentials it is straightforward to show that,

▶2. $C_n = \tfrac{1}{2}(A_n - iB_n)$ and $C_{-n} = \tfrac{1}{2}(A_n + iB_n)$. (8.9)

These equations and their inverses establish the link between the two commonest sets of functions.

The next requirement is that of obtaining explicit expressions for A_n and B_n once $f(t)$ is given. To do this we multiply (8.7) through by $\cos (m\omega t)$ and integrate it from $-T/2$ to $T/2$. [This will 'project out' all parts of both f and the series which are orthogonal to $\cos (m\omega t)$.]

$$\int_{-T/2}^{T/2} f(t) \cos (m\omega t) \, \mathrm{d}t = \int_{-T/2}^{T/2} \left\{ \frac{A_0}{2} \cos (m\omega t) \right.$$
$$\left. + \sum_{n=1}^{\infty} [A_n \cos (n\omega t) \cos (m\omega t) + B_n \sin (n\omega t) \cos (m\omega t)] \right\} \mathrm{d}t. \tag{8.10}$$

The first term on the right-hand side is zero unless $m = 0$ when it has the value $\frac{1}{2}A_0 T$, and, in view of the previously shown mutual orthogonality of the set of functions $\{\cos n\omega t, \sin m\omega t\}$, the only non-zero term appearing in the summation is that from the cosine series for which $n = m \neq 0$; this term has the value $\frac{1}{2}A_m T$.

With this enormous simplification of the right-hand side, (8.10) can be rewritten to give an explicit expression for A_m

$$A_m = \frac{2}{T} \int_{-T/2}^{T/2} f(t) \cos (m\omega t) \, dt, \quad m = 0, 1, \ldots, \infty. \quad \text{(8.11 a)}$$

▶3. Follow a similar procedure to show that

$$B_m = \frac{2}{T} \int_{-T/2}^{T/2} f(t) \sin (m\omega t) \, dt, \quad m = 1, 2, \ldots, \infty. \quad \text{(8.11 b)}$$

▶4. Deduce that

$$C_m = \frac{1}{T} \int_{-T/2}^{T/2} f(t) \exp (-im\omega t) \, dt, \quad -\infty < m < \infty. \quad \text{(8.11 c)}$$

▶5. Show that if the function $f(t)$ is real, then $C_{-n} = C_n^*$.

Using these explicit formulae it is now possible to write any given periodic function $f(t)$ of period $T = 2\pi/\omega$ in the form of a series of sinusoidal terms, containing the fundamental frequency ω and some or all of its harmonics (together with a possible constant). The amplitudes of the various harmonics are obtainable from the original function by means of equations (8.11).

As an example (which will be used for illustration throughout this chapter) consider the square-wave function $f(t)$ illustrated in fig. 8.2 and given by

$$
\begin{aligned}
f(t) &= -1, \quad -T/2 \leqslant t < 0, \\
&= +1, \quad 0 \leqslant t < T/2.
\end{aligned}
\quad \text{(8.12)}
$$

This function is to be represented as a Fourier series [perhaps so that the effect of a frequency dependent integrating circuit on an electrical input of this form can be determined] and is therefore assumed to be representable in the form of expression (8.7). The required coefficients are obtained by substituting (8.12) in (8.11 a, b). Since $f(t)$ is in this

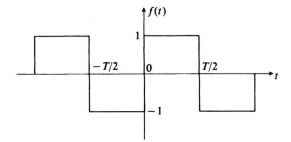

Fig. 8.2 The square-wave function $f(t)$.

case an odd function of t whereas $\cos(m\omega t)$ is an even one, all A_m given by (8.11 a) are zero. Recalling that $\omega T = 2\pi$, we have also that

$$B_m = \frac{2}{T} \int_{-T/2}^{T/2} f(t) \sin(m\omega t)\, dt$$

$$= \frac{4}{T} \int_{0}^{T/2} \sin(m\omega t)\, dt$$

$$= \frac{2}{\pi m} (1 - (-1)^m). \tag{8.13}$$

This is zero if m is even and equals $4/\pi m$ for m odd.

Thus, instead of (8.12), we can write $f(t)$ as

$$f(t) = \frac{4}{\pi} \left[\sin(\omega t) + \frac{\sin(3\omega t)}{3} + \frac{\sin(5\omega t)}{5} + \cdots \right], \tag{8.14}$$

i.e. expressed in terms of its component frequencies with simple sinusoidal functions [but at the price of having introduced an infinite series]. As the number of terms in the series is increased the function $f(t)$ given by (8.14) approaches more and more closely the original function given by (8.12) and fig. 8.2.

The general question as to under what circumstances a series generated by the above procedures (expanding according to (8.7) with coefficients given by (8.11)) is a valid representation of the original function is mathematically complicated, but for practical purposes the answer may be summarized as:

> If (i) there exists only a finite number of maxima, minima and discontinuities of the function $f(t)$ in one period, and (ii) $\int_{-T/2}^{T/2} |f(t)|\, dt$ is convergent, then the Fourier series converges to $f(t)$ as the number of terms $\to \infty$. (8.15)

At the discontinuities (if any) of $f(t)$ a further result is needed. If $f(t)$ is discontinuous at $t = t_1$ then the series converges to $\frac{1}{2}[f(t_1-) + f(t_1+)]$

at that point. Here the quantity $f(t_1 +)$ is the limit of $f(t)$ as $t \to t_1$ from values greater than t_1, and $f(t_1 -)$ is the corresponding limit as $t \to t_1$ from below. It should be remembered that in the series, t is fixed at the value t_1 and it is the increasing number of terms that brings about the convergence.

Referred specifically to our example, these results mean that in the range $0 < t < T/2$ the series (8.14) tends to a limit in value of 1 for all t, that in $-T/2 < t < 0$, series (8.14) tends to -1 for all t, but that at $t = 0$ we must expect it to yield the value $\frac{1}{2}[(-1) + (+1)] = 0$. Likewise at $t = T/2$, we expect the series to give $\frac{1}{2}[(+1) + (-1)] = 0$. Since at $t = 0$ or $T/2$ every term of the calculated series individually vanishes, its values at the discontinuities are immediately verified. For values of t outside the range $(-T/2, T/2)$, the series repeats periodically the values it takes inside the range.

In order to illustrate pictorially the convergence of the series (8.14) to the original function, the value of the series is shown graphically superimposed upon the original function in fig. 8.3, after various numbers of terms have been added. In each picture except the last, the next contribution to be made (but not yet included) is shown dashed. It will be seen that after only a few terms the series is a reasonable approximation except near the discontinuities of $f(t)$.

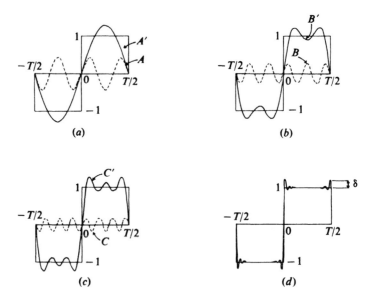

Fig. 8.3 The evaluation of equation (8.14) after various numbers of terms have been added. The next term to be added is shown dashed. In (d) δ gives the Gibbs overshoot. For an explanation of A, A', etc. see the text.

It will be noticed how in any particular range of t values and after any particular number of terms have been included, the next term generally has, at around that value of t, the correct sign to (over-) compensate the discrepancy between the sum so far and the original function. This is not true for every t value [or consequently for every range of t value], but can be illustrated by examining, for example, the values near the points A, B, C of the next terms to be added and the corresponding discrepancies A', B', C' between the sum so far and $f(t)$.

As the number of terms in the series is increased the 'overshoot' nearest a point of discontinuity moves closer to the discontinuity (see 8.3 (d)), but in fact never disappears even in the limit of infinitely many terms, although it moves in position arbitrarily close to the discontinuity. This behaviour is known as the *Gibbs phenomenon*. It does not contradict our previous results since for any t *inside* the range 0 to $T/2$ for example, the partial sum to N terms of the series can be made arbitrarily close to $+1$ for all $N >$ some N_0, by taking N_0 large enough. [Roughly speaking, the Gibbs overshoot has to be squeezed into the non-zero gap between t and the position of the discontinuity, 0 or $T/2$.] The Gibbs phenomenon is characteristic of Fourier series at a discontinuity, its size being proportional to the magnitude of the discontinuity. For the unit square-wave function discussed here, its value $\delta = 0.179\ldots$ (see fig. 8.3 (d)).

8.3   Symmetry considerations

In the example of the previous section all the coefficients A_m were shown to be zero on account of the antisymmetry of the particular function $f(t)$. In general, if the given function $f(t)$ possesses some degree of symmetry about the point $t = 0$ or $t = T/4$, then some economy of labour can be affected in the calculation of the expansion coefficients.

Since cos $(m\omega t)$ and sin $(m\omega t)$ are even and odd functions of t respectively it is apparent by (8.11 a, b), that as general results:

If $f(t)$ is an odd function of t, i.e. $f(-t) = -f(t)$, then
$A_m = 0$ for all m.   (8.16)

If $f(t)$ is an even function of t, i.e. $f(-t) = f(t)$, then
$B_m = 0$ for all m.   (8.17)

Nothing beyond the oddness or evenness of $f(t)$ is required for these results, since they each depend only on the fact that the integral from $-T/2$ to $T/2$ of an odd and an even function multiplied together is necessarily zero. A general function $f(t)$ will usually be neither even nor odd and so both A_m and B_m are in general non-zero.

The consequences of symmetry or antisymmetry of $f(t)$ about $t = T/4$

are a little harder to see, but the following argument gives the required result.

Suppose $f(t)$ has either even or odd symmetry about $t = T/4$, i.e. $f(-s) = \pm f(s)$, where s has been written for $t - T/4$. Then according to (8.11 b)

$$B_m = \frac{2}{T} \int_{-T/2}^{T/2} f(t) g_m(t)\, dt, \qquad (8.18)$$

where $g_m(t) = \sin(m\omega t)$. Expressed in terms of s, g_m becomes $G_m(s) = \sin(m\omega s + m\omega T/4)$, and, recalling that $\omega T = 2\pi$, we have

$$G_m(s) = \sin(m\omega s)\cos(m\pi/2) + \cos(m\omega s)\sin(m\pi/2).$$

If m is even then $\sin(m\pi/2) = 0$ and $G_m(s)$ is an odd function of s, whilst if m is odd $\cos(m\pi/2) = 0$ and G_m is an even function. For the evaluation of B_m, the independent variable in the integral can be changed from t to s. The limits of integration can in fact be left unaltered since f is of course periodic in s as well as t. Two particular combinations of circumstances of interest now arise,

(i) if $f(-s) = f(s)$ and m is even [i.e. G_m is odd] then the integral is zero,
(ii) if $f(-s) = -f(s)$ and m is odd [i.e. G_m is even] then the integral is zero.

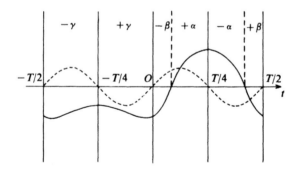

Fig. 8.4 Contributions to B_2 for a function (solid line) with symmetry about $t = T/4$. The contributions are $\pm\alpha$, etc., and the dashed curve is $\sin(2\omega t)$.

As a pictorial way of visualizing these results, fig. 8.4 illustrates case (i) for some periodic function (solid line) which is symmetric about $t = T/4$ (but has no other particular symmetry). The function $\sin(2\omega t)$ is also shown (dashed line). The range $-T/2$ to $T/2$ has been marked off into sections and labelled with the contributions each section would make to the integral (8.18) for B_2. Where the function and $\sin(2\omega t)$ have opposite signs the contribution is negative. The quantities α, β, γ are positive but their actual values are irrelevant for our purpose, the important point being that because of the symmetry of the function about $T/4$ the contributions cancel in pairs.

▶6. By making a similar sketch for say $\sin(3\omega t)$ verify that the integral does not necessarily vanish for odd values of m.

Corresponding results about the A_m for functions with some symmetry about $t = T/4$ can be obtained by considering the symmetry properties of $\cos(m\omega s + m\omega T/4)$. These and the other results of this section are summarized in the following table, where those coefficients A_m and B_m which are necessarily zero are indicated.

	A_{2n}	A_{2n+1}	B_{2n}	B_{2n+1}
f even about $t = 0$			0	0
f odd about $t = 0$	0	0		
f even about $t = T/4$		0	0	
f odd about $t = T/4$	0			0	(8.19)

All of the above results follow automatically when (8.11 a, b) are evaluated in any particular case, but a prior knowledge of them will often enable some coefficients to be set to zero on inspection and so substantially reduce the computational labour. As an example the square-wave function of equation (8.12) and fig. 8.2, is

(i) an odd function of t, and therefore all $A_m = 0$, and
(ii) even about $T/4$ and therefore $B_{2n} = 0$.

Thus we can say immediately that only sine terms of odd harmonics will be present in the series and therefore need to be calculated; this is confirmed in expansion (8.14).

▶7. Find, without calculation, which terms will be present in the Fourier series for the periodic functions $f(t)$ of period T, given in the range $-T/2$ to $T/2$ by,

(i) $f = 2$ for $0 \leqslant |t| < T/4$, $f = 1$ for $T/4 \leqslant |t| < T/2$;
(ii) $f(t) = \exp[-(t - T/4)^2]$;
(iii) $f = -1$ for $-T/2 \leqslant t < -3T/8$ and $3T/8 \leqslant t < T/2$, $f = 1$ for $-T/8 \leqslant t < T/8$, and the graph of f is completed by two straight lines in the remaining ranges so as to form a continuous function.

8.4   Function continuation for fixed intervals

Even when a physically occurring function is not periodic, Fourier analysis can in some cases still be usefully employed. If a function is defined only over a fixed finite interval [for example, the displacement of a violin string] and the physical behaviour of the system is required only in that interval, then, by assuming that the function is mathematically defined outside that interval in a suitably periodic way, a Fourier representation of the function can be used. This is so, since (provided the function satisfies the conditions of statement (8.15)) everywhere within the interval the Fourier series will certainly converge to the function, and it will also do so at the end points if the choice of mathematical continuation makes the function continuous there.

Rather than discuss the choice of continuation in general terms we will use a specific example and treat it in some detail.

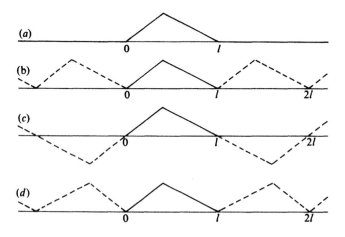

Fig. 8.5   Plucked string with fixed ends: (*a*) the physical situation; (*b*)–(*d*) show possible mathematical continuations; (*c*) is antisymmetric about 0 and (*d*) symmetric.

Consider a string of length *l* which is plucked at some point along its length, as in fig. 8.5 (*a*). Inside the range 0 to *l* [here distance x replaces time] the displacement is given by $f(x)$, whilst outside it is undefined. The most natural assumption is that it is zero outside the range, but in order to express the displacement as a Fourier series we must make the function periodic. The choice of continuation is by no means unique. Figure 8.5 (*b*) shows the most obvious periodic continuation, producing a series of period *l*, but with no particular symmetry properties.

A second choice is shown in (*c*). Here the period is 2*l* and so in the resulting series there will be twice as many wave numbers [spatial analogue of frequencies] below any given wave number, but on the other hand the function is now an odd function of x and so only sine terms will be present. Since for analytically tractable cases all the required coefficients are thus found from a single formula, this means a saving in computational labour. Choice (d) is similar to (c) except that here only cosine terms appear.

Example 8.1. Find Fourier series to represent the displacement of the string when it is pulled aside by y_0 at the point $x = l/4$.

The function to be represented is

$$y(x) = 4y_0 x/l, \quad 0 \leqslant x < l/4,$$

$$= \frac{4y_0}{3}\left(1 - \frac{x}{l}\right), \quad l/4 \leqslant x \leqslant l.$$

Solution 1. Take the continuation as in fig. 8.5 (b) and then the series is

$$y(x) = \frac{a_0}{2} + \sum_{n=1}^{\infty} [a_n \cos (nkx) + b_n \sin (nkx)],$$

where $k = 2\pi/l$ and is the fundamental wave number.

Then, from (8.11) we have (taking the integral from 0 to l rather than $-l/2$ to $l/2$ for convenience),

$$a_n = \frac{2}{l}\int_0^{l/4} \frac{4y_0 x}{l} \cos (nkx)\,\mathrm{d}x + \frac{2}{l}\int_{l/4}^{l} \frac{4y_0}{3}\left(1 - \frac{x}{l}\right)\cos (nkx)\,\mathrm{d}x.$$

Using integration by parts the student should show that this gives

▶8. $a_n = y_0$ for $n = 0$,
        $= -8y_0/3n^2\pi^2$ for n odd,
        $= -16y_0/3n^2\pi^2$ for $n = 4m + 2$,
        $= 0$ for $n = 4m$ $(m \neq 0)$.

The values of b_n are similarly obtained from

$$b_n = \frac{2}{l}\int_0^{l/4} \frac{4y_0 x}{l} \sin (nkx)\,\mathrm{d}x + \frac{2}{l}\int_{l/4}^{l} \frac{4y_0}{3}\left(1 - \frac{x}{l}\right)\sin (nkx)\,\mathrm{d}x,$$

and are

▶9. $b_n = 8y_0(-1)^{\frac{1}{2}(n-1)}/3n^2\pi^2$,   for n odd,
       $= 0$,   for n even.

The first few terms of the Fourier series thus obtained are

$$y(x) = \frac{y_0}{2} - \frac{8y_0}{3\pi^2}\left[\cos\left(\frac{2\pi x}{l}\right) - \sin\left(\frac{2\pi x}{l}\right) + \tfrac{1}{4}\cos\left(\frac{4\pi x}{l}\right)\right.$$

$$\left. + \tfrac{1}{9}\cos\left(\frac{6\pi x}{l}\right) + \tfrac{1}{9}\sin\left(\frac{6\pi x}{l}\right) + \tfrac{1}{25}\cos\left(\frac{10\pi x}{l}\right) - \cdots\right]. \quad (8.20)$$

Solution 2. This time we continue the function as in fig. 8.5 (*c*) so that the series contains only sine terms

$$y(x) = \sum_{n=1}^{\infty} B_n \sin(nk'x),$$

where here $k' = 2\pi/2l$. The coefficients B_n are given by

$$B_n = \frac{2}{2l}\int_{-l}^{l} y(x)\sin(nk'x)\,\mathrm{d}x$$

$$= 2\frac{1}{l}\int_{0}^{l} y(x)\sin(nk'x)\,\mathrm{d}x,$$

since the integrand is symmetric about $x = 0$.

Proceeding as before the integral is found to be

▶10. $B_n = 32y_0 \sin(n\pi/4)/3n^2\pi^2.$

This solution thus gives a series

$$y(x) = \frac{32y_0}{3\pi^2}\left[(2)^{-1/2}\sin\left(\frac{\pi x}{l}\right) + \tfrac{1}{4}\sin\left(\frac{2\pi x}{l}\right)\right.$$

$$\left. + \frac{(2)^{-1/2}}{9}\sin\left(\frac{3\pi x}{l}\right) - \frac{(2)^{-1/2}}{25}\sin\left(\frac{5\pi x}{l}\right) - \cdots\right]. \quad (8.21)$$

▶11. Show that the series corresponding to the continuation in fig. 8.5 (*d*) is

$$y(x) = \frac{y_0}{2} + \frac{8y_0}{3\pi^2}\sum_{n=1}^{\infty}\frac{4\cos(n\pi/4) - 3 - (-1)^n}{n^2}\cos\left(\frac{n\pi x}{l}\right). \quad (8.22)$$

It will be apparent to the reader who has carried through the calculations needed to obtain equations (8.20)–(8.22), that the introduction of some symmetry, in the latter two, produces a significant saving in computational labour. It should be emphasized that all choices of continuation give series which converge to the function in fig. 8.5 (*a*) in the interval

0 to l, but of course differ widely outside that range. A further consideration in the choice of continuation is the convergence of the resultant
series. We will again show this by considering a specific example.

Example 8.2. Find Fourier series to represent the function $y(x) = x$
in the range $0 \leqslant x < \pi$.

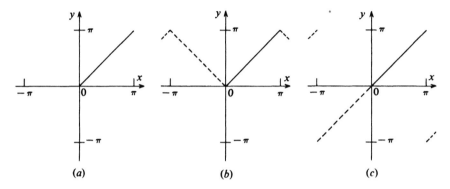

Fig. 8.6   The function $y = x$ in $0 \leqslant x < \pi$ and two possible continuations.

The function is shown in fig. 8.6 (*a*). Two possible continuations with
period 2π and fundamental wave number $= 1$ are shown in fig. 8.6 (*b*)
and (*c*). Continuation (*b*) is symmetric and yields a cosine series with
coefficients

$$A_n = \frac{2}{2\pi} \times 2 \int_0^\pi x \cos (nx) \, dx$$

$$= 0 \qquad \text{for } n \text{ even } (n \neq 0),$$
$$= \pi \qquad \text{for } n = 0,$$
$$= -4/\pi n^2 \text{ for } n \text{ odd},$$

giving

$$y(x) = \frac{\pi}{2} - \frac{4}{\pi} \sum_{m=0}^{\infty} \frac{\cos (2m + 1)x}{(2m + 1)^2}. \tag{8.23}$$

On the other hand continuation (*c*) yields a sine series

▶12. $$y(x) = 2 \sum_{n=1}^{\infty} (-1)^{n+1} \frac{\sin (nx)}{n}. \tag{8.24}$$

Again both series converge to $y(x) = x$ in $0 \leqslant x < \pi$, but (8.23) is
clearly preferable for numerical computation since (*a*) it has an n^{-2}

convergence whereas (8.24) converges only as n^{-1}, and (b) it does not show the Gibbs phenomenon which accompanies the discontinuous function produced by the continuation in fig. 8.6 (c). The choice of continuation or period is not always as wide as this section may have suggested. For some situations additional physical conditions at the ends of the given interval limit or determine the allowed forms or periods. Examples of this occur, for example, in connection with the Fourier solutions of flow and diffusion problems considered in chapter 10.

8.5   Differentiation and integration

Since in a Fourier expansion both sides are functions of the independent variable, it is natural to consider differentiation and integration with respect to that variable. We will not prove the relevant properties but merely state them and give an illustration.

(i) *Differentiation*. The Fourier series for the derivative $f'(t)$ can be obtained from the FS for $f(t)$ by differentiating term by term.
(ii) *Integration*. The Fourier series for the indefinite integral $F(t) = \int^t f(s)\,ds$ can be obtained (to within a constant) by term by term integration of the FS for $f(t)$.

In terms of formulae, for $f(t)$ given by (8.7),

$$f'(t) = \sum_{n=1}^{\infty} [-n\omega A_n \sin(n\omega t) + n\omega B_n \cos(n\omega t)], \tag{8.25}$$

and

$$F(t) = \frac{A_0 t}{2} + \sum_{n=1}^{\infty} \left[\frac{B_n}{n\omega}(1 - \cos n\omega t) + \frac{A_n}{n\omega} \sin(n\omega t) \right], \tag{8.26}$$

where the constant of integration is $\sum (B_n/n\omega)$ and is chosen here so as to make $F(0) = 0$.

As shown in (8.25) each differentiation produces an additional factor $n\omega$ in the numerator of each term and consequently the convergence of the series becomes less and less rapid. In any case, the process can only be used as many times as $f(t)$ is itself differentiable.

As an example we may use the function already considered in fig. 8.6 (b) and given by (8.23) as

$$y(x) = \frac{\pi}{2} - \frac{4}{\pi} \sum_{m=0}^{\infty} \frac{\cos(2m+1)x}{(2m+1)^2}. \tag{8.23 bis}$$

Differentiating term by term yields

$$\frac{dy}{dx} = \frac{4}{\pi} \sum_{m=0}^{\infty} \frac{\sin(2m+1)x}{2m+1}. \tag{8.27}$$

This is just the series given in (8.14) with, as in this case, $T = 2\pi$, $\omega = 1$ and a change of notation. The series (8.27) is thus that for the periodic function $g(x) = -1$, for $-\pi \leqslant x < 0$ and $g(x) = 1$, for $0 \leqslant x < \pi$; but this is precisely the value of the derivative of the original function, thus verifying the differentiation property for this case. The validity of the integration procedure for this example worked backwards is obvious, the constant of integration having to be determined separately.

▶13. Using a set of tables evaluate the series (8.23) for the function $y(x) = x$, by adding on successive terms. Use values of x of (say) $r\pi/6$ for $r = (0, 6)$. Note how the factor $(2m + 1)^{-2}$ produces rapid convergence.

▶14. Illustrate the limitations of the differentiation property by 'deducing' from (8.24) that, if the differentiability or otherwise of the original function is ignored, then

$$1 + 1 + 1 + \cdots = -\tfrac{1}{2}.$$

▶15. As far as possible, determine by inspection the form of the functions of which the following are the Fourier series,

(i) $\cos\theta + \tfrac{1}{9}\cos 3\theta + \tfrac{1}{25}\cos 5\theta + \cdots$;

(ii) $\sin\theta + \tfrac{1}{27}\sin 3\theta + \tfrac{1}{125}\sin 5\theta + \cdots$;

(iii) $\dfrac{l^2}{3} - \dfrac{4l^2}{\pi^2}\left[\cos\left(\dfrac{\pi x}{l}\right) - \tfrac{1}{4}\cos\left(\dfrac{2\pi x}{l}\right) + \tfrac{1}{9}\cos\left(\dfrac{3\pi x}{l}\right) - \cdots\right].$

[You may find it helpful to deduce from (8.23) that

$$S_0 \equiv \textstyle\sum_{m=0}^{\infty}(2m + 1)^{-2} = \pi^2/8,$$

and other summation relationships derivable from this.]

8.6   Fourier transforms

So far we have considered physically periodic functions and functions defined in a fixed finite interval, in so far as representing them as superpositions of sinusoidal functions is concerned. For the reasons discussed at the beginning of the chapter it is desirable to obtain such a representation even for functions defined over an infinite interval and with no periodicity. The representation we will obtain in this case is called the **Fourier integral** or **transform** of the particular function, and is obtained formally by letting the interval T of the Fourier series become infinite.

Before carrying out this procedure properly, we give a qualitative outline of what is involved.

The frequencies present in a Fourier series are all those of the form $\omega = 2\pi n/T$ for all integral values of n. If T tends to infinity the 'quantum' of frequency $2\pi/T$ becomes vanishingly small and the spectrum of frequencies allowed becomes a continuum. With this continuum of admitted frequencies the Fourier sum goes over into an integral and the expansion coefficients A_n, B_n or C_n become functions of the *continuous* variable ω. In previous parts of this chapter the symbol ω has been used to indicate the *fixed* quantity $2\pi/T$, but in connection with Fourier transforms (FT) it will be used as the continuous variable analogous to the integral variable n of Fourier series. For passage from the series to the transform it will be more convenient to use the exponential base set $\exp(i2\pi nt/T)$ for the series, rather than the sine and cosine set.

Turning the above discussion into equations we expect that on letting $T \to \infty$, expansion (8.8) will take on the general form

$$f(t) = K_1 \int_{-\infty}^{\infty} g(\omega) \exp(i\omega t) \, d\omega, \tag{8.28}$$

and that (8.11 c) for the series coefficients C_m will become an equation for $g(\omega)$ of the form

$$g(\omega) = K_2 \int_{-\infty}^{\infty} f(t) \exp(-i\omega t) \, dt. \tag{8.29}$$

Here K_1 and K_2 are constants and their relative values are clearly arbitrary as any increase in K_1 can always be compensated by a decrease in K_2. However since (8.29) can be substituted back into (8.28) and hence $f(t)$ obtained on both sides of an equation, the product $K_1 K_2$ is not arbitrary.

Having sketched in outline the procedure to be followed in making the transition to continuous frequencies and the Fourier integral, we now carry it through in a quantitative way. We first note, for any function $f(t)$ given by a series, the trivial result that

$$f(t) = \sum_{n=-\infty}^{\infty} F_n = \frac{T}{2\pi} \sum_{n=-\infty}^{\infty} \frac{2\pi}{T} F_n. \tag{8.30}$$

For our particular case fig. 8.7 shows that $\sum (2\pi/T)F_n$ has a readily interpretable meaning. Plotting F_n as a function of n and $\omega = 2\pi n/T$ simultaneously, it is clear that $(2\pi/T)F_n$ is just the area of the nth rectangle $(-\infty < n < \infty)$ and that the sum approximates the area under the solid curve. Expressed in terms of ω it is $\int_{-\infty}^{\infty} F(\omega) \, d\omega$ where now F is treated as a function of ω. This now is not merely an approximate relationship

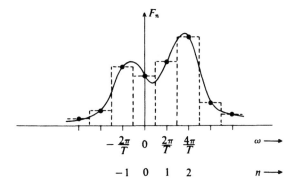

Fig. 8.7 The relationship between the Fourier terms F_n for a function of period T and the Fourier integral $\int F(\omega)\,d\omega$ of the function. The solid line shows $F(\omega)$.

but is, as $T \to \infty$ and consequently the widths of the rectangles tend to zero, the mathematical definition of the integral. Recalling that in the Fourier series $F_n = C_n \exp(i2\pi nt/T) = c(\omega)\exp(i\omega t)$, we obtain from (8.8) that

$$f(t) = \frac{T}{2\pi} \int_{-\infty}^{\infty} c(\omega)\exp(i\omega t)\,d\omega,$$

whilst from (8.11 c) we get

$$c(\omega) = \frac{1}{T} \int_{-T/2}^{T/2} f(t)\exp(-i\omega t)\,dt.$$

Finally writing $Tc(\omega)/(2\pi)^{1/2} = g(\omega)$ and letting $T \to \infty$ we obtain the two defining relationships for the Fourier transform

$$f(t) = \frac{1}{(2\pi)^{1/2}} \int_{-\infty}^{\infty} g(\omega)\exp(i\omega t)\,d\omega \qquad (8.31)$$

and

$$g(\omega) = \frac{1}{(2\pi)^{1/2}} \int_{-\infty}^{\infty} f(t)\exp(-i\omega t)\,dt. \qquad (8.32)$$

Including the $(2\pi)^{-1/2}$ in the definition of $g(\omega)$, whose mathematical existence as $T \to \infty$ is here assumed without proof, amounts to choosing the relative values of K_1 and K_2. The actual choice made is aimed at making (8.31) and (8.32) as symmetric as possible. The function $g(\omega)$ is often called the (amplitude) spectrum of $f(t)$. It is assumed of course that $\int_{-\infty}^{\infty} |f(t)|\,dt$ exists.

To illustrate these definitions we consider the following simple example.

Example 8.3. Find the Fourier transform of the exponential decay function $f(t) = 0$ for $t < 0$ and $f(t) = \exp(-\lambda t)$ for $t \geqslant 0$ ($\lambda > 0$).

Using definition (8.32)

$$g(\omega) = (2\pi)^{-1/2} \int_0^\infty \exp(-\lambda t) \exp(-i\omega t) \, dt$$

$$= (2\pi)^{-1/2} (\lambda + i\omega)^{-1},$$

which is the required transform. [This may be verified by resubstitution of the result into (8.31) to recover $f(t)$, but evaluation of the integral requires the use of complex variable contour integration (chapter 16).]

An important function which appears in many areas of physical science, either precisely or as an approximation to a physical situation, is the Gaussian or normal distribution. Its Fourier transform is of importance both for itself and because, when interpreted statistically, it readily illustrates a form of 'Uncertainty Principle', independent of the introduction of quantum mechanics.

We take the Gaussian distribution in the normalized form

$$f(t) = \frac{1}{(2\pi)^{1/2}} \frac{1}{\tau} \exp\left(-\frac{t^2}{2\tau^2}\right), \quad -\infty < t < \infty, \tag{8.33}$$

which, interpreted as a distribution, has zero mean and a root-mean-square deviation $\Delta t = \tau$. To find its Fourier transform or spectrum we evaluate

$$g(\omega) = (2\pi\tau)^{-1} \int_{-\infty}^\infty \exp(-t^2/2\tau^2) \exp(-i\omega t) \, dt$$

$$= (2\pi\tau)^{-1} \int_{-\infty}^\infty \exp\left\{-\frac{1}{2\tau^2} [t^2 + 2\tau^2 i\omega t \right.$$

$$\left. + (\tau^2 i\omega)^2 - (\tau^2 i\omega)^2]\right\} dt,$$

where the quantity $-(\tau^2 i\omega)^2/2\tau^2$ has been both added and subtracted in the exponent in order to allow the factors involving the variable of integration t to be expressed as a complete square. On bringing the last [t-independent] factor outside the integral sign, the expression can be written as

$$g(\omega) = \frac{\exp(-\tau^2\omega^2/2)}{(2\pi)^{1/2}} \frac{1}{(2\pi)^{1/2}\tau} \int_{-\infty}^\infty \exp\left[-\frac{(t + i\tau^2\omega)^2}{2\tau^2}\right] dt.$$

The last factor is in fact the normalization integral for the Gaussian distribution (8.33) and equals unity, although to show this strictly, needs

results from complex variable theory (chapter 16). That it is equal to
unity can be made plausible by a change of variable to $s = t + i\tau^2\omega$
and assuming the imaginary parts introduced into the integration path
and limits (where the integrand goes rapidly to zero anyway) make no
difference.

We are thus left with the result that

$$g(\omega) = \frac{1}{(2\pi)^{1/2}} \exp\left(-\frac{\tau^2\omega^2}{2}\right), \tag{8.34}$$

which is another Gaussian distribution with zero mean and an R.M.S.
deviation $\Delta\omega = 1/\tau$. The R.M.S. deviation in t was τ, and so it is seen
that the deviations or 'spreads' in t and ω are inversely related by

$$\Delta\omega \cdot \Delta t = 1, \tag{8.35}$$

independent of the value of τ. In the physical terms of time and frequency,
the narrower an electrical impulse (say) is in time the greater the spread
of frequency components it must contain. Similar physical statements
are valid for other pairs of Fourier related variables, such as spatial position and wave number. In an obvious notation $\Delta k \cdot \Delta x = 1$.

Uncertainty relationships, as usually expressed in quantum mechanics,
can be related to the above if the de Broglie and Einstein relationships
for momentum p and energy E are introduced,

$$p = \hbar k \quad \text{and} \quad E = \hbar\omega.$$

Here $\hbar$ is Planck's constant divided by 2π. In quantum mechanics $f(t)$
is a wave function and the distribution of the wave intensity in time is
given by $|f|^2$ (also a Gaussian). Similarly the intensity distribution in
frequency is given by $|g|^2$. These two distributions have R.M.S. devia-
tions of $\tau/\sqrt{2}$ and $1/(\sqrt{2}\tau)$, giving together with the above relations

$$\Delta E \cdot \Delta t = \tfrac{1}{2}\hbar \quad \text{and} \quad \Delta p \cdot \Delta x = \tfrac{1}{2}\hbar.$$

The factors of $\tfrac{1}{2}$ which appear are specific to the Gaussian form, but any
distribution $f(t)$ produces for the product $\Delta\omega \cdot \Delta t$ a quantity $\lambda\hbar$ in which
λ is strictly positive [in fact the Gaussian value of $\tfrac{1}{2}$ is the minimum pos-
sible].

Our second example, the *diffraction grating* is taken from optics and
illustrates a spatial Fourier transform. The pattern of transmitted light
produced by a partially opaque (or phase changing) object upon which a
coherent beam of radiation falls, is called its diffraction pattern, and in
particular, when the cross-section of the object is small compared with the
distance at which the light is observed the pattern is known as the Fraun-
hofer diffraction pattern. The essential quantity is the dependence of the
light amplitude (and hence intensity) on the angle the viewing direction

makes with the incident beam. As will be seen, this is entirely determined by the amplitude and phase of the light at the object, the intensity in a particular direction being determined by the Fourier component of this spatial distribution corresponding to a particular wave number directly related to that direction.

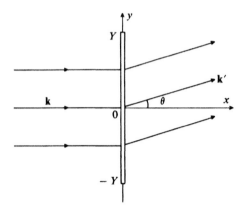

Fig. 8.8   Diffraction grating of width $2Y$ with light of wave length $2\pi/k$ being diffracted through an angle θ.

We consider a simple one-dimensional screen of width $2Y$ on which light of wave number k ($=2\pi/\lambda$) is incident normally (see fig. 8.8), and suppose at the position $(0, y)$ the amplitude of transmitted light is $f(y)$ per unit length in the y-direction [$f(y)$ may be complex]. Both the screen and beam are assumed infinite in the z-direction.

At a position $\mathbf{r}' = (x', y')$ with $x' > 0$ the total light amplitude will be the superposition of all the (Huygen) wavelets originating from the various parts of the screen. For large r', these can be treated as plane waves to give†

$$A(\mathbf{r}') = \int_{-Y}^{Y} \frac{f(y) \exp\left[i\mathbf{k}' \cdot (\mathbf{r}' - y\mathbf{j})\right]}{|\mathbf{r}' - y\mathbf{j}|} \, dy, \tag{8.36}$$

where $\mathbf{j}$ is the unit vector in the y-direction. The factor $\exp\left[i\mathbf{k}' \cdot (\mathbf{r}' - y\mathbf{j})\right]$ represents the phase change undergone by the light in travelling from the point $y\mathbf{j}$ on the screen to the point $\mathbf{r}'$, and the denominator represents the reduction in amplitude with distance. [Recall that the system is infinite

† This is the approach first used by Fresnel. For simplicity we have omitted from the integral a multiplicative inclination factor which depends on angle θ, and decreases as θ increases.

in the z-direction and so the 'spreading' is effectively in two dimensions only.]

If the medium is the same on both sides of the screen $\mathbf{k}' = (k \cos \theta, k \sin \theta, 0)$, and if $r' \gg Y$, expression (8.36) can be approximated by

$$A(\mathbf{r}') = \frac{\exp(\mathrm{i}\mathbf{k}' \cdot \mathbf{r}')}{r'} \int_{-\infty}^{\infty} f(y) \exp(-\mathrm{i}k \sin \theta\, y) \, \mathrm{d}y. \qquad (8.37)$$

We have used that $f(y) = 0$ for $|y| > Y$, to extend the integral to infinite limits. The intensity in direction θ is then given by

$$I(\theta) = |A|^2 = \frac{2\pi}{r'^2} |\tilde{f}(k \sin \theta)|^2 . \qquad (8.38)$$

The function $\tilde{f}(\omega)$ is an alternative notation for $g(\omega)$, as defined by (8.32).

The amplitude in a direction θ is thus directly proportional to the Fourier component of the light amplitude distribution at the screen corresponding to wave number $k \sin \theta$. This result is general and does not depend upon any periodicity in the screen, but if the screen transmits light in such a way that the amplitude is periodic in y, further, more specific, results can be obtained.

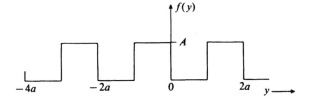

Fig. 8.9 The amplitude of light transmitted through a uniformly illuminated diffraction grating consisting of alternate transparent and opaque strips, all of width a.

As a particular case consider a grating consisting of $4N$ equal strips of width a (in the y-direction), alternately opaque and transparent. Then, if the grating is uniformly illuminated

$$f(y) = A \text{ when } (2n+1)a \leqslant y \leqslant (2n+2)a, \quad -N \leqslant n < N,$$
$$= 0 \text{ otherwise,}$$

as is illustrated in fig. 8.9. Writing $k \sin \theta = \mu$, we require to evaluate

$$\tilde{f}(\mu) = (2\pi)^{-1/2} \int_{-Y}^{Y} f(y) \exp(-\mathrm{i}\mu y) \, \mathrm{d}y$$

$$= (2\pi)^{-1/2} \sum_{r=-N}^{N-1} \int_{-\infty}^{\infty} g(y - 2ar) \exp(-\mathrm{i}\mu y) \, \mathrm{d}y,$$

where $g(u) = A$ for $a \leqslant u < 2a$ and $g(u) = 0$ elsewhere. Now in the integral for the rth term replace $y - 2ar$ by u to give

$$\tilde{f}(\mu) = (2\pi)^{-1/2} \sum_{r=-N}^{N-1} \exp(-i2ar\mu) \int_{a}^{2a} A \exp(-i\mu u) \, du$$

▶16.
$$= (2\pi)^{-1/2} \exp(-i\mu a/2) \frac{A \sin(2a\mu N)}{\mu \cos(a\mu/2)}.$$

Hence,

$$I(\theta) = \frac{|A|^2}{r'^2} \frac{\sin^2(2a\mu N)}{\mu^2 \cos^2(a\mu/2)} \quad \text{with} \quad \mu = k \sin\theta = \frac{2\pi}{\lambda} \sin\theta.$$

This distribution has maxima and minima at those values of μ which make $2a\mu N$ an odd or even multiple of $\pi/2$. For large N these are very closely spaced and effectively give a low intensity background. More pronounced maxima of intensity occur where θ is such that the denominator vanishes, namely $\frac{1}{2}a\mu = \frac{1}{2}(2m + 1)\pi$, i.e. $a \sin\theta = \frac{1}{2}(2m + 1)\lambda$ with m integral. The value of $|\tilde{f}(\mu)|$ under these conditions is

$$\lim_{\mu \to (2m+1)\pi/a} \left| \frac{A}{(2\pi)^{1/2}\mu} \frac{\sin(2a\mu N)}{\cos(a\mu/2)} \right|$$

$$= \lim_{\mu \to (2m+1)\pi/a} \left| \frac{2aNA}{(2\pi)^{1/2}} \frac{\cos(2a\mu N)}{\cos(a\mu/2) - \frac{1}{2}a\mu \sin(a\mu/2)} \right|$$

$$= \frac{4aNA}{(2\pi)^{1/2}(2m + 1)\pi}, \tag{8.39}$$

and as expected $I(\theta)$ is proportional to N^2, i.e. for these values of θ the light from the $2N$ transparent slits interferes constructively.

The principal maximum occurs at $\mu = 0$, with (using the procedure of (8.39)) the value $|\tilde{f}(0)|$ given by $2aNA/(2\pi)^{1/2}$. Hence the distribution has a central maximum ($\theta = 0$) with subsidiary maxima at angles arcsin $[(2m + 1)\lambda/2a]$, with intensities reduced by factors of $(2m + 1)^2\pi^2/4$ relative to the central one. This is illustrated schematically in fig. 8.10 for a moderate value of N. Naturally, for a diffracted spectrum to be observable the corresponding value of $|\theta|$ must be $< \pi/2$.

8.7   Properties of Fourier transforms

As would be expected, Fourier transforms have many properties analogous to those of Fourier series (section 8.5) with regard to the connection between transforms of related functions. Here they will only be listed, but the reader should verify them directly by working from the definition of

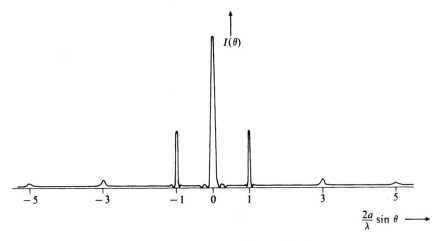

Fig. 8.10 Diffraction pattern from the grating discussed in the text.

the transform. As previously we denote $FT\{f(t)\}$ by $g(\omega)$. The unfamiliar last term in (8.41) is discussed below.

▶17. Differentiation $\qquad$ $FT\{f'(t)\} = i\omega g(\omega).$ $\qquad$ (8.40)

▶18. Integration $\qquad$ $FT\{\int^t f(s)ds\} = -i\omega^{-1}g(\omega) + 2\pi C\delta(\omega).$

(8.41)

▶19. Translation $\qquad$ $FT\{f(t + a)\} = \exp(ia\omega)g(\omega).$ (8.42)

▶20. Exponential multiplication $\quad$ $FT\{\exp(\alpha t)f(t)\} = g(\omega + i\alpha).$ (8.43)

In (8.43) α may be real, imaginary or complex. The last term $2\pi C\delta(\omega)$ in (8.41) represents the FT of the constant of integration associated with the definition of the indefinite integral. The function involved, $\delta(\omega)$, is fully discussed in the next section.

To illustrate both a use, and the proof, of one of the above relations, we may consider (8.43) in connection with an amplitude-modulated radio wave. Suppose a message to be broadcast is represented by $f(t)$. The message can be added electronically to a constant signal a of such a magnitude that $a + f(t)$ is never negative, and the sum then used to modulate the amplitude of a carrier signal of frequency ω_c. Using a complex exponential notation, the transmitted amplitude is now

$$F(t) = A(a + f(t)) \exp(i\omega_c t). \qquad (8.44)$$

Ignoring in the present context the effect of the term $Aa \exp(i\omega_c t)$

which gives a contribution to the transmitted spectrum only at $\omega = \omega_c$, we obtain for the new spectrum

$$G(\omega) = (2\pi)^{-1/2} A \int_{-\infty}^{\infty} f(t) \exp(i\omega_c t) \exp(-i\omega t) \, dt$$

$$= (2\pi)^{-1/2} A \int_{-\infty}^{\infty} f(t) \exp[-i(\omega - \omega_c)t] \, dt$$

$$\doteq A g(\omega - \omega_c), \tag{8.45}$$

which is simply a shift of the whole spectrum by the carrier frequency. The use of different carrier frequencies enables signals to be separated.

8.8   The δ-function

In the previous section the δ-function was referred to without having been either introduced or defined; this was done so that related properties of the FT could be presented together. In this section we remedy this omission.

Referring back to the defining equations (8.31) and (8.32) and substituting from one into the other we obtain the equation

$$f(t) = (2\pi)^{-1/2} \int_{-\infty}^{\infty} d\omega \exp(i\omega t)$$

$$\times (2\pi)^{-1/2} \int_{-\infty}^{\infty} dt' \exp(-i\omega t') f(t')$$

$$= \int_{-\infty}^{\infty} dt' f(t') \times (2\pi)^{-1} \int_{-\infty}^{\infty} d\omega \exp[i\omega(t - t')]. \tag{8.46}$$

Here we have written the differentials immediately following the integral signs to which they refer and in obtaining the second line from the first we have assumed that the order of the integrations can be reversed.

Now, if it is recalled that $f(t)$ is an arbitrary [but sufficiently well-behaved] function, and also noted that in (8.46) the left-hand side refers to a value of f at a *particular* value of t, whilst the right-hand side contains an integral over *all* values of the argument of f, then it is clear that the expression

$$(2\pi)^{-1} \int_{-\infty}^{\infty} d\omega \exp[i\omega(t - t')], \tag{8.47}$$

considered as a function of t', has some remarkable properties. The expression is known as the **Dirac δ-function** and is denoted by $\delta(t - t')$.

Qualitatively speaking, since the left-hand side of (8.46) is independent of the value of $f(t')$ for all $t' \neq t$ and f itself is an arbitrary function, the δ-function must have the effect of making the integral over t' receive zero contribution from all t' except in the immediate neighbourhood of $t' = t$, where the contribution is so large that a finite value for the integral results. We are thus able to rewrite (8.46) in the following (somewhat unrigorous) form

$$f(t) = \int_{-\infty}^{\infty} dt' \, f(t') \, \delta(t - t')$$

$$= f(t) \int_{-\infty}^{\infty} \delta(t - t') \, dt' = f(t) \int_{t-\epsilon}^{t+\epsilon} \delta(t - t') \, dt',$$

where ϵ is any quantity > 0. Expressed in another way,

$$\int \delta(t - t') \, dt' = 1$$

if the range of integration includes $t' = t$.

The δ-function is a less obviously valid mathematical function than most which are encountered in physical science, but its admissibility has by now been well justified by pure mathematical methods and its utility demonstrated in many applications, both idealized and practical. Its formal properties may be summarized as,

(i) $\delta(x) = 0$ for $x \neq 0$,

(ii) $\int_{-a}^{b} \delta(x) \, dx = 1$, all $a, b > 0$,

(iii) $\int f(y) \, \delta(x - y) \, dy = f(x)$ if the range of integration includes the point $y = x$,

(iv) $\delta(h(x)) = \sum_{i} \dfrac{\delta(x - x_i)}{|h'(x_i)|}$.

$$(8.48)$$

The last of these in which the x_i are those values of x for which $h(x) = 0$ and $h'(x)$ stands for dh/dx, may be verified by considering an integral of the form $\int f(x) \, \delta(h(x)) \, dx$ and making a change of variable to $z = h(x)$.

The δ-function can be visualized as a very sharp narrow pulse (in space, time, density, etc.) producing an integrated effect of definite magnitude. For many practical purposes, effects which are not strict δ-functions may be analysed as such, if they take place in an interval which is much shorter than the response interval of the system on which they act. The idealized notions of an impulse at time t_0, $j(t) = J \, \delta(t - t_0)$, or of a point charge

q at position $\mathbf{r}_0$, $\rho(\mathbf{r}) = q\,\delta(\mathbf{r} - \mathbf{r}_0) = q\,\delta(x - x_0)\,\delta(y - y_0)\,\delta(z - z_0)$, are examples of cases in which the δ-function can also be employed, to present a discrete quantum as if it were a continuum distribution.

Mention was made in the discussion of chapter 7 of the use of δ-functions in enabling the general solution of linear differential equations with arbitrary driving terms, to be built up in terms of Green's functions. As has been indicated, chapter 7 should be restudied after the methods of this and the next two chapters have been assimilated.

We have already obtained one representation of the δ-function, as an integral

$$\delta(t) = (2\pi)^{-1} \int_{-\infty}^{\infty} \exp(i\omega t)\,d\omega. \tag{8.49}$$

Considered as a Fourier transform it shows that a very narrow time peak at the origin results from the superposition of a complete spectrum of harmonic waves, all frequencies having the same amplitude, and all waves being in phase at $t = 0$. This suggests that the δ-function may also be represented as the limit of the transform of a uniform distribution of unit height as the width of the distribution becomes infinite.

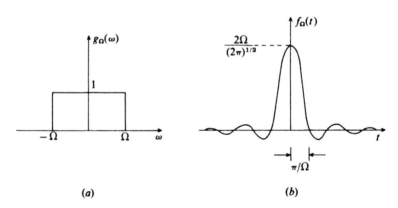

Fig. 8.11 (a) A Fourier transform showing a rectangular distribution of frequencies between $\pm\Omega$; (b) the function of which it is the transform is proportional to $t^{-1}\sin\Omega t$.

Consider the distribution shown in fig. 8.11 (a). From (8.31)

$$f_\Omega(t) = (2\pi)^{-1/2} \int_{-\Omega}^{\Omega} 1 \times \exp(i\omega t)\,d\omega \tag{8.50}$$

► 21.
$$= \frac{2\Omega}{(2\pi)^{1/2}} \frac{\sin(\Omega t)}{\Omega t}.$$

This function is illustrated in fig. 8.11 (*b*) and it is apparent that, for large Ω, it both gets very large at $t = 0$ and also becomes very narrow about $t = 0$, as we qualitatively expect and require. We also note that in the limit $\Omega \to \infty$, $f_\Omega(t)$, as defined in (8.50), tends to $(2\pi)^{1/2}\,\delta(t)$ by virtue of (8.49). Hence we may conclude that the δ-function can also be represented by

$$\delta(t) = \lim_{\Omega \to \infty} \frac{\sin(\Omega t)}{\pi t}. \tag{8.51}$$

Several other limit function representations are equally valid, e.g. the limiting cases of rectangular or triangular distributions; the only essential requirements are a knowledge of the area under such a curve and that undefined operations such as differentiation are not inadvertently carried out on the δ-function whilst some non-explicit representation is being employed.

8.9 Parseval's theorem

Using the result of the previous section we can now prove Parseval's theorem,† which relates the integral of $|f(t)|^2$ over all t to the integral of $|g(\omega)|^2$ over all ω. When f and g are physical amplitudes these integrals relate to the total intensity involved in some physical process.

From the complex conjugate of the definition of the inverse FT

$$f^*(t) = (2\pi)^{-1/2} \int_{-\infty}^{\infty} g^*(\omega') \exp(-i\omega' t)\, d\omega'.$$

Substituting this and the definition into

$$I = \int_{-\infty}^{\infty} |f(t)|^2\, dt = \int_{-\infty}^{\infty} f(t)f^*(t)\, dt,$$

we obtain (all integrals are from $-\infty$ to $+\infty$)

$$I = \int dt \times (2\pi)^{-1/2} \int d\omega\, g(\omega) \exp(i\omega t)$$

$$\times (2\pi)^{-1/2} \int d\omega'\, g^*(\omega') \exp(-i\omega' t)$$

$$= \int d\omega\, g(\omega) \int d\omega'\, g^*(\omega') \times (2\pi)^{-1} \int dt\, \exp[i(\omega - \omega')t]$$

$$= \int d\omega\, g(\omega) \int d\omega'\, g^*(\omega')\, \delta(\omega - \omega'), \qquad \text{using (8.47),}$$

† This is a particular case of the more general Parseval's theorem, which, for two functions $f_1(t)$ and $f_2(t)$, relates $\int f_1^* f_2\, dt$ to $\int g_1^* g_2\, d\omega$.

$$= \int \mathrm{d}\omega \, g(\omega)g^*(\omega), \qquad\qquad \text{using (8.48 iii),}$$

$$= \int |g(\omega)|^2 \, \mathrm{d}\omega.$$

Hence we have Parseval's theorem, that

$$\int_{-\infty}^{\infty} |f(t)|^2 \, \mathrm{d}t = \int_{-\infty}^{\infty} |g(\omega)|^2 \, \mathrm{d}\omega. \tag{8.52}$$

▶22. Prove the corresponding result for Fourier series

$$\int_{-\pi}^{\pi} |f(\theta)|^2 \, \mathrm{d}\theta = 2\pi \sum_{n} |C_n|^2. \tag{8.53}$$

It should be noted that the constants which appear in these relationships depend upon the normalizations assumed in the definitions of the series and transforms.

As an example of Parseval's theorem applied to series we may take our previous example of a plucked string. Here $f(x)$ is the displacement of the string at position x, and $|f(x)|^2$ is proportional to the stored energy per unit length. On the other hand $|C_n|^2$ is proportional to the stored energy in the nth mode. It is apparent that both $\int_0^l |f(x)|^2 \, \mathrm{d}x$ and $\sum_n |C_n|^2$ represent (to within a constant) the total work done in producing the given configuration.

For an example from Fourier transforms, we may consider a damped oscillatory function

$$
\begin{aligned}
f(t) &= 0, \quad t < 0, \\
&= \exp(-t/\tau) \sin(\omega_0 t), \quad t \geq 0.
\end{aligned}
$$

This could represent the displacement of a damped harmonic oscillator, the current in an antenna, or the electric field of a radiated wave. The transform is easily obtained

$$g(\omega) = \int_{-\infty}^{\infty} f(t) \exp(-i\omega t) \, \mathrm{d}t$$

▶23.
$$= \frac{1}{2} \left[\frac{1}{\omega + \omega_0 - (i/\tau)} - \frac{1}{\omega - \omega_0 - (i/\tau)} \right]$$

The physical interpretation of $|g(\omega)|^2$ as the energy radiated or dissipated per unit frequency interval is immediate, as is the identification of Parseval's theorem as a statement about the total energy involved.

As a by-product it may also be noted that, if $\tau\omega_0 \gg 1$, then for $\omega \approx \omega_0$,

$$|g(\omega)|^2 \propto \frac{1}{(\omega - \omega_0)^2 + (1/\tau^2)},$$

which is the familiar form of the response of a damped oscillator to driving frequencies near to its resonant frequency. This indicates that the radiated intensity spectrum from such a system is very similar to its 'response spectrum'.

8.10   Convolution and deconvolution

It is apparent that an attempt to measure the value of any physical quantity as a function of some independent variable is, to some extent, rendered less precise than is intrinsically possible by the finite resolution of the measuring apparatus employed. In this section we will show how one of the properties of Fourier transforms can be used to make allowance for the resolution of the apparatus.

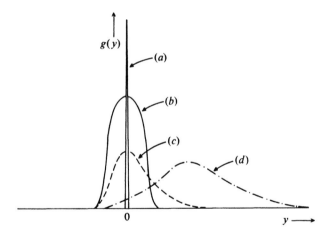

Fig. 8.12 Resolution functions: (a) ideal δ-function; (b) typical unbiased resolution; (c) and (d) show biases tending to shift observations to higher values than the true one.

Let $g(y)$ be the resolution function of the apparatus. By this we mean that the probability that a reading which should have been at $y = 0$ but is recorded instead as having been between y and $y + dy$, is given by $g(y)\,dy$. Figure 8.12 shows qualitatively some possible functions of this sort. It is obvious that a resolution such as (a), as much like a δ-function as possible, is desirable. Curve (b) is typical of the effective resolution of many measuring systems, having a finite spread but no significant bias. The other two curves (c) and (d) show bias to various degrees.

Suppose that a physical quantity has a *true* distribution $f(x)$ as a function of an independent variable x, but is measured by an instrument with resolution function $g(y)$; then we wish to calculate what the *observed* distribution $h(z)$ will be. The symbols x, y and z all refer to the same physical variable (e.g. length or angle), but are denoted differently because the variable appears in the analysis in three different roles.

We first require the probability that a true reading lying between x and $x + dx$ (which has a probability $f(x) dx$ of being selected) is moved by the instrumental resolution through an interval $z - x$ into an interval dz wide; this is $g(z - x) dz$. Hence the combined probability that the interval dx will give rise to an observation appearing in the interval dz is $f(x) dx g(z - x) dz$, and combining together all values of x which can lead to an observation in the range z to $z + dz$, we obtain

$$h(z) \, dz = dz \int_{-\infty}^{\infty} dx \, f(x) g(z - x). \qquad (8.54)$$

The integral on the right-hand side is called the **convolution** of the functions f and g and is often written $f * g$ [not to be confused with multiplication as denoted in some computer programming languages]. In words, our result is that the observed distribution is the convolution of the true distribution and the experimental resolution function. This can only result in the observed distribution being broader and smoother than the true one and, if $g(y)$ has a bias, in maxima normally being displaced from their true positions.

It is obvious from (8.54) that if the resolution function can be represented by a δ-function, $g(y) = \delta(y)$, then $h(z) = f(z)$ and the observed distribution is the true one.

▶24. Show that convolution is commutative, i.e. $f * g = g * f$. (8.55)

The next step is to obtain a result from Fourier transform theory which also involves convolution. To do this we consider the FT of a function $f(t)$ which can be written as a product $f(t) = f_1(t) f_2(t)$. For such a function its FT is given by (all integrals are from $-\infty$ to ∞)

$$g(\omega) = (2\pi)^{-1/2} \int f_1(t) f_2(t) \exp(-i\omega t) \, dt$$

$$= (2\pi)^{-1/2} \int dt \, f_2(t) \exp(-i\omega t)$$

$$\times (2\pi)^{-1/2} \int d\omega' g_1(\omega') \exp(i\omega' t)$$

$$= (2\pi)^{-1} \int d\omega' g_1(\omega') \int dt \, f_2(t) \exp\left[-i(\omega - \omega')t\right]$$

$$= (2\pi)^{-1/2} \int d\omega' g_1(\omega') g_2(\omega - \omega')$$

$$= (2\pi)^{-1/2} g_1 * g_2. \tag{8.56}$$

Hence the FT of a product is equal to the convolution of the separate Fourier transforms multiplied by $(2\pi)^{-1/2}$ (the *Convolution theorem*).

The similarity between (8.54) and (8.56) enables us to find a solution to the following problem.

Example 8.4. An experimental quantity $f(x)$ is measured by an instrument with known resolution function $g(y)$ to give an observed distribution $h(z)$. How may $f(x)$ be extracted from the measured distribution?

To save lengthy explicit formulae, we will use a symbolic notation. As on a previous occasion denote the FT of a function $k(t)$ [in our problem t could be x, y or z] by $\tilde{k}$, a function of ω, and the inverse FT (equation (8.31)) of $j(\omega)$ by $\check{j} = \check{j}(t)$.

Consider now, for the function $k(t)$,

$$\hat{\tilde{k}} = (2\pi)^{-1/2} \int d\omega \exp(i\omega t) \times (2\pi)^{-1/2} \int dt' k(t') \exp(-i\omega t')$$

$$= (2\pi)^{-1} \int dt' k(t') \int d\omega \exp[i\omega(t - t')]$$

$$= \int dt' k(t') \, \delta(t - t')$$

$$= k(t),$$

i.e. $\check{\tilde{k}} = k$.

▶25. Prove $\check{\check{j}} = j$.

For the problem we have from (8.54) that

$$h = f * g.$$

Let us define F, G, H, all functions of ω, such that

$$\tilde{f} = F, \qquad \tilde{g} = G, \qquad \tilde{h} = H, \tag{8.57 a}$$

and consequently that

$$\check{F} = \check{\tilde{f}} = f, \text{ etc.} \tag{8.57 b}$$

Then, comparing the above with (8.56) (with the roles of t and ω inter-changed, i.e. in the form 'that if $g(\omega) = g_1(\omega)g_2(\omega)$, then $f(t) = (2\pi)^{-1/2}f_1(t)*f_2(t)$') we see that

$$(2\pi)^{-1/2}H(\omega) = F(\omega)G(\omega)$$

or $$F(\omega) = (2\pi)^{-1/2}\frac{H(\omega)}{G(\omega)} = (2\pi)^{-1/2}\frac{\bar{h}}{\tilde{g}}.$$

Finally we obtain an expression for the required f in terms of the meas-ured h and the known g as

$$f = \hat{F} = (2\pi)^{-1/2}\left(\widehat{\frac{\bar{h}}{\tilde{g}}}\right). \tag{8.58}$$

In words, 'to extract the true distribution, divide the FT of the observed distribution by that of the resolution function for each value of ω, and then take the inverse FT of the function so generated'.

This explicit method of extracting true distributions is straightforward for exact functions, but in practice, because of experimental and statis-tical uncertainties in experimental data or because data over only a limited range is available, it is often not very precise, involving as it does three (numerical) transforms each requiring (in principle) an integral over an infinite range.

▶26. Treating the modulated signal of (8.44) as a product, use the Con-volution theorem to obtain result (8.45).

8.11   Examples for solution

1. Represent a periodic rectangular voltage of magnitude V, period T, and mark-space ratio 1/3, by a Fourier series. (The length of the pulse is $T/3$ and assume it occurs symmetrically about $t = 0$.) Show that every third harmonic is absent.

2. A given function $f(t)$ is periodic in t with period $2\pi/\omega$. It is re-quired to approximate $f(t)$ by a sine and cosine series $f_N(t)$ with a finite number of terms N, each of frequency ω or one of its harmonics. Show that for any N the mean square error is least when the coeffi-cients of the finite series are equal to the corresponding Fourier coefficients.

3. A function $f(x) = \exp(-x^2)$ in $0 \leqslant x \leqslant 1$. Show how it should be continued to give as its Fourier series, a series (the actual form is not wanted) (i) with only cosine terms, (ii) with only sine terms, (iii) with period 1, (iv) with period 2. Would there be any difference

between the values of the series in (iii) and (iv), (a) at $x = 0$, (b) at $x = 1$?

4. Electricity from the mains is supplied to a rectifying device in order to produce a d.c. supply. The output voltage of the rectifier is $f(t) = \cos \omega t$ if $\cos \omega t > 0$, and $f(t) = 0$ otherwise. This voltage is fed into a linear smoothing device which reduces any sinusoidal wave of frequency $n\omega$ by a factor $1/10n^2$ ($n \neq 0$). What will be the approximate amplitude of the voltage fluctuations in the final output?

5. Laplace's equation $\nabla^2 u = 0$ is to be solved in two-dimensional polar coordinates (r, θ) for the interior of the unit circle, with the boundary conditions

$$u(r = 1, \theta) = \sin \theta, \quad 0 \leqslant \theta \leqslant \pi,$$
$$= 0, \quad \pi \leqslant \theta \leqslant 2\pi.$$

The method of separation of variables (chapter 10) leads to a solution of the form

$$u(r, \theta) = \sum_{n=0}^{\infty} r^n [A_n \cos (n\theta) + B_n \sin (n\theta)].$$

Find the actual solution and compare it with that found in example 4 above. In both cases, by sketching the function to be represented by the sum of harmonics of the basic frequency, verify that the series found has the correct symmetry properties.

6. Demonstrate explicitly for the function defined in (8.12) that Parseval's theorem is valid, namely

$$\int_{-T/2}^{T/2} |f(t)|^2 \, dt = T \sum_{n=-\infty}^{\infty} |C_n|^2.$$

[Use the result (8.14) and the one quoted in ▶15.] Show that a filter which will pass frequencies only up to $8\pi/T$ will still transmit more than 90 per cent of the power in such a square-wave voltage signal.

7. [To be treated by the reader without familiarity with quantum mechanics as an exercise in Fourier transforms.]
   In quantum mechanics two equal mass particles of momenta and energies $(\mathbf{p}_j, E_j, j = 1, 2)$ and represented by plane wave functions†
$\phi_j = \exp[i(\mathbf{p}_j \cdot \mathbf{r}_j - E_j t)]$ interact through a potential $V = V(|\mathbf{r}_1 - \mathbf{r}_2|)$. The probability of scattering to a state with momenta and energies $(\mathbf{p}_j', E_j')$ is determined by the modulus squared of the

---

† Planck's constant h has been set equal to 2π.

quantity $M = \int_{-\infty}^{\infty}\int_{-\infty}^{\infty}\int_{-\infty}^{\infty} \psi_f^* V \psi_i \, dr_1 \, dr_2 \, dt$. The initial state $\psi_i = \phi_1\phi_2$ and the final state $\psi_f = \phi_1'\phi_2'$.

(i) By writing $r_1 + r_2 = 2R$ and $r_1 - r_2 = r$ and assuming $dr_1 \, dr_2 =$ constant $\times$ $(dR \, dr)$, show that M can be written as the product of three 1-variable integrals.

(ii) From two of the integrals deduce energy and momentum conservation in the form of δ-functions.

(iii) Show that M is proportional to the Fourier transform of V, i.e. $\tilde{V}(\mathbf{k})$ where $2\mathbf{k} = (\mathbf{p}_2 - \mathbf{p}_1) - (\mathbf{p}_2' - \mathbf{p}_1')$.

8. In the previous example, for some ion–atom scattering, V may be approximated by $V = |r_1 - r_2|^{-1} \exp(-\mu|r_1 - r_2|)$. Show that the probability of scattering is proportional to $(\mu^2 + k^2)^{-2}$ where $k = |\mathbf{k}|$. [Use the direction of $\mathbf{k}$ as the polar axis for the $\mathbf{r}$-integration.]

9. [For those who have access to a computer.] Evaluate the series of (8.24) for a number of values of x (say 20 values in $0 \leqslant x \leqslant \pi$) and for an increasing number of terms, so as to demonstrate the Gibbs phenomenon discussed in section 8.2.

9
Partial differential equations

In this and the following chapter the solution of differential equations of the types typically encountered in physical science and engineering is extended to situations involving more than one independent variable. Only linear equations will be considered.

The most commonly occurring independent variables are those describing position and time, and we will couch our discussion and examples in notations which suggest these variables; of course the methods are not restricted to such cases. To this end we will, in discussing partial differential equations (p.d.e.), use the symbols u, v, w, for the dependent variables and reserve x, y, z, t for independent ones. For reasons explained in the preface, the partial equations will be written out in full rather than employ a suffix notation, but unless they have a particular significance at any point in the development, the arguments of the dependent variables, once established, will be generally omitted.

As in other chapters we will concentrate most of our attention on second-order equations since these are the ones which arise most often in physical situations. The solution of first-order p.d.e. will necessarily be involved in treating these, and some of the methods discussed can be extended without difficulty to third- and higher-order equations.

The method of 'separation of variables' has been placed in a separate chapter, but the division is rather arbitrary and has really only been made because of the general usefulness of that method. It will be readily apparent that some of the results of the present chapter are in fact solutions in the form of separated variables, but arrived at by a different approach.

9.1 Arbitrary functions

As with ordinary differential equations (chapter 5), it is instructive to study how p.d.e. may be formed from a potential set of solutions. Such study can provide an indication of how equations obtained, not from potential solutions but from physical arguments, might be solved.

For definiteness let us suppose we have a group of functions $\{u_i(x, y)\}$ involving two independent variables x and y. Without further specifica-

tion this is of course a very wide set of functions and we could not expect
to obtain a useful equation which they all satisfy. However among these
functions are those in which x and y appear in a particular way; either
the function is a simple one, or it can be written as a function [however
complicated] of a single variable p which is itself a simple function of
x and y.

Let us illustrate this latter case by considering the three functions of
x and y

$$\text{(i) } u_1(x, y) = x^4 + 4(x^2y + y^2 + 1),$$

$$\text{(ii) } u_2(x, y) = \sin x^2 \cos 2y + \cos x^2 \sin 2y,$$

$$\text{(iii) } u_3(x, y) = \frac{x^2 + 2y + 2}{3x^2 + 6y + 5}. \tag{9.1}$$

All are fairly complicated functions of x and y and a single differential
equation of which each one is a solution is not obvious. However, if we
observe that in fact each is a function of a variable p given by $p = x^2 + 2y$,
then a great simplification takes place. [It is not suggested that the existence and form of p are apparent from the equations – the examples are
only given in order that possible methods of solving future given p.d.e.
can be indicated.]

Written in terms of p the equations become,

$$\text{(i) } u_1(x, y) = (x^2 + 2y)^2 + 4 = p^2 + 4 = f_1(p),$$

$$\text{(ii) } u_2(x, y) = \sin (x^2 + 2y) = \sin p = f_2(p),$$

$$\text{(iii) } u_3(x, y) = \frac{(x^2 + 2y) + 2}{3(x^2 + 2y) + 5} = \frac{p + 2}{3p + 5} = f_3(p). \tag{9.2}$$

Now let us form, for each u_i, the partial differentials $\partial u_i/\partial x$ and $\partial u_i/\partial y$.
In each case these are (writing both the form for general p and the one
appropriate to our particular value of $p = x^2 + 2y$)

$$\frac{\partial u_i}{\partial x} = \frac{df_i(p)}{dp} \frac{\partial p}{\partial x} = 2xf_i', \quad i = 1, 2, \text{ or } 3,$$

$$\frac{\partial u_i}{\partial y} = \frac{df_i(p)}{dp} \frac{\partial p}{\partial y} = 2f_i', \quad i = 1, 2, \text{ or } 3. \tag{9.3}$$

From these all reference to the form of the f_i can be eliminated thus,

$$\frac{\partial p}{\partial y} \frac{\partial u_i}{\partial x} = \frac{\partial p}{\partial x} \frac{\partial u_i}{\partial y}, \tag{9.4}$$

or, for the specific form of p,

$$\frac{\partial u_1}{\partial x} = x \frac{\partial u_1}{\partial y}. \tag{9.5}$$

It is thus apparent that not only the three functions u_1, u_2, u_3 of (9.1) are solutions of p.d.e. (9.5), but so also is *any arbitrary function* $f(p)$ in which its argument p is given the value $x^2 + 2y$.

▶1. Verify by direct differentiation with respect to x and y that u_1, u_2, u_3 are solutions of (9.5).

▶2. For the following functions,

(i) $x^2(x^2 - 4) + 4y(x^2 - 2) + 4(y^2 - 1)$,

(ii) $x^4 + 2x^2y + y^2$,

(iii) $\dfrac{x^4 + 4x^2y + 4y^2 + 4}{2x^4 + x^2(8y + 1) + 8y^2 + 2y}$,

(a) determine whether they can be written as functions of $p = x^2 + 2y$ only,
(b) determine whether they satisfy p.d.e. (9.5) by direct substitution.

In summary, the first order p.d.e. (9.5) has as a solution *any* function of the variable $x^2 + 2y$; this points the way for the solution of p.d.e. of other orders as follows. It is *not* generally true that an nth order p.d.e. can always be considered as resulting from the elimination of n arbitrary *functions* from its solution. However, given specific equations, we may try to solve them by seeking combinations of variables in terms of which the solutions may be expressed as arbitrary functions. Where this is possible we may expect n combinations to be involved in the solution.

Now let us start afresh and suppose we are presented with a first-order p.d.e. as in (9.5), namely

$$\frac{\partial u}{\partial x} - x \frac{\partial u}{\partial y} = 0. \tag{9.6}$$

As a possible solution try a form $u(x, y) = f(p)$ where p is some, at present unknown, combination of x and y. Then, just as in (9.3), $\partial u/\partial x = f'(p)(\partial p/\partial x)$ and $\partial u/\partial y = f'(p)(\partial p/\partial y)$, or, on substitution into (9.6),

$$\frac{\partial p}{\partial x} - x \frac{\partial p}{\partial y} = 0. \tag{9.7}$$

But now consider also the necessary condition for $f(p)$ to remain constant as x and y vary; this is that p remains constant. In mathematical form therefore, $f = $ constant implies that x and y vary in such a way that

$$\mathrm{d}p = \frac{\partial p}{\partial x}\,\mathrm{d}x + \frac{\partial p}{\partial y}\,\mathrm{d}y = 0. \tag{9.8}$$

Now the forms of (9.7) and (9.8) are very alike, and become the same if we require that

$$\frac{\mathrm{d}x}{1} = \frac{\mathrm{d}y}{-x},$$

or, on integration, that

$$-\tfrac{1}{2}x^2 = y + c. \tag{9.9}$$

Identifying the constant of integration c with the constant $-\tfrac{1}{2}p$, we conclude in a deductive way that if x and y satisfy

$$p = x^2 + 2y \tag{9.10}$$

(but are otherwise arbitrary), then $u(x, y) = f(p)$ is a constant.

Since (9.7) was obtained directly from (9.6), and condition (9.10) makes (9.7) and (9.8) identical, we conclude that any $u(x, y)$ which can be expressed in this way, as $f(p)$ with p given by (9.10), must also be a solution of (9.6).

Naturally, the exact functional form of f for any particular situation must be determined by some set of boundary conditions. For complete determination the conditions will have to take a form equivalent to specifying $u(x, y)$ along a suitable continuum of points in the xy-plane [usually along a line].

Finally, to go round the cycle once more, we can confirm this form of solution by changing to a new variable

$$\xi = x^2 + 2y \tag{9.11}$$

when (9.6) becomes

► 3. $$2x\,\frac{\partial u}{\partial \xi} - 2x\,\frac{\partial u}{\partial \xi} = 0,$$

i.e. an identity – again showing that u is an arbitrary function of ξ.

9.2   General solutions and boundary conditions

Although most of the p.d.e. we will wish to consider in a physical context will be of second order (i.e. containing $\partial^2 u/\partial x^2$ or $\partial^2 u/\partial x\,\partial y$, etc.), we will use first-order equations to illustrate the general considerations involved in the form of solution and in the matching of given boundary conditions. The algebra can then be kept very elementary without losing any of the essential features. A simple example will first be worked in order to give a concrete illustration for the subsequent discussion.

Example 9.1. Find a solution of

$$x \frac{\partial u}{\partial x} - 2y \frac{\partial u}{\partial y} = 0, \tag{9.12}$$

(a) which takes the value $2y + 1$ on the line $x = 1$, and (b) which has value 4 at the point (1, 1) [separately].

Comparing the given equation with that of (9.6) and following the same arguments as in (9.7)–(9.9), we see that $u(x, y)$ will be constant along lines of x, y which satisfy

$$\frac{dx}{x} = \frac{dy}{-2y},$$

i.e. $c = x^2 y$.

Thus the general solution of the given equation is an arbitrary function of x^2y and the boundary value condition (a) shows that the particular function required is

$$u(x, y) = 2(x^2y) + 1 = 2x^2y + 1. \tag{9.13}$$

For boundary condition (b) some obvious acceptable solutions are

$$u(x, y) = x^2y + 3, \tag{9.14 a}$$
$$u(x, y) = 4x^2y, \tag{9.14 b}$$
$$u(x, y) = 4. \tag{9.14 c}$$

Each is a valid solution [the freedom of choice of form arising from the fact that u is specified only at one point (1, 1), and not (say) along a continuum as in boundary condition (a)] and all three are particular examples of the general solution which may be written, for example,

$$u(x, y) = x^2y + 3 + g(x^2y), \tag{9.15}$$

where $g = g(p)$ is an arbitrary function subject only to $g(1) = 0$. For example, the forms of g corresponding to the particular solutions (9.14 a–c) are $g(p) = 0$, $g(p) = 3p - 3$, $g(p) = 1 - p$.

▶4. Find the forms of g corresponding to the three solutions of (9.14), if the general form is written as

(i) $4 + g(x^2y)$,   (ii) $x^4y^2 + g(x^2y)$.

Let us now discuss these particular results in a more general form. It is clear that so far as the equation is concerned, if $u(x, y)$ is a solution then so is any multiple of $u(x, y)$ or any linear sum of separate solutions $u_1(x, y) + u_2(x, y)$. However, when it comes to fitting the boundary conditions this is not so – for example, although $u(x, y)$ of (9.13) satisfies (9.12) and takes the value $2y + 1$ on the line $x = 1$, $u_1(x, y) = 4u(x, y) = 8x^2y + 4$, whilst satisfying the p.d.e., has the value $8y + 4$ on $x = 1$. Likewise $u_2(x, y) = u(x, y) + f_1(x^2y)$, for arbitrary f_1, satisfies (9.12) but takes the value $2y + 1 + f_1(y)$ on the line $x = 1$ and this is not the required form unless f_1 is identically zero.

Thus we see that when treating the superposition of solutions of p.d.e., two considerations arise, one concerning the equation itself, the other connected with the boundary conditions. The *equation* is said to be *homogeneous* if '$u(x, y)$ is a solution implies that $\lambda u(x, y)$, for any constant λ, is also a solution'. The *problem* is said to be homogeneous if, in addition, the boundary conditions are such that if they are satisfied by $u(x, y)$ then they are also satisfied by $\lambda u(x, y)$. This last requirement by itself is referred to as *homogeneous boundary conditions*.

As examples, (9.6) is homogeneous, but

$$\frac{\partial u}{\partial x} - x \frac{\partial u}{\partial y} + au = g(x, y) \tag{9.16}$$

would not be unless $g(x, y)$ were zero. Boundary condition (*a*) of example 9.1 is not homogeneous, but a boundary condition

$$u(x, y) = 0 \text{ on the line } y = 4x^{-2} \tag{9.17}$$

would be, since

$$u(x, y) = \lambda(x^2y - 4) \tag{9.18}$$

satisfies (9.17) for any λ [and being a function of x^2y satisfies (9.12)].

The reason for discussing the homogeneity of the equation and its boundary conditions is that there is a close parallel in linear p.d.e. of the complementary function and the particular integral property of ordinary differential equations. This is that the general solution of the inhomogeneous problem can be written as the sum of *any* particular solution of the problem and the general solution of the corresponding homogeneous problem. Thus, for example, the general solution of (9.16) subject to, say, the boundary condition $u(x, y) = h(y)$ on the line $x = 0$, is given by

$$u(x, y) = v(x, y) + w(x, y), \tag{9.19}$$

where $v(x, y)$ is any solution (however simple) satisfying (9.16) and such that $v(0, y) = h(y)$, and $w(x, y)$ is the general solution of

$$\frac{\partial w}{\partial x} - x \frac{\partial w}{\partial y} + aw = 0, \qquad (9.20)$$

with $w(0, y) = 0$. If the boundary conditions are sufficiently completely specified then the only possible solution of (9.20) will be $w(x, y) \equiv 0$ and $v(x, y)$ will be the complete solution by itself as in part (a) of the example already worked.

▶5. Find (the most general) solutions $u(x, y)$ of the following equations consistent with the boundary values given.

(i) $y \dfrac{\partial u}{\partial x} - x \dfrac{\partial u}{\partial y} = 0 : u(x, 0) = 1 + \sin x.$

(ii) $i \dfrac{\partial u}{\partial x} = 3 \dfrac{\partial u}{\partial y} : u = (4 + 3i)x^2$ on the line $x = y.$

(iii) $\sin x \sin y \dfrac{\partial u}{\partial x} + \cos x \cos y \dfrac{\partial u}{\partial y} = 0:$
$$u = \cos 2y \text{ on } x + y = \tfrac{1}{2}\pi.$$

(iv) $\dfrac{\partial u}{\partial x} + 2x \dfrac{\partial u}{\partial y} = 0 : u = 2$ on parabola $y = x^2.$

(v) $y \dfrac{\partial u}{\partial x} - x \dfrac{\partial u}{\partial y} = 3x : (a)\ u = x^2$ on the line $y = 0,$
$$(b)\ u = 2 \text{ at the point } (1, 0).$$

(vi) $y^2 \dfrac{\partial u}{\partial x} + x^2 \dfrac{\partial u}{\partial y} = x^2 y^2(x^3 + y^3):$

9.3   Second-order equations

The class of differential equations to be considered next, that of second-order linear p.d.e., contains most of the important partial equations of physics. Just to gain familiarity with their general form, some of the more important ones will now be briefly discussed.

As indicated at the beginning of the chapter, commonly occurring independent variables are those of position and time, and equations governing oscillations and waves are ready examples of the type to be considered.

The transverse vibrations of a uniform string held under a uniform tension T satisfy the equation

$$\frac{\partial^2 u}{\partial x^2} - \frac{\rho}{T} \frac{\partial^2 u}{\partial t^2} = -\frac{f(x, t)}{T}, \qquad (9.21\,a)$$

where ρ is the mass per unit length of the string and $f(x, t)$ is the external transverse force per unit length acting on the string at time t. This equation is second-order (since the highest derivative present is the second), but unless $f(x, t)$ is zero, it is not homogeneous. As is probably familiar to the student, the quantity T/ρ gives the square of the velocity of propagation c of a transverse disturbance on a string when no external force acts. In this circumstance the vibrations satisfy

$$\frac{\partial^2 u}{\partial x^2} = \frac{\rho}{T} \frac{\partial^2 u}{\partial t^2} = \frac{1}{c^2} \frac{\partial^2 u}{\partial t^2}. \tag{9.21 b}$$

Free longitudinal vibrations of an elastic rod obey a very similar equation,

$$\frac{\partial^2 u}{\partial x^2} = \frac{\rho}{E} \frac{\partial^2 u}{\partial t^2} = \frac{1}{c^2} \frac{\partial^2 u}{\partial t^2}. \tag{9.22}$$

The only difference is that here ρ is the normal density (mass per unit volume) and E is Young's modulus.

Closely similar examples, but involving two or three spatial dimensions rather than one, are provided by the second-order equations governing the vibrations of a stretched membrane subject to an external force density $f(x, y, t)$,

$$T\left(\frac{\partial^2 u}{\partial x^2} + \frac{\partial^2 u}{\partial y^2}\right) + f(x, y, t) = \rho(x, y) \frac{\partial^2 u}{\partial t^2} \tag{9.23}$$

(with obvious meanings for ρ and T), and by Schrodinger's equation for a non-relativistic quantum mechanical particle in a potential $V(x, y, z)$,

$$-\frac{\hbar^2}{2m_0}\left(\frac{\partial^2 u}{\partial x^2} + \frac{\partial^2 u}{\partial y^2} + \frac{\partial^2 u}{\partial z^2}\right) + V(x, y, z)u = i\hbar \frac{\partial u}{\partial t}. \tag{9.24}$$

The function $u = u(x, y, z)$ in this latter equation is more usually denoted by ψ the quantum mechanical wave function; $\hbar$ is Planck's constant divided by 2π and m_0 is the mass of the particle. This equation contains four independent variables x, y, z and t, being second-order in the three spatial ones and first-order in the time t. Further, the equation is homogeneous, since if $u(x, y, z, t)$ is a solution then so is λu for any constant λ.

An important equation of the same general form as Schrodinger's but occurring in the classical theory of flow processes is the heat diffusion equation (or with a change of interpretation of the physical parameters, the general diffusion equation). For a material of uniform thermal conductivity k, specific heat s and density ρ, the temperature u satisfies

$$k\left(\frac{\partial^2 u}{\partial x^2} + \frac{\partial^2 u}{\partial y^2} + \frac{\partial^2 u}{\partial z^2}\right) + f(x, y, z, t) = s\rho \frac{\partial u}{\partial t}. \tag{9.25}$$

[For the more general case the first term is div $(k\nabla u)$ and $s\rho$ may depend on x, y, z.] The second term $f(x, y, z, t)$, representing the density of heat sources, is often missing in practical applications, whilst in the simplest cases the heat flow is one-dimensional resulting in the two-variable homogeneous p.d.e.

$$\frac{\partial^2 u}{\partial x^2} = \frac{s\rho}{k}\frac{\partial u}{\partial t}. \qquad (9.26)$$

9.4 Equation and boundary condition types

For the solution of actual physical problems any of the above equations must be supplemented by boundary conditions, as we have seen in connection with first-order p.d.e. Since the equations are of second order we might expect that relevant boundary conditions would involve specifying u, or some of its first derivatives, or both, along a suitable set of boundaries bordering or enclosing the region (of independent variables) over which a solution is sought.

Three common types of boundary conditions occur and are associated with the names of Dirichlet, Neumann and Cauchy. They are:

Dirichlet. u is given at each point of the boundary.

Neumann. $\partial u/\partial n$, the normal component of the gradient of u, is given at each point of the boundary.

Cauchy. Both u and $\partial u/\partial n$ are given at each point of the boundary.

We will not go into details,† but merely note that whether the various types of boundary conditions are appropriate (in that they give a unique, sometimes to within a constant, well-defined solution) depends upon the type of second-order equation under consideration and on whether the region of solution is bounded by a closed or open curve [or surface if there are more than two independent variables]. A part of a closed boundary may be at infinity if conditions are imposed on u or $\partial u/\partial n$ there.

The classification of the types of second-order equation involving two independent variables is carried out as follows. Let the second-order linear p.d.e. be written as

$$A\frac{\partial^2 u}{\partial x^2} + 2B\frac{\partial^2 u}{\partial x\,\partial y} + C\frac{\partial^2 u}{\partial y^2} = \text{function of } x, y, \frac{\partial u}{\partial x}, \frac{\partial u}{\partial y}. \qquad (9.27)$$

† For a discussion the reader is referred, for example, to Morse and Feshbach, *Methods of theoretical physics, Part I* (McGraw-Hill, 1953) chapter 6.

Then the equation type is said to be,

(i)   *hyperbolic if $B^2 > AC$,*
(ii) *parabolic if $B^2 = AC$,*   (9.28)
(iii) *elliptic if $B^2 < AC$.*

This classification holds good even in cases where A, B and C are functions of x and y (and not restricted to be constants), but in these cases the equation may be of a different type in different parts of the region. Furthermore the equation type is not changed if a new pair of independent variables ξ and η, given by a one-to-one mapping with continuous first and second derivatives, is used instead of x and y (see example 4 of section 9.14).

It can be shown (for example in the reference given) that the appropriate boundary condition and equation type pairings are

Equation type	Boundary	Conditions
Hyperbolic	Open	Cauchy
Parabolic	Open	Dirichlet or Neumann
Elliptic	Closed	Dirichlet or Neumann.

As examples, the one-dimensional wave equation (9.21 b) is hyperbolic $[B^2 - AC = -(1)(-1/c^2) > 0]$, the diffusion equation (9.26) is parabolic $[B = C = 0]$, whilst Laplace's equation in two dimensions,

$$\frac{\partial^2 u}{\partial x^2} + \frac{\partial^2 u}{\partial y^2} = 0,$$ (9.29)

is clearly elliptic. Extensions can be made to cover equations containing more than two independent variables; a more detailed discussion is called for but we will content ourselves with stating that the equations just mentioned do not change their type when extended to two or three spatial dimensions. Thus the 3D-Laplace equation is elliptic and requires either Dirichlet or Neumann boundary conditions and a closed boundary which, as we have already noted, may be at infinity if the behaviour of u is specified there [most often u or $\partial u/\partial n \to 0$ at infinity].

▶6. Consider the following equations and situations in an entirely qualitative way and satisfy yourself in each case that the boundary curve in the xt- or $r\theta$-plane should be as given, and that the stated boundary conditions are appropriate.

(a) Equation (9.21 b) for transverse waves on a semi-infinite string the end of which is made to move in a prescribed way. Open, Cauchy.

(b) Equation (9.26) for one-dimensional heat diffusion in a bar of length L which is given an initial temperature distribution and then thermally isolated. Open, Dirichlet and Neumann.

(c) Laplace's equation in two-dimensional polars

$$\frac{\partial^2 u}{\partial r^2} + \frac{1}{r}\frac{\partial u}{\partial r} + \frac{1}{r^2}\frac{\partial^2 u}{\partial \theta^2} = 0,$$

for two long conducting concentric cylinders of radii a and b ($b > a$). Closed, Dirichlet and Neumann, but not both simultaneously on any part of the boundary.

9.5   Uniqueness theorem

Although we have merely stated the appropriate boundary types and conditions for the general case, one particular example is sufficiently important that a proof that a unique solution is obtained will now be given.

This is Poisson's equation in three dimensions,

$$\nabla^2 u = -\rho/\epsilon\epsilon_0, \tag{9.30}$$

(for the electrostatic case) with either Dirichlet or Neumann conditions on a closed boundary appropriate to such an elliptic equation. What will be shown is that (to within an unimportant constant) the solution of (9.30) is *unique* if either the potential u or its normal derivative is specified on all surfaces bounding a given region of space (including, if necessary, a hypothetical spherical surface of indefinitely large radius on which u or $\partial u/\partial n$ is prescribed to have an arbitrarily small value).

Stated somewhat more formally this is:

Uniqueness theorem. If u is real and its first and second partial derivatives are continuous in a region V and on its boundary S, and if $\nabla^2 u = f$ and either $u = g$ or $\partial u/\partial n = h$ on S, where f, g and h are prescribed functions, then u is unique (at least to within a constant).

To prove the theorem suppose that two solutions $u = u_1(x, y, z)$ and $u = u_2(x, y, z)$ both satisfy the conditions given above. Then we will show that $u_1 = u_2 +$ constant.

Denote the function $u_1 - u_2$ by v. Then since $\nabla^2 u_1 = f = \nabla^2 u_2$ in V and either $u_1 = g = u_2$ or $\partial u_1/\partial n = h = \partial u_2/\partial n$ on S we must have

$$\nabla^2 v = 0 \text{ in } V, \tag{9.31}$$

and,

$$\text{either } v = 0, \text{ or } \partial v/\partial n = 0 \text{ on } S. \tag{9.32}$$

Now if we use Green's first theorem (equation (4.12) of chapter 4) for the case where both scalar functions are taken as v we have

$$\int_S v \frac{\partial v}{\partial n} \, dS = \int_V \{v\nabla^2 v + (\nabla v)^2\} \, dV. \tag{9.33}$$

But either part of (9.32) makes the left-hand side vanish, whilst the first term on the right is zero because of (9.31). Thus we are left with

$$\int_V (\nabla v)^2 \, dV = 0.$$

Since $(\nabla v)^2$ can never be negative, this can only be satisfied if

$$\nabla v = 0$$

or v, and hence $u_1 - u_2$, is a constant in V.

If Dirichlet conditions are given, then $u_1 \equiv u_2$ on (some part of) S and hence everywhere in V. For Neumann conditions u_1 and u_2 can differ throughout V by an arbitrary (but unimportant) constant.

The importance of the Uniqueness theorem lies in the fact that it shows that if a solution to a Poisson (or Laplace) equation can be found which fits a given set of Dirichlet or Neumann conditions, by any means whatever, then that solution is the correct one, since only one exists.

This result is the mathematical justification for the method of treating some problems in electrostatics by what is called the '*method of images*'. We will only discuss briefly the most familiar and elementary of such problems (others will be found at the end of the chapter).

Example 9.2. Find the electrostatic potential associated with a charge e placed at a point P a distance h above an infinite earthed conducting plane.

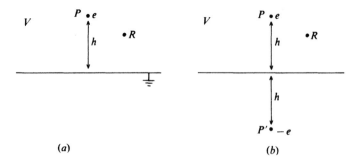

(a)   (b)

Fig. 9.1   A charge e placed at a distance h from an infinite earthed conducting plane: (a) the physical system; (b) the image charge system.

The physical situation is shown in fig. 9.1 (a). The potential u_1, due to the charge, at the general point R (with position $\mathbf{r}$) is straightforwardly

$$u_1 = \frac{e}{4\pi\epsilon_0|\mathbf{r} - \mathbf{r}_P|},$$

but the potential at R due to the induced charge on the plate is difficult to calculate.

However posed as a problem in solving

$$\nabla^2 u = -e \, \delta(\mathbf{r} - \mathbf{r}_P)/\epsilon_0 \text{ (above the plane)} \tag{9.34}$$

with the boundary conditions $u = 0$ on the plane and at infinity in the upper half-space, the solution can be obtained by another approach and justified by the Uniqueness theorem.

Consider the system of charges shown in fig. 9.1 (b); the charge $-e$ is called the image charge† of the original. In the region V above the plane, only the charge $+e$ is present and so (9.34) is still satisfied; furthermore the potential due to both charges $\to 0$ at infinity in the upper half-space [actually in all directions] and on the plane is zero everywhere, since at all points it receives equal and opposite contributions from the two equal and opposite charges.

The potential is in fact

$$u(\mathbf{r}) = \frac{e}{4\pi\epsilon_0|\mathbf{r} - \mathbf{r}_P|} - \frac{e}{4\pi\epsilon_0|\mathbf{r} - \mathbf{r}_{P'}|}, \tag{9.35}$$

and although it has been obtained by unusual methods, it does satisfy all the required boundary conditions as well as (9.34) in the appropriate region and must therefore by the Uniqueness theorem be the required solution.

Since the potential is unique, so also is the force field derived from it, so that the force experienced by an infinitesimal test charge anywhere in the upper half-space of the physical system is the same as it would experience in the presence of the original charge and its image [but of course with the plane ignored].

It should be noted that (9.35) in no way represents the correct potential below the plane in the physical situation of fig. 9.1 (a).

---

† The name 'method of images' is somewhat misleading since no actual method is involved. The appropriate set of image charges for any given set of conductors is obtained either from experience or in an *ad hoc* way. The only methodical aspect is the subsequent justification of the solution by means of the Uniqueness theorem.

9.6   Equations with constant coefficients

We now return to the problem of finding specific solutions to some types of second-order p.d.e. and begin by seeking solutions which are arbitrary functions of particular combinations of the independent variables† (just as in section 9.2 for first-order p.d.e.).

Clearly we can only hope to find such solutions if all the terms of the equation involve the same total number of differentiations, i.e. all terms are of the same order [although the number of differentiations with respect to the individual independent variables may be different]. For our present investigation this means that all terms are second order and that no terms in $\partial u/\partial x$, $\partial u/\partial y$ or u are present, and in addition the equation is homogeneous.

Both the wave equation

$$\frac{\partial^2 u}{\partial x^2} - \frac{1}{c^2}\frac{\partial^2 u}{\partial t^2} = 0, \tag{9.21 bis}$$

and the 2-D Laplace equation

$$\frac{\partial^2 u}{\partial x^2} + \frac{\partial^2 u}{\partial y^2} = 0, \tag{9.29 bis}$$

meet this requirement, but the diffusion equation

$$\frac{\partial^2 u}{\partial x^2} - \frac{s\rho}{k}\frac{\partial u}{\partial t} = 0, \tag{9.26 bis}$$

does not.

With both terms in each of (9.21) and (9.29) involving two differentiations, by assuming a solution of the form $u(x, y) = f(p)$, where p is some at present unknown function of x and y (or t), we may be able to obtain a common factor $f''(p)$ as the only appearance of f on the left-hand side. Then, because of the zero on the right-hand side, all reference to the form of f can be cancelled out.

We can gain some guidance as to suitable forms for the combination $p = p(x, y)$ by considering $\partial u/\partial x$ when u is given by $u(x, y) = f(p)$, for then

$$\frac{\partial u}{\partial x} = f'(p)\frac{\partial p}{\partial x}.$$

Clearly a second differentiation with respect to x (or y) is going to lead to a single term on the right-hand side containing f only as f'', only if the

---

† In order to keep the level of presentation elementary, we have deliberately omitted any explicit discussion of characteristics.

factor $\partial p/\partial x$ is a constant so that $\partial^2 p/\partial x^2$ (or $\partial^2 p/\partial y\, \partial x$) is necessarily zero. This shows that p must be a linear function of x. In an exactly similar way p must be a linear function of y.

It is now possible to obtain a general form of solution for the restricted type of second-order p.d.e. we have discussed, namely

$$A\frac{\partial^2 u}{\partial x^2} + 2B\frac{\partial^2 u}{\partial x\, \partial y} + C\frac{\partial^2 u}{\partial y^2} = 0, \tag{9.36}$$

A, B and C being constants. As in the preliminary discussion, we assume a solution

$$u(x, y) = f(p), \tag{9.37}$$

where p is a *linear* function of x and y,

$$p = ax + by. \tag{9.38}$$

Now proceeding to evaluate the terms ready for substitution in (9.36) we obtain

$$\frac{\partial u}{\partial x} = af'(p), \qquad \frac{\partial u}{\partial y} = bf'(p),$$

$$\frac{\partial^2 u}{\partial x^2} = a^2 f''(p), \qquad \frac{\partial^2 u}{\partial x\, \partial y} = abf''(p), \qquad \frac{\partial^2 u}{\partial y^2} = b^2 f''(p),$$

and on substituting,

$$(Aa^2 + 2Bab + Cb^2)f''(p) = 0. \tag{9.39}$$

This is the form we have been seeking, since now a solution independent of the form of f can be obtained if we require that a and b satisfy

$$Aa^2 + 2Bab + Cb^2 = 0. \tag{9.40}$$

From this quadratic, two values for the ratio [all that matters] of the two constants a and b are obtained,

$$b/a = [-B \pm (B^2 - AC)^{1/2}]/C. \tag{9.41}$$

If we denote these two ratios by λ_1 and λ_2, then *any* functions of the two variables

$$\begin{aligned} p_1 &= x + \lambda_1 y, \\ p_2 &= x + \lambda_2 y, \end{aligned} \tag{9.42}$$

will be solutions of the original equation (9.36). The omission of the constant factor a from p_1 and p_2 is of no consequence since this can always be absorbed into the particular form of any chosen function – only the relative weighting of x and y in p is important.

With p_1 and p_2 being different (in general), we can thus write the general solution of (9.36) as

$$u(x, y) = f_1(x + \lambda_1 y) + f_2(x + \lambda_2 y),\tag{9.43}$$

where f_1 and f_2 are arbitrary functions.

▶7. Show that the alternative solution $f''(p) = 0$ of (9.39) only leads to the trivial solution $u(x, y) = gx + hy + k$, for which all second derivatives are individually zero.

▶8. Show that the requirement that neither λ_1 nor λ_2 is a function of x or y (so as to maintain the linearity of p in x and y) means that the type of solution discussed is only appropriate to A, B and C all being constants (or all having the same dependence on x and y).
By considering

$$6x \frac{\partial^2 u}{\partial x^2} - 2y \frac{\partial^2 u}{\partial x\, \partial y} - (y + \tfrac{3}{2}x) \frac{\partial^2 u}{\partial y^2} = 0,$$

show that if only λ_1 were required to be a constant, this restriction would not apply.

Example 9.3. Find the general solution of the one-dimensional wave equation (9.21 b).

$$\frac{\partial^2 u}{\partial x^2} - \frac{1}{c^2} \frac{\partial^2 u}{\partial t^2} = 0.$$

This equation is that of (9.36) with $A = 1$, $B = 0$ and $C = -1/c^2$ and so the two values of λ_1 and λ_2 are, from (9.41), $-c$ and c. This means that any arbitrary functions of the quantities

$$p_1 = x - ct, \qquad p_2 = x + ct,$$

will be satisfactory solutions of (9.21 b), and that the general solution will be

$$u(x, t) = f_1(x - ct) + f_2(x + ct).\tag{9.44}$$

Since $u(x, t) = f_1(x - ct)$ represents the displacement of a string at time t and position x, it is clear that all positions x and times t for which '$x - ct$ is the same' will have the same instantaneous displacement. But '$x - ct$ is the same' is exactly the relation between the time and position of an observer travelling with velocity c along the positive x-direction. Consequently a moving observer sees a constant displacement of the string, i.e. to a stationary observer, the initial profile $u(x, 0)$ moves with velocity c along the x-axis as if it were a rigid system. Thus $f_1(x - ct)$ represents a wave form of constant shape travelling along the positive x-axis with velocity c, the actual form of the wave depending upon the function f_1.

Similarly the term $f_2(x + ct)$ is a constant wave form travelling with velocity c in the negative x-direction. The general solution (9.44) represents a superposition of these.

If the functions f_1 and f_2 are the same, then the complete solution (9.44) represents identical progressive waves going in opposite directions. This may result in a wave pattern whose profile does not progress and is described as a *standing wave*.

As a simple example, suppose both $f_1(p)$ and $f_2(p)$ have the form†

$$f_i(p) = A \cos (kp + \epsilon).$$

Then (9.44) can be written as

$$u(x, t) = A[\cos (kx - kct + \epsilon) + \cos (kx + kct + \epsilon)]$$
$$= 2A \cos (kx + \epsilon) \cos (kct).$$

The important thing to notice is that the shape of the wave pattern is the same at all times but that its amplitude depends upon the time. At some points x which satisfy

$$\cos (kx + \epsilon) = 0,$$

there is no displacement at any time. Such points are called *nodes*.

The type of solution obtained here is discussed further in chapter 10 on the method of separation of variables.

▶9. Verify by direct differentiation and substitution that (9.44) is indeed a solution of the wave equation (9.21 b).

Example 9.4. Find the general solution of the two-dimensional Laplace equation (9.29),

$$\frac{\partial^2 u}{\partial x^2} + \frac{\partial^2 u}{\partial y^2} = 0.$$

Following the established procedure, we look for a solution which is a function $f(p)$ of $p = x + \lambda y$. Substituting this we obtain directly (or by means of (9.41)) that

$$1 + \lambda^2 = 0.$$

† In the usual notation, k is the wave number ($= 2\pi/$wavelength) and $kc = \omega$, the angular frequency of the wave.

This requires that $\lambda = \pm i$, and thus the satisfactory variables p are

$$p = x \pm iy.$$

The general solution required is therefore, in terms of arbitrary functions f_1 and f_2,

$$u(x, y) = f_1(x + iy) + f_2(x - iy). \tag{9.45}$$

It will be apparent from these two worked examples that the nature of the appropriate linear combination of x and y depends upon whether or not $B^2 > AC$. This is exactly the same criterion as determines whether the p.d.e. is hyperbolic or elliptic. Hence, as a general result, hyperbolic and elliptic equations of the form (9.36) have as solutions functions whose arguments have the form $x + \alpha y$ and $x + i\beta y$ respectively, where α and β themselves are real.

The one case not taken care of by this result is that in which $B^2 = AC$, i.e. a parabolic equation. In this case λ_1 and λ_2 are not different and only one suitable combination of x and y results,

$$u(x, y) = f_1[x - (B/C)y]. \tag{9.46}$$

To find the second part of the general solution we try [in analogy with the corresponding situation in ordinary differential equations] a solution of the form

$$u(x, y) = g(x, y)f_2[x - (B/C)y].$$

Substituting this and using $A = B^2/C$ results in

▶10. $\left(A \dfrac{\partial^2 g}{\partial x^2} + 2B \dfrac{\partial^2 g}{\partial x\, \partial y} + C \dfrac{\partial^2 g}{\partial y^2} \right) f_2 = 0.$ (9.47)

There are several simple solutions of this equation for g, but as only one is required we take the simplest non-trivial one

$$g(x, y) = x,$$

to give as the general solution of the parabolic equation

$$u(x, y) = f_1[x - (B/C)y] + xf_2[x - (B/C)y]. \tag{9.48}$$

▶11. Convince yourself that alternative solutions of (9.47), e.g. $g(x, y) = y$, do not lead to a solution which is not represented by (9.48).

To complete this section we will give alternative derivations of the general solutions (9.43) and (9.48) by expressing the original p.d.e. in terms of new variables before solving it. The actual solution will then become almost trivial; but, of course, it will be recognized that suitable new variables could hardly have been guessed if it were not for the work already done. This does not detract from the validity of the derivation to be described, only from the likelihood that it would be discovered by inspection.

We start again with (9.36) and change to new variables

$$\xi = x + \lambda_1 y, \qquad \eta = x + \lambda_2 y. \tag{9.49}$$

With this change of variable, we can replace the differentiations in (9.36) by

$$\frac{\partial}{\partial x} = \frac{\partial}{\partial \xi} + \frac{\partial}{\partial \eta},$$

$$\frac{\partial}{\partial y} = \lambda_1 \frac{\partial}{\partial \xi} + \lambda_2 \frac{\partial}{\partial \eta}. \tag{9.50}$$

When this is carried out, and the fact that

$$A + 2B\lambda_i + C\lambda_i^2 = 0, \quad i = 1, 2, \tag{9.51}$$

is used, (9.36) becomes

▶12. $$[2A + 2B(\lambda_1 + \lambda_2) + 2C\lambda_1\lambda_2] \frac{\partial^2 u}{\partial \xi \, \partial \eta} = 0. \tag{9.52}$$

Then, providing the factor in square brackets does not vanish, we obtain

$$\frac{\partial^2 u}{\partial \xi \, \partial \eta} = 0, \tag{9.53}$$

which has as its successive integrals

$$\frac{\partial u}{\partial \eta} = F(\eta) \tag{9.54}$$

and

$$u(\xi, \eta) = f(\eta) + g(\xi). \tag{9.55}$$

This is just the same as (9.43)

$$u(x, y) = g(x + \lambda_1 y) + f(x + \lambda_2 y). \tag{9.56}$$

▶13. Verify that the condition that the factor in square brackets of (9.52) does not vanish is that $B^2 \neq AC$, i.e. that the original equation is not parabolic.

If the equation is parabolic ($B^2 = AC$), instead of (9.49) we use new variables

$$\xi = x + \lambda y, \qquad \eta = x, \tag{9.57}$$

and (9.36) reduces to [recall $\lambda = -(B/C)$]

▶14. $$A \frac{\partial^2 u}{\partial \eta^2} = 0. \tag{9.58}$$

Two straightforward integrations give as the general solution of (9.58)

$$u(\xi, \eta) = \eta f(\xi) + g(\xi), \tag{9.59}$$

which in terms of x and y has exactly the form of (9.48)

$$u(x, y) = xf(x + \lambda y) + g(x + \lambda y). \tag{9.60}$$

This concludes the alternative and independent derivations of the general solution of homogeneous p.d.e. with constant coefficients of the type given by (9.36).

Example 9.5. $u(x, y)$ satisfies

$$\frac{\partial^2 u}{\partial x^2} - 3 \frac{\partial^2 u}{\partial x \, \partial y} + 2 \frac{\partial^2 u}{\partial y^2} = 0.$$

Find the value of $u(0, 1)$ if $u = -x^2$ and $\partial u/\partial y = 0$ for $y = 0$ and all x.

From our general result, functions of $p = x + \lambda y$ will be solutions provided that

$$1 - 3\lambda + 2\lambda^2 = 0,$$

i.e. $\lambda = 1$ or $\tfrac{1}{2}$.

The general solution is thus

$$u(x, y) = f(x + y) + g(x + \tfrac{1}{2}y). \tag{9.61}$$

The boundary condition $u(x, y) = -x^2$ on $y = 0$ implies

$$-p^2 = f(p) + g(p), \quad \text{for all } p, \tag{9.62}$$

whilst $\partial u/\partial y = 0$ when $y = 0$ yields

$$0 = f'(p) + \tfrac{1}{2}g'(p), \quad \text{for all } p. \tag{9.63}$$

Differentiating (9.62) with respect to p and subtracting from (9.63) gives

$$2p = -\tfrac{1}{2}g'(p),$$

or that

$$g(p) = -2p^2 + k.$$

Equation (9.62) then gives that

$$f(p) = p^2 - k$$

and thus that

$$\begin{aligned} u(x, y) &= f(x + y) + g(x + \tfrac{1}{2}y) \\ &= (x + y)^2 - k - 2(x + \tfrac{1}{2}y)^2 + k \\ &= -x^2 + \tfrac{1}{2}y^2. \end{aligned}$$

Finally at the point $(0, 1)$,

$$u(0, 1) = \tfrac{1}{2}.$$

►15. Solve example 9.5 if the boundary condition is $u = \partial u/\partial y = 1$, when $y = 0$ (for all x).

►16. Solve

$$\frac{\partial^2 u}{\partial x^2} + 2\frac{\partial^2 u}{\partial x \, \partial y} + \frac{\partial^2 u}{\partial y^2} = 0,$$

subject to $u = 0$ when $x = 0$, and $u = x^2$ when $y = 1$.

9.7   The wave equation

We have already found the general solution of the one-dimensional wave equation as (9.44)

$$u(x, t) = f(x - ct) + F(x + ct), \tag{9.64}$$

with f and F arbitrary functions. However, the equation is of such general importance that further discussion will not be out of place. Let us see how to obtain its solution if we are given initial conditions (boundary conditions) in the general form:

initial displacement $u(x, 0) = \phi(x),$ (9.65 a)

initial velocity $\dfrac{\partial u\,(x, 0)}{\partial t} = \psi(x),$ (9.65 b)

for all parts of a string whose transverse displacement is $u(x, t)$. The functions $\phi(x)$ and $\psi(x)$ are given and describe the displacement and velocity of each part of the string at the arbitrary time $t = 0$.

It is clear that what is needed are the particular forms of the functions

f and F in (9.64) which takes the values given by (9.65) at $t = 0$. This means that

$$\phi(x) = u(x, 0) = f(x - 0) + F(x + 0) \tag{9.66 a}$$

and $\quad \psi(x) = \dfrac{\partial u(x, 0)}{\partial t} = -cf'(x - 0) + cF'(x + 0). \tag{9.66 b}$

It should be noted that $f'(x - 0)$ stands for $df(p)/dp$ evaluated, after the differentiation, at $p = x - c \times 0$ and likewise with $F'(x + 0)$.

Looking on the two left-hand sides as functions of $p = x \pm ct$, but everywhere evaluated at $t = 0$, we may integrate (9.66 b) between an arbitrary [and irrelevant] lower limit p_0 and an indefinite upper limit p and obtain

$$\frac{1}{c} \int_{p_0}^{p} \psi(q) \, dq + K = -f(p) + F(p), \tag{9.67}$$

the constant K depending upon p_0.

Adding and subtracting (9.66 a) [with x replaced by p] and (9.67), we can establish the forms of the functions f and F as

$$f(p) = \frac{\phi(p)}{2} - \frac{1}{2c} \int_{p_0}^{p} \psi(q) \, dq - \frac{K}{2} \tag{9.68 a}$$

and $\quad F(p) = \dfrac{\phi(p)}{2} + \dfrac{1}{2c} \displaystyle\int_{p_0}^{p} \psi(q) \, dq + \dfrac{K}{2}. \tag{9.68 b}$

Adding (9.68 a), with $p = x - ct$, to (9.68 b), with $p = x + ct$, gives as the solution to the original problem (equation (9.64))

$$u(x, t) = \tfrac{1}{2}[\phi(x - ct) + \phi(x + ct)] + \frac{1}{2c} \int_{x - ct}^{x + ct} \psi(q) \, dq. \tag{9.69}$$

Notice that all dependence on p_0 has disappeared.

Each of the terms in (9.69) has a fairly straightforward physical interpretation. In each case the $\frac{1}{2}$ represents the fact that only half of a displacement profile which starts at any particular point on the string travels towards any other position x, the other half travelling away from it. The first term $\frac{1}{2}\phi(x - ct)$ arises from the initial displacement at a distance ct to the left (negative direction) of x which travels forward arriving at x at time t. Similarly the second contribution is due to the initial displacement at a distance ct to the right of x.

The interpretation of the final term is a little less obvious, but viewed in the form

$$\frac{1}{2} \int_{t' = x/c - t}^{t' = x/c + t} \text{(initial transverse velocity evaluated at a distance } ct' \text{ from } x) \, dt', \tag{9.70}$$

it clearly represents the accumulated transverse displacement at position x, due to the passage past x of all parts of the initial motion whose effects can reach x in a time t, both backward and forward travelling.

The extension to the three-dimensional wave equation of solutions of the type we have so far considered presents no serious difficulty. In Cartesian coordinates the 3-D wave equation is

$$\frac{\partial^2 u}{\partial x^2} + \frac{\partial^2 u}{\partial y^2} + \frac{\partial^2 u}{\partial z^2} - \frac{1}{c^2} \frac{\partial^2 u}{\partial t^2} = 0. \tag{9.71}$$

In close analogy with the 1-D case we try solutions which are functions of linear combinations of all four variables

$$p = lx + my + nz + \mu t. \tag{9.72}$$

It is clear that a solution of (9.71), $u(x, y, z, t) = f(p)$, will be acceptable provided that

$$(l^2 + m^2 + n^2 - \mu^2/c^2) f''(p) = 0. \tag{9.73}$$

Thus, as in the 1-D case, f can be arbitrary provided that

$$l^2 + m^2 + n^2 = \mu^2/c^2.$$

Using the obvious normalization, we take $\mu = \pm c$ and l, m, n three numbers such that

$$l^2 + m^2 + n^2 = 1.$$

In other words (l, m, n) are the Cartesian components of a unit vector $\hat{n}$, which points along the direction of propagation of the wave. The quantity p can be written in terms of vectors as the scalar expression $p = \hat{n} \cdot \mathbf{r} \pm ct$, and the general solution of (9.71) as

$$u(x, y, z, t) = u(\mathbf{r}, t) = f_1(\hat{n} \cdot \mathbf{r} - ct) + f_2(\hat{n} \cdot \mathbf{r} + ct), \tag{9.74}$$

where $\hat{n}$ is *any* unit vector. It would perhaps be more transparent to write $\hat{n}$ explicitly as one of the arguments of u.

9.8   Particular integrals

As was discussed in section 9.2, in order to obtain the general solution of a p.d.e. a particular integral is needed. [For the homogeneous problem –

equation and boundary conditions – it is formally a solution which is identically zero.]

We will not discuss at length general methods for obtaining such particular integrals, but merely note that some of those available for ordinary differential equations can be suitably extended,† and give an example using D-operator methods.

Example 9.6. Find the general solution of the equation

$$\frac{\partial^2 u}{\partial x^2} + \frac{\partial^2 u}{\partial y^2} = 6(x + y). \tag{9.75}$$

Following our previous methods and results, the complementary function is $u(x, y) = f_1(x + iy) + f_2(x - iy)$, and only a particular integral now remains to be found.

One solution [perfectly adequate in itself] is obvious by inspection, namely

$$u(x, y) = x^3 + y^3. \tag{9.76}$$

However we will pretend this has not been noticed and proceed by a formal D-operator method.

Let the operator D_x stand for $\partial/\partial x$ and D_y for $\partial/\partial y$. Then a solution of (9.75) can be written formally as

$$u(x, y) = \frac{1}{D_x^2 + D_y^2} 6(x + y).$$

Using methods analogous to those of chapter 5 this can be evaluated as follows [recall D^{-1} represents an integration]

$$u(x, y) = \frac{6}{D_x^2} [1 + (D_y^2/D_x^2)]^{-1}(x + y)$$

$$= 6 \left[\frac{1}{D_x^2} - \frac{D_y^2}{D_x^4} + \cdots \right](x + y)$$

$$= \frac{6}{D_x^2} (x + y) - 0 + 0 - \cdots$$

$$= \frac{3}{D_x} (x^2 + 2yx)$$

$$= x^3 + 3yx^2.$$

† See for example Piaggio, *Differential equations* (Bell, 1954), pp. 175 *et seq.*

Since we require only one particular solution, all arbitrary functions of y have been set equal to zero when integrating with respect to x.

Thus the general solution of (9.75) can be written

$$u(x, y) = f_1(x + iy) + f_2(x - iy) + x^3 + 3yx^2.$$

▶17. Show that this result is not in contradiction with the result (9.76) obtained by inspection; i.e. show that the difference between $x^3 + y^3$ and $x^3 + 3yx^2$ can be written in the form $g(x + iy) + h(x - iy)$.

9.9   The diffusion equation

One important class of second-order p.d.e. which we have not yet considered is that in which the second derivative with respect to one independent variable appears, but only the first derivative with respect to another [usually time]. This is exemplified by the one-dimensional diffusion equation of section 9.3

$$K\frac{\partial^2 u(x, t)}{\partial x^2} = \frac{\partial u}{\partial t}, \tag{9.77}$$

in which K is a constant with the dimensions of length^2 × time^{-1}. The physical constants which go to make up K in a particular case depend upon the nature of the process (e.g. solute diffusion, heat flow, etc.) and the material being described.

With (9.77) we cannot hope to successfully repeat the method of section 9.6, since now $u(x, t)$ is differentiated a different number of times on the two sides of the equation. Any attempted solution in the form $u(x, t) = f(p)$ with $p = ax + bt$, therefore, will only lead to an equation in which the form of f cannot be cancelled out. Clearly we must try other methods.

Solutions may be obtained using the standard method of separation of variables and this is discussed in the next chapter.

A simple solution is also given if both sides of (9.77), as it stands, are separately set equal to a constant, say α,

$$\frac{\partial^2 u}{\partial x^2} = \frac{\alpha}{K}, \tag{9.78 a}$$

$$\frac{\partial u}{\partial t} = \alpha. \tag{9.78 b}$$

These have general solutions

$$u = \frac{1}{2}\frac{\alpha}{K}x^2 + xg(t) + h(t)$$

and $u = \alpha t + k(x)$,

which may be made compatible with each other if $g(t)$ is taken as a constant (or zero), $h(t)$ taken as αt, and $k(x)$ as $(\alpha/2K)x^2 + gx$. An acceptable solution is thus

$$u(x, t) = \frac{1}{2} \frac{\alpha}{K} x^2 + gx + \alpha t + \text{constant}. \qquad (9.79)$$

For an example of the use of this form of solution see example 15 of section 9.14.

Let us now return to seeking solutions of equations by combining the independent variables in particular ways. Having seen that a linear combination of x and t will be of no value, we must search for other possible combinations. It has already been noted that K has the dimensions of length$^2 \times$ time^{-1} and so the combination of variables

$$\eta = \frac{x^2}{Kt}, \qquad (9.80)$$

will be dimensionless. Let us see if we can mathematically satisfy (9.77) with a solution $u(x, t)$ which has the form $u(x, t) = f(\eta)$ first, and leave its interpretation until afterwards.

Evaluating the necessary derivatives we have

$$\frac{\partial u}{\partial x} = f'(\eta) \frac{\partial \eta}{\partial x} = f'(\eta) \frac{2x}{Kt},$$

$$\frac{\partial^2 u}{\partial x^2} = \frac{2}{Kt} f'(\eta) + f''(\eta) \left(\frac{2x}{Kt} \right)^2,$$

$$\frac{\partial u}{\partial t} = -f'(\eta) \frac{x^2}{Kt^2}. \qquad (9.81)$$

Substituting from (9.81) into (9.77) we find the new equation can be written entirely in terms of η.

▶18. $4\eta f''(\eta) + (2 + \eta)f'(\eta) = 0. \qquad (9.82)$

This is a straightforward ordinary differential equation which can be integrated (with a minimum of explanation) as follows,

$$\frac{f''(\eta)}{f'(\eta)} = -\frac{1}{2\eta} - \frac{1}{4},$$

$$\ln [\eta^{1/2} f'(\eta)] = -\tfrac{1}{4}\eta + c,$$

$$f'(\eta) = \frac{A}{\eta^{1/2}} \exp{(-\tfrac{1}{4}\eta)},$$

$$f(\eta) = A \int_{\eta_0}^{\eta} \mu^{-1/2} \exp\left(-\tfrac{1}{4}\mu\right) d\mu, \tag{9.83}$$

with η given by (9.80).

If we now write this in terms of a slightly different variable

$$\xi = \frac{\eta^{1/2}}{2} = \frac{x}{2(Kt)^{1/2}}, \tag{9.84}$$

then $d\xi = \tfrac{1}{4}\eta^{-1/2} d\eta$, and (9.83) takes the form

$$u(x, t) = f(\eta) = g(\xi) = B \int_{\xi_0}^{\xi} \exp\left(-\nu^2\right) d\nu. \tag{9.85}$$

Here B is a constant and it should be noticed that x and t only appear on the right-hand side in the indefinite upper limit ξ, and then only in the combination $xt^{-1/2}$. This kind of function of x and t may be slightly unfamiliar but is in essence no different from (say) $\sin\left(x/K^{1/2}t^{1/2}\right)$. If ξ_0 is chosen as zero, it is, to within a constant factor,[†] the error function Erf $\left(x/2(Kt)^{1/2}\right)$, which is tabulated in many books of tables. Only non-negative values of x and t are to be considered here so $\xi \geq \xi_0$.

Next let us try to determine what kind of (say) temperature distribution and flow this represents. For definiteness we take $\xi_0 = 0$. Firstly since $u(x, t)$ in (9.85) only depends upon the product $xt^{-1/2}$, it is clear that all points x at times t such that $xt^{-1/2}$ have the same value, have the same temperature. Put another way, at any specific time t, the region having a particular temperature has moved along the positive x-axis a distance proportional to the square root of the time. This is. a typical *diffusion* process.

It will also be noticed that at $t = 0$, $\xi \to \infty$, and u becomes quite independent of x (except perhaps at $x = 0$); the solution then represents a uniform spatial temperature distribution. On the other hand at $x = 0$, $u(x, t)$ is identically $= 0$ for all t.

▶19. In order to get a feel for the temperature distribution described by (9.85), use axes as illustrated in fig. 9.2 and roughly sketch some of the contours of equal temperature. (Do not worry about the overall normalization.)

A calculation about a resistive cable involving this general kind of solution will be found in the exercises at the end of this chapter.

† Take $B = 2\pi^{-1/2}$ to give the usual error function normalized so that Erf $(\infty) = 1$.

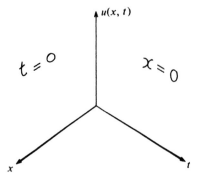

Fig. 9.2   Axes for exercise ✦19.

Example 9.7. An infrared laser delivers a pulse of energy (heat) E to a point P on a large insulated sheet of thickness b, thermal conductivity k, specific heat s and density ρ. The sheet is initially at a uniform temperature. If $u(r, t)$ is the excess temperature, a time t later, at a point a distance $r\,(\gg b)$ from P, then show (by substitution in the appropriate equation) that a suitable expression for u is

$$u(r, t) = \frac{\alpha}{t}\exp{(-r^2/2\beta t)} \tag{9.86}$$

where α and β are constants.

(i) Show that $\beta = 2k/s\rho$.
(ii) Show that, as expected, the total heat in the sheet is independent of time and hence evaluate α.
(iii) Show that the total heat which ultimately flows past any radius r is E.

The equation to be solved is the heat diffusion equation

$$k\nabla^2 u(\mathbf{r}, t) = s\rho\,\frac{\partial u(\mathbf{r}, t)}{\partial t}. \tag{9.87}$$

Since we only require the solution for $r \gg b$ we can treat the problem as two-dimensional with obvious circular symmetry. We thus need only the r-derivative term in the expression for ∇^2 giving

$$\frac{k}{r}\frac{\partial}{\partial r}\left(r\frac{\partial u}{\partial r}\right) = s\rho\,\frac{\partial u}{\partial t}, \tag{9.88}$$

where now $u(\mathbf{r}, t) = u(r, t)$.

(i) Substitution of the given expression (9.86) into (9.88) shows that it is a solution provided that

▶20. $\beta = 2k/s\rho$.

(ii) The total heat in the system at any time t is

$$\int_0^\infty 2\pi r \, dr \cdot b \cdot \rho s \cdot u(r, t) = 2\pi b \rho s \alpha \int_0^\infty \frac{r}{t} \exp\left(-r^2/2\beta t\right) dr$$

▶21. $= 2\pi b \rho s \alpha \beta$.

This is independent of t and must equal the total heat input E. Thus

$$\alpha = \frac{E}{2\pi b \rho s \beta} = \frac{E}{4\pi b k}.$$

(iii) The total heat flow past a radius r is

$$-\int_0^\infty k \frac{\partial u(r, t)}{\partial r} \cdot 2\pi r b \cdot dt = -\int_0^\infty 2\pi r b k \cdot \frac{E}{4\pi b k t} \cdot \frac{(-r)}{\beta t}$$
$$\times \exp\left(-r^2/2\beta t\right) \cdot dt$$

▶22. $= E[\exp\left(-r^2/2\beta t\right)]_0^\infty$

$$= E \text{ for all } r.$$

9.10   Superposing solutions

The solution of this last example and of the result (9.85) are worth a little further discussion. It will be clear from the way it was derived and from the results (ii) and (iii) of example 9.7, that (9.86) gives the temperature response of the sheet to a heat impulse of magnitude E at position $r = 0$ and time $t = 0$. In fact, in mathematical terms it is the response to a δ-function input, i.e. it is the solution of (see (9.25)),

$$k\nabla^2 u + E\,\delta(\mathbf{r})\,\delta(t) = s\rho\,\partial u/\partial t.$$

In the theory of ordinary differential equations, such a response function was called the *Green's function* and the same name is used here [although now we have two independent variables].

  The point of this observation is that, since the diffusion equation without internal heat sources is linear, the temperature distribution $u(\mathbf{r}, t)$ in space and time can be built up for an *arbitrary* energy input pattern by superposing the corresponding Green's function solutions. This is an exact parallel of the procedures of chapter 7. In practice, the labour of computing superposed solutions is generally rather large and makes the method unattractive.

Equation (9.85) does not give the response to a δ-function initial condition, but rather that to a 'step function' initial condition. As we have noted, it corresponds to the situation of a long one-dimensional conducting bar which is initially everywhere (all x) at the same temperature [$\propto B$], but one end of which ($x = 0$) is maintained permanently (all t) at zero temperature, $u(0, t) = 0$. The heat subsequently flows out of the bar at the $x = 0$ end and ultimately the temperature at all points of the bar falls to zero ($\xi \to 0$ as $t \to \infty$ for all $x > 0$).

It is clear that in principle any initial temperature distribution in the bar can be represented as a superposition of such step functions (using different lower limits ξ_0 corresponding to different 'lower ends' x_0 of each uniform temperature distribution). Like our other superposition method this approach is usually cumbersome in practice [the superposition becomes an integral for any smoothly varying temperature distribution] and we will not attempt to take it any further in this book.

▶23. Confirm that your sketched contours of ▶19 are in agreement with the physical description in the last paragraph but one.

▶24. Confirm that $u(x, t)$ given by (9.85) with $\xi_0 = 0$, $B = 2\pi^{-1/2}T$, and x no longer restricted to be positive, is the appropriate temperature distribution to describe the situation in which two semi-infinite conducting rods, initially at temperatures $+T$ and $-T$, are brought together, end to end, at time $t = 0$.

9.11   Laplace transform methods

To conclude our treatment of the diffusion equation for this chapter, we will demonstrate, by means of an example and a brief discussion, its solution using integral transform methods – in particular the Laplace transform.

Example 9.8. A semi-infinite tube of unit cross-section initially contains pure water. At time $t = 0$, one end of the tube is put into contact with a salt solution and maintained at a constant concentration u_0. Find the total amount of salt that has diffused into the tube at time t, if the diffusion constant is K.

The concentration $u(x, t)$ at time t and distance x from the end of the tube satisfies

$$K\frac{\partial^2 u}{\partial x^2} = \frac{\partial u}{\partial t}. \tag{9.89}$$

This equation has to be solved subject to the boundary conditions $u(0, t) = u_0$ for all t and $u(x, 0) = 0$ for all $x > 0$.

It will be recalled from chapter 5 that one of the major virtues of Laplace transformations is the possibility of replacing derivatives of functions by simple multiplication by a scalar. In our equation, if the derivative with respect to time were so removed, it would contain only differentiation with respect to a single variable; let us therefore take the Laplace transform of (9.89) with respect to t,

$$\int_0^\infty K \frac{\partial^2 u}{\partial x^2} \exp\left(-st\right) dt = \int_0^\infty \frac{\partial u}{\partial t} \exp\left(-st\right) dt.$$

Denoting the transform of $u(x, t)$ by $U(x, s)$ and using result (5.65) to evaluate the right-hand side, together with $u(x, 0) = 0$ [or by integrating directly by parts], we then obtain

$$K \frac{\partial^2 U}{\partial x^2} = sU. \tag{9.90}$$

Notice that, as indicated, the differentiation $\partial^2/\partial x^2$ can be brought outside the integral sign since the integration is with respect to t and does not affect x.

The solution of (9.90) is immediate

$$U(x, s) = A \exp\left[(s/K)^{1/2}x\right] + B \exp\left[-(s/K)^{1/2}x\right].$$

At $x = \infty$, $u(x, t)$ must $= 0$ for all finite t and so $U(\infty, s)$ must $= 0$ also; consequently we must require that $A = 0$. The value of B is determined by the need for $u(0, t) = u_0$ and hence that [see table 5.1, result 1] $U(0, s) = u_0 s^{-1}$.

We thus conclude that the appropriate expression for the Laplace transform of $u(x, t)$ is

$$U(x, s) = \frac{u_0}{s} \exp\left[-\left(\frac{s}{K}\right)^{1/2} x\right]. \tag{9.91}$$

To obtain $u(x, t)$ from this would require the inversion of this transform – a task which we have seen in chapter 5 is generally difficult and requires a contour integration. However in the present problem an alternative method is available.

Let $w(t)$ be the amount of salt that has diffused into the tube in time t, then

$$w(t) = \int_0^\infty u(x, t) \, dx,$$

and its transform

$$W(s) = \int_0^\infty \exp(-st)\,dt \int_0^\infty u(x, t)\,dx,$$

$$= \int_0^\infty dx \int_0^\infty u(x, t) \exp(-st)\,dt$$

$$= \int_0^\infty U(x, s)\,dx$$

▶ 25. $= u_0 K^{1/2} s^{-3/2}.$

Referring to result 11 of table 5.1, this shows that $w(t) = 2(K/\pi)^{1/2} \times u_0 t^{1/2}$, which is thus the required expression for the amount of diffused salt at time t.

This example shows that in some circumstances the use of a Laplace transformation can greatly simplify the solution of a p.d.e. However it will have been observed that (as with ordinary d.e.) the easy elimination of some derivatives is usually paid for by the introduction of a difficult inverse transformation. It need hardly be pointed out that the boundary conditions must be capable of being incorporated into the transform in a manageable way.

9.12   The telegraphy equation

So far we have considered only equations in which there is a single dependent variable, but in the theory of transmission lines, for example, two dependent variables are present, the voltage and the current, both being functions of position and time.

Suppose a line has resistance R, inductance L, capacitance C, and leakage conductance G, all per unit length. We may construct the (coupled) p.d.e. satisfied by the voltage $v(x, t)$ and current $i(x, t)$, by considering a small length δx of the line (AB in fig. 9.3) and working to first order in δx. Fig. 9.3 shows the equivalent electrical network for this small (infinitesimal) length of line and the currents and voltages at points A and B [the conductor FD may be arbitrarily assumed at zero voltage].

Considering the voltage change along AB immediately leads to

$$Ri\,\delta x + L\frac{\partial i}{\partial t}\delta x = -\frac{\partial v}{\partial x}\delta x.$$

The currents flowing (towards FD) through the capacitance and leakage

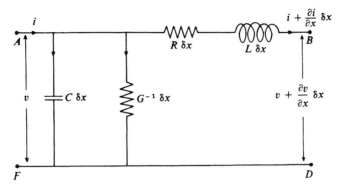

Fig. 9.3 Equivalent circuit for a length δx of a transmission line of resistance R, inductance L, capacitance C and leakage conductance G, all per unit length.

conductance are $(\partial v/\partial t)C\,\delta x$ and $vG\,\delta x$ respectively and so current conservation gives

$$i = C\,\frac{\partial v}{\partial t}\,\delta x + Gv\,\delta x + i + \frac{\partial i}{\partial x}\,\delta x.$$

In the limit of $\delta x \to 0$ these two physical statements lead to the coupled first order p.d.e.

$$Ri + L\,\frac{\partial i}{\partial t} = -\frac{\partial v}{\partial x}, \tag{9.92 a}$$

$$Gv + C\,\frac{\partial v}{\partial t} = -\frac{\partial i}{\partial x}. \tag{9.92 b}$$

In general these equations will have to be solved subject to initial conditions that $i(x, 0)$ and $v(x, 0)$ are certain specified functions of x, and also subject to boundary conditions [e.g. the line being short circuited ($v = 0$) or open circuited ($i = 0$) at certain points].

In some circumstances they can be reduced to a p.d.e. containing only one of the dependent variables. For example, in a 'lossless' transmission line in which R and G can be taken as zero, by differentiating one of equations (9.92) with respect to t and the other with respect to x, either v or i can be eliminated, yielding for the other the second-order equation

▶26.
$$\frac{\partial^2 u}{\partial x^2} - LC\,\frac{\partial^2 u}{\partial t^2} = 0, \tag{9.93}$$

where u stands for either $i(x, t)$ or $v(x, t)$. This is exactly the one-dimensional wave equation of example 9.3 and has the solutions discussed in that example. The velocity of propagation of the current or voltage waves is $(LC)^{-1/2}$ [L and C are defined per unit length and so this expression has the correct dimensions for a velocity].

As a second special case consider a submarine cable which has no inductance or leakage conductance, $L = G = 0$. Again differentiation of the two equations (9.92) with respect to different independent variables allows the elimination of either i or v and results in an equation identical in form to the diffusion equation

▶27.
$$\frac{1}{RC}\frac{\partial^2 u}{\partial x^2} = \frac{\partial u}{\partial t}.$$
(9.94)

As before, u stands for either $i(x, t)$ or $v(x, t)$.

▶28. Check that $(RC)^{-1}$ has the appropriate dimensions for a diffusion constant.

In less specialized cases, when the elimination of i or v is not practicable, solutions of (9.92) can be found using Laplace transform methods. This removes the time differentiations and produces ordinary coupled differential equations. Initial conditions on the current and voltage are usually straightforward to incorporate in this approach.

If we denote the specified values of $i(x, 0)$ and $v(x, 0)$ by $i_0(x)$ and $v_0(x)$ respectively, then Laplace transforming equations (9.92) yield

$$(R + Ls)I(x, s) = -\frac{\partial V(x, s)}{\partial x} + Li_0(x)$$
(9.95 a)

and
$$(G + Cs)V(x, s) = -\frac{\partial I(x, s)}{\partial x} + Cv_0(x).$$
(9.95 b)

These two coupled equations for the transforms $I(x, s)$ and $V(x, s)$ of the current and voltage [together with the transform of any imposed boundary conditions, e.g. $V(a, s) = 0$ (short circuit) or $I(a, s) = 0$ (open circuit) at some point $x = a$] are then used to obtain I and V and hence, by means of tables† or contour integration, the solution $i(x, t)$ and $v(x, t)$.

For a continuation of this approach the reader is referred to example 18 of section 9.14 at the end of this chapter.

† And a certain trust that the solution is unique.

9.13   Some other equations

We will conclude our treatment of p.d.e. by considering briefly two further equations of physical science; one taken from the realm of engineering, and one from quantum physics.

The first, which approximately describes the transverse vibrations of a rod, is

$$a^4 \frac{\partial^4 u}{\partial x^4} + \frac{\partial^2 u}{\partial t^2} = 0. \qquad (9.96)$$

The value of the constant a depends upon the geometry of the rod, and the density and Young's modulus of the material from which it is made.

This is a new type of p.d.e. for which we have no particular method. However, with its simplicity of form we might be able to guess a suitable solution. In view of the different number of differentiations in the two terms, we are led to seek a function which retains its essential form on repeated differentiation. Perhaps the simplest such function to try is

$$u(x, t) = A \exp (\lambda x + \mu t). \qquad (9.97)$$

[This is at one and the same time both a 'combination' and a 'separation' of the variables x and t.]

Substituting in (9.96) gives

$$(a^4\lambda^4 + \mu^2)u(x, t) = 0,$$

which can be satisfied non-trivially if λ is taken as $(\mu^{1/2}/a) \exp [\frac{1}{4}(2m + 1)\pi i]$ with $m = 0, 1, 2, 3$ (taking μ as real and > 0). The same sort of solutions result if $\mu = in$, as is appropriate to the case of (9.97) describing periodic vibrations.

▶29. Verify that the solution just obtained is equivalent to the 'separation of variables' result that 'if the solution of (9.96) is written $u(x, t) = f(x)g(t)$ and $g(t) = \cos (nt + \epsilon)$, then

$$f(x) = A \sin \lambda x + B \cos \lambda x + C \sinh \lambda x + D \cosh \lambda x, \qquad (9.98)$$

where $\lambda = n^{1/2}/a$'.

From the field of quantum mechanics we may consider the Schrodinger equation (9.24). The general solution of this equation forms a major part of the whole of non-relativistic quantum mechanics, but we will do no more than find its simplest solution for the potential-free case, $V(x, y, z) = 0$. For this case the equation becomes

$$-\frac{\hbar^2}{2m_0} \left(\frac{\partial^2 u}{\partial x^2} + \frac{\partial^2 u}{\partial y^2} + \frac{\partial^2 u}{\partial z^2} \right) = i\hbar \frac{\partial u}{\partial t}. \qquad (9.99)$$

As has been previously noted this has the appearance of a diffusion equation, but the presence of the i on the right-hand side produces different physical consequences, e.g. changing decaying solutions into oscillatory ones.

As on previous occasions when faced with a p.d.e. containing derivatives of different orders, but with constant coefficients, we try as a solution an exponential function with the independent variables appearing linearly in the exponent. Thus, putting

$$u(x, y, z, t) = A \exp{(lx + my + nz + \lambda t)}, \qquad (9.100)$$

we obtain, after substitution in (9.99) [and cancellation of common factors], that

$$-\frac{\hbar^2}{2m_0} (l^2 + m^2 + n^2) = i\hbar\lambda. \qquad (9.101)$$

Several possibilities for l, m, n and λ will satisfy this equation, but one of particular interest is that in which λ is taken as $-iE/\hbar$ and l, m and n are taken as $ip_x/\hbar$, $ip_y/\hbar$ and $ip_z/\hbar$ respectively. As presented here, this is nothing but playing with nomenclature, but put in this form (9.101) takes the very familiar form of the relationship between the kinetic energy and momentum of a non-relativistic particle, namely

▶ 30. $$\frac{1}{2m_0} (p_x^2 + p_y^2 + p_z^2) = E. \qquad (9.102)$$

Readers familiar with elementary modern physics will recognize the identification of l, m, n and λ as essentially the content of de Broglie's and Einstein's relations.

Solution (9.100) now has the form of a plane wave

$$u(x, y, z, t) = u(\mathbf{r}, t) = A \exp{[i(\mathbf{p} \cdot \mathbf{r} - Et)/\hbar]}, \qquad (9.103)$$

describing a free quantum mechanical particle of mass m_0 and kinetic energy E travelling with momentum $\mathbf{p} = (p_x, p_y, p_z)$ in a direction given by the unit vector $\mathbf{p}/p$.

9.14   Examples for solution

1. Find solutions of $\dfrac{1}{x} \dfrac{\partial u}{\partial x} + \dfrac{1}{y} \dfrac{\partial u}{\partial y} = 0$,

   for which (i) $u(0, y) = y$; (ii) $u = 1$ at the point $(1, 1)$.

2. Solve $\sin x \dfrac{\partial u}{\partial x} + \cos x \dfrac{\partial u}{\partial y} = \cos x$,

   subject to (i) $u(\pi/2, y) = 0$; (ii) $u(\pi/2, y) = y(y + 1)$.

3. A function $u(x, y)$ satisfies $2\partial u/\partial x + 3\partial u/\partial y = 10$, and on the line $y = 4x$ has the value 3. Evaluate $u(2, 4)$.

4. Show that if a change of variables $\xi = \xi(x, y)$, $\eta = \eta(x, y)$ is made in (9.27) so that it reads

$$A' \frac{\partial^2 u}{\partial \xi^2} + 2B' \frac{\partial^2 u}{\partial \xi \, \partial \eta} + C' \frac{\partial^2 u}{\partial \eta^2} = \text{function of } \xi, \eta, \frac{\partial u}{\partial \xi}, \frac{\partial u}{\partial \eta},$$

then

$$B'^2 - A'C' = (B^2 - AC)\left[\frac{\partial(\xi, \eta)}{\partial(x, y)}\right]^2,$$

and hence that the equation type is unchanged by the change of variable. [The algebra is somewhat lengthy, but straightforward.]

5. A point charge e is placed at point P a distance b from the origin O and a second charge of magnitude $-ea/b$ $(a < b)$ at a point Q between O and P and at a distance a^2/b from O. Show by considering similar triangles QOS and SOP, where S is any point on the surface of a sphere, centre O and radius a, that the net potential anywhere on the sphere due to the two charges is zero.

Use this result (backed up by the Uniqueness theorem) to find the force with which a point charge e placed a distance b from the centre of a spherical conductor of radius a $(< b)$ is attracted to the sphere (i) if the sphere is earthed, and (ii) if the sphere is uncharged and insulated.

6. By putting $\xi = x + iy$, $\eta = x - iy$ reduce the Laplace equation in two dimensions to

$$4 \frac{\partial^2 v}{\partial \xi \, \partial \eta} = 0,$$

where $v(\xi, \eta) = u(x, y)$. Integrate this twice to obtain result (9.45) directly.

7. In ▶15 (page 245), if the boundary condition were $u = \partial u/\partial y = 1$ when $y = 0$ for all $x > 0$, in which region of the xy-plane would u be determined?

8. Solve $6 \dfrac{\partial^2 u}{\partial x^2} - 5 \dfrac{\partial^2 u}{\partial x \, \partial y} + \dfrac{\partial^2 u}{\partial y^2} = 14$,

subject to $u = 2x + 1$ and $\partial u/\partial y = 4 - 6x$ both on $y = 0$.

9. Solve $\dfrac{\partial^2 u}{\partial x \, \partial y} + 3 \dfrac{\partial^2 u}{\partial y^2} = x(2y + 3x)$.

10. An incompressible fluid of density ρ and negligible viscosity flows with velocity v along a long thin straight tube, perfectly light and flexible, of cross-section A and held under a tension T. Assume that small transverse displacements u of the tube are governed by

$$\frac{\partial^2 u}{\partial t^2} + 2v \frac{\partial^2 u}{\partial x\, \partial t} + \left(v^2 - \frac{T}{\rho A}\right) \frac{\partial^2 u}{\partial x^2} = 0.$$

Show that the general solution consists of a superposition of two wave forms travelling with different velocities.

The tube initially has a small transverse displacement $a \cos kx$ and is suddenly released from rest. Find its subsequent motion.

11. Find the general solution of $\partial^2 u/\partial x^2 + \partial^2 u/\partial y^2 = x^2 + y^2$, (i) by inspection, (ii) by using D-operator methods. By considering $(x + iy)^4 + (x - iy)^4$ [or some similar function] verify that they are equivalent.

12. Find the most general solution of $\partial^2 u/\partial x^2 + \partial^2 u/\partial y^2 = x^2 y^2$.

13. Obtain the general solution of the one-dimensional wave equation by factorizing the differential operator.

14. Obtain a solution to example 9.6 (page 248) by changing variables to $\xi = x + iy$ and $\eta = x - iy$ in equation (9.75) and solving in terms of ξ and η before resubstituting for x and y.

15. A sheet of material of thickness w, specific heat s, density ρ and thermal conductivity k, is isolated in a vacuum but its two sides are exposed to sources of heat radiation of intensities J_1 and J_2 W m^{-2}. Ignoring short-term transients and assuming a solution of the form (9.79), show that:

(i) its rate of temperature rise is, as expected, $(J_1 + J_2)/s\rho w$ deg. s^{-1},
(ii) the temperature difference between its two surfaces is $(J_2 - J_1)w/2k$ deg.

16. In a cable of resistance R and capacitance C per unit length, voltage signals obey the equation $\partial^2 V/\partial x^2 = RC\, \partial V/\partial t$. This has solutions of the form given in (9.85) and also of the form $V = Ax + D$. Find a combination of these which represents a steady voltage V_0 applied at $x = 0$ at time $t = 0$. Obtain a solution describing the propagation of the voltage signal resulting from the application of the signal $V = V_0$ for $0 < t < T$, $V = 0$ otherwise, to the $x = 0$ end of an infinite cable. Show that for $t \gg T$ the maximum signal occurs at a value of x proportional to $t^{1/2}$ and has a magnitude proportional to t^{-1}.

17. The daily and annual variations of temperature at the surface of the earth may be represented by sine-wave oscillations with equal amplitude and periods of 1 and 365 days respectively. Assume that

for (angular) frequency ω the temperature at depth x in the earth is given by $u(x, t) = A \sin(\omega t + \mu x) \exp(-\lambda x)$, where λ and μ are constants (to be found).

Find the ratio of the depths below the surface at which the amplitudes have dropped to $1/20$ of their surface values.

At what time of year is the soil coldest at the greater of these depths, assuming the smoothed annual variation at the surface has a minimum at 25 January?

18. *Transmission line using Laplace transforms* (cont).
(a) Eliminate $I(x, s)$ from (9.95) to obtain a second-order equation for $V(x, s)$.
(b) For the simplest case in which $i_0(x)$ and $v_0(x)$ are identically zero and the end $x = 0$ of a semi-infinite line is maintained at a constant voltage E for all $t > 0$, show that the general solution for $V(x, s)$ is

$$V(x, s) = A(s) \exp(-\lambda x) + B(s) \exp(\lambda x),$$

with the boundary conditions $V(0, s) = Es^{-1}$, $V(\infty, s)$ is finite, and $\lambda = \lambda(s)$ given by

$$\lambda^2 = (R + Ls)(G + Cs).$$

(c) From (b) the voltage transform $V(x, s)$ is given by

$$V(x, s) = \frac{E}{s} \exp\{-x[(R + Ls)(G + Cs)]^{1/2}\}.$$

The general inversion of this to give $v(x, t)$ is difficult, but, using table 5.1 of transforms, establish the given solutions for the following special cases:

(i) The lossless line, $R = G = 0$ for which $v(x, t) = EU(t - (x/c))$ with $c = (LC)^{-1/2}$, i.e. the voltage is zero until the effect of the applied voltage, travelling with velocity c, reaches the point x, after which it always has the value E.
(ii) The 'distortionless' line in which the geometry and materials are so arranged that $(R/L) = (G/C) = \alpha$. For such a line

$$v(x, t) = E \exp(-\alpha x/c) U(t - x/c),$$

i.e. the effect of the applied voltage still propagates with velocity $c = (LC)^{-1/2}$, but suffers an attenuation e^{-1} in a distance c/α. [For general values of R, L, G and C, attenuation always occurs with the amount of it depending upon the frequency of the signal. For the particular arrangement $(R/L) = (G/C)$ all frequencies are attenuated by the same factor and so any signal – in the above case a step function – is transmitted undistorted.]

(iii) The submarine cable, $L = G = 0$. Here the transform of the voltage $v(x, t)$ is given by

$$V(x, s) = \frac{E}{s} \exp\left[-x(CR)^{1/2}s^{1/2}\right].$$

The inverse transform of this is not given in table 5.1, but is in fact

$$v(x, t) = E\{1 - \mathrm{Erf}\,[x(CR)^{1/2}/2t^{1/2}]\}.$$

[In view of the discussion leading to (9.94) and the form (9.85) of a solution to the diffusion equation, this general type of solution is not surprising.]

19. A rod is clamped at both ends, $x = 0$ and $x = l$. Show that if it executes transverse vibrations as described in section 9.13 with angular frequency $n = a^2\lambda^2$, then λ must satisfy

$$\cosh \lambda l = \sec \lambda l$$

and that the vibration is described by

$$u(x, t) = A[(\sinh \lambda x - \sin \lambda x)(\cosh \lambda l - \cos \lambda l)$$
$$- (\cosh \lambda x - \cos \lambda x)(\sinh \lambda l - \sin \lambda l)]\cos(a^2\lambda^2 t + \epsilon).$$

[At a clamped point both u and $\partial u/\partial x$ must vanish.]

20. By forming and solving its Laplace transform find the solution of the one-dimensional wave equation

$$\frac{\partial^2 u}{\partial x^2} = \frac{1}{c^2}\frac{\partial^2 u}{\partial t^2},$$

subject to the conditions

$$u(x, 0) = \frac{\partial u(x, 0)}{\partial t} = 0 \quad \text{and} \quad u(0, t) = f(t).$$

Find in particular the displacement at time t and position x of a string which is initially at rest and one end of which is raised, starting at $t = 0$, through a height h at a constant velocity v.

21. Consider a lossless transmission line ($R = G = 0$) of length l, which is initially at zero voltage and carries no current. The end $x = l$ is shorted and a constant voltage E is applied at $x = 0$ at time $t = 0$. Obtain an expression for the Laplace transform $V(x, s)$ of the voltage and by expanding it in a (convergent) sum of exponentials show that the voltage at the point x has the form shown in fig. 9.4.

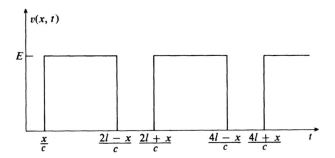

Fig. 9.4   The voltage $v(x, t)$ at position x as a function of time t for the transmission line of example 21.

10
Separation of variables

In the previous chapter we demonstrated methods by which some of the simplest partial differential equations may be solved. In particular, solutions containing the independent variables in definite combinations were sought, thus reducing the effective number of them. Alternatively one of the independent variables was eliminated (at least from differential coefficients) by making a Laplace transformation.

In the present chapter the opposite policy will be pursued, namely that of trying to keep the independent variables as separate as possible. By this we mean that if we are seeking a solution $u(x, y, z, t)$ to some p.d.e. then we attempt to obtain one which has the product form†

$$u(x, y, z, t) = X(x) Y(y) Z(z) T(t). \tag{10.1}$$

A solution which has this form is said to be *separable* in x, y, z and t and seeking solutions of this form is called the method of **separation of variables**.

As simple examples, we may observe that of the functions

(i) $xyz^2 \sin bt$,   (ii) $xy + zt$,   (iii) $(x^2 + y^2)z \cos \omega t$,

(i) is completely separable, (ii) is inseparable in that no single variable can be separated out from it and written as a multiplicative factor, whilst (iii) is separable in z and t but not in x and y.

When seeking p.d.e. solutions of the form (10.1), we are not requiring that there is no connection at all between the functions X, Y, Z and T (for example, certain parameters may appear in two or more of them), only that the function X does not depend upon y, z, t, that Y does not depend upon x, z, t, etc.

---

† It should be noted that the use of upper-case (capital) letters here to denote functions of the corresponding lower-case variable is intended to enable an easy correspondence between a function and its argument to be made. In the present chapter, upper-case letters have no connection at all with the Laplace transform of the corresponding lower case variable.

10.1   The general method

For a general partial equation it is likely that a separable solution is impossible, but certainly some common and important equations do have physically applicable solutions of this form and we will illustrate the method of solution by studying the three-dimensional wave equation. [A closely related result has already been derived in section 9.7.]

The wave equation in three dimensions is

$$\nabla^2 u = \frac{1}{c^2} \frac{\partial^2 u}{\partial t^2}. \tag{10.2}$$

We will work in Cartesian coordinates for the present and assume a solution of the form (10.1). (The solution in alternative coordinate systems, e.g. spherical polars, is considered later in the chapter.) Expressed in Cartesian coordinates (10.2) takes the form

$$\frac{\partial^2 u}{\partial x^2} + \frac{\partial^2 u}{\partial y^2} + \frac{\partial^2 u}{\partial z^2} = \frac{1}{c^2} \frac{\partial^2 u}{\partial t^2}, \tag{10.3}$$

and substituting (10.1) gives

$$X''YZT + XY''ZT + XYZ''T = c^{-2}XYZT''. \tag{10.4}$$

It will be noticed that we have here reverted to a dashed notation for derivatives even though (10.3) is expressed in partial derivatives. This is to emphasize the fact that each of the functions X, Y, Z and T has only one independent variable and thus its only derivative is its total derivative. For the same reason, in each term of (10.4) three of the four functions are unaltered by the differentiation and behave exactly as constant multipliers.

If we now divide (10.4) throughout by $u = XYZT$ we obtain the equation as

$$\frac{X''}{X} + \frac{Y''}{Y} + \frac{Z''}{Z} = \frac{1}{c^2} \frac{T''}{T}. \tag{10.5}$$

This form shows the particular characteristic which is the basis of the method of separation of variables, namely that, of the four terms, the first is a function of x only, the second of y only, the third of z only, and the right-hand side a function of t only, and yet there is an equation connecting them. This can only be so if *each* of the terms does not in fact [despite appearances] depend upon the corresponding independent variable, but is equal to a constant, the four constants being such that (10.5) is satisfied.

Since there is only one equation to be satisfied and four constants involved, there is considerable freedom in the values they may take. For the purposes of continuing our illustrative example let us make the choice of $-l^2$, $-m^2$, $-n^2$, for the first three constants. The constant associated with $c^{-2}T''/T$ must then necessarily have the value $-\mu^2 = -(l^2 + m^2 + n^2)$.

Having recognized that each term of (10.5) is individually equal to a constant (parameter), we can now replace it by four separate ordinary differential equations,

$$\frac{X''}{X} = -l^2, \quad \frac{Y''}{Y} = -m^2, \quad \frac{Z''}{Z} = -n^2, \quad \frac{1}{c^2}\frac{T''}{T} = -\mu^2. \quad (10.6)$$

These can be solved without difficulty; the general solution of the first one for example being

$$X(x) = A \exp(ilx) + B \exp(-ilx), \quad (10.7)$$

and similarly with the others.

The important point to notice however is not the simplicity of the equations (10.6) [the corresponding ones for a general p.d.e. are usually far from simple], but that, by the device of assuming a separable solution, a *partial* differential equation (10.3), which contained derivatives with respect to four independent variables all in one equation, has been reduced to four *separate ordinary* differential equations (10.6). These ordinary equations are connected only through four constant parameters which satisfy an algebraic equation. The constants are called the *separation constants*.

To finish off our example by explicitly constructing a solution, suppose that we take as particular solutions of (10.6) the four functions

$$X(x) = \exp(ilx), \quad Y(y) = \exp(imy),$$
$$Z(z) = \exp(inz), \quad T(t) = \exp(-ic\mu t). \quad (10.8)$$

Then the particular solution of (10.3) so constructed is

$$u(x, y, z, t) = \exp(ilx)\exp(imy)\exp(inz)\exp(-ic\mu t) \quad (10.9\ a)$$
$$= \exp[i(lx + my + nz - c\mu t)]. \quad (10.9\ b)$$

This is a special case of the general solution (9.74) obtained in the previous chapter and represents a plane wave [of unit amplitude] propagating in a direction given by the vector (l, m, n) with velocity c. In the conventional notation of wave theory, l, m and n are the components of the wave number $\mathbf{k}$ [whose magnitude $k = 2\pi/(\text{wavelength of the wave})$] and $c\mu$ is the angular frequency ω of the wave. This gives the equation in the form

$$u(x, y, z, t) = \exp\left[i(k_x x + k_y y + k_z z - \omega t)\right]$$
$$= \exp\left[i(\mathbf{k} \cdot \mathbf{r} - \omega t)\right], \tag{10.10}$$

and makes the exponent in the exponential dimensionless.

It has been noted in the previous chapter that exponential solutions of p.d.e. have at one and the same time both a 'combination' and a 'separation' of variables aspect; they are therefore perhaps somewhat misleading to use in illustrative examples. However, the point is that so far as the method of separation of variables is concerned, the above problem is finished at equation (10.9 a) and the remainder of the manipulation is only for the purpose of tying the result in with our previous work.

10.2   Some particular equations (revisited)

In this section we will use the separation method to obtain further solutions to some of the equations which were considered in the last chapter, restricting ourselves for the moment to Cartesian coordinates. Some results will be given as worked examples and others left for the reader to obtain from the ▶ exercises.

Example 10.1. Use the method of separation of variables to obtain for the one-dimensional diffusion equation

$$K \frac{\partial^2 u}{\partial x^2} = \frac{\partial u}{\partial t}, \tag{10.11}$$

a solution which $\to 0$ as $t \to \infty$ for all x.

Here we have only two independent variables x and t, and therefore assume a solution of the form

$$u(x, t) = X(x)T(t). \tag{10.12}$$

Substituting in (10.11) and dividing through by $u = XT$ we obtain

$$K \frac{X''}{X} = \frac{T'}{T}.$$

Now, arguing exactly as before that the left-hand side is a function of x only and the right-hand side of t only, we conclude that each side must equal a constant which we will take as $-\lambda$. This gives us two ordinary equations

$$KX'' + \lambda X = 0,$$
$$T' + \lambda T = 0,$$

with solutions

$$X(x) = A \sin [(\lambda/K)^{1/2}x] + B \cos [(\lambda/K)^{1/2}x], \qquad (10.13 \text{ a})$$

$$T(t) = \exp(-\lambda t). \qquad (10.13 \text{ b})$$

Combining these to give the assumed solution (10.12) yields

$$u(x, t) = [A \sin (\lambda/K)^{1/2}x + B \cos (\lambda/K)^{1/2}x] \exp(-\lambda t). \quad (10.14)$$

In order to satisfy the requirement that $u \rightarrow 0$ as $t \rightarrow \infty$, λ must be > 0 [and since $K > 0$ the solution is sinusoidal in x – and not a disguised hyperbolic function].

It will be noticed in this example that any value of λ, provided it is positive, will suffice. The only requirement is that the function (10.13 a) appearing in the solution (10.14) is multiplied by the function (10.13 b) corresponding to the same value of λ. In view of the linearity of (10.11), this opens up the possibility that further solutions might be obtained which are sums (superpositions) of functions like (10.14), but involving different values of λ in each term. This point will be taken up again in a later section.

▶1. For the case of no external force density, find solutions of (9.23), describing the vibrations of a uniform stretched membrane, which are separable in x, y and t.

In particular find those solutions which describe the vibration of a membrane stretched on a rectangular frame of length a and width b, so determining the natural frequencies of such a membrane as

$$\omega^2 = \frac{\pi^2 T}{\rho} \left(\frac{n^2}{a^2} + \frac{m^2}{b^2} \right),$$

where n and m are any positive integers.

▶2. Obtain solution (9.103) for the Schrodinger equation (9.99) using separation of variables and show that the separation constants must be such that equation (9.102),

$$p_x^2 + p_y^2 + p_z^2 = 2m_0 E,$$

is satisfied.

▶3. Obtain a different solution to the Schrodinger equation describing a particle confined to a cubical box of side a [u must vanish at the walls of the box]. For this situation show the quantum mechanical result that the energy of the system can take only the (quantized) values

$$E = \frac{\hbar^2 \pi^2}{2m_0 a^2}(n_x^2 + n_y^2 + n_z^2),$$

where n_x, n_y, n_z are positive integers.

As a final example, before passing on to non-Cartesian coordinate systems, we will investigate the separable solutions of the two-dimensional Laplace equation in the form

$$\frac{\partial^2 u}{\partial x^2} + \frac{\partial^2 u}{\partial y^2} = 0. \tag{10.15}$$

If we assume a solution $u(x, y) = X(x)Y(y)$ then we must have

▶4. $X'' = \lambda X$ and $Y'' = -\lambda Y$. $\tag{10.16}$

Taking λ as > 0, the general solution, for that particular λ, becomes

$$u(x, y) = (A \cosh \mu x + B \sinh \mu x)(C \cos \mu y + D \sin \mu y), \tag{10.17}$$

where $\mu^2 = \lambda$. If λ is < 0 then the roles of x and y interchange. The particular combinations of sinusoidal and hyperbolic functions and the values of λ allowed will be determined by the geometrical properties of any specific problem, together with any prescribed or necessary boundary conditions.

A particular case of (10.17) links up with the 'combination' result $u(x, y) = f(x + iy)$ of the previous chapter, namely when

▶5. $A = B,$ $D = iC,$ and $f(p) = AC \exp(\mu p).$

10.3   Laplace's equation in polar coordinates

So far we have considered the solution of p.d.e. only in Cartesian coordinates, but many systems in two and three dimensions are more naturally expressed in some form of polar coordinates, in which full advantage can be taken of inherent symmetries. For example, the potential associated with an isolated charge has a very simple expression $e/4\pi\epsilon_0 r$ when polar coordinates are used, but involves all three coordinates [and square roots] when Cartesians are employed. For these reasons we now turn to the use of separation of variables in two-dimensional polar, spherical polar and cylindrical polar coordinates.

Most of the partial equations we have considered so far have involved the operator ∇^2, e.g. the wave equation, the diffusion equation, Schrodinger's and Poisson's equations. It is therefore appropriate that we recall the expressions for ∇^2 when expressed in these polar coordinate systems. From section 4.7 they are:

$$\nabla^2 = \frac{1}{r}\frac{\partial}{\partial r}\left(r\frac{\partial}{\partial r}\right) + \frac{1}{r^2}\frac{\partial^2}{\partial\phi^2} \quad \text{(two-dimensional polars)}, \quad (10.18\,\text{a})$$

$$\nabla^2 = \frac{1}{r^2}\frac{\partial}{\partial r}\left(r^2\frac{\partial}{\partial r}\right) + \frac{1}{r^2\sin\theta}\frac{\partial}{\partial\theta}\left(\sin\theta\frac{\partial}{\partial\theta}\right) + \frac{1}{r^2\sin^2\theta}\frac{\partial^2}{\partial\phi^2}$$
$$\text{(spherical polars)}, \quad (10.18\,\text{b})$$

$$\nabla^2 = \frac{1}{r}\frac{\partial}{\partial r}\left(r\frac{\partial}{\partial r}\right) + \frac{1}{r^2}\frac{\partial^2}{\partial\phi^2} + \frac{\partial^2}{\partial z^2}$$
$$\text{(cylindrical polars)}. \quad (10.18\,\text{c})$$

The first of these is not given explicitly in chapter 4, but may be obtained from the last by taking z identically zero, or from the general expression (4.29), or by substituting $x = r\cos\phi$, $y = r\sin\phi$ into the Cartesian formula for ∇^2.

The simplest of the equations containing ∇^2 is Laplace's equation

$$\nabla^2 u = 0. \tag{10.19}$$

It contains most of the essential features of the other more complicated equations and so we will consider its solution first.

1. *Two-dimensional polars.* Suppose that we need to find a solution of (10.19) which has a prescribed behaviour on the circle $r = a$ (e.g. to find the shape taken up by a circular drumskin when its rim is slightly deformed from being planar). Then we may seek solutions of (10.19) which are separable in r and ϕ (measured from some arbitrary radius as $\phi = 0$), and hope to accommodate the boundary condition by examining the solution found for r set equal to a.

Thus, writing $u(r, \phi) = R(r)\Phi(\phi)$ and using (10.18 a), Laplace's equation (10.19) becomes

$$\frac{\Phi}{r}\frac{\partial}{\partial r}\left(r\frac{\partial R}{\partial r}\right) + \frac{R}{r^2}\frac{\partial^2\Phi}{\partial\phi^2} = 0.$$

Now, employing the same device as previously, that of dividing through by $u = R\Phi$ (and multiplying through by r^2), results in the separated equation

$$\frac{r}{R}\frac{\partial}{\partial r}\left(r\frac{\partial R}{\partial r}\right) + \frac{1}{\Phi}\frac{\partial^2\Phi}{\partial\phi^2} = 0. \tag{10.20}$$

Following our earlier argument this must mean that the two *ordinary* equations

$$\frac{r}{R}\frac{d}{dr}\left(r\frac{dR}{dr}\right) = n^2,\tag{10.21 a}$$

$$\frac{1}{\Phi}\frac{d^2\Phi}{d\phi^2} = -n^2,\tag{10.21 b}$$

are valid. We have chosen the separation constants to have the form $\pm n^2$ for later convenience; for the present, n is a general (complex) number.

The second of the equations (10.21 b) has the immediate general solution

$$\Phi(\phi) = A \exp(in\phi) + B \exp(-in\phi).\tag{10.22}$$

Equation (10.21 a) is the homogeneous equation

$$r^2R'' + rR' - n^2R = 0,$$

which can be solved either by trying a power solution in r, or, more deductively, by making the substitution $r = \exp(t)$ as described in section 5.10 and so reducing it to an equation with constant coefficients.

▶6. Carry out this latter procedure to show that

$$R(r) = Cr^n + Dr^{-n}.\tag{10.23}$$

Returning to the solution (10.22) of the azimuthal equation, we can see that if Φ, and hence u, is to be single-valued and not change when ϕ increases by 2π, then we must have that n is a real integer. Mathematically other values of n are permissible, but for the description of real physical situations it is clear that this limitation must be imposed. Having thus restricted the possible values of n, the same limitations must be carried over into (10.23) – the same n must be in both if they are to be compatible factors in a solution of (10.20).

We may thus write a particular solution of the two-dimensional Laplace equation as

$$u(r, \phi) = [A \exp(in\phi) + B \exp(-in\phi)][Cr^n + Dr^{-n}],\tag{10.24}$$

where A, B, C, D are arbitrary constants and n is a real integer.

As an (artificial) example of matching a boundary condition, we may reconsider the deformed circular drumskin and suppose that its supporting rim $r = a$ is twisted so that it is not planar by a small amount $\epsilon \sin \phi$

[roughly sketched and greatly exaggerated in fig. 10.1]. We require to choose n and the constants so that

$$\epsilon \sin \phi = u(a, \phi) = [A \exp (in\phi) + B \exp (-in\phi)][Ca^n + Da^{-n}],$$

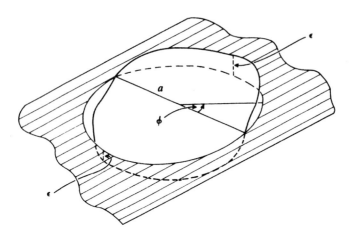

Fig. 10.1 Sketch of the deformed circular drumskin discussed in section 10.3.

and, in addition, to ensure that the deformation of the skin remains finite everywhere. This second condition requires that $D = 0$, and the first that $n = 1$ and $AC = -BC = \epsilon a^{-n}/2i$.

Hence the appropriate shape for the drumskin [valid over the whole skin, not just on the rim] is

$$u(r, \phi) = \frac{\epsilon r}{a} \sin \phi.$$

2. *Spherical polars.* Passing now to three dimensions, we come to possibly the most widely applicable single equation in physical science, namely the solution of $\nabla^2 u = 0$ in spherical polar coordinates,

$$\frac{1}{r^2} \frac{\partial}{\partial r}\left(r^2 \frac{\partial u}{\partial r}\right) + \frac{1}{r^2 \sin \theta} \frac{\partial}{\partial \theta}\left(\sin \theta \frac{\partial u}{\partial \theta}\right) + \frac{1}{r^2 \sin^2 \theta} \frac{\partial^2 u}{\partial \phi^2} = 0.$$

(10.25)

Our method of procedure will be as before; we try a solution of the form

$$u(r, \theta, \phi) = R(r)\Theta(\theta)\Phi(\phi).$$

(10.26)

Substituting this in (10.25), dividing through by $u = R\Theta\Phi$, and multiplying by r^2, we obtain

▶7. $\qquad \dfrac{1}{R}\dfrac{\partial}{\partial r}\left(r^2\dfrac{\partial R}{\partial r}\right) + \dfrac{1}{\Theta\sin\theta}\dfrac{\partial}{\partial\theta}\left(\sin\theta\dfrac{\partial\Theta}{\partial\theta}\right) + \dfrac{1}{\Phi\sin^2\theta}\dfrac{\partial^2\Phi}{\partial\phi^2} = 0.$

$$(10.27)$$

The first term depends only on r and the second and third (taken together) only on θ and ϕ. Thus (10.27) is equivalent to the two equations

$$\frac{1}{R}\frac{d}{dr}\left(r^2\frac{dR}{dr}\right) = \lambda, \qquad\qquad (10.28\ a)$$

$$\frac{1}{\Theta\sin\theta}\frac{\partial}{\partial\theta}\left(\sin\theta\frac{\partial\Theta}{\partial\theta}\right) + \frac{1}{\Phi\sin^2\theta}\frac{\partial^2\Phi}{\partial\phi^2} = -\lambda. \qquad (10.28\ b)$$

Equation (10.28 a) is a homogeneous equation

$$r^2R'' + 2rR' - \lambda R = 0, \qquad\qquad (10.29)$$

which can be reduced by the substitution $r = \exp(t)$ and $R(r) = S(t)$ to

▶8. $\qquad S'' + S' - \lambda S = 0.$

This has the general solution

$$S(t) = A\exp(m_1 t) + B\exp(m_2 t),$$

i.e. $\qquad R(r) = Ar^{m_1} + Br^{m_2}, \qquad\qquad (10.30)$

where

$$m_1 + m_2 = -1, \qquad\qquad (10.31\ a)$$

$$m_1 m_2 = -\lambda. \qquad\qquad (10.31\ b)$$

In view of (10.31 a) we can take m_1 and m_2 as given by l and $-(l + 1)$ and then λ has the form $l(l + 1)$. [It should be noted that at this stage nothing has been either assumed or proved about l being an integer.]

Hence we have obtained some information about the first factor in the separated variable solution which will now have the form

$$u(r, \theta, \phi) = (Ar^l + Br^{-(l+1)})\Theta(\theta)\Phi(\phi), \qquad (10.32)$$

where Θ and Φ must satisfy (10.28 b) with $\lambda = l(l + 1)$.

The next step is to take this latter equation further. Multiplying it through by $\sin^2\theta$ and substituting for λ, it too takes a separated form

$$\left[\frac{\sin\theta}{\Theta}\frac{\partial}{\partial\theta}\left(\sin\theta\frac{\partial\Theta}{\partial\theta}\right) + l(l+1)\sin^2\theta\right] + \frac{1}{\Phi}\frac{\partial^2\Phi}{\partial\phi^2} = 0. \qquad (10.33)$$

Taking the separation constants as m^2 and $-m^2$, the equation in the azimuthal angle ϕ has the by now very familiar solution

$$\Phi(\phi) = C \sin m\phi + D \cos m\phi. \tag{10.34}$$

As in the two-dimensional case, single-valuedness of u requires that m is a non-zero integer. The particular case $m = 0$ yields the solution

$$\Phi(\phi) = C\phi + D. \tag{10.35}$$

This form is appropriate to a solution with axial symmetry ($C = 0$) or one, such as the magnetic scalar potential associated with a wire carrying a current I ($C = I/2\pi$ and D arbitrary), which is multivalued [but manageably so].

Having settled the form of $\Phi(\phi)$, we are left only with the equation satisfied by $\Theta(\theta)$ which is (from (10.33))

$$\frac{\sin \theta}{\Theta} \frac{d}{d\theta} \left(\sin \theta \frac{d\Theta}{d\theta}\right) + l(l + 1) \sin^2 \theta = m^2. \tag{10.36}$$

A change of the independent variable from θ to $\mu = \cos \theta$ [we avoid $x = \cos \theta$ in case confusion with the spatial coordinate arises] will reduce this to a form for which solutions are known, and of which some study has been made in an earlier chapter (chapter 6).

Putting

$$\mu = \cos \theta, \quad \frac{d\mu}{d\theta} = -\sin \theta, \quad \frac{d}{d\theta} = -(1 - \mu^2)^{1/2} \frac{d}{d\mu},$$

gives the following equation for $M(\mu) \equiv \Theta(\theta)$,

▶ 9.
$$\frac{d}{d\mu}\left[(1 - \mu^2) \frac{dM}{d\mu}\right] + \left[l(l + 1) - \frac{m^2}{1 - \mu^2}\right] M = 0. \tag{10.37}$$

This equation is the *associated Legendre equation* which was mentioned (equation (7.23)) in the discussion of Sturm–Liouville equations.

For the case $m = 0$, (10.37) reduces to the Legendre equation which was studied at length in section 6.7 and has the solution (finite at $\mu = 0$)

$$M(\mu) = P_l(\mu). \tag{10.38}$$

We have not explicitly solved (10.37) for general m, but the solution to it is the so-called *associated Legendre function*†

$$\Theta(\theta) \equiv M(\mu) = P_l^m(\mu) = (1 - \mu^2)^{\frac{1}{2}|m|} \frac{d^{|m|}}{d\mu^{|m|}} P_l(\mu). \tag{10.39}$$

---

† The reader may refer to Morse and Feshbach, *Methods of theoretical physics* (McGraw-Hill, 1953) for example, for full details. For a proof that (10.39) satisfies (10.37) see example 4 of section 10.8.

Here m must be an integer with $0 \leqslant |m| \leqslant l$.

It is one of the important conditions for obtaining finite polynomial solutions of Legendre's equation that l is an integer $\geqslant 0$. This condition therefore also applies to the solutions (10.38) and (10.39) and is reflected back to the radial part of the general solution, as given in (10.32).

Now that the solutions of each of the three ordinary differential equations governing R, Θ and Φ have been obtained, either explicitly or in terms of tabulated functions, we may finally assemble the complete separated variable solution of Laplace's equation in spherical polars. It is

$$u(r, \theta, \phi) = [Ar^l + Br^{-(l+1)}]$$
$$\times P_l^m(\cos \theta)[C \sin m\phi + D \cos m\phi], \quad (10.40)$$

where the three multiplicative functions are connected only [but crucially] through the integer parameters l and m with $l \geqslant |m| \geqslant 0$.

▶10. Denoting the three terms of ∇^2 as given in (10.18 b) by ∇_r^2, ∇_θ^2, ∇_ϕ^2, in an obvious way, evaluate directly for the two functions given, $\nabla_r^2 u$, $\nabla_\theta^2 u$, $\nabla_\phi^2 u$, and verify in each case that, although the individual terms are not necessarily zero, their sum $\nabla^2 u$ is. Identify the corresponding values of l and m.

(i) $u(r, \theta, \phi) = \left(Ar^2 + \dfrac{B}{r^3}\right) \dfrac{3\cos^2 \theta - 1}{2}.$

(ii) $u(r, \theta, \phi) = \left(Ar + \dfrac{B}{r^2}\right) \sin \theta \exp(i\phi).$

Example 10.2. An uncharged conducting sphere of radius a is placed at the origin in an initially uniform electrostatic field E. Show that it behaves as an electric dipole.

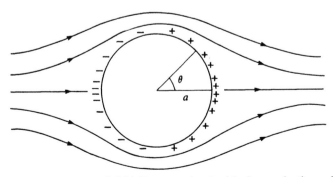

Fig. 10.2 Induced charge and field lines associated with the conducting sphere of example 10.2.

The uniform field, taken in the direction of the polar axis, has an electrostatic potential

$$u = Ez = Er \cos \theta,$$

where u is arbitrarily taken as zero at $r = 0$. This satisfies the Laplace equation $\nabla^2 u = 0$, as must the potential v when the sphere is present; for large r the asymptotic form of v must still be $Er \cos \theta$.

The $\cos \theta$ dependence of v [and the absence of any azimuthal dependence] indicates that the θ, ϕ dependence of $v(r, \theta, \phi)$ is given by $P_1^0(\cos \theta) = \cos \theta$. Thus the r dependence of v must also correspond to an $l = 1$ solution. The most general such solution is $Ar^1 + Br^{-2}$. The asymptotic form immediately gives A as E and so yields for the solution (outside the sphere, $r \geqslant a$),

$$v(r, \theta, \phi) = \left(Er + \frac{B}{r^2}\right) \cos \theta.$$

Since the sphere is conducting, v must not depend on θ for any r equal to a. This can only be if $B/a^2 = -Ea$, thus fixing B. The final solution is therefore

$$v = E\left(r - \frac{a^3}{r^2}\right) \cos \theta,$$

showing that the sphere behaves as a dipole of moment $-Ea^3$ [physically because of the effect of the charge distribution induced in the sphere, as roughly indicated in fig. 10.2].

3. *Cylindrical polars.* Having dealt with the spherical polar case at some length, only an outline of the solution in cylindrical polars will be given, leaving the details for the reader to fill in.

The equation to be solved is

$$\frac{1}{r} \frac{\partial}{\partial r}\left(r \frac{\partial u}{\partial r}\right) + \frac{1}{r^2} \frac{\partial^2 u}{\partial \phi^2} + \frac{\partial^2 u}{\partial z^2} = 0. \qquad (10.41)$$

Substituting $u(r, \phi, z) = R(r)\Phi(\phi)Z(z)$ and using separation constants $\pm n^2$ and $\pm m^2$ yields the three ordinary equations

$$Z'' - n^2 Z = 0, \qquad (10.42\ a)$$

$$\Phi'' + m^2 \Phi = 0, \qquad (10.42\ b)$$

▶11. $$r \frac{d}{dr}\left(r \frac{dR}{dr}\right) + (n^2 r^2 - m^2)R = 0. \qquad (10.42\ c)$$

The first two equations have straightforward solutions and the third can be transformed by writing

$$\mu = nr, \qquad r\frac{\mathrm{d}}{\mathrm{d}r} = \mu\frac{\mathrm{d}}{\mathrm{d}\mu}, \qquad R(r) = M(\mu), \tag{10.43}$$

into Bessel's equation of order m:

▶12.
$$\frac{\mathrm{d}^2 M}{\mathrm{d}\mu^2} + \frac{1}{\mu}\frac{\mathrm{d}M}{\mathrm{d}\mu} + \left(1 - \frac{m^2}{\mu^2}\right)M = 0. \tag{10.44}$$

The solutions $J_m(\mu)$ of Bessel's equation were investigated in chapter 6 and will not be pursued here.

The complete separated variable solution in cylindrical polars of $\nabla^2 u = 0$ is therefore

$$\begin{aligned} u(r, \phi, z) = J_m(nr)[A\cos m\phi &+ B\sin m\phi] \\ &\times [C\exp(nz) + D\exp(-nz)]. \end{aligned} \tag{10.45}$$

10.4   Spherical harmonics

In obtaining solutions in polar coordinates of $\nabla^2 u = 0$, we found that the angular part of the solution was given by

$$\Theta(\theta)\Phi(\phi) = P_l^m(\cos\theta)[C\sin m\phi + D\cos m\phi]. \tag{10.46}$$

This general form is sufficiently common that particular functions of θ and ϕ called spherical harmonics are defined and tabulated. The spherical harmonics $Y_l^m(\theta, \phi)$ are defined by

$$Y_l^m(\theta, \phi) = (-1)^m \left[\frac{2l+1}{4\pi}\frac{(l-m)!}{(l+m)!}\right]^{1/2}$$
$$\times P_l^m(\cos\theta)\exp(\mathrm{i}m\phi), \quad \text{for } m \geqslant 0. \tag{10.47}$$

For values of $m < 0$ the relation

$$Y_l^{-|m|}(\theta, \phi) = (-1)^{|m|} Y_l^{|m|*}(\theta, \phi) \tag{10.48}$$

defines the spherical harmonic. The asterisk denotes complex conjugation.

Since they contain as their θ-dependent part the solution to the associated Legendre equation, and this is a Sturm–Liouville equation, the Y_l^m are mutually orthogonal when integrated from -1 to $+1$ over $\mathrm{d}(\cos\theta)$. Their mutual orthogonality with respect to ϕ $(0 \leqslant \phi \leqslant 2\pi)$ is even more obvious. The numerical factor in (10.47) is chosen to make the Y_l^m an orthonormal set, i.e.

$$\int_{-1}^{1}\int_{0}^{2\pi} Y_l^{m*}(\theta, \phi) Y_{l'}^{m'}(\theta, \phi)\,\mathrm{d}\phi\,\mathrm{d}(\cos\theta) = \delta_{ll'}\,\delta_{mm'}. \tag{10.49}$$

As well as forming an orthonormal set they form a complete set in that any reasonable (likely to be met in a physical situation) function of θ and ϕ can be expanded as a sum of such functions

$$f(\theta, \phi) = \sum_{l=0}^{\infty} \sum_{m=-l}^{l} a_{lm} Y_l^m(\theta, \phi), \qquad (10.50)$$

with the constants a_{lm} given by

$$a_{lm} = \int_{-1}^{1} \int_{0}^{2\pi} Y_l^{m*}(\theta, \phi) f(\theta, \phi) \, d\phi \, d(\cos \theta). \qquad (10.51)$$

This is an exact analogy with Fourier series and a particular example of the general property of Sturm–Liouville solutions (chapter 7).

▶13. Starting with the information (chapter 6) that $P_0(\mu) = 1$, $P_1(\mu) = \mu$, $P_2(\mu) = \frac{1}{2}(3\mu^2 - 1)$, obtain expressions as functions of θ and ϕ for all $Y_l^m(\theta, \phi)$ up to $l = 2$.
   Verify the normalization of (say) Y_2^1 and the orthogonality of (say) Y_2^1 and Y_1^1.

▶14. Without worrying about the normalization, make rough polar sketches of the modulus squared $|Y_l^m(\theta, \phi)|^2$ of the spherical harmonics calculated in ▶13. (In a polar plot the length of the radius vector in a particular direction θ gives the value of the function for that θ.)
   [These plots, which are independent of ϕ, give the probability distributions for finding an atomic electron in different polar directions – for a fixed r – when the electron is in a state with angular momentum quantum numbers l and m.]

10.5   The wave and Schrodinger equations in spherical polars

The development of solutions of $\nabla^2 u = 0$ carried out in the previous two sections can be readily employed in solving some other equations in which the ∇^2 operator appears. In particular, partial differential equations expressed in polar coordinates in which the variables θ and ϕ appear only in the ∇^2 term can be solved in an almost identical way. What is more, the angular parts of the solutions $\Theta(\theta)\Phi(\phi)$ are identical to those of the Laplace solutions.
   As a specific example, consider the three-dimensional wave equation

$$\nabla^2 u = \frac{1}{c^2} \frac{\partial^2 u}{\partial t^2}. \qquad (10.52)$$

Expressed in spherical polar coordinates and with a separated variable trial solution $u(r, \theta, \phi, t) = R(r)\Theta(\theta)\Phi(\phi)T(t)$ this becomes in the usual way,

$$\frac{1}{Rr^2}\frac{\partial}{\partial r}\left(r^2\frac{\partial R}{\partial r}\right) + \frac{1}{\Theta r^2 \sin\theta}\frac{\partial}{\partial\theta}\left(\sin\theta\frac{\partial\Theta}{\partial\theta}\right) + \frac{1}{\Phi r^2 \sin^2\theta}\frac{\partial^2\Phi}{\partial\phi^2}$$
$$= \frac{1}{c^2 T}\frac{\partial^2 T}{\partial t^2}. \quad (10.53)$$

Setting both sides equal to a separation constant $-k^2$ (k^2 must be >0 to give a physically oscillating solution), T has the general solution

$$T(t) = A\exp(ikct) + B\exp(-ikct). \quad (10.54)$$

The quantity kc is usually written as ω, the angular frequency of the wave; k itself is the wave number.

After this separation off of the t dependence, (10.53) with the right-hand side now set equal to $-k^2$, is identical to (10.25) except for the $-k^2$ term. On multiplying through by r^2 an additional term $k^2 r^2 R$ is added to the equation for R, but Θ and Φ are identical to those obtained in (10.39) and (10.34) respectively.

With the additional term $k^2 r^2 R$, instead of (10.29) we have as the equation satisfied by R,

$$r^2 R'' + 2rR' + [k^2 r^2 - l(l+1)]R = 0. \quad (10.55)$$

This looks much like Bessel's equation (cf. equation (10.44)) and can in fact be reduced to it by writing $R(r) = r^{-1/2}S(r)$. The function $S(r)$ then satisfies

▶15. $\quad r^2 S'' + rS' + [k^2 r^2 - (l + \tfrac{1}{2})^2]S = 0,$ (10.56)

which, after changing the variable to $\mu = kr$, is Bessel's equation of order $l + \tfrac{1}{2}$ and has as its solutions $S(\mu) = J_{l+1/2}(\mu)$ or $Y_{l+1/2}(\mu)$, the latter being the second independent solution (infinite at $\mu = 0$ and not to be confused with a spherical harmonic).

One possible complete solution of the wave equation is thus

$$u(r, \theta, \phi, t) = r^{-1/2}J_{l+1/2}(kr)P_l^m(\cos\theta)$$
$$\times [A\cos m\phi + B\sin m\phi]\exp(ikct). \quad (10.57)$$

▶16. It was shown in (6.34) that $J_{1/2}(\mu) \propto \mu^{-1/2}\sin\mu$, thus giving the dependence of (10.57) as

$$\propto \frac{\sin(kr)}{r}\exp(ikct)$$

when $l = 0$. Show directly that this is a solution of (10.53) – one which has no θ or ϕ dependence.

The solutions $\propto r^{-1/2}J_{l+1/2}(kr)$ of (10.55), when suitably normalized are called *spherical Bessel functions* and denoted by $j_l(kr)$. The normalization is

$$j_l(\mu) = \left(\frac{\pi}{2\mu}\right)^{1/2} J_{l+1/2}(\mu). \tag{10.58}$$

They are trigonometric functions of μ and for $l = 0$ and 1 are given by

$$j_0(\mu) = \sin \mu,$$

$$j_1(\mu) = \frac{\sin \mu}{\mu} - \cos \mu. \tag{10.59}$$

The independent solutions of (10.55), $n_l(\mu)$ are derived similarly from the $Y_{l+1/2}(\mu)$.

It will be noticed that although the solution (10.57) corresponds to a definite frequency solution $\omega = kc$, except for the case $l = 0$ involving $j_0(\mu)$, the zeros of the radial function ($j_l(kr)$) are not equally spaced in r and so there is no precise wavelength associated with the solution.

To conclude this section, let us mention briefly the Schrodinger equation applicable to the electron in a hydrogen atom, the nucleus of which is taken at the origin and is assumed massive compared to the electron. Under these circumstances the equation is (equation (9.24))

$$-\frac{\hbar^2}{2m_0} \nabla^2 u - \frac{e^2}{4\pi\epsilon_0} \frac{u}{r} = i\hbar \frac{\partial u}{\partial t}. \tag{10.60}$$

Assuming a 'stationary state' solution for which the energy is E and the time dependent factor T in u is given by $T(t) = \exp(-iEt/\hbar)$, this also becomes an equation similar to (10.25), except that again the r-dependent part is modified.[†] However, as with the wave equation, the angular parts of the solution are identical with those of equations (10.34) and (10.39) and are expressed in terms of spherical harmonics (see the note to ▶14).

The important point to make is that for any equation involving ∇^2, so long as θ and ϕ do not appear in the equation other than as part of ∇^2, then a separated variable solution in spherical polars will always lead to spherical harmonic solutions.

[†] For the solution by series of the r-equation the reader may consult, e.g. Schiff, *Quantum mechanics* (McGraw-Hill, 1955) p. 82.

10.6   Superposition methods

As has been hinted at earlier in this chapter, the solutions we have obtained so far in separated variable form have by no means been used to their full capabilities. In fact, to illustrate them physically we have contrived situations which fit the solutions, rather than the other way about. The contrivance has been in choosing problems whose boundary conditions are immediately met by the choice of a single constant.

However, we have seen that in obtaining the separated solutions, there is in general a considerable freedom in the values of the separation constants; sometimes they must be integers, or must be $\geqslant 0$, or some such conditions, but even such restrictions normally leave a lot of latitude. The only essential condition is that we associate the correct function of one independent variable with the appropriate functions of the others – the correct one being the one with the same values of the separation constants.

If the original partial equation is linear (as are Laplace's, Schrodinger's, the diffusion and the wave equations) then mathematically acceptable solutions can be formed by superposing (adding) solutions corresponding to different values of the separation constants. To take a two-variable example, if

$$u_{\lambda_1}(x, y) = X_{\lambda_1}(x) Y_{\lambda_1}(y) \tag{10.61}$$

is a solution of a linear p.d.e. obtained by giving the separation constant the value λ_1, then the superposition

$$u(x, y) = a_1 X_{\lambda_1}(x) Y_{\lambda_1}(y) + a_2 X_{\lambda_2}(x) Y_{\lambda_2}(y) + \cdots$$
$$= \sum_i a_i X_{\lambda_i}(x) Y_{\lambda_i}(y) \tag{10.62}$$

is also a solution of the same p.d.e. for any constants a_i.

The value of this is that a boundary condition, say that $u(x, y)$ takes a particular form $f(x)$ when $y = 0$, might be met by choosing the constants a_i so that

$$f(x) = \sum_i a_i Y_{\lambda_i}(0) X_{\lambda_i}(x). \tag{10.63}$$

This will generally be possible provided the functions $X_{\lambda_i}(x)$ form a complete set – as do the sinusoidal functions of Fourier series or the spherical harmonics of section 10.4.

To illustrate this approach by a concrete example and at the same time indicate that generally some boundary conditions must be obtained by physical arguments, we will apply it to the following example (with more explanation than would be usual in a normal solution).

Example 10.3. A bar of length l is initially at a temperature of 0 °C. One end of the bar ($x = 0$) is held at 0 °C and the other is supplied with heat at a constant rate per unit area of H. Find the temperature within the bar after time t.

With the notation we have used several times before, the heat diffusion equation satisfied by the temperature $u(x, t)$ is

$$K \frac{\partial^2 u}{\partial x^2} = \frac{\partial u}{\partial t},\tag{10.64}$$

with $K = k/s\rho$.

It is clear that ultimately ($t = \infty$), when all the transients have died away, the end $x = l$ will attain a temperature u_0 such that

$$\frac{k u_0}{l} = H,$$

and there will be a constant temperature gradient,

$$u(x, \infty) = \frac{u_0 x}{l}.\tag{10.65}$$

In example 10.1 we obtained a separated variable solution for the one-dimensional diffusion equation

$$u(x, t) = [A \sin (\lambda/K)^{1/2} x \\ + B \cos (\lambda/K)^{1/2} x] \exp (-\lambda t),\quad \text{(10.14 bis)}$$

corresponding to a separation constant λ. If we restrict λ to be positive then all of these solutions are transient ones decaying to zero as $t \to \infty$. These are just what is needed for adding to (10.65) to give the correct solution as $t \to \infty$, but at the same time allowing the possibility of satisfying the initial condition that the bar is everywhere at zero temperature,

$$u(x, 0) = 0 \quad \text{for all } x.\tag{10.66}$$

One further boundary condition to be met is that the $x = 0$ end is permanently at zero temperature, i.e. $u(0, t) = 0$ for all t. Looking at the sum of (10.65) and (10.14 bis) for $x = 0$, this implies that $B = 0$ for each λ and so reduces the proposed solution to

$$u(x, t) = \frac{u_0 x}{l} + \sum_{\text{some } \lambda > 0} A_\lambda \sin \left[\left(\frac{\lambda}{K} \right)^{1/2} x \right] \exp (-\lambda t).\tag{10.67}$$

As we have noted [and contrived] this satisfies all physical conditions at $t = \infty$ and at $x = 0$, as well as being a solution of (10.64); all that remains is to satisfy (10.66). This requires that

$$0 = \frac{u_0 x}{l} + \sum_{\text{some }\lambda > 0} A_\lambda \sin\left(\frac{\lambda}{K}\right)^{1/2} x, \quad 0 \leqslant x \leqslant l. \tag{10.68}$$

That the A_λ can be chosen to make this true for all x in $0 \leqslant x \leqslant l$ follows from the work of chapter 8 on Fourier series. The physical function $-u_0 x/l$ for which a Fourier series is needed is shown in fig. 10.3 (a). Equation (10.68) shows that we want a series, which is odd in x [sine terms only] and continuous at $x = 0$ and l [no discontinuities, since the series must converge at the end points]. This leads to the continuation shown in fig. 10.3 (b) with a period of $4l$.

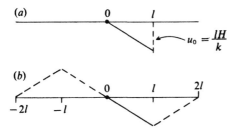

Fig. 10.3 (a) The equilibrium temperature distribution of the bar in example 10.3. (b) The appropriate continuation for a Fourier series containing only sine functions.

The corresponding Fourier series can be found in the usual way as

►17.
$$-\frac{u_0 x}{l} = -\frac{8u_0}{\pi^2} \sum_{n=0}^{\infty} \frac{(-1)^n}{(2n + 1)^2} \sin\left[\frac{(2n + 1)\pi x}{2l}\right].$$

Comparing this with (10.68) shows that the only values of λ needed are those given by

$$\left(\frac{\lambda}{K}\right)^{1/2} = \frac{(2n + 1)\pi}{2l}, \quad n = 0, 1, 2, \ldots, \infty,$$

and that then

$$A_\lambda = -\frac{8u_0}{\pi^2} \frac{(-1)^n}{(2n + 1)^2}.$$

The final formula for $u(x, t)$ is thus

$$u(x, t) = \frac{Hx}{k} - \frac{8Hl}{k\pi^2} \sum_{n=0}^{\infty} \frac{(-1)^n}{(2n+1)^2} \sin\left[\frac{(2n+1)\pi x}{2l}\right]$$
$$\times \exp\left[-\frac{k(2n+1)^2\pi^2 t}{4l^2 s\rho}\right],$$

giving the temperature for all positions $0 \leqslant x \leqslant l$ and for all times $t \geqslant 0$.

10.7   Solutions by expansion

It is sometimes possible to use the Uniqueness theorem of the previous chapter together with the results we obtained in section 10.3, where Laplace's equation was considered in polar coordinates, to obtain the solution of the equation appropriate to a particular physical situation.

The essence of the method is to assume that the required solution of the Laplace equation, $\nabla^2 u = 0$, can be written as a superposition of (say) the solutions found in (10.40), namely

$$u(r, \theta, \phi) = \sum_{l=0}^{\infty} \sum_{m=-l}^{l} [Ar^l + Br^{-(l+1)}]P_l^m(\cos\theta)$$
$$\times [C \sin m\phi + D \cos m\phi], \quad (10.69)$$

where all the constants, A, B, C, D, may depend upon l and m.

Boundary conditions of a physical nature will then fix or eliminate some of the constants, e.g.

(i) u is finite at the origin implies all $B = 0$,
(ii) $u \to 0$ as $r \to \infty$ implies all $A = 0$,
(iii) axial symmetry of the solution implies that only $m = 0$ terms are present,
(iv) $u = 0$ at $r = a$ relates the constants A and B.

The remaining constants are then found by determining u at values of r, θ, ϕ for which it can be evaluated by other means, e.g. by direct calculation on an axis of symmetry. Once the remaining constants have been fixed by these special considerations to have particular values, the Uniqueness theorem can be invoked to establish that they must have these values in general.

To illustrate the method consider the problem of calculating the gravitational potential at a general point in space due to a uniform ring of matter. Everywhere except on the ring the potential $u(\mathbf{r})$ satisfies the Laplace equation and so if we use polar coordinates with the normal to the ring as polar axis (fig. 10.4) a solution of the form (10.69) can be assumed.

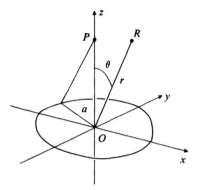

Fig. 10.4 The polar axis Oz is taken normal to the plane of the ring of matter and passing through its centre.

We expect the potential $u(r, \theta, \phi)$ to $\rightarrow 0$ as $r \rightarrow \infty$ and also to be finite at $r = 0$. At first sight this might seem to imply that all A and B and hence u must be identically zero – an unacceptable result. In fact, what it means is that different expressions must apply to different regions of space. On the ring itself we no longer have $\nabla^2 u = 0$ and so it is not surprising that the form of expression changes there. Let us therefore take two separate regions.

1. $r > a$. In this region

(i) we must have $u \rightarrow 0$ as $r \rightarrow \infty$ implying all $A = 0$,
(ii) the system is axially symmetric and so only $m = 0$ terms appear.

With these restrictions we can write as the trial form

$$u(r, \theta, \phi) = \sum_{l=0}^{\infty} B_l \, r^{-(l+1)} P_l^0(\cos \theta), \tag{10.70}$$

with the constants B_l still to be determined. This we do by calculating directly the potential where it can be done simply – clearly, in this case, on the polar axis.

Considering the point P a distance z from the plane of the ring (taken as $\theta = \pi/2$), all parts of the ring are a distance $(z^2 + a^2)^{1/2}$ from it, where a is the radius of the ring. The potential there is thus straightforwardly

$$u(z, 0, \phi) = \frac{GM}{(z^2 + a^2)^{1/2}}, \tag{10.71}$$

with G the gravitational constant and M the mass of the ring of matter. This has to be the same as (10.70) for the particular values $r = z$, $\theta = 0$, and ϕ undefined. Recalling that $P_l^0(\cos\theta) = P_l(\cos\theta)$ with $P_l(1) = 1$, putting $r = z$ in (10.70) gives

$$u(z, 0, \phi) = \sum_{l=0}^{\infty} \frac{B_l}{z^{l+1}}. \tag{10.72}$$

On the other hand, expanding (10.71) for $z > a$ (as it is in this region of space) we obtain

$$u(z, 0, \phi) = \frac{GM}{z}\left[1 - \frac{1}{2}\left(\frac{a}{z}\right)^2 + \frac{3}{8}\left(\frac{a}{z}\right)^4 - \cdots\right]. \tag{10.73}$$

Comparing (10.72) and (10.73) shows that†

$$B_0 = GM,$$

$$B_{2l} = GMa^{2l}\binom{-\frac{1}{2}}{l} = \frac{GMa^{2l}(-1)^l(2l-1)!!}{2^l l!}, \quad (l \geqslant 1), \tag{10.74}$$

$$B_{2l+1} = 0.$$

We can now conclude the argument by saying that if a solution for a general point (r, θ, ϕ) exists at all [which of course we very much expect on physical grounds] then it must be (10.70) with the B_l given by (10.74). This is so because it is a function with no arbitrary constants which satisfies all the boundary conditions and the Uniqueness theorem states that there is only one such function.

The expression for the potential in the region $r > a$ is therefore

$$u(r, \theta, \phi) = \frac{GM}{r}\left[1 + \sum_{n=1}^{\infty}\frac{(-1)^n(2n-1)!!}{2^n n!}\left(\frac{a}{r}\right)^{2n}P_{2n}(\cos\theta)\right]. \tag{10.75}$$

2. $r < a$. In a similar way an expression valid for $r < a$ can be found. Without much explanation, the main steps in its derivation are as follows. The finiteness of u at $r = 0$ and the axial symmetry give

$$u(r, \theta, \phi) = \sum_{l=0}^{\infty} A_l r^l P_l^0(\cos\theta). \tag{10.76}$$

Comparing (10.76) for $r = z$, $\theta = 0$ with the $z < a$ expansion of (10.71) [which is valid for any z] establishes $A_{2l+1} = 0$, $A_0 = GM/a$ and

▶ 18. $$A_{2l} = \frac{GM}{a^{2l+1}}\binom{-\frac{1}{2}}{l}.$$

† $(2l - 1)!! = 1 \times 3 \times \cdots \times (2l - 1)$.

The final expression valid [and convergent] for $r < a$ is thus

$$u(r, \theta, \phi) = \frac{GM}{a}\left[1 + \sum_{n=1}^{\infty} \frac{(-1)^n(2n-1)!!}{2^n n!}\left(\frac{r}{a}\right)^{2n} P_{2n}(\cos\theta)\right].$$

$$(10.77)$$

▶19. Check that the solution obtained (i) has the expected physical value for large r and for $r = 0$, (ii) is continuous at $r = a$.

10.8   Examples for solution

1. Solve the first-order equations

$$\text{(i)} \quad \frac{\partial u}{\partial x} - x\frac{\partial u}{\partial y} = 0, \qquad \text{(ii)} \quad x\frac{\partial u}{\partial x} - 2y\frac{\partial u}{\partial y} = 0,$$

by separating the variables. Verify in each case that the solution is compatible with the general forms deduced for these equations in sections 9.1 and 9.2.

2. A conducting cube has as its six faces the planes $x, y, z = 0, a$ and contains no internal sources of heat. Verify that the temperature distribution

$$u(x, y, z, t) = A\cos\left(\frac{\pi x}{a}\right)\sin\left(\frac{\pi z}{a}\right)\exp\left(-\frac{2K\pi^2}{a^2}t\right),$$

where K has the same meaning as in example 10.1, obeys the appropriate diffusion equation.

On a rough sketch indicate the temperature distribution and heat flow pattern to which this solution corresponds.

3. As described in section 9.13 the free transverse vibrations of a thick rod satisfy the equation

$$a^4\frac{\partial^4 u}{\partial x^4} + \frac{\partial^2 u}{\partial t^2} = 0.$$

Obtain a solution in the form of separated variables, and for a rod clamped at one end $x = 0$ and free at the other $x = l$ show that the angular frequency ω of the vibrations satisfies

$$\cosh\left(\frac{\omega^{1/2}l}{a}\right) = -\sec\left(\frac{\omega^{1/2}l}{a}\right).$$

[At a free end $\partial^2 u/\partial x^2$ and $\partial^3 u/\partial x^3$ both vanish.]

4. Prove (10.39) satisfies the associated Legendre equation (10.37) as follows:

(i) Evaluate $dP_l^m/d\mu$ and $d^2P_l^m/d\mu^2$ with P_l^m given by (10.39).

(ii) Substitute these results in the left-hand side of (10.37) in the form

$$(1 - \mu^2)\frac{d^2 P_l^m}{d\mu^2} - 2\mu\frac{d P_l^m}{d\mu} + \left[l(l + 1) - \frac{m^2}{1 - \mu^2}\right] P_l^m \equiv A \text{ (say)}.$$

(iii) Differentiate the Legendre equation m times using Leibniz' theorem to obtain

$$B \equiv (1 - \mu^2)\frac{d^{m+2}P_l}{d\mu^{m+2}} - 2\mu(m + 1)\frac{d^{m+1}P_l}{d\mu^{m+1}}$$

$$+ [l(l + 1) - m(m + 1)]\frac{d^m P_l}{d\mu^m} = 0.$$

(iv) Verify that $A = (1 - \mu^2)^{m/2}B$, and therefore $A = 0$; this proves that P_l^m satisfies (10.37).

5. For $l = 0, 1, 2$ use the results of ▶13 to evaluate

$$\sum_{m=-l}^{l} |Y_l^m(\theta, \phi)|^2,$$

showing that whatever the values of θ and ϕ, the above expression is independent of them. Use this result to reconcile the note to ▶14 with the clearly arbitrary choice of polar axis in describing a (unpolarized) collection of atoms.

  [The above expression is independent of θ and ϕ for any l, but a general proof is more involved.]

6. Express the function $f(\theta, \phi) = \sin\theta[\sin^2(\theta/2)\cos\phi + i\cos^2(\theta/2)$ $\times \sin\phi] + \sin^2(\theta/2)$ as a sum of spherical harmonics.

7. Find the form† assumed by a membrane stretched on a circular drum of unit radius, when one half of the rim is held flat and the other given a sinusoidal distortion. Mathematically, solve Laplace's equation for the interior of the unit circle with the boundary condition

$$u(r = 1, \phi) = \sin\phi \quad \text{for } 0 \leqslant \phi \leqslant \pi,$$
$$= 0 \quad \text{for } \pi \leqslant \phi \leqslant 2\pi.$$

[See example 5 of section 8.11.]

---

† Consider the possibility of $n = 0$ terms arising from (10.21).

8. Find the form† of a membrane stretched between two concentric rings of radii a and b $(b > a)$ if the smaller one is distorted out of being planar by an amount $c|\phi|$, $(-\pi \leqslant \phi \leqslant \pi)$.

9. A string of length l, fixed at its two ends is plucked at its midpoint by an amount A and then released. Assume (but prove if you wish) that the subsequent displacement is given by

$$u(x, t) = \sum_{n=0}^{\infty} \frac{8A}{\pi^2} \frac{1}{(2n+1)^2} \sin \frac{(2n+1)\pi x}{l} \cos \frac{(2n+1)\pi ct}{l},$$

where, in the usual notation, $c^2 = T/\rho$.

Calculate the total kinetic energy of the string when it passes through its unplucked position, by finding it for each mode (each n) and then summing. $[\sum_0^{\infty} (2n+1)^{-2} = \pi^2/8.]$

Show that it is equal to the work done in plucking the string initially.

10. (i) By integrating the relation

$$(1 - 2\mu t + t^2)^{-1/2} = \sum_{n=0}^{\infty} t^n P_n(\mu),$$

show that,

$$\int_0^1 P_{2n+1}(\mu) \, d\mu = \frac{(-1)^n (2n)!}{2^{2n+1} n! (n+1)!}.$$

(ii) A conducting spherical shell of radius a is cut round its equator and the two halves connected to voltages of $+V$ and $-V$. Find an expression for the potential anywhere inside the two hemispheres. $[\int_{-1}^{1} P_n^2(\mu) \, d\mu = 2/(2n+1).]$

Answer: $u(r, \theta, \phi) = V \sum_{n=0}^{\infty} \frac{(-1)^n (2n)!(4n+3)}{2^{2n+1} n! (n+1)!}$

$$\times \left(\frac{r}{a}\right)^{2n+1} P_{2n+1}(\cos \theta).$$

11. Develop the problem corresponding to the previous one, for cylindrical coordinates; namely find a potential $u(r, \phi)$ which satisfies:

(a) $\nabla^2 u = 0$ inside the cylinder $r = b$;
(b) $u = V$ on the half-cylinder $r = b$, $\cos \phi > 0$;
(c) $u = -V$ on the half-cylinder $r = b$, $\cos \phi < 0$.

[The geometry here is three-dimensional, but the problem is effectively two-dimensional with no z-dependence.]

† Consider the possibility of $n = 0$ terms arising from (10.21).

12. A slice of biological material of thickness L is placed into a solution of a radioactive isotope of concentration C_0. Find the concentration of radioactive ions at a distance x from either surface after a time t. [Very closely related to example 10.3.]

13. Two identical copper bars each of length a are initially one at $0\,°C$ and the other at $100\,°C$. They are joined end to end and thermally isolated at time $t = 0$. Obtain in the form of a Fourier series an expression $u(x, t)$ for the temperature at any point distance x from the join at a later time t. [Bear in mind the heat flow conditions at the free ends of the bars.]

Verify that your series gives the obvious answer for the total heat ultimately flowing across the junction. [$\sum_0^\infty (2n + 1)^{-2} = \pi^2/8$.]

Taking $a = 0.5\,m$ estimate the time it takes for one of the free ends to attain a temperature of $55\,°C$. [Thermal conductivity of copper $= 3.8 \times 10^2\,J\,m^{-1}\,K^{-1}\,s^{-1}$, $s\rho$ for copper $= 3.4 \times 10^6\,J\,m^{-3}$.]

14. (i) Show that the gravitational potential due to a uniform disc of material of radius a and mass M is given by

$$\frac{2GM}{a}\left[1 - \frac{r}{a}P_1(\cos\theta) + \frac{1}{2}\left(\frac{r}{a}\right)^2 P_2(\cos\theta)\right.$$

$$\left. - \frac{1}{8}\left(\frac{r}{a}\right)^4 P_4(\cos\theta) + \cdots\right]\quad\text{for } r < a,$$

$$\frac{GM}{r}\left[1 - \frac{1}{4}\left(\frac{a}{r}\right)^2 P_2(\cos\theta) + \frac{1}{8}\left(\frac{a}{r}\right)^4 P_4(\cos\theta) - \cdots\right]\quad\text{for } r > a,$$

where the polar axis is normal to the plane of the disc.
(ii) Reconcile the presence of a $P_1(\cos\theta)$ term for which $P_1(-\cos\theta) = -P_1(\cos\theta)$ with the symmetry, with respect to the plane of the disc, of the physical system.
(iii) Deduce the gravitational field near an infinite sheet of matter of constant density ρ per unit area.

11
Numerical methods

It frequently happens that the end product of a calculation or piece of analysis is one or more equations, algebraic or differential (or an integral), which cannot be evaluated in closed form or in terms of available tabulated functions. From the point of view of the physical scientist or engineer, who needs numerical values for prediction or comparison with experiment, the calculation or analysis is thus incomplete. The present chapter on numerical methods indicates (at the very simplest levels) some of the ways in which further progress towards extracting numerical values might be made.

In the restricted space available in a book of this·nature it is clearly not possible to give anything like a full discussion, even of the elementary points that will be made in this chapter. The limited objective adopted is that of explaining and illustrating by very simple examples some of the basic principles involved. The examples used can in many cases be solved in closed form anyway, but this 'obviousness' of the answer should not detract from their illustrative usefulness, and it is hoped that their transparency will help the reader to appreciate some of the inner workings of the methods described.

The student who proposes to study complicated sets of equations or make repeated use of the same procedures by, for example, writing computer programmes to carry out the computations, will find it essential to acquire a good understanding of topics hardly mentioned here. Amongst these are the sensitivity of the procedures adopted to errors introduced by the limited accuracy with which a numerical value can be stored in a computer (rounding errors), and to the errors introduced as a result of the approximations made in setting up the numerical procedures (truncation errors). For this scale of application, books specifically devoted to numerical analysis, data analysis and computer programming should be consulted.

So far as is possible the method of presentation here is that of indicating and discussing in a qualitative way the main steps in the procedure, and then to follow this with an elementary worked example. The examples have been restricted in complexity to a level at which they can be carried out with a set of mathematical tables and a desk calculator. Naturally it will not be possible for the student to check all the numerical values presented unless he has a calculator or computer readily available [and even then it might well be tedious to do so]. However, he is advised to check the initial step and at least one step in the middle of each repetitive calculation given, in order to be sure that he understands correctly how the symbolic equations are used with actual numbers. There is clearly some advantage in choosing a step at a point in the calculation where the values involved are changing sufficiently that whatever calculating device he is using will have the accuracy to show this.

Where alternative methods for solving the same type of problem are discussed, for example, in finding the roots of a polynomial equation, we have usually used the same example to illustrate each method. This could give the wrong impression that the methods are very restricted in applicability, but it is felt by the author that using the same examples repeatedly has sufficient advantages in terms of illustrating the *relative* characteristics of competing methods, as to justify doing so. Once the principles are clear, little is to be gained by using new examples each time and, in fact, having some prior knowledge of the 'correct answer' should allow the reader to make for himself some evaluation of the efficiency and dangers of particular methods as he follows the successive steps through.

Two other points remain to be mentioned. The first of these is the location of a chapter on numerical methods in a book of this kind. To the extent that large parts of the book can be read in any order, the position of an individual chapter does not matter. However, some methods are discussed here which relate to matrices and simultaneous linear algebraic equations, but matrices themselves are not discussed in any detail until chapter 14. On the other hand it was felt that the present chapter should follow closely after those on the solution of differential equations and this accounts for its actual location. The reader who has no familiarity with matrix equations presumably has no need to solve them numerically, and so can with equanimity omit the corresponding sections of the present chapter until he has need of them.

Finally, unlike the case with every other chapter, the value of a reasonably large selection of examples at the end of this one for the student to use as practice or for self-testing, is not too clear cut. The reader with sufficient computing resources available to tackle them can easily devise for himself algebraic or differential equations to be solved, or functions to be integrated (perhaps ones which have arisen in other contexts). Further, their solutions for the most part will be self-checking. Consequently, although a few simple examples are included, no attempt has been made to test the full range of ideas which may have been learned from reading the chapter.

11.1 Algebraic and transcendental equations

The problem of finding the real roots of an equation of the form $f(x) = 0$, where $f(x)$ is an algebraic or transcendental function of x, is one which can sometimes be treated numerically even if explicit solutions in closed form are not feasible. Examples of the types of equations referred to are the quartic equation

$$ax^4 + bx + c = 0,$$

and the transcendental equation

$$x - 3 \tanh x = 0.$$

The latter type is characterized by the fact that it effectively contains an infinite polynomial on the left-hand side.

We will discuss four methods which in various circumstances, can be used to obtain the real roots of equations of the above types. In all cases we will take as the specific equation to be solved the fifth order polynomial equation

$$f(x) \equiv x^5 - 2x^2 - 3 = 0. \tag{11.1}$$

The reasons for using the same equation each time are discussed in the previous section.

For future reference and so that the reader may, with the help of a ruler, follow some of the calculations leading to the evaluation of the real root of (11.1), a reasonably accurate graph of $f(x)$ for the range $0 \leqslant x \leqslant 1.9$ is shown in fig. 11.1.

Equation (11.1) is one for which no solution can be found with x in closed form, that is in the form $x = a$, where a does not explicitly contain x. The general scheme to be employed will be an iterative one in which successive approximations to a real root of (11.1) will be obtained, hopefully with each approximation better than the preceding one, but certainly with the requirement that the approximations converge and that they have as their limit the sought-for root. Let us denote the required root by ξ and the values of successive approximations by $x_1, x_2, \ldots, x_n, \ldots$. Then for any particular method to be successful

$$\lim_{n \to \infty} x_n = \xi, \quad \text{where } f(\xi) = 0. \tag{11.2}$$

However, success as used here is not the only criterion. Since, in practice, only a finite number of iterations will be possible, it is important that the values of x_n be close to that of ξ for all $n > N$, where N is a relatively low number. Exactly how low naturally depends upon the computing resources available and the accuracy required in the final answer.

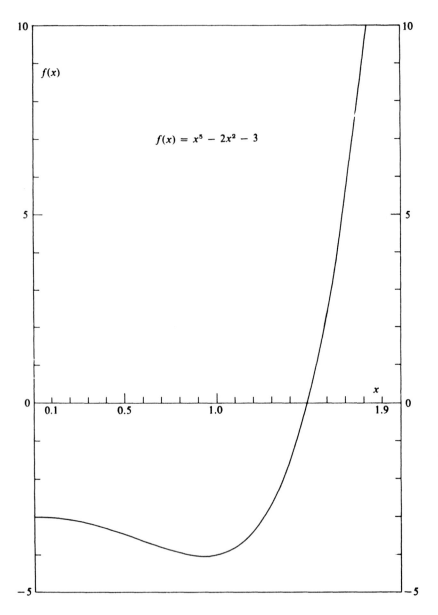

Fig. 11.1 A graph of the function $f(x) = x^5 - 2x^2 - 3$ for x in the range $0 \leqslant x \leqslant 1.9$.

So that the reader may assess the progress of the calculations which follow, we record that to 9 significant figures the real root of (11.1) has the value

$$\xi = 1.495\ 106\ 40. \tag{11.3}$$

We now consider in turn each of four methods for determining this value.

1. *Rearrangement of the equation.* If the equation $f(x) = 0$ can be recast into the form

$$x = \phi(x), \tag{11.4 a}$$

where $\phi(x)$ is a *slowly* varying function of x, then an iteration scheme

$$x_{n+1} = \phi(x_n) \tag{11.4 b}$$

will often produce a fair approximation to ξ after a few iterations. Clearly $\xi = \phi(\xi)$ since $f(\xi) = 0$, and thus when x_n is close to ξ the next approximation x_{n+1} will differ little from x_n, the actual size of the difference giving an order of magnitude indication of the inaccuracy in x_{n+1} (as compared to ξ).

In the present case the equation can be written

$$x = (2x^2 + 3)^{1/5}. \tag{11.5 a}$$

Because of the presence of the 1/5th power, the right-hand side is insensitive to the value of x and so the form (11.5 a) fits the general requirements for the method to work satisfactorily. It only remains to choose a starting approximation. It is relatively easy to see from fig. 11.1 that the value 1.5 would be a good starting value, but so that the behaviour of the procedure at values some way from the actual root can be studied, we will make the poorer choice of $x_1 = 1.7$.

With this starting value and the general recurrence relationship

$$x_{n+1} = (2x_n^2 + 3)^{1/5}, \tag{11.5 b}$$

successive values can be found. These are recorded in table 11.1. Although not strictly necessary, the value of $f(x_n) \equiv x_n^5 - 2x_n^2 - 3$ is also shown at each stage.

It is seen that x_7 and all later x_n agree with the precise answer (equation (11.3)) to within 1 part in 10^4. On the other hand $f(x_n)$ and $x_n - \xi$ are both reduced by a factor of only about 4 for each iteration; thus a large number of them would be required to produce a very accurate answer. The factor of 4 is of course specific to this particular problem and would be different for a different equation. Although they do not illustrate a great deal in this case, the successive values of x_n are shown in graph (a) of fig. 11.2.

2. *Linear interpolation.* In this approach two values A_1 and B_1 of x are chosen with $A_1 < B_1$ and such that $f(A_1)$ and $f(B_1)$ have opposite signs.

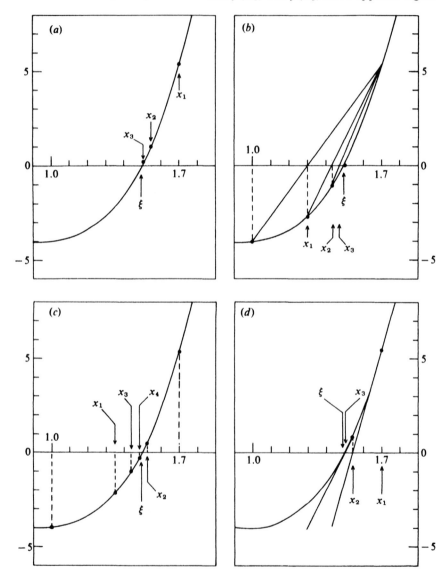

Fig. 11.2 Graphical illustrations of the iteration methods discussed in the text: (a) rearrangement; (b) linear interpolation; (c) binary chopping; (d) Newton–Raphson.

Table 11.1

n	x_n	$f(x_n)$
1	1.7	5.42
2	1.544 18	1.01
3	1.506 86	2.28×10^{-1}
4	1.497 92	5.37×10^{-2}
5	1.495 78	1.28×10^{-2}
6	1.495 27	3.11×10^{-3}
7	1.495 14	7.34×10^{-4}
8	1.495 12	1.76×10^{-4}

The chord joining the two points $(A_1, f(A_1))$ and $(B_1, f(B_1))$ is then notionally constructed, as illustrated in graph (b) of fig. 11.2, and the value x_1 at which the chord cuts the x-axis determined by the interpolation formula

$$x_n = \frac{A_n f(B_n) - B_n f(A_n)}{f(B_n) - f(A_n)}, \qquad (11.6)$$

with $n = 1$. Next $f(x_1)$ is evaluated and the process repeated after replacing with x_1 either A_1 or B_1, according as $f(x_1)$ has the same sign as $f(A_1)$ or as $f(B_1)$ respectively. [In fig. 11.2 (b), A_1 is the one replaced.]

As in our particular example, there is a tendency (if the curvature of $f(x)$ is of constant sign near the root) for one of the two ends of the successive chords to be fixed.

Starting with the initial values $A_1 = 1$ and $B_1 = 1.7$, the results of the first five iterations using (11.6) are given in table 11.2 and indicated in graph (b) of fig. 11.2.

Table 11.2

n	A_n	$f(A_n)$	B_n	$f(B_n)$	x_n	$f(x_n)$
1	1.0	-4.0000	1.7	5.4186	1.2973	-2.6916
2	1.2973	-2.6916	1.7	5.4186	1.4310	-1.0957
3	1.4310	-1.0957	1.7	5.4186	1.4762	-0.3482
4	1.4762	-0.3482	1.7	5.4186	1.4897	-0.1016
5	1.4897	-0.1016	1.7	5.4186	1.4936	-0.0289
6	1.4936	-0.0289	1.7	5.4186	1.4947	-0.0082

As with the rearrangement method the improvement in accuracy is a fairly constant factor at each iteration (approximately 3 in this case), and for our particular example there is little to choose between the two. Both tend to their limiting value of ξ monotonically, from either higher or lower values, and this makes it difficult to estimate limits within which ξ can safely be presumed to lie. The next method to be described gives at any stage a range of values within which ξ is known to lie.

3. *Binary chopping.* Again two values of x, A_1 and B_1 which straddle the root are chosen, with $A_1 < B_1$ and $f(A_1)$ and $f(B_1)$ having opposite signs. The interval between them is then halved by forming

$$x_n = \tfrac{1}{2}(A_n + B_n), \tag{11.7}$$

with $n = 1$, and $f(x_1)$ evaluated. [Notice that x_1 is determined solely by A_1 and B_1 and not by the values of $f(A_1)$ and $f(B_1)$, as in the linear interpolation method.] Now x_1 is used to replace either A_1 or B_1, depending on which of $f(A_1)$ or $f(B_1)$ has the same sign as $f(x_1)$, i.e. if $f(A_1)$ and $f(x_1)$ have the same sign, x_1 replaces A_1. The process is then repeated to obtain x_2, x_3, etc.

This has been carried through below for our standard equation (11.1) and is illustrated in fig. 11.2 (c). The entries in table 11.3 have been rounded to 4 places of decimals. It is suggested that the reader follows through the sequential replacements of the A_n and B_n in the table and correlates the first few of these with graph (c) of fig. 11.2 and also with the schematic tree in fig. 11.3.

Table 11.3

n	A_n	$f(A_n)$	B_n	$f(B_n)$	x_n	$f(x_n)$
1	1.0000	-4.0000	1.7000	5.4186	1.3500	-2.1610
2	1.3500	-2.1610	1.7000	5.4186	1.5250	0.5968
3	1.3500	-2.1610	1.5250	0.5968	1.4375	-0.9946
4	1.4375	-0.9946	1.5250	0.5968	1.4813	-0.2573
5	1.4813	-0.2573	1.5250	0.5968	1.5031	0.1544
6	1.4813	-0.2573	1.5031	0.1544	1.4922	-0.0552
7	1.4922	-0.0552	1.5031	0.1544	1.4977	0.0487
8	1.4922	-0.0552	1.4977	0.0487	1.4949	-0.0085

Clearly the accuracy with which ξ is known in this approach increases by only a factor of 2 at each step, but this accuracy is predictable at the outset of a calculation and (unless $f(x)$ has very violent behaviour near

$x = \xi$) a range of x in which ξ lies can be safely stated at any stage. At the stage reached above it may be stated that $1.4949 < \xi < 1.4977$. Binary chopping thus gives a simple (less multiplication than linear interpolation, for example), predictable, and relatively safe method of solving algebraic or transcendental equations although its convergence is slow.

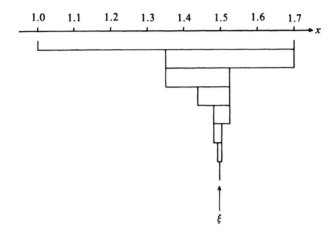

Fig. 11.3 Schematic tree of the binary chopping procedure used to obtain table 11.3.

4. *Newton–Raphson method.* The Newton–Raphson (N–R) procedure is somewhat similar to the interpolation method but, as will be seen, has one distinct advantage over it. Instead of constructing the chord between two points on the curve of $f(x)$ against x, the tangent to the curve is notionally constructed at each successive value of x_n and the next value x_{n+1} taken at the point at which the tangent cuts the axis $f(x) = 0$. This is illustrated in graph (d) of fig. 11.2.

If the nth value is x_n then the tangent to the curve of $f(x)$ at that point has slope $f'(x_n)$ and passes through the point $x = x_n$, $y = f(x_n)$. It is thus

$$y(x) = (x - x_n)f'(x_n) + f(x_n).\tag{11.8}$$

The value of x at which $y = 0$ is then taken as x_{n+1}; thus the condition $y(x_{n+1}) = 0$ yields from (11.8) the iteration scheme

$$x_{n+1} = x_n - \frac{f(x_n)}{f'(x_n)}.\tag{11.9}$$

This is the **Newton–Raphson iteration formula.** Clearly when x_n is close to ξ, x_{n+1} is close to x_n, as it should be. It is also apparent that if any of the x_n comes close to a stationary point of f so that $f'(x_n)$ is close to zero, then the scheme is not going to work well.

For our standard example, (11.9) becomes

$$x_{n+1} = x_n - \frac{x_n^5 - 2x_n^2 - 3}{5x_n^4 - 4x_n}$$

$$= \frac{4x_n^5 - 2x_n^2 + 3}{5x_n^4 - 4x_n}. \tag{11.10}$$

Again taking a starting value of $x_1 = 1.7$ we obtain in succession the entries in table 11.4. The different values are given to an increasing number of decimal places as the calculation proceeds and $f(x_n)$ is recorded also.

Table 11.4

n	x_n	$f(x_n)$
1	1.7	5.42
2	1.545 01	1.03
3	1.498 87	7.20×10^{-2}
4	1.495 13	4.49×10^{-4}
5	1.495 106 40	2.6×10^{-8}
6	1.495 106 40	—

It is apparent that this method is unlike the previous ones in that the increase in accuracy of the answer is not constant throughout the iterations, but improves dramatically as the required root is approached. Away from the root the behaviour of the series is less satisfactory and from its geometrical interpretation it can be seen that if, for example, there were a maximum or minimum near the root then the series could oscillate between values on either side of it (instead of 'homing' on the root). The reason for the good convergence near the root is discussed in the next section.

Of the four methods mentioned, no single one is ideal and in practice some mixture of them is usually to be preferred. The particular combination of methods selected will depend a great deal on how easily the progress of the calculation may be monitored, but some combination of the first three methods mentioned, followed by the Newton–Raphson scheme if great accuracy is required, would be suitable for most circumstances.

11.2   Convergence of iteration schemes

For iteration schemes in which x_{n+1} can be expressed as a differentiable function of x_n, e.g. the rearrangement or Newton–Raphson methods of the previous section, a partial analysis of the conditions necessary for a successful scheme can be made as follows.

Suppose the general iteration formula is expressed as

$$x_{n+1} = F(x_n) \qquad (11.11)$$

[(11.5 b) and (11.10) are examples]; then the sequence of values x_1, $x_2, \ldots, x_n, \ldots$ is required to converge to the value ξ, which satisfies both

$$f(\xi) = 0, \qquad (11.12 \text{ a})$$

and

$$\xi = F(\xi). \qquad (11.12 \text{ b})$$

If the error in the solution at the nth stage is ϵ_n, i.e. $x_n = \xi + \epsilon_n$, then

$$\xi + \epsilon_{n+1} = x_{n+1} = F(x_n) = F(\xi + \epsilon_n). \qquad (11.13 \text{ a})$$

For the iteration process to converge, a decreasing error is required, that is $|\epsilon_{n+1}| < |\epsilon_n|$. To see what this implies about F, we expand the right-hand term of (11.13 a) by means of a Taylor series to replace (11.13 a) by

$$\xi + \epsilon_{n+1} = \xi + \epsilon_n F'(\xi) + \tfrac{1}{2}\epsilon_n^2 F''(\xi) + \cdots. \qquad (11.13 \text{ b})$$

This shows that for small ϵ_n,

$$\epsilon_{n+1} \simeq F'(\xi)\epsilon_n,$$

and that a necessary (but not sufficient) condition for convergence is that

$$|F'(\xi)| < 1. \qquad (11.14)$$

[Notice that this is a condition on $F'(\xi)$ and not one on $f'(\xi)$, which may have any finite value.]

Figure 11.4 illustrates in a graphical way how the convergence proceeds for the case $0 < F'(\xi) < 1$.

▶1. Sketch corresponding graphs showing

(i) the convergence for the case $-1 < F'(\xi) < 0$,
(ii) the divergence when $F'(\xi) > 1$.

Equation (11.13 b) suggests that if $F(x)$ can be chosen so that $F'(\xi) = 0$ then the ratio $|\epsilon_{n+1}/\epsilon_n|$ could be made very small, of order ϵ_n in fact.

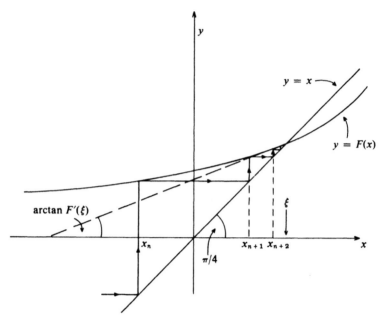

Fig. 11.4 Illustration of the convergence of the iteration scheme $x_{n+1} = F(x_n)$ when $0 < F'(\xi) < 1$, where $\xi = F(\xi)$.

Or to go even further, if it can be arranged that the first several derivatives of F vanish at $x = \xi$, then the convergence, once x_n has become close to ξ, could become very rapid indeed. If the first $N - 1$ derivatives of F vanish at $x = \xi$, i.e.

$$F'(\xi) = F''(\xi) = \cdots = F^{(N-1)}(\xi) = 0, \qquad (11.15\ a)$$

and consequently

$$\epsilon_{n+1} = O(\epsilon_n^N), \qquad (11.15\ b)$$

then the scheme is said to have **Nth order convergence**.

This is the cause of the significant difference in convergence between the Newton–Raphson scheme and the others discussed (judged by reference to (11.15 b), so that the differentiability of the function F is not a prerequisite). The N–R procedure has second-order convergence as is shown by the following analysis. Since

$$F(x) = x - f(x)/f'(x),$$
$$F'(x) = 1 - f'(x)/f'(x) + [f(x)f''(x)]/[f'(x)]^2$$
$$= f(x)f''(x)/[f'(x)]^2.$$

Now, provided $f'(\xi) \neq 0$, that $F'(\xi) = 0$ follows because $f(x) = 0$ at $x = \xi$.

▶2. A common iteration scheme for finding the square root of the number X is

$$x_{n+1} = \frac{1}{2}\left(x_n + \frac{X}{x_n}\right). \tag{11.16}$$

Show that it has second-order convergence and illustrate its efficiency by finding say $\sqrt{16}$ starting with the very bad guess $\sqrt{16} = 1$.

11.3   Simultaneous linear equations

Many situations in physical science can be described approximately or exactly by means of a number N of simultaneous linear equations in N variables x_i ($i = 1, 2, \ldots, N$). They take the general form

$$
\begin{aligned}
a_{11}x_1 + a_{12}x_2 + \cdots + a_{1N}x_N &= y_1, \\
a_{21}x_1 + a_{22}x_2 + \cdots + a_{2N}x_N &= y_2,
\end{aligned}
$$

$$a_{N1}x_1 + a_{N2}x_2 + \cdots + a_{NN}x_N = y_N, \tag{11.17}$$

where the a_{ij} are constants and form the elements of a square matrix A. The y_i are given and form a vector y.

If A is non-singular, i.e. the determinant of A is not zero, then the inverse matrix A^{-1} can be formed and the values of the x_i obtained in terms of it as

$$\mathbf{x} = A^{-1}\mathbf{y}, \tag{11.18 a}$$

or, using subscripts and the summation convention,

$$x_i = (A^{-1})_{ij}y_j. \tag{11.18 b}$$

This approach is discussed in chapter 14 in much more detail and will not be pursued here.

The idea of eliminating the variables one at a time (*Gaussian elimination*) from equations like (11.17) is probably very familiar to the reader and so a specific example will not be given to illustrate this alone. Instead we will show how a calculation along such lines might be arranged, so that the errors due to the inherent lack of perfect precision in any calculating equipment do not become excessive. This can happen if the value of N is large and particularly (we will merely state this) if the elements on the leading diagonal of the matrix $(a_{11}, a_{22}, \ldots, a_{NN}$, in (11.17)) are small compared with the off-diagonal ones.

The process to be described is known as *Gaussian elimination with interchange*. The only difference from straightforward elimination is that before each variable x_i is eliminated (to reduce the number of variables and equations both by one) the equations are reordered to put the largest (in modulus) remaining coefficient of x_i on the leading diagonal.

We will take as an illustration a straightforward three-variable example, which can in fact be solved perfectly well without any interchange, since with simple numbers and only two eliminations to perform, rounding errors do not have a chance to build up. However the important thing is that the reader appreciates how this would be applied in say a computer programme for a 100-variable case, perhaps with unforseeable zeros or very small numbers appearing on the leading diagonal.

Example 11.1. Solve the simultaneous equations

$$\begin{aligned}
x_1 + 6x_2 - 4x_3 &= 8, \text{ (a)} \\
3x_1 - 20x_2 + x_3 &= 12, \text{ (b)} \\
-x_1 + 3x_2 + 5x_3 &= 3. \text{ (c)}
\end{aligned} \qquad (11.19)$$

Firstly interchange rows (a) and (b) to bring the $3x_1$ onto the leading diagonal

$$\begin{aligned}
3x_1 - 20x_2 + x_3 &= 12, \text{ (i)} \\
x_1 + 6x_2 - 4x_3 &= 8, \text{ (d)} \\
-x_1 + 3x_2 + 5x_3 &= 3. \text{ (e)}
\end{aligned}$$

For $j = (d)$ and (e), replace row (j) by

$$\text{row } (j) - \frac{a_{j1}}{3} \times \text{row (i)},$$

to give the two equations

$$\begin{aligned}
(6 + 20/3)x_2 + (-4 - 1/3)x_3 &= 8 - 12/3, \text{ (ii)} \\
(3 - 20/3)x_2 + (5 + 1/3)x_3 &= 3 + 12/3. \text{ (}f\text{)}
\end{aligned}$$

Now $|(6 + 20/3)| > |(3 - 20/3)|$ and so no interchange is needed before the next elimination. To eliminate x_2, replace row (f) by

$$\text{row } (f) - \frac{(-11/3)}{(38/3)} \times \text{row (ii)}.$$

This gives

$$\left[\frac{16}{3} + \frac{11}{38} \times \frac{(-13)}{3} \right] x_3 = 7 + \frac{11}{38} \times 4. \text{ (iii)}$$

Collecting together and tidying up the final equations we have

$$3x_1 - 20x_2 + x_3 = 12, \text{ (i)}$$
$$38x_2 - 13x_3 = 12, \text{ (ii)}$$
$$x_3 = 2. \text{ (iii)}$$

Starting with (iii) and working backwards it is now a simple matter to obtain

$$x_1 = 10, \qquad x_2 = 1, \qquad x_3 = 2,$$

the required solution.

This example gives an explicit way of solving (11.19) to an accuracy that is limited only by the rounding errors in the calculating facilities available, and the calculation has been planned to minimize these. However, in some cases it may be that only an approximate solution is needed, and then for large numbers of variables an iterative method may produce a satisfactory degree of precision with less calculation. The method we will describe is the **Gauss–Seidel iteration** and is based upon the following analysis.

The problem is, given the matrix and vector equation

$$A\mathbf{x} = \mathbf{y}, \tag{11.20}$$

where A and $\mathbf{y}$ are known, to find the components of vector $\mathbf{x}$. The procedure is:

(i) Rearrange the equations (usually by simple division on both sides of each equation) so that all diagonal elements of the new matrix B are unity, that is (11.20) becomes

$$B\mathbf{x} = \mathbf{z}, \tag{11.21}$$

where

$$B = I - C \tag{11.22}$$

and C has zeros as its diagonal elements.
(ii) Putting (11.22) into (11.21) produces

$$C\mathbf{x} + \mathbf{z} = I\mathbf{x} = \mathbf{x}, \tag{11.23}$$

which forms the basis of an iteration scheme

$$\mathbf{x}^{(m+1)} = C\mathbf{x}^{(m)} + \mathbf{z}, \tag{11.24}$$

where the vector $\mathbf{x}^{(m)}$ is the mth approximation to the required solution of (11.20), which we denote by $\boldsymbol{\xi}$. [$\mathbf{x}^{(m)}$ is not to be confused with the mth derivative of $\mathbf{x}$, but is used here instead of $\mathbf{x}_m$ because, as is customary, we are denoting the ith element of $\mathbf{x}$ by x_i.]

(iii) To improve the convergence, the matrix C (which has zeros on its diagonal) can be written as the sum of two matrices L and U which have non-zero elements only below and above the leading diagonal respectively;

$$L_{ij} = C_{ij} \quad (i > j), \qquad L_{ij} = 0 \quad (i \leqslant j), \qquad \text{(11.25 a)}$$

$$U_{ij} = C_{ij} \quad (i < j), \qquad U_{ij} = 0 \quad (i \geqslant j). \qquad \text{(11.25 b)}$$

This enables the latest values of the components of $\mathbf{x}$ to be used at each stage and an improved form of (11.24) to be obtained

$$\mathbf{x}^{(m+1)} = L\mathbf{x}^{(m+1)} + U\mathbf{x}^{(m)} + \mathbf{z}. \qquad \text{(11.26)}$$

To see why this is possible we may notice, for example, that when calculating say $x_4^{(m+1)}$, the quantities $x_3^{(m+1)}$, $x_2^{(m+1)}$ and $x_1^{(m+1)}$ are already known and that these are the only ones needed to evaluate $(L\mathbf{x}^{(m+1)})_4$ because of the structure of L.

Example 11.2. Obtain an approximate solution to the equations of example 11.1.

The equations are

$$\begin{aligned} x_1 + 6x_2 - 4x_3 &= 8, \\ 3x_1 - 20x_2 + x_3 &= 12, \\ -x_1 + 3x_2 + 5x_3 &= 3. \end{aligned} \right\} \qquad \text{(11.19 bis)}$$

Divide the equations by 1, -20 and 5 respectively to give

$$\begin{aligned} x_1 + 6x_2 - 4x_3 &= 8, \\ -0.15x_1 + x_2 - 0.05x_3 &= -0.6, \\ -0.2x_1 + 0.6x_2 + x_3 &= 0.6. \end{aligned}$$

Thus, set out in matrix form, (11.26) is in this case

$$\begin{bmatrix} x_1 \\ x_2 \\ x_3 \end{bmatrix}_{(m+1)} = \begin{bmatrix} 0 & 0 & 0 \\ 0.15 & 0 & 0 \\ 0.2 & -0.6 & 0 \end{bmatrix} \begin{bmatrix} x_1 \\ x_2 \\ x_3 \end{bmatrix}_{(m+1)}$$

$$+ \begin{bmatrix} 0 & -6 & 4 \\ 0 & 0 & 0.05 \\ 0 & 0 & 0 \end{bmatrix} \begin{bmatrix} x_1 \\ x_2 \\ x_3 \end{bmatrix}_{m} + \begin{bmatrix} 8 \\ -0.6 \\ 0.6 \end{bmatrix}. \qquad \text{(11.27)}$$

Suppose initially we make the guess that $x_1^{(1)} = x_2^{(1)} = x_3^{(1)} = 2$. Then the successive sets of values of the three quantities generated by (11.27) are as shown in table 11.5.

Table 11.5

m	$x_1^{(m)}$	$x_2^{(m)}$	$x_3^{(m)}$
1	2	2	2
2	4	0.1	1.34
3	12.76	1.381	2.323
4	9.008	0.867	1.881
5	10.321	1.042	2.039
6	9.902	0.987	1.988
7	10.029	1.004	2.004

Thus it is seen that, even with the rather poor initial guess, a close approximation to the exact result $x_1 = 10$, $x_2 = 1$, $x_3 = 2$ is obtained in only a few iterations.

Although for the solution of most matrix equations $A\mathbf{x} = \mathbf{y}$ the number of operations involved increases rapidly with the size of the matrix (roughly as the cube of the number of variables, N) for one particularly simple kind of matrix the computing required is only proportional to N. This type often occurs in physical situations in which an ordered set of objects interact only with their nearest neighbours, and is one in which only the leading diagonal and the diagonals immediately above and below it contain non-zero entries. Such matrices are known as **tridiagonal** matrices.

A typical matrix equation involving a tridiagonal matrix is thus

$$
\begin{bmatrix}
b_1 & c_1 & & & & & \\
a_2 & b_2 & c_2 & & & \mathbf{0} & \\
& a_3 & b_3 & c_3 & & & \\
& & \cdot & \cdot & \cdot & & \\
& & & \cdot & \cdot & \cdot & \\
& \mathbf{0} & & & \cdot & \cdot & \\
& & & & a_{N-1} & b_{N-1} & c_{N-1} \\
& & & & & a_N & b_N
\end{bmatrix}
\begin{bmatrix}
x_1 \\ x_2 \\ x_3 \\ \cdot \\ \cdot \\ \cdot \\ x_{N-1} \\ x_N
\end{bmatrix}
=
\begin{bmatrix}
y_1 \\ y_2 \\ y_3 \\ \cdot \\ \cdot \\ \cdot \\ y_{N-1} \\ y_N
\end{bmatrix}
$$

(11.28)

In such an equation the first and last rows involve x_1 and x_N respectively, and so a solution could be found by letting x_1 be unknown and then solving each row of the equation in turn in terms of x_1, and finally determining x_1 by requiring the next to last line to generate for x_N an equation compatible with that given by the last line. However, if the matrix is large this becomes a very cumbersome operation and a simpler method is to assume a form of solution (no summation convention in this section)

$$x_{i-1} = \theta_{i-1}x_i + \phi_{i-1}. \tag{11.29}$$

Then, since the ith line of (11.28) is

$$a_i x_{i-1} + b_i x_i + c_i x_{i+1} = y_i,$$

we must have, by substituting for x_{i-1}, that

$$(a_i \theta_{i-1} + b_i)x_i + c_i x_{i+1} = y_i - a_i \phi_{i-1}.$$

This again is of the form (11.29) but with i replaced by $i + 1$.

Thus the recurrence formulae for the θ_i and ϕ_i are†

$$\theta_i = \frac{-c_i}{a_i \theta_{i-1} + b_i}, \qquad \phi_i = \frac{y_i - a_i \phi_{i-1}}{a_i \theta_{i-1} + b_i}, \tag{11.30}$$

and from the first row of (11.28), $\theta_1 = -c_1/b_1$ and $\phi_1 = y_1/b_1$. The equation may therefore be solved for the x_i in two stages without carrying an unknown quantity throughout. First, all the θ_i and ϕ_i are generated using (11.30) and the values of θ_1 and ϕ_1, and then after these are obtained, (11.29) is used to evaluate the x_i starting with x_N ($= \phi_N$) and working backwards.

As a simple worked example consider the equation

$$\begin{bmatrix} 1 & 0 & 0 & 0 & 0 & 0 \\ -1 & 2 & 1 & 0 & 0 & 0 \\ 0 & 2 & -1 & 2 & 0 & 0 \\ 0 & 0 & 3 & 1 & 1 & 0 \\ 0 & 0 & 0 & 3 & 4 & 2 \\ 0 & 0 & 0 & 0 & 0 & 2 \end{bmatrix} \begin{bmatrix} x_1 \\ x_2 \\ x_3 \\ x_4 \\ x_5 \\ x_6 \end{bmatrix} = \begin{bmatrix} 2 \\ 3 \\ -3 \\ 10 \\ 7 \\ 2 \end{bmatrix}, \tag{11.31}$$

and its solution by means of table 11.6, in which the arrows indicate the general flow of the calculation.

It will be seen later that such equations can be used to solve certain types of differential equations by numerical approximation.

† The method fails if the matrix elements are such as to make $a_i \theta_{i-1} + b_i = 0$ for any i.

Table 11.6

i	a_i	b_i	c_i	$a_i\theta_{i-1}+b_i$	θ_i	y_i	$a_i\phi_{i-1}$	ϕ_i	x_i
1	—	1	0	—	0	2	—	2	2 (check)
2	-1	2	1	2	$-1/2$	3	-2	5/2	1
3	2	-1	2	-2	1	-3	5	4	3
4	3	1	1	4	$-1/4$	10	12	$-1/2$	-1
5	3	4	2	13/4	$-8/13$	7	$-3/2$	34/13	2
6	0	2	—	2	—	2	0	1	1

11.4 Numerical integration

Sometimes, for example, when a function is presented in the form of a tabulation, it is not possible to evaluate a required integral in closed form, that is to give an explicit expression equal to the integral

$$I = \int_a^b f(x)\,dx; \qquad (11.32)$$

then a numerical evaluation becomes necessary. This is done by regarding I as representing the area under the curve of $f(x)$, as discussed in chapter 1, and attempting to estimate this area.

The simplest methods of doing this involve dividing up the interval $a \leqslant x \leqslant b$ into N equal sections each of length $h = (b - a)/N$. The

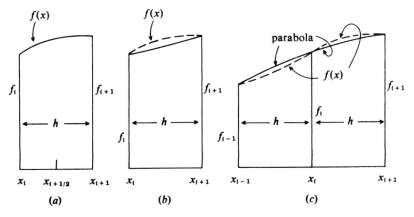

Fig. 11.5 (a) Definition of nomenclature. (b) The approximation in using the trapezium rule. (c) The Simpson's rule approximation.

dividing points are labelled x_i with $x_0 = a$, $x_N = b$ and i running from 0 to N. The point x_i is at a distance ih from a. The central value of x in a strip (at $x_i + \frac{1}{2}h$) is denoted for brevity by $x_{i+1/2}$, and for the same reason $f(x_i)$ is written as f_i. This nomenclature is indicated graphically in fig. 11.5 (a).

So that we may compare later estimates of the area under the curve with the true value, we next calculate exactly an expression [even though we cannot evaluate it] for I. To do this we need consider only one strip, say that between x_i and x_{i+1}. For this strip the area is

$$\int_{-h/2}^{h/2} f(x_{i+1/2} + y) \, dy = \int_{-h/2}^{h/2} \sum_{n=0}^{\infty} f^{(n)}(x_{i+1/2}) \frac{y^n}{n!} \, dy$$

$$\text{(Taylor's expansion)}$$

$$= \sum_{n=0}^{\infty} f^{(n)}(x_{i+1/2}) \int_{-h/2}^{h/2} \frac{y^n}{n!} \, dy$$

$$= \sum_{n \text{ even}} f_{i+1/2}^{(n)} \frac{2}{(n+1)!} \left(\frac{h}{2}\right)^{n+1} \quad \text{(exact)}.$$

$$(11.33)$$

Notice that only even derivatives of f survive the integration and that all derivatives are evaluated at $x_{i+1/2}$.

Now we turn to various ways of approximating I, given the values of, or a means to calculate, f_i for $i = 0, 1, \ldots, N$.

1. *Trapezium rule.* In this simple case the area shown in fig. 11.5 is approximated by that drawn in (b), i.e. by a trapezium. The area A_i of the trapezium is clearly

$$A_i = \frac{1}{2}(f_i + f_{i+1})h, \tag{11.34}$$

and if such contributions for all strips are added together the estimate of the total area, and hence of I is

$$I \text{ (estim.)} = \sum_{i=0}^{N-1} A_i = \frac{h}{2} (f_0 + 2f_1 + 2f_2 + \cdots + 2f_{N-1} + f_N). \tag{11.35}$$

This provides a very simple expression for estimating integral (11.32); its accuracy is limited only by the extent to which h can be made very small (and hence N very large) without making the calculation excessively long. Clearly the estimate provided is only exact if $f(x)$ is a linear function of x.

The error made in calculating the area of a strip when the trapezium rule is used may be estimated as follows. The values used are (equation

(11.34)) f_i and f_{i+1}. These can be (accurately) expressed in terms of $f_{i+1/2}$ and its derivatives by a Taylor series

$$f_{i+1/2 \pm 1/2} = f_{i+1/2} \pm \frac{h}{2} f'_{i+1/2} + \frac{1}{2!}\left(\frac{h}{2}\right)^2 f''_{i+1/2} \pm \frac{1}{3!}\left(\frac{h}{2}\right)^3 \times f^{(3)}_{i+1/2} + \cdots. \quad (11.36)$$

Thus

$$A_i(\text{estim.}) = \tfrac{1}{2}h(f_i + f_{i+1})$$
$$= h\left[f_{i+1/2} + \frac{1}{2!}\left(\frac{h}{2}\right)^2 f''_{i+1/2} + O(h^4)\right], \quad (11.37)$$

whilst, from the first few terms of the exact result (11.33),

$$A_i(\text{exact}) = hf_{i+1/2} + \frac{2}{3!}\frac{h^3}{8} f''_{i+1/2} + O(h^5). \quad (11.38)$$

Thus the error

$$\Delta A_i = A_i(\text{estim.}) - A_i(\text{exact}) = \left(\frac{1}{8} - \frac{1}{24}\right) h^3 f''_{i+1/2} + O(h^5)$$
$$\simeq \frac{h^3}{12} f''_{i+1/2}. \quad (11.39)$$

The total error in $I(\text{estim.})$ is thus approximately given by

$$\frac{nh^3}{12} \langle f'' \rangle = \frac{h^2(b-a)}{12} \langle f'' \rangle, \quad (11.40)$$

where $\langle f'' \rangle$ represents an average value for the second derivative of f over the interval a to b.

2. *Simpson's rule.* Whereas the trapezium rule makes a linear interpolation of f, Simpson's rule effectively mimics the local variation of $f(x)$ with parabolas. The strips are treated two at a time (fig. 11.5 (c)) and therefore the number of them should be made even (that is N should be even).

In the neighbourhood of x_i (where i is odd) it is supposed that $f(x)$ can be adequately represented by a quadratic form

$$f(x_i + y) = f_i + ay + by^2. \quad (11.41)$$

Applying this in particular to $y = \pm h$ yields an expression for b,

$$f_{i+1} = f(x_i + h) = f_i + ah + bh^2,$$
$$f_{i-1} = f(x_i - h) = f_i - ah + bh^2.$$

Thus $bh^2 = \frac{1}{2}(f_{i+1} + f_{i-1} - 2f_i).$ (11.42)

Now if representation (11.41) is assumed, the area of the double strip from x_{i-1} to x_{i+1} is given by

$$A_i(\text{estim.}) = \int_{-h}^{h} (f_i + ay + by^2)\, dy = 2hf_i + \frac{2b}{3} h^3.$$

Substituting for bh^2 from (11.42) then yields for the estimated area

$$\begin{aligned} A_i(\text{estim.}) &= 2hf_i + \tfrac{2}{3}h \cdot \tfrac{1}{2}(f_{i+1} + f_{i-1} - 2f_i) \\ &= \tfrac{1}{3}h(4f_i + f_{i+1} + f_{i-1}), \end{aligned}$$ (11.43)

an expression involving only given quantities. [Note that the value of b need never be calculated.]

For the full integral,

$$I(\text{estim.}) = \frac{h}{3}\left(f_0 + f_N + 4 \sum_{m\,\text{odd}} f_m + 2 \sum_{m\,\text{even}} f_m \right).$$ (11.44)

▶3. Follow the same procedure as in the trapezium rule case to show that the error in the estimated area is approximately

$$\frac{(b-a)}{180} h^4 \langle f^{(4)} \rangle.$$

[In (11.33) and (11.36) replace $h/2$ by h and $i + \frac{1}{2}$ by i.]

In the two cases considered, the function f was mimicked by linear and quadratic functions. These yield exact answers if f is itself a linear or quadratic function (respectively) of x. This process could be continued by increasing the order of the polynomial mimicking function so as to increase the accuracy with which more complicated functions f could be numerically integrated; but the same effect can be achieved with less effort by not insisting upon equally spaced points x_i.

The detailed analysis of methods of numerical integration, in which the integration points are not equally spaced and the weightings given to the values at each point do not fall into a few simple groups, is too long to be given here. The reader is referred to books devoted specifically to the theory of numerical analysis, where he will find details of the integration points and weights for many possible schemes.†

We will content ourselves here with mentioning only **Gaussian integration** which is based upon the orthogonality properties, in the interval

† The points and weights may be found in, e.g. Abramowitz and Stegun, *Handbook of mathematical functions* (Dover, 1965) p. 887.

$-1 \leqslant x \leqslant 1$, of the Legendre polynomials $P_l(x)$ [discussed in chapter 6]. In order to use these properties the integral between limits a and b in (11.32) has to be changed to one between limits -1 and $+1$. This is easily done by a change of variable from x to z given by

$$z = \frac{2x - b - a}{b - a}, \tag{11.45 a}$$

when I becomes

$$I = \frac{(b - a)}{2} \int_{-1}^{1} g(z) \, dz, \tag{11.45 b}$$

in which $g(z) \equiv f(x)$.

The integration points x_i for an n-point Gaussian integration are given by the zeros of $P_n(x)$, i.e. the x_i are such that $P_n(x_i) = 0$. For example, the x_i for a 3-point integration are at 0 and $\pm 0.774\,60$. The corresponding weightings w_i (also obtainable from the theory of Legendre polynomials) are $0.888\,89$ and $0.555\,56$. The value of the integral of $f(x)$ between -1 and $+1$ is given by $\sum_i w_i f(x_i)$. It can be shown that an n-point Gaussian integration can evaluate exactly the integral of a $2n - 1$ degree polynomial.

We will finish this section by evaluating the same integral

$$\int_0^1 f(x) \, dx = \int_0^1 \frac{1}{1 + x^2} \, dx \tag{11.46}$$

by each of the three methods (using in each case a three-point formula, i.e. using x_0, x_1 and x_2) and also exactly, so that they may be compared.

(i) The *exact* evaluation.

$$\int_0^1 \frac{dx}{1 + x^2} = [\arctan x]_0^1 = \frac{\pi}{4} = 0.785\,40.$$

(ii) *Trapezium* rule.

$$I = \tfrac{1}{2} \cdot \tfrac{1}{2} [f(0) + 2f(\tfrac{1}{2}) + f(1)]$$
$$= \tfrac{1}{4} [1 + 8/5 + 1/2] = 0.7750.$$

(iii) *Simpson's* rule.

$$I = \tfrac{1}{3} \cdot \tfrac{1}{2} [f(0) + 4f(\tfrac{1}{2}) + f(1)]$$
$$= \tfrac{1}{6} [1 + 16/5 + 1/2] = 0.7833.$$

(iv) *Gaussian* integration.

$$I = \frac{1 - 0}{2} \int_{-1}^{1} \frac{dz}{1 + \frac{1}{4}(z + 1)^2}$$

$$= \tfrac{1}{2}\{0.555\,56[f(-0.774\,60) + f(0.774\,60)] + 0.888\,89f(0)\}$$

$$= \tfrac{1}{2}\{0.555\,56[0.987\,458 + 0.559\,503] + 0.888\,89[0.8]\}$$

$$= 0.785\,27.$$

These results are sufficient to indicate that in practice a compromise has to be struck between the accuracy of the result achieved and the calculational labour which goes into obtaining it.

11.5   Finite differences

In several places in the previous section comparison was made between, on the one hand, sums and differences of sequential values of f_i, and, on the other, the derivatives of f at one of the points at which the f_i were evaluated. Here, by way of preparation for the numerical treatment of differential equations in the next and subsequent sections, we will do this in a more systematic way.

Again we consider a set of values f_i of a function $f(x)$ evaluated at equally spaced points x_i, with common separation h. The basis for our discussion will again be the Taylor series expansion, but on this occasion about the point x_i. It is

$$f_{i\pm1} = f_i \pm h \frac{df}{dx} + \frac{h^2}{2!} \frac{d^2 f}{dx^2} \pm \frac{h^3}{3!} \frac{d^3 f}{dx^3} + \cdots, \qquad (11.47 \pm)$$

where all the derivatives are evaluated at $x = x_i$ and a general one will be denoted by $f_i^{(n)}$.

From (11.47), three different expressions which approximate $f_i^{(1)}$ can be derived. The first of these, obtained by subtracting the two equations is

$$f_i^{(1)} \equiv \frac{df}{dx}\bigg|_{x_i} = \frac{f_{i+1} - f_{i-1}}{2h} - \frac{h^2}{3!} \frac{d^3 f}{dx^3} - \cdots. \qquad (11.48)$$

The quantity $(f_{i+1} - f_{i-1})/2h$ is known as the *central difference* approximation to $f_i^{(1)}$ and can be seen from (11.48) to be in error by approximately $(h^2/6)f_i^{(3)}$.

An alternative approximation, obtained from (11.47 +) alone, is given by

$$f_i^{(1)} \equiv \frac{df}{dx}\bigg|_{x_i} = \frac{f_{i+1} - f_i}{h} - \frac{h}{2!}\frac{d^2 f}{dx^2} - \cdots. \qquad (11.49)$$

The *forward difference*, $(f_{i+1} - f_i)/h$, is clearly a poorer approximation since it is in error by approximately $(h/2)f_i^{(2)}$ (as compared with $(h^2/6)f_i^{(3)}$), and similarly for the *backward difference*, $(f_i - f_{i-1})/h$, obtained from (11.47 −), although the sign of the error is reversed in this latter case.

This type of differencing approximation can be continued to the higher derivatives of f in an obvious manner. By adding the two equations (11.47) a central difference approximation to $f_i^{(2)}$ can be obtained as

$$f_i^{(2)} \equiv \frac{d^2 f}{dx^2}\bigg|_{x_i} \approx \frac{f_{i+1} - 2f_i + f_{i-1}}{h^2}. \qquad (11.50)$$

The error in this approximation (also known as the second difference of f) is easily shown to be $(h^2/12)f_i^{(4)}$.

Of course if the function $f(x)$ is a sufficiently simple function of x, all derivatives beyond a particular one may vanish anyway, and then there is no error in taking the difference to give the derivative.

▶4. The following is copied from a table of entries for the values of a second-order polynomial $f(x)$ at values of x from 1 to 12 inclusive,

2, 2, ?, 8, 14, 22, 32, 46, ?, 74, 92, 112.

The entries marked ? were illegible and in addition an error was made in the transcription. Find the illegible entries and locate and correct the copying error.

Would your procedure have worked if the copying error had been in $f(6)$?

11.6 Difference schemes for differential equations

For the remaining sections of this chapter our attention will be on the solution of differential equations by numerical methods. We consider first the simplest kind of equation – one of first order, represented typically by

$$\frac{dy}{dx} = f(x, y), \qquad (11.51)$$

where y is taken as the dependent variable and x the independent one. If this equation can be solved analytically then this is the best course to adopt; but sometimes this is not possible and a numerical approach becomes necessary. Some of the examples we will use can in fact be solved easily by an explicit integration, but, for the purposes of illustration, this is an advantage rather than the reverse since useful comparisons can then be made between the numerically derived solution and the exact one.

The first method to be described is not so much numerical as graphical, but as it is sometimes useful it is included here. The method, known as that of **isoclines**, is to sketch for a number of values of a constant c those curves (the isoclines) in the xy-plane along which $f(x, y) = c$, that is those curves along which dy/dx (where y is the required solution) is a constant of known value. [Notice that these are not generally straight lines.] Since a straight line of slope dy/dx at and through any particular point is a tangent to the curve $y = y(x)$ at that point, small elements of straight lines of slopes appropriate to the isoclines they cut, effectively form the curve $y = y(x)$.

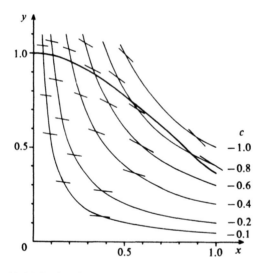

Fig. 11.6 The isocline method. The cross lines on each isocline show the slopes that solutions of $dy/dx = -2xy$ must have at the points where they cross the isoclines. The heavy line is the solution with $y(0) = 1$, viz. $\exp(-x^2)$.

Figure 11.6 illustrates in outline the method as applied to the solution of

$$\frac{dy}{dx} = -2xy. \tag{11.52}$$

The thinner curves (rectangular hyperbolae) are a selection of the isoclines along which $-2xy$ is constant and equal to the corresponding value of c. The small cross lines on each curve show the slopes ($=c$) solutions of (11.52) must have if they cross the curve. The thick curve is the solution which has value $y = 1$ at $x = 0$ and takes the slope dictated by the value of c on each isocline it crosses. [It is the function $y = \exp(-x^2)$.]

We now turn to more directly numerical methods and, so as to illustrate some of the difficulties in applying them to differential equations, carry through one or two elementary calculations based upon representations of derivatives by differences, as discussed in the previous section.

Consider the differential equation

$$\frac{dy}{dx} = -y, \quad y(0) = 1, \tag{11.53}$$

and the possibility of solving it numerically by approximating dy/dx by the forward difference

$$\left.\frac{dy}{dx}\right|_{x_i} \approx \frac{y_{i+1} - y_i}{h} \tag{11.54}$$

using the notation of section 11.5, but with f now replaced by y. This would lead to the recurrence relation

$$\begin{aligned}
y_{i+1} &= y_i + h(dy/dx)_i \\
&= y_i - hy_i \quad \text{[from (11.53)]} \\
&= (1 - h)y_i.
\end{aligned} \tag{11.55}$$

Thus since $y_0 = y(0) = 1$ is given, $y_1 = y(0 + h) = y(h)$ can be calculated and so on (this is the *Euler* method). Table 11.7 shows the values of $y(x)$ obtained if this is done using various values of h, for selected values of x. The exact value [$y(x) = \exp(-x)$] is also shown.

Table 11.7

x	$h = 0.01$	$h = 0.1$	$h = 0.5$	$h = 1$	$h = 1.5$	$h = 2$	$h = 3$	Exact
0	(1)	(1)	(1)	(1)	(1)	(1)	(1)	(1)
0.5	0.605	0.590	0.500	0	-0.500	-1	-2	0.607
1.0	0.366	0.349	0.250	0	0.250	1	4	0.368
1.5	0.221	0.206	0.125	0	-0.125	-1	-8	0.223
2.0	0.134	0.122	0.063	0	0.063	1	16	0.135
2.5	0.081	0.072	0.032	0	-0.032	-1	-32	0.082
3.0	0.049	0.042	0.016	0	0.016	1	64	0.050

It is clear that to maintain anything like a reasonable accuracy only very small steps h can be used. Indeed, if h is taken large enough, not only is the accuracy bad but, as can be seen, for $h > 1$ the calculated solution oscillates and for $h > 2$ it diverges. Equation (11.55) is of the form $y_{i+1} = \lambda y_i$, and a necessary condition for non-divergence is $|\lambda| < 1$, i.e. $0 < h < 2$. [Clearly the satisfaction of this condition in no way guarantees accuracy.]

Part of this difficulty arises because of the poor approximation in (11.54); its right-hand side is a closer approximation to dy/dx evaluated at $x = x_i + \frac{1}{2}h$ rather than at $x = x_i$. This is the result of using a forward difference approximation rather than the more accurate [but of course still approximate] central difference.

If a central difference is used (*Milne's* method), equation (11.48) *et seq.* give the recurrence relation

$$y_{i+1} = y_{i-1} + 2h(dy/dx)_i \tag{11.56 a}$$

in general, or

$$y_{i+1} = y_{i-1} - 2hy_i \tag{11.56 b}$$

in this case. An additional difficulty then arises, since two initial values of y are needed. These can be estimated by other means (e.g. a Taylor series as discussed later) but for illustration we will take the accurate value of $y(-h) = \exp(h)$ as giving y_{-1}. Taking for the sake of example $h = 0.5$, and using (11.56 b), gives the results shown in table 11.8.

Table 11.8

x	$y(x)$
-0.5	(1.648)
0	(1.000)
0.5	0.648
1.0	0.352
1.5	0.296
2.0	0.056
2.5	0.240
3.0	-0.184

Although some improvement is noticeable (as compared to the corresponding column in table 11.7) in the early values of the calculated $y(x)$, this scheme runs into difficulties as is obvious from the last two lines.

Some part of this poor performance is not really attributable to the approximations made in estimating dy/dx but to the form of the equation itself. *Any* rounding error occurring in the evaluation effectively introduces a small amount of the solution of

$$\frac{dy}{dx} = +y$$

into y. This equation has solution $y(x) = \exp(x)$ and will ultimately render the calculations totally inaccurate. [The numbers given in tables 11.7 and 11.8 are not sufficiently precise to show this effect for the small amount of calculation carried out.]

We have only illustrated rather than analysed some of the difficulties associated with simple (finite) difference iteration schemes for differential equations, but they may be summarized as (i) insufficiently precise approximations to the derivatives, and (ii) inherent instability due to rounding errors.

Difference schemes for partial differential equations are discussed briefly later.

11.7 Taylor series solutions

Consider again the first-order equation

$$\frac{dy}{dx} = f(x, y), \quad y(x_0) = y_0. \tag{11.57}$$

Since a Taylor series expansion is exact if all its terms are included and the limits of convergence are not exceeded, we may seek to use it to evaluate y_1, y_2, etc. for the above equation and boundary condition.

The Taylor series is

$$y(x + h) = y(x) + hy'(x) + \frac{h^2}{2!} y''(x) + \frac{h^3}{3!} y^{(3)}(x) + \cdots. \tag{11.58}$$

Turning this into the present notation at the point $x = x_i$,

$$y_{i+1} = y_i + hy_i^{(1)} + \frac{h^2}{2} y_i^{(2)} + \frac{h^3}{6} y_i^{(3)} + \cdots. \tag{11.59}$$

But for the required solution $y(x)$,

$$y_i^{(1)} = \left. \frac{dy}{dx} \right|_{x_i} = f(x_i, y_i) \tag{11.60 a}$$

is given, and the second derivative can be obtained from it,

$$y_i^{(2)} = \frac{\partial f}{\partial x} + \frac{\partial f}{\partial y} \frac{dy}{dx} = \frac{\partial f}{\partial x} + f \frac{\partial f}{\partial y}, \tag{11.60 b}$$

evaluated at $x = x_i$, $y = y_i$. This process can be continued for the third and higher derivatives all of which are to be evaluated at (x_i, y_i).

Having obtained expressions for the derivatives $y_i^{(n)}$ in (11.59) two alternative ways of proceeding are open.

(i) To use (11.59) to evaluate y_{i+1} and then repeat the whole process to obtain y_{i+2}, and so on.

(ii) To apply (11.59) several times but using a different value of h each time and so obtain the corresponding values of $y(x + h)$.

It is clear that approach (i) does not require so many terms of (11.59) to be kept, but on the other hand the $y_i^{(n)}$ have to be recalculated at each step. With approach (ii) fairly accurate results for y may be obtained for values of x close to the given starting value, but for large values of h a large number of terms of (11.59) must be retained.

If the original equation is, say, second order,

$$\frac{d^2 y}{dx^2} = g\left(x, y, \frac{dy}{dx}\right), \tag{11.61}$$

rather than first order, then values of dy/dx will need to be calculated at each x_i in order to adopt approach (i). A given initial value for $y'(x_0)$ will be required for either method. A Taylor series expansion for $y_{i+1}^{(1)}$ is easily obtained from the original Taylor expansion (11.58) by differentiating it with respect to x; it is

$$y_{i+1}^{(1)} = y_i^{(1)} + h y_i^{(2)} + \frac{h^2}{2} y_i^{(3)} + \frac{h^3}{6} y_i^{(4)} + \cdots. \tag{11.62}$$

As an example of a Taylor series solution using approach (ii) referred to above, the following problem will be solved.

Example 11.3. Evaluate the solution of the equation

$$\frac{dy}{dx} = 2y^{3/2}, \quad y(0) = 1, \tag{11.63}$$

for $x = 0.1$ to 0.5 in steps of 0.1.

The necessary derivatives will be calculated without commentary, since the origin of each factor is fairly clear.

$$
\begin{aligned}
y(0) &= 1, & y(0) &= 1. \\
y' &= 2y^{3/2}, & y'(0) &= 2. \\
y'' &= \tfrac{3}{2} \cdot 2y^{1/2} \cdot 2y^{3/2} = 6y^2, & y''(0) &= 6. \\
y^{(3)} &= 12y \cdot 2y^{3/2} = 24y^{5/2}, & y^{(3)}(0) &= 24. \\
y^{(4)} &= 60y^{3/2} \cdot 2y^{3/2} = 120y^3, & y^{(4)}(0) &= 120. \\
y^{(5)} &= 360y^2 \cdot 2y^{3/2} = 720y^{7/2}, & y^{(5)}(0) &= 720.
\end{aligned}
$$

Thus the Taylor expansion of the solution about the origin (more correctly a Maclaurin series) is

$$y(x) = 1 + 2x + \frac{6}{2!}x^2 + \frac{24}{3!}x^3 + \frac{120}{4!}x^4 + \frac{720}{5!}x^5 + \cdots.$$

Hence, $y(\text{estim.}) = 1 + 2x + 3x^2 + 4x^3 + 5x^4 + 6x^5$.

Table 11.9

x	$y(\text{estim.})$	$y(\text{exact})$
0	1.0000	1.0000
0.1	1.2346	1.2346
0.2	1.5619	1.5625
0.3	2.0331	2.0408
0.4	2.7254	2.7778
0.5	3.7500	4.0000

Comparison with the exact value (table 11.9) shows that using the first six terms gives a value which is correct to 1 part in 100 up to $x = 0.3$.

▶5. Integrate (11.63) analytically and hence verify the exact values quoted.

11.8 Prediction and correction

An improvement in the accuracy obtainable using difference methods is possible if steps are taken, sometimes retrospectively to allow for the inaccuracies in approximating derivatives by differences. We shall describe only the simplest schemes of this kind and begin with a *prediction* method usually called the Adams method.

The forward difference estimate of y_{i+1}, namely

$$y_{i+1} = y_i + hy_i^{(1)} = y_i + hf(x_i, y_i) \tag{11.64}$$

(where the equation to be solved is again $dy/dx = f(x, y)$), would give exact results if y were a linear function of x in the range $x_i \leqslant x \leqslant x_i + h$. The idea behind the Adams method is to allow some relaxation of this and suppose that y can be adequately approximated by a parabola over the interval $x_{i-1} \leqslant x \leqslant x_{i+1}$. Then dy/dx can be approximated by a linear function in the same interval. That is

$$f(x, y) = \frac{dy}{dx} \approx a + b(x - x_i), \quad \text{for } x_i - h \leqslant x \leqslant x_i + h. \tag{11.65}$$

The values of a and b are fixed by the calculated values of f at x_{i-1} and x_i, which we may denote by f_{i-1} and f_i, as

$$a = f_i, \quad b = \frac{f_i - f_{i-1}}{h}. \tag{11.66}$$

Thus

$$y_{i+1} - y_i \approx \int_{x_i}^{x_i+h} \left[f_i + \frac{(f_i - f_{i-1})}{h} (x - x_i) \right] dx,$$

which yields

▶6. $$y_{i+1} = y_i + hf_i + \tfrac{1}{2}h(f_i - f_{i-1}). \tag{11.67}$$

The last term of this expression is seen to be a correction to result (11.64). That it is in some sense the second-order correction $[\tfrac{1}{2}h^2 y^{(2)}_{i-1/2}]$ to a first-order formula is apparent.

For using such a procedure, in addition to the usual requirement of the value of y_0, a value must be found for y_1 and hence f_1. This has to be obtained by other methods, e.g. a Taylor series expansion.

Improvements to simple difference formulae can also be obtained by using *correction* methods. Here a rough prediction of the next value y_{i+1} is first made and then this is used in a better predicting formula [not originally usable, since the formula itself requires a (rough) value of y_{i+1} for its evaluation]. The value of y_{i+1} is then recalculated using this better formula.

Such a scheme based on the forward difference formula might be,

(i) predict y_{i+1} using $y_{i+1} = y_i + hf_i$,
(ii) calculate f_{i+1} using this value,
(iii) recalculate y_{i+1} using $y_{i+1} = y_i + \tfrac{1}{2}h(f_i + f_{i+1})$. Here $\tfrac{1}{2}(f_i + f_{i+1})$ has replaced the f_i used in (i), since it better represents the average value of dy/dx in the interval $x_i \leqslant x \leqslant x_{i+1}$.

Steps (ii) and (iii) can be iterated if more accuracy is required.

Many more complex schemes of prediction and correction, in most cases combining the two in the same process, have been devised, but the reader is referred to more specialist texts for discussions of them.

11.9   Runge–Kutta methods

The Runge–Kutta method of integrating

$$\frac{dy}{dx} = f(x, y) \tag{11.68}$$

is a step-by-step process of obtaining an approximation for y_{i+1} starting from the value y_i. Among its advantages are that no functions other than f are used, no subsidiary differentiation is needed, and no subsidiary starting values need be calculated.

To be set against these advantages is the fact that f is evaluated using somewhat complicated arguments and that this has to be done several times for each increase in the value of i. However, once a procedure has been established, for example using a computer programme, the method usually gives good results.

The basis of the method is to simulate the (accurate) Taylor series for $y(x_i + h)$, not by calculating all the higher derivatives of y at the point x_i, but by taking a particular combination of the values of the first derivative of y evaluated at a number of carefully chosen points. Equation (11.68) is used to evaluate these first derivatives. The accuracy of the simulation can be made to be up to whatever power of h is desired, but naturally the greater the accuracy the more complex the calculation and, in any case, rounding errors cannot ultimately be avoided.

The setting up of the calculational scheme may be illustrated by considering the particular case in which second-order accuracy in h is required. To second order, the Taylor expansion is

$$y_{i+1} = y_i + hf_i + \frac{h^2}{2}\frac{df}{dx}\bigg|_{x_i}, \tag{11.69}$$

where

$$\frac{df}{dx}\bigg|_{x_i} = \left(\frac{\partial f}{\partial x} + \frac{\partial f}{\partial y}f\right)\bigg|_{x_i} \equiv \frac{\partial f_i}{\partial x} + \frac{\partial f_i}{\partial y}f_i, \tag{11.70}$$

the last step being merely the definition of the abbreviated notation.

We assume that this can be simulated by a form

$$y_{i+1} = y_i + \alpha_1 hf_i + \alpha_2 hf(x_i + \beta_1 h, y_i + \beta_2 hf_i), \tag{11.71}$$

that is, effectively using a weighted mean of the value of dy/dx at x_i and its value at some point yet to be determined. The object will be to choose values of α_1, α_2, β_1 and β_2, so that (11.71) coincides with (11.69) up to the coefficient of h^2.

Expanding the function f in the last term of (11.71) in a Taylor series of its own we obtain

$$f(x_i + \beta_1 h, y_i + \beta_2 hf_i)$$
$$= f(x_i, y_i) + \beta_1 h\frac{\partial f_i}{\partial x} + \beta_2 hf_i\frac{\partial f_i}{\partial y} + O(h^2), \tag{11.72}$$

where $\partial f_i/\partial x$ and $\partial f_i/\partial y$ are as defined by (11.70). Putting this result into (11.71) and rearranging in powers of h we obtain

$$y_{i+1} = y_i + (\alpha_1 + \alpha_2)hf_i + \alpha_2 h^2 \left(\beta_1 \frac{\partial f_i}{\partial x} + \beta_2 f_i \frac{\partial f_i}{\partial y} \right). \qquad (11.73)$$

Comparing this with (11.69) and (11.70) shows that there is in fact some freedom remaining in the choice of the α's and β's. In terms of an arbitrary α_1 (but $\neq 1$)

$$\alpha_2 = 1 - \alpha_1, \qquad \beta_1 = \beta_2 = \frac{1}{2(1 - \alpha_1)}. \qquad (11.74)$$

One possible choice is $\alpha_1 = \frac{1}{2}$, and then $\alpha_2 = \frac{1}{2}$, $\beta_1 = \beta_2 = 1$. In this case the procedure (equation (11.71)) can be summarized by

$$\begin{aligned} y_{i+1} &= y_i + \tfrac{1}{2}(a_1 + a_2), \\ a_1 &= hf(x_i, y_i), \\ a_2 &= hf(x_i + h, y_i + a_1). \end{aligned} \qquad (11.75)$$

Similar schemes giving higher-order accuracy in h can be devised. Two such schemes, which we give without derivation, are

(i) To order h^3,

$$\begin{aligned} y_{i+1} &= y_i + \tfrac{1}{6}(b_1 + 4b_2 + b_3), \\ b_1 &= hf(x_i, y_i), \\ b_2 &= hf(x_i + \tfrac{1}{2}h, y_i + \tfrac{1}{2}b_1), \\ b_3 &= hf(x_i + h, y_i + 2b_2 - b_1). \end{aligned} \qquad (11.76)$$

(ii) To order h^4,

$$\begin{aligned} y_{i+1} &= y_i + \tfrac{1}{6}(c_1 + 2c_2 + 2c_3 + c_4), \\ c_1 &= hf(x_i, y_i), \\ c_2 &= hf(x_i + \tfrac{1}{2}h, y_i + \tfrac{1}{2}c_1), \\ c_3 &= hf(x_i + \tfrac{1}{2}h, y_i + \tfrac{1}{2}c_2), \\ c_4 &= hf(x_i + h, y_i + c_3). \end{aligned} \qquad (11.77)$$

11.10   Higher-order equations

The discussion of numerical solutions of differential equations has so far been in terms of one dependent and one independent variable related by a first-order equation. It is straightforward to carry out the extension to the case of several dependent variables $y^{[r]}$, governed by R first-order equations

$$\frac{dy^{[r]}}{dx} = f^{[r]}(x, y^{[1]}, y^{[2]}, \ldots, y^{[R]}), \quad r = 1, 2, \ldots, R. \qquad (11.78)$$

The integration of these by the methods discussed for the one variable case presents no particular difficulty, provided that all of the equations are advanced through any particular step before any of them is taken through the following one.

Higher-order equations can be reduced to a set of simultaneous equations provided that they can be written in the form

$$\frac{d^R y}{dx^R} = f(x, y, y', \ldots, y^{(R-1)}), \tag{11.79}$$

where R is the order of the equation. To do this a new set of variables p_r is defined by

$$p_r = \frac{d^r y}{dx^r}, \quad r = 1, 2, \ldots, R - 1. \tag{11.80}$$

Equation (11.79) is then equivalent to the set of simultaneous first-order equations

$$\frac{dy}{dx} = p_1,$$

$$\frac{dp_r}{dx} = p_{r+1}, \quad r = 1, 2, \ldots, R - 2,$$

$$\frac{dp_{R-1}}{dx} = f(x, y, p_1, \ldots, p_{R-1}). \tag{11.81}$$

These can then be treated in the way indicated in the previous paragraph.

In practical problems it often happens that boundary conditions applicable to a higher-order equation do not consist of the function and all its derivatives at one particular point, but rather of (say) the value of the function at two separate (end-) points. In these cases an explicit step-by-step 'marching' scheme of solution is not available and other methods have to be tried.

One obvious method is to treat the problem as a 'marching one', but using a number of (intelligently guessed) initial values for the derivatives at the starting point. The aim then is to find by interpolation or some other form of iteration, those starting values for the derivatives which produce the given value of the function at the finishing point.

In some cases, for example, that of a second-order equation for $y(x)$ with constant coefficients and with the value of y given at the two end-points, the problem can be reduced by a differencing scheme to a matrix equation.

Consider the second-order equation

$$y'' + 2ky' + \mu y = f(x), \tag{11.82 a}$$

with the boundary conditions

$$y(0) = A, \qquad y(1) = B. \tag{11.82 b}$$

If (11.82 a) is replaced by a central difference equation

$$\frac{y_{i+1} - 2y_i + y_{i-1}}{h^2} + 2k \frac{y_{i+1} - y_{i-1}}{2h} + \mu y_i = f(x_i), \tag{11.83}$$

we obtain from it the recurrence relation

$$(1 + kh)y_{i+1} + (\mu h^2 - 2)y_i + (1 - kh)y_{i-1} = h^2 f(x_i). \tag{11.84}$$

For $h = 1/(N - 1)$ this is in exactly the form of the $N \times N$ tridiagonal matrix equation (11.28) with

$$b_1 = b_N = 1, \qquad c_1 = a_N = 0, \qquad a_i = (1 - kh),$$
$$b_i = (\mu h^2 - 2), \qquad c_i = (1 + kh), \qquad i = 2, 3, \ldots, N - 1,$$

and the y_i of that equation replaced here by $y_1 = A$, $y_N = B$ and the other $y_i = h^2 f(x_i)$. The solutions can be obtained as in (11.29) and (11.30).

11.11   Partial differential equations

The extension of previous methods to partial differential equations thus involving two or more independent variables, proceeds in a more or less obvious way. Rather than an interval divided into equal steps by the points at which solutions to the equations are to be found, a mesh of points in two or more dimensions has to be set up and all the variables given an increased number of subscripts.

Considerations of the stability, accuracy and feasibility of particular calculational schemes are in principle the same as for the one-dimensional case, but in practice are too complicated to be discussed here.

Rather than note generalities we are unable to pursue in any quantitative way, we will conclude by indicating in outline how two familiar partial equations of physical science can be set up for numerical solution. The first of these is Laplace's equation in two dimensions

$$\frac{\partial^2 \phi}{\partial x^2} + \frac{\partial^2 \phi}{\partial y^2} = 0, \tag{11.85}$$

with the value of ϕ given on the perimeter of a closed domain.

A grid with spacings Δx and Δy in the two directions is first chosen, so that, for example, x_i stands for the point $x_0 + i \Delta x$ and $\phi_{i,j}$ for the value $\phi(x_i, y_j)$. Next, using a second central difference formula, (11.85) is turned into

$$\frac{\phi_{i+1,j} - 2\phi_{i,j} + \phi_{i-1,j}}{(\Delta x)^2} + \frac{\phi_{i,j+1} - 2\phi_{i,j} + \phi_{i,j-1}}{(\Delta y)^2} = 0, \tag{11.86}$$

for $(i = 0, 1, \ldots, N)$ and $(j = 0, 1, \ldots, M)$. If $(\Delta x)^2 = \lambda(\Delta y)^2$, this then becomes a recurrence relationship

$$\phi_{i+1, j} + \phi_{i-1, j} + \lambda(\phi_{i, j+1} + \phi_{i, j-1}) = 2(1 + \lambda)\phi_{i, j}. \tag{11.87}$$

The boundary conditions in their simplest form (rectangular domain) mean that

$$(\phi_{0, j}), \quad (\phi_{N, j}), \quad (\phi_{i, 0}), \quad (\phi_{i, M}), \tag{11.88}$$

have predetermined values. [Clearly non-rectangular boundaries can be accommodated, either by more complex boundary value prescriptions or by using non-Cartesian coordinates.]

To find a set of values of $\phi_{i, j}$ satisfying (11.87), an initial guess is made, subject to the quantities listed in (11.88) having their fixed values, and then the values not on the boundary are iteratively adjusted in order to try to bring about condition (11.87) everywhere. We will not go into particular methods of doing this here, but when it has been achieved to within some required accuracy, the values $\phi_{i, j}$ give the solution of (11.85).

Our final example is based upon the diffusion equation in one dimension

$$\frac{\partial \phi}{\partial t} = K \frac{\partial^2 \phi}{\partial x^2}. \tag{11.89}$$

If $\phi_{i, j}$ stands for $\phi(x_0 + i \Delta x, t_0 + j \Delta t)$, then a forward difference representation of the time derivative and a central difference one of the spatial derivative lead to the following relationship

$$\frac{\phi_{i, j+1} - \phi_{i, j}}{\Delta t} = K \frac{\phi_{i+1, j} - 2\phi_{i, j} + \phi_{i-1, j}}{(\Delta x)^2}. \tag{11.90}$$

This allows the construction of an explicit scheme for generating the temperature distribution at later times, given it at an earlier one, namely

$$\phi_{i, j+1} = \alpha(\phi_{i+1, j} + \phi_{i-1, j}) + (1 - 2\alpha)\phi_{i, j}, \tag{11.91}$$

where $\alpha = K \Delta t/(\Delta x)^2$.

Although this scheme is explicit it is not a good one because of the asymmetric way the differences were formed. However, the effect of this can be minimized if we study the errors introduced in the following way.

From Taylor's series in time

$$\phi_{i, j+1} = \phi_{i, j} + \Delta t \frac{\partial \phi_{i, j}}{\partial t} + \tfrac{1}{2}(\Delta t)^2 \frac{\partial \phi_{i, j}}{\partial t^2} + \cdots, \tag{11.92}$$

(still using the notation that $\partial \phi_{i,j}/\partial t = \partial \phi(x_i, t_j)/\partial t$, etc.). Thus the first correction term to the left-hand side of (11.90) is

$$-\tfrac{1}{2}(\Delta t)\frac{\partial^2 \phi_{i,j}}{\partial t^2}.$$
(11.93 a)

The first omitted term on the right-hand side of the same equation is, by a similar argument,

$$-K\frac{2(\Delta x)^2}{4!}\frac{\partial^4 \phi_{i,j}}{\partial x^4}.$$
(11.93 b)

But using the fact that ϕ satisfies (11.89) we obtain

$$\frac{\partial^2 \phi}{\partial t^2} = \frac{\partial}{\partial t}\left(K\frac{\partial^2 \phi}{\partial x^2}\right) = K\frac{\partial^2}{\partial x^2}\left(\frac{\partial \phi}{\partial t}\right) = K^2\frac{\partial^4 \phi}{\partial x^4},$$
(11.94)

and so, to this accuracy, the two errors (11.93 a, b) can be made to cancel if α is chosen so that

$$-\frac{K^2(\Delta t)}{2} = -\frac{2K(\Delta x)^2}{4!},$$

i.e. $\alpha = \tfrac{1}{6}$.

11.12   Examples for solution

1. Use an iteration procedure to find to 4 significant figures the root of the equation $40x = \exp(x)$.

2. Using the Newton–Raphson procedure, find, correct to three decimal places, the root nearest to 7.0 of the equation $4x^3 + 2x^2 - 200x - 50 = 0$.

3. Show that the rearrangement method discussed in section 11.1 has only first-order convergence.

4. The square root of a number N is to be determined by means of the iteration process $x_{n+1} = x_n[1 - (N - x_n^2)f(N)]$. Find how to choose $f(N)$ so that the process has second-order convergence.
    Given that $\sqrt{7} = 2.65$ approximately, calculate $\sqrt{7}$ as accurately as a single application of the formula will allow.

5. The following table of values of a polynomial of low degree contains an error. Identify and correct the erroneous value and extend the table up to $x = 1.2$.

x	$p(x)$	x	$p(x)$
0.0	0.000	0.5	0.157
0.1	0.011	0.6	0.216
0.2	0.040	0.7	0.245
0.3	0.081	0.8	0.256
0.4	0.128	0.9	0.243

6. Use a Taylor series to solve the equation

$$\frac{dy}{dx} + xy = 0, \quad y(0) = 1,$$

evaluating $y(x)$ for $x = 0.0$ to 0.5 in steps of 0.1.

12
Calculus of variations

In a previous chapter it was shown how to find stationary values of functions of a single variable $f(x)$, of several variables $f(x, y, \ldots)$ and of constrained variables $f(x, y, \ldots)$ subject to $g_i(x, y, \ldots) = 0$, $(i = 1, 2, \ldots, m)$. In all these cases the forms of the functions f and g_i were known and the problem was one of finding suitable values of the variables $x, y, \ldots$.

We now turn to a different kind of problem, one in which there are not free variables which must be chosen in order to bring about a particular condition for a given function, but in which the functions are free and must be chosen to bring about a particular condition for a given expression which depends upon these functions.

To give a more concrete example of the type of question to be answered, we may ask the following. 'Why does a uniform rope suspended between two points take up the shape it does? Why doesn't it hang in an arc of a circle or in the form of three sides of a rectangle? Is it possible to predict the shape in which it will hang, that is to find a *function*, $y = y(x)$, that gives the vertical height of the rope as a function of horizontal position?'

The answers to the first two questions lie directly in the realm of physics and the 'umbrella' answer would be that the rope takes up the shape it does because for that shape the gravitational potential energy is the lowest possible consistent with having the rope of a certain length and with having its ends fixed.† On the basis of the general physical statement that 'the gravitational potential energy is ... ends fixed', the calculus of variations, which forms the subject of this chapter, aims to answer the third question affirmatively and to produce the explicit function $y = y(x)$.

Two other more transparent examples may also be given of the type of question involved. 'Along what curve joining two fixed points is the total path-length of the curve a minimum?' 'In what shape should a fixed length of fencing be arranged so as to enclose the largest possible area?' The answers to the questions are physically obvious, but the important

---

† In this particular case an alternative analysis based upon there being no net force acting upon any small part of the rope, is available.

point to notice here is that the *questions* do not contain any particular *functions*. It is the *answers* which give the functions [and the calculus of variations which provides the answers].

What the questions do provide are the general principles by which the particular functions are to be determined; these principles must be expressible in mathematical form. The mathematical form common to the three examples cited is an *integral*. In each case the quantity which has to be maximized or minimized by an appropriate choice of a function may be expressed as an integral involving the function and the variables describing the geometry of the situation [which are at the same time the variables upon which the function depends].

In the case of the rope, each elementary piece of the rope has a gravitational potential energy proportional to its vertical height above an arbitrary but common zero and to the length of the piece. The total potential energy is thus given by an integral for the whole rope of such elementary contributions. For the shortest-path and fencing problems the quantities to be minimized and maximized (respectively) are even more obviously expressed by integrals whose integrands involve the functions giving the shape of the shortest path or the lay-out of the fencing.

So we are led by this different type of question to study the value of an integral, for which the integrand has a specified form in terms of a function and its derivatives, and how that value changes when the form of the function is varied. Specifically we aim to find the function which makes the integral *stationary*, that is the function which makes the value of the integral a (local) maximum or minimum.

12.1   Euler's equation

As has been our practice elsewhere, we will assume that all the functions we need to deal with are sufficiently smooth and differentiable.

Let us take as the integral for study

$$I = \int_a^b F(y, y', x)\, \mathrm{d}x, \tag{12.1}$$

where a, b and the form of F are fixed by given considerations, e.g. the physics of the problem, but the curve $y = y(x)$ has to be chosen so as to give a stationary value to I, which clearly is a function [more technically a functional] of the curve, i.e. $I = I(y)$.

As an example for the form F, we could consider the total energy of a particle moving in a harmonic potential well $V = \frac{1}{2}ky^2$,

$$E = \tfrac{1}{2}m\dot{y}^2 + \tfrac{1}{2}ky^2.$$

Here the independent variable x of (12.1) has been replaced by time, but the form $F(u_1, u_2, u_3) = \frac{1}{2}mu_2^2 + \frac{1}{2}ku_1^2$, illustrates what is meant by $F(y, y', x)$ in (12.1).

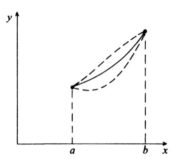

Fig. 12.1 Possible paths for the integral of equation (12.1). The solid line is the curve along which the integral is assumed stationary. The dashed curves represent small variations of the path from this.

Referring to fig. 12.1, we now want to choose a function $y = y(x)$ (given, say, by the solid line) such that first-order small changes in it (for example the two dashed paths) will make only second-order changes in the values of I. Put in other words, we require that the changes in the contributions to I of the various factors and parts in F when the path is varied, should cancel each other to a first approximation.

Writing this in more mathematical form, let us suppose that $y = y(x)$ is the required function for making I stationary, and consider replacing $y(x)$ by $y(x) + \alpha\eta(x)$ where $\eta(x)$ is another function and α is small. Here $\eta(x)$ is arbitrary, but like $y(x)$ it is assumed to have sufficiently amenable mathematical properties. Our requirement that the change in I is $O(\alpha^2)$ can thus be expressed as

$$\left.\frac{\mathrm{d}I}{\mathrm{d}\alpha}\right|_{\alpha=0} = 0 \quad \text{for all } \eta(x). \tag{12.2}$$

Putting the new form into (12.1) explicitly gives

$$I(y, \alpha) = \int_a^b F(y + \alpha\eta, y' + \alpha\eta', x)\,\mathrm{d}x,$$

and expanding in a Taylor series in α, this becomes

$$I(y, \alpha) = \int_a^b F(y, y', x)\,\mathrm{d}x + \int_a^b \left(\frac{\partial F}{\partial y}\alpha\eta + \frac{\partial F}{\partial y'}\alpha\eta'\right)\mathrm{d}x + O(\alpha^2).$$

With this form for $I(y, \alpha)$, (12.2) implies that for all $\eta(x)$,

$$\int_a^b \left(\frac{\partial F}{\partial y} \eta + \frac{\partial F}{\partial y'} \eta' \right) dx = 0.$$

Integrating the second term by parts, yields

►1.
$$\int_a^b \left\{ \frac{\partial F}{\partial y} - \frac{d}{dx} \left(\frac{\partial F}{\partial y'} \right) \right\} \eta(x) \, dx + \left[\eta \frac{\partial F}{\partial y'} \right]_a^b = 0. \tag{12.3}$$

If we restrict ourselves to cases with fixed end-points, i.e. not only a and b are given but also $y(a)$ and $y(b)$, this result can be simplified. Such a restriction means that only $\eta(x)$ such that $\eta(a) = \eta(b) = 0$ are to be considered, in which case the last term on the left-hand side of (12.3) vanishes since it equals zero at both end-points. It is then easy to see that, since η is arbitrary (12.3) requires

$$\frac{\partial F}{\partial y} = \frac{d}{dx} \left(\frac{\partial F}{\partial y'} \right), \tag{12.4}$$

i.e. a differential equation for $y = y(x)$ since the form of F is known. This equation is known as the Euler or Euler–Lagrange equation.

12.2   Special cases and simple examples

In certain cases a first integral of the Euler equation can be obtained for a general form of F.

(i) *F does not contain y explicitly.* In this case $\partial F/\partial y = 0$, and (12.4) can be integrated immediately giving

$$\frac{\partial F}{\partial y'} = \text{constant}. \tag{12.5}$$

(ii) *F does not contain x explicitly.* Multiplying both sides of (12.4) by y' and using

$$\frac{d}{dx} \left(y' \frac{\partial F}{\partial y'} \right) = y' \frac{d}{dx} \left(\frac{\partial F}{\partial y'} \right) + y'' \frac{\partial F}{\partial y'},$$

we obtain

$$y' \frac{\partial F}{\partial y} + y'' \frac{\partial F}{\partial y'} = \frac{d}{dx} \left(y' \frac{\partial F}{\partial y'} \right).$$

But since F is a function of y and y' only, and not of x, the left-hand side of this is the total derivative of F, namely dF/dx. Hence

$$F - y' \frac{\partial F}{\partial y'} = \text{constant}. \tag{12.6}$$

With these two special cases many results of physical and geometrical interest can be obtained [and of course (12.4) may often be solved on an *ad hoc* basis once F is given]. We begin with two very simple examples for which the answers are obvious before the calculation is done, so that the working of the method will be transparent at all points.

Example 12.1. Show that the shortest path between two points is a straight line.

Let the two points be A and B (fig. 12.2).

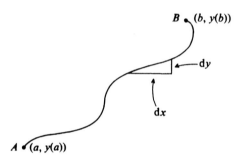

Fig. 12.2 The shortest-path problem of example 12.1.

The length of an element of path ds is given by

$$\mathrm{d}s = [(\mathrm{d}x)^2 + (\mathrm{d}y)^2]^{1/2} = (1 + y'^2)^{1/2}\,\mathrm{d}x,$$

and hence the total path length along a curve joining A to B is given by

$$L = \int_a^b (1 + y'^2)^{1/2}\,\mathrm{d}x. \tag{12.7}$$

This is the case whatever path is chosen, and we are now going to apply our previous results to select a path which makes L stationary [clearly a minimum].

Since the integral contains neither y nor x explicitly, we may use either (12.5) or (12.6). We will use (12.5) and obtain

$$\text{constant} = k = \frac{\partial F}{\partial y'} = \frac{y'}{(1 + y'^2)^{1/2}}.$$

This is easily rearranged and integrated to give

▶2. $$y = \frac{k}{(1 - k^2)^{1/2}}\, x + c, \tag{12.8}$$

i.e. the expected straight line in the form $y = mx + c$.

▶3. Use (12.7) and (12.8) to show that, as expected, the minimum value of L is given by

$$L^2 = [y(b) - y(a)]^2 + (b - a)^2.$$

▶4. Use (12.6) to obtain the result (12.8).

Example 12.2. Find the closed convex curve of given circumference which encloses the greatest area.

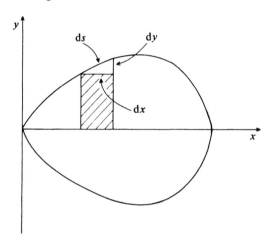

Fig. 12.3 The largest enclosed area problem of example 12.2.

Let the curve pass through the origin and suppose it is symmetric with respect to the x-axis [this assumption is not essential]. We will use s the distance along the curve (measured from the origin) as the independent variable and y as the dependent one. Then $y = y(s)$ is such that $y(0) = y(l/2) = 0$. The element of area dA (shaded in fig. 12.3) is given by

$$dA = y\, dx = y[(ds)^2 - (dy)^2]^{1/2} = y(1 - y'^2)^{1/2}\, ds,$$

and the total area by

$$A = 2\int_0^{l/2} y(1 - y'^2)^{1/2}\, ds. \tag{12.9}$$

This is an example of case (ii) and so using (12.6) we obtain as a first integral of the differential equation for y,

▶5. $y(1 - y'^2)^{1/2} + yy'^2(1 - y'^2)^{-1/2} = k,$

giving, on rearrangement,

▶6. $ky' = \pm(k^2 - y^2)^{1/2}.$

On integration and use of $y(0) = 0$, this gives

$$y/k = \sin(s/k),$$

and the other end-point condition, $y(l/2) = 0$, fixes the value of k as $l/2\pi$ to yield

$$y = \frac{l}{2\pi} \sin\frac{2\pi s}{l}. \qquad (12.10)$$

From this $dy = \cos(2\pi s/l)\, ds$ and since $(ds)^2 = (dx)^2 + (dy)^2$ we obtain also that $dx = \pm \sin(2\pi s/l)\, ds$. This in turn can be integrated and, using $x(0) = 0$, gives x in terms of s as,

$$x - \frac{l}{2\pi} = -\frac{l}{2\pi}\cos\frac{2\pi s}{l}.$$

We thus obtain the expected result that x and y lie on the circle

$$\left(x - \frac{l}{2\pi}\right)^2 + y^2 = \frac{l^2}{4\pi^2},$$

of radius $l/2\pi$.

▶7. Verify directly from (12.9) and (12.10) that $A = l^2/4\pi$.

These two examples have been carried through at some length even though the answers are more easily obtainable in other ways, expressly so that the method is apparent and so that the way it works can be filled in mentally in terms of the known answer at almost every step. Our next example does not have such an intuitively obvious answer.

Example 12.3. Two rings, each of radius a, are placed parallel with their centres $2b$ apart and on a common normal. An axially symmetric soap film is formed between them but does not cover the ends of the rings. Find the shape assumed by the film.

Creating a soap film surface requires an energy γ per unit area [numerically equal to the surface tension of the soap solution], and so the stable shape of a soap film, being the one which minimizes the energy, is also the one minimizing the surface area [neglecting gravitational effects]. It is obvious that any surface shape such as the one shown dashed in fig. 12.4 (a) cannot be a minimum, but it is not clear that some shape intermediate between the solid curve in (a) [with area $4\pi ab$ – or twice this for the double surface of the film] and the form shown in (b) [area – $2\pi a^2$]

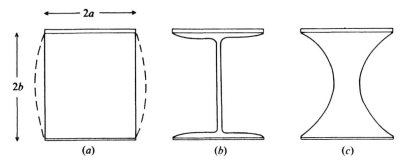

Fig. 12.4 Possible soap films between two parallel circular rings. See text for explanation.

will produce a lower total area than both of these extremes.† If there is such a shape (fig. 12.4 (c)), it will be that which best compromises between the criteria of the minimum ring-to-ring distance on the film surface of (a), and the minimum waist measurement of the surface in (b).

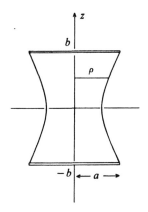

Fig. 12.5 Coordinate system for the minimization of the soap film area.

We take cylindrical polar coordinates as in fig. 12.5, and, with no azimuthal dependence of ρ, let the radius at height z be $\rho(z)$ with $\rho(\pm b) = a$.

Counting only one side of the film, the element of surface area between z and $z + dz$ is

$$2\pi\rho[(dz)^2 + (d\rho)^2]^{1/2},$$

† With no pressure difference between the inside and outside of the film, such shapes are not in fact possible, but are used here to illustrate the kind of reasoning involved.

and so the total surface

$$S = 2\pi \int_{-b}^{b} \rho(1 + \rho'^2)^{1/2} \, dz. \tag{12.11}$$

This expression does not contain z explicitly, so to obtain an equation for ρ which minimizes S we apply (12.6), obtaining

$$\rho(1 + \rho'^2)^{1/2} - \rho\rho'^2(1 + \rho'^2)^{-1/2} = k,$$

which on rearrangement gives an explicit form for ρ' and hence for ρ,

▶8. $\operatorname{arcosh} \dfrac{\rho}{k} = \dfrac{z}{k} + c.$

Using the boundary conditions $\rho(\pm b) = a$, requires that $c = 0$ and that k is such that $a/k = \cosh b/k$. [If b/a is too large, no such k can be found.] Thus we see that a minimizing curve is possible,

$$\rho/k = \cosh z/k, \tag{12.12}$$

and in profile the soap film is a catenary with the minimum distance from the axis equal to k. This analysis is taken a little further in example 3 of section 12.7.

12.3   Fermat's principle

Fermat's principle of geometrical optics states that a ray of light travelling in a region of variable refractive index follows a path such that the total optical path length [physical length × refractive index] is a minimum.

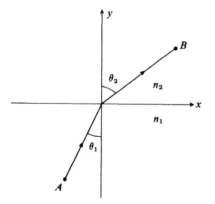

Fig. 12.6 Path of light ray at the plane interface between media of refractive indices n_1 and n_2 ($< n_1$).

From this, Snell's law concerning refraction at an interface can be deduced. Let the interface be $y =$ constant (fig. 12.6) and let it separate two regions of refractive indices n_1 and n_2. For a ray which passes through A and B, its element of physical path length is $ds = (1 + y'^2)^{1/2} \, dx$ and its total optical path length

$$P = \int_A^B n(y)(1 + y'^2)^{1/2} \, dx.$$

Applying (12.6) we obtain, after some rearrangement,

$$n(y)(1 + y'^2)^{-1/2} = k, \tag{12.13}$$

where k is a constant. Recalling that y' is the tangent of the angle ϕ between the instantaneous direction of the ray and the x-axis, this *general* result [not dependent on the configuration presently under consideration] can be put in the form,

$$n \cos \phi = \text{constant} \tag{12.14}$$

along a ray, even though n and ϕ individually vary.

For our particular configuration n is constant in each medium and therefore by (12.13) so is y'. Thus the rays travel in straight lines in each medium [as we have anticipated in fig. 12.6, but not assumed in our working], and since k is the same along the *whole* path we have $n_1 \cos \phi_1 = n_2 \cos \phi_2$, or, in terms of the conventional angles in the figure,

$$n_1 \sin \theta_1 = n_2 \sin \theta_2. \tag{12.15}$$

A more complicated example is given in section 12.7.

12.4   Some extensions

It is quite possible to relax many of the restrictions we have imposed hitherto, for example, not to use fixed end-points but merely end-points constrained to lie on given curves. We will not pursue this particular aspect, but will list briefly some other extensions:

(i) *More than one dependent variable.* Here $F = F(y_1, y_1', y_2, y_2', \ldots, y_n, y_n', x)$ where each $y_i = y_i(x)$. The analysis proceeds as before leading to n separate and simultaneous equations,

$$\frac{\partial F}{\partial y_i} = \frac{d}{dx}\left(\frac{\partial F}{\partial y_i'}\right), \quad i = 1, \ldots, n. \tag{12.16}$$

(ii) *Higher derivatives involved.*

▶9. Show by the original method and repeated integration by parts, that, if $F = F(y, y', y'', \ldots, y^{(n)}, x)$, then the required minimizing function $y = y(x)$ satisfies†

$$\frac{\partial F}{\partial y} - \frac{\mathrm{d}}{\mathrm{d}x}\left(\frac{\partial F}{\partial y'}\right) + \frac{\mathrm{d}^2}{\mathrm{d}x^2}\left(\frac{\partial F}{\partial y''}\right) - \cdots + (-1)^n \frac{\mathrm{d}^n}{\mathrm{d}x^n}\left(\frac{\partial F}{\partial y^{(n)}}\right) = 0.$$

(12.17)

(iii) *Several independent variables.* For cases in which the integral is multiple

$$I = \iint \cdots \int F\left(y, \frac{\partial y}{\partial x_1}, \frac{\partial y}{\partial x_2}, \ldots, \frac{\partial y}{\partial x_n}, x_1, x_2, \ldots, x_n\right) \mathrm{d}x_1 \, \mathrm{d}x_2 \ldots \mathrm{d}x_n,$$

the same kind of analysis as used before leads to an equation for y,

$$\frac{\partial F}{\partial y} = \sum_1^n \frac{\partial}{\partial x_i}\left(\frac{\partial F}{\partial y_{x_i}}\right).$$

(12.18)

Here y_{x_i} stands for $\partial y/\partial x_i$.

12.5   Hamilton's principle and Lagrange's equations

The mechanics of systems whose configuration can be uniquely specified by a number of coordinates q_i [usually distances and angles] together with time, is a subject which can be approached from different directions. Much of the original work and many of the equations and definitions in this field are associated with the names of Hamilton and Lagrange. The choice of a postulatory starting point is not unique, but in order to illustrate the use of the calculus of variations we will take as a postulate Hamilton's principle and restrict our attention to cases where the forces involved are derivable from a potential V.

Hamilton's principle states that if a system moves from one configuration at time t_0 to another at time t_1 then the path of the motion (in configuration space q_i, $\dot{q}_i$) is such as to make

$$\mathscr{L} = \int_{t_0}^{t_1} L \, \mathrm{d}t$$

(12.19)

stationary. The Lagrangian L is defined in terms of the kinetic energy T and the potential energy V by $L = T - V$. Here V is a function of the q_i only, not of the $\dot{q}_i$.

---

† Equation (12.17) holds only if the end-point conditions are such that the contribution from the definite integrals vanishes – otherwise it appears on the right-hand side.

In this spirit we apply the Euler equation to $\mathcal{L}$ and obtain as in (12.16)

$$\frac{d}{dt}\left(\frac{\partial L}{\partial \dot{q}_i}\right) = \frac{\partial L}{\partial q_i}, \quad i = 1, \ldots, n, \tag{12.20}$$

or more explicitly

▶10.
$$\frac{d}{dt}\left(\frac{\partial T}{\partial \dot{q}_i}\right) - \frac{\partial T}{\partial q_i} = -\frac{\partial V}{\partial q_i}, \quad i = 1, \ldots, n, \tag{12.21}$$

These equations, one for each $\dot{q}_i$, are called **Lagrange's equations**. A proof that they lead to the conservation of energy, $T + V = $ constant, at least in the case of T being a homogeneous quadratic in the $\dot{q}_i$ is constructed in example 6 of section 12.7.

As a specific illustration, we will derive the velocity of small wave motions on a string, using this approach. We are in fact here considering an extension of the above to a case involving one isolated independent coordinate t, together with a continuum [the q_i become the continuous variable x], the expressions for T and V becoming integrals over x rather than sums over i. Specifically (referring to fig. 12.7), we have

$$T = \int_0^l \frac{\rho}{2}\left(\frac{\partial y}{\partial t}\right)^2 dx, \quad V = \int_0^l \frac{P}{2}\left(\frac{\partial y}{\partial x}\right)^2 dx,$$

where ρ and P are the local density and tension of the string – both may depend on x.

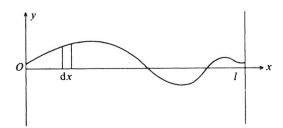

Fig. 12.7 Transverse displacement of a string of length l.

Expression (12.19) then becomes

$$\mathcal{L} = \tfrac{1}{2}\int_{t_0}^{t_1} dt \int_0^l \left[\rho\left(\frac{\partial y}{\partial t}\right)^2 - P\left(\frac{\partial y}{\partial x}\right)^2\right] dx,$$

and since y does not appear explicitly, by following (12.18) we obtain

$$-\frac{\partial}{\partial x}\left[P\frac{\partial y}{\partial x}\right] + \frac{\partial}{\partial t}\left[\rho\frac{\partial y}{\partial t}\right] = 0. \tag{12.22}$$

If, in addition, ρ and P do not depend on x or t then

$$\frac{\partial^2 y}{\partial x^2} = \frac{1}{c^2} \frac{\partial^2 y}{\partial t^2} \quad \text{with } c^2 = \frac{P}{\rho}. \tag{12.23}$$

Other examples for solution will be found at the end of the chapter and Lagrange's equations are mentioned again in other chapters.

12.6   Conditional variation

Just as the problem of finding stationary values of $f(x, y)$, where x, y are not independent but constrained by $g(x, y) = 0$, is solved by means of Lagrange's undetermined multipliers, so the corresponding problem in the variational calculus is solved by analogous methods. Use of the multipliers is discussed more fully in section 1.14, but in outline is as follows.

Suppose that we require to find stationary values of

$$I = \int_a^b F(y, y', x) \, dx, \tag{12.24}$$

but subject to the constraint that the value of

$$J = \int_a^b G(y, y', x) \, dx \tag{12.25}$$

is given. The method of attack is to take a new variable $K = I + \lambda J$ and find the stationary values of that. [We know $\delta J = 0$, and if $\delta K = 0$ then we will have $\delta I = 0$, where δX stands for the first-order 'variation' in X when a first-order change is made in the function $y = y(x)$.]

This process will yield

$$\frac{\partial F}{\partial y} - \frac{d}{dx}\left(\frac{\partial F}{\partial y'}\right) + \lambda\left[\frac{\partial G}{\partial y} - \frac{d}{dx}\left(\frac{\partial G}{\partial y'}\right)\right] = 0, \tag{12.26}$$

and this, together with constraint (12.25), will, in principle, yield the required solution $y(x)$.

We will not carry the discussion of this general approach very far, but rather illustrate it with a full example. The reader should identify for himself the quantities and conditions in this particular example corresponding to those just mentioned in connection with the more general problem.

Example 12.4. To find the shape assumed by a uniform rope when suspended by its ends from two points at equal heights.

This is a problem which can be solved straightforwardly using previously described methods, by taking the distance from one end of the rope as the independent variable. It is also one for which the solution – a catenary – is widely known even among students unfamiliar with the calculus of variations. For these two reasons we choose it to illustrate the method of conditional variation, using x (see fig. 12.8) as the independent variable.

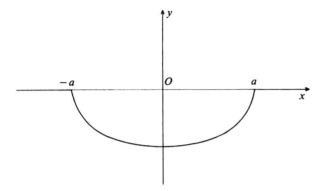

Fig. 12.8 Coordinate system for a uniform rope with fixed ends suspended under gravity.

Let the rope of length $2L$ be suspended between the points $x = \pm a$, $y = 0$ ($L > a$), and have uniformity density ρ per unit length. Then we require to find a stationary value of its gravitational potential energy

$$I = -\rho g \int y \, ds = -\rho g \int_{-a}^{a} y(1 + y'^2)^{1/2} \, dx,$$

for small changes in the form of the rope, but subject to its total length $\int ds$ remaining constant,

$$J = \int_{-a}^{a} (1 + y'^2)^{1/2} \, dx = 2L. \tag{12.27}$$

As we have already indicated, we take $K = I + \lambda J$ (omitting the constant $-\rho g$ from I for brevity) and apply the Euler equation to that. With

$$K = \int_{-a}^{a} (y + \lambda)(1 + y'^2)^{1/2} \, dx,$$

the independent variable is not present and so by (12.6) a first integral is

$$(y + \lambda)(1 + y'^2)^{1/2} - (y + \lambda)y'^2(1 + y'^2)^{-1/2} = k,$$

which reduces to

▶11. $y'^2 = [(y + \lambda)/k]^2 - 1.$

Making the substitution $(y + \lambda) = k \cosh z$, this can be integrated immediately to

$$k \operatorname{arcosh} \frac{y + \lambda}{k} = x + c, \qquad (12.28)$$

c being the constant of integration.

We now have three unknown constants λ, k, c, and they must be evaluated using the two end conditions $y(\pm a) = 0$, and the constraint (12.27).

The end conditions give

$$\cosh \frac{a + c}{k} = \frac{\lambda}{k} = \cosh \frac{-a + c}{k}, \qquad (12.29)$$

and since $a \neq 0$, these imply $c = 0$ and $\lambda/k = \cosh (a/k)$.

Putting $c = 0$, the constraint, in which $y' = \sinh (x/k)$, takes the form

$$2L = \int_{-a}^{a} [1 + \sinh^2 (x/k)]^{1/2} \, dx$$
$$= 2k \sinh (a/k). \qquad (12.30)$$

Collecting together (12.28), (12.29) and (12.30), the form adopted by the free hanging rope is thus

$$y(x) = k \cosh \left(\frac{x}{k} \right) - k \cosh \left(\frac{a}{k} \right), \qquad \text{[a catenary]}$$

where k is the solution of $\sinh (a/k) = L/k$.

12.7   Examples for solution

1. A surface of revolution, whose equation in cylindrical polar coordinates is $\rho = \rho(z)$ $(\rho = (x^2 + y^2)^{1/2})$, is bounded by the circles $\rho = a$, $z = \pm c$ $(a > c)$. The function $\rho(z)$ is chosen so that the surface integral $I = \int \rho^{-1/2} \, dA$ is stationary for small variations, dA being the element of area. Show that $\rho(z) = K + z^2/4K$, where $K = \frac{1}{2}[a \pm (a^2 - c^2)^{1/2}]$.

2. Show that the least value of the integral

$$\int_{P_1}^{P_2} \frac{(1 + y'^2)^{1/2}}{y} \, dx,$$

where P_1 is $(-1, 1)$ and P_2 is $(1, 1)$, is $2 \ln (1 + \sqrt{2})$. Assume that the Euler equation gives a minimizing curve.

3. Reference example 12.3 and fig. 12.4 (page 338).
(a) Putting $b = kx$ and $a = \lambda b$, show that the necessary condition

$a/k = \cosh b/k$ only has a solution if λ is greater than a certain value λ_m, and that when $\lambda = \lambda_m$, x satisfies $x = \coth x$.

(b) From tables verify that $x \approx 1.20$ and $\lambda_m \approx 1.51$ and hence that the minimum radius of the film is $0.83b$.

(c) Show that the stationary value of the surface area $S = 2\pi \times [kb + a(a^2 - k^2)^{1/2}]$.

(d) For $b \ll a$ find an approximate expression for k, and show that the corresponding value of S is less than that $(4\pi ab)$ of fig. 12.4 (a) [i.e. that a curve like fig. 12.4 (c) does reduce the total surface area]. [Comparison with fig. 12.4 (b) involves detailed physics considerations outside the scope of this book. See also the footnote to example 12.3.]

4. The refractive index n of a medium is a function of the distance r from a fixed point O only. Prove that the equation of a light ray (assumed to lie in a plane through O) travelling in the medium satisfies (in polar coordinates)

$$\frac{1}{r^2}\left(\frac{dr}{d\theta}\right)^2 = \frac{r^2}{a^2}\frac{n^2(r)}{n^2(a)} - 1,$$

where a is the distance of the ray from O at the point at which $dr/d\theta = 0$.

If $n = [1 + (\alpha^2/r^2)]^{1/2}$ and the ray starts and ends far from O, find the angle through which the ray is turned if its minimum distance from O is a.

5. The Lagrangian $L(\mathbf{x}, t)$ for a pi-meson is given by $\frac{1}{2}(\dot{\phi}^2 - |\nabla\phi|^2 - \mu^2\phi^2)$, where μ is the meson mass and $\phi(\mathbf{x}, t)$ its wave function. Assuming Hamilton's principle find the equation satisfied by ϕ.

6. (a) For a system described in terms of coordinates q_i and t, show that if t does not appear explicitly in the expressions for x, y, z ($x = x(q_i, t)$, etc.), then the kinetic energy T is a homogeneous quadratic function of the $\dot{q}_i$ (it may also involve the q_i). Deduce that $\dot{q}_i (\partial T/\partial \dot{q}_i) = 2T$.

(b) By multiplying (12.21) by $\dot{q}_i$ and summing, and expressing dT/dt in terms of q_i and $\dot{q}_i$, show that $d(T + V)/dt = 0$.

7. For a system specified by a coordinate q and t, show that the equation of motion is unchanged if the Lagrangian $L(q, \dot{q}, t)$ is replaced by

$$L_1 = L + \frac{d\phi(q, t)}{dt},$$

ϕ being an arbitrary function. Deduce that the equation of motion of a particle, which moves in one dimension x subject to a force $= -dV(x)/dx$ (x being measured from a point O), is unchanged if O

is forced to move with constant velocity v (x still being measured from O).

8. Derive the differential equations for the polar coordinates r, θ of a particle of unit mass moving in a field of potential $V(r)$. Find the form of V if the path of the particle is given by $r = a \sin \theta$.

9. You are provided with a light line of length $\pi a/2$ and some lead shot of total mass M. Use a variational method to determine how the lead shot must be distributed along the line if the loaded line is to hang in an arc of a circle of radius a when its ends are attached to two points at the same height. [Measure the distance s along the line from its centre.]

13
General eigenvalue problem

We have seen in chapter 12 that the problem of finding a curve which makes the value of a given integral stationary when the integral is taken along the curve, results in each case in a differential equation for the curve. It is not a great extension to ask whether this may be used to solve differential equations, by setting up a suitable variational problem and then seeking ways other than the Euler equation of finding or estimating stationary solutions.

13.1   Laplace's equation

Let us consider again probably the most familiar of all equations in physical mathematics, Laplace's equation

$$\nabla^2 \phi = 0,$$

and ask if this can be related to a variational problem. With x, y, z as independent variables it can be written

$$\frac{\partial}{\partial x}\left(\frac{\partial \phi}{\partial x}\right) + \frac{\partial}{\partial y}\left(\frac{\partial \phi}{\partial y}\right) + \frac{\partial}{\partial z}\left(\frac{\partial \phi}{\partial z}\right) = 0.$$

Reference to chapter 12 shows that Laplace's equation has the same form as (12.18), provided F is chosen so that $\partial F/\partial \phi = 0$ and $\partial F/\partial \phi_x = \phi_x$, where ϕ_x stands for $\partial \phi/\partial x$, and similarly for y and z. It follows immediately that the required form for F is $F = \frac{1}{2}(\phi_x^2 + \phi_y^2 + \phi_z^2)$, or, as an integral,

$$I = \frac{1}{2}\iiint (\nabla \phi)^2 \, dx \, dy \, dz.$$

Interpreted physically, the solutions of Laplace's equation are thus the ones which make the mean squared gradient of the potential stationary [a minimum].

This is an example resulting in an unconstrained variation problem. Our next example will be based on restricted variation.

13.2   Sturm–Liouville equation

This equation has already been discussed in chapter 7, equation (7.10), and the same notation will be used here.

Suppose we search for stationary values of the integral

$$I = \int_a^b (p(x)y'^2(x) + q(x)y^2(x))\, dx, \tag{13.1}$$

with $y(a) = y(b) = 0$ and p and q any sufficiently smooth and differentiable functions of x. However, in addition we impose a normalization condition

$$J = \int_a^b \rho(x)y^2(x)\, dx = \text{constant.} \tag{13.2}$$

Here $\rho(x)$ is a positive weight function defined in $a \leqslant x \leqslant b$, but which may in particular cases be a constant.

Then, as in section 12.6, we use undetermined Lagrange multipliers,† and consider $K = I - \lambda J$ given by

$$K = \int_a^b [py'^2 + (q - \lambda\rho)y^2]\, dx.$$

On application of the Euler equation this yields

$$\frac{d}{dx}\left(p\frac{dy}{dx}\right) - qy + \lambda\rho y = 0, \tag{13.3}$$

which is exactly the Sturm–Liouville equation (7.10), with eigenvalue λ. Now since both I and J are quadratic in y and its derivative, finding stationary values of K is equivalent to finding stationary values of I/J. Thus we have the important result that 'finding functions y which minimize I/J is equivalent to finding functions y which are solutions of the Sturm–Liouville equation'.

Of course this does not tell us how to find such functions y, and to have to do it by solving (13.3) directly, naturally defeats the purpose of the exercise. We will see in the next section how some progress can be made. It is worth recalling that the functions $p(x)$, $q(x)$ and $\rho(x)$ can have many different forms, and so (13.3) represents quite a wide variety of equations.

Finally in this section we recall some properties of the solutions of the Sturm–Liouville equation and deduce one further result concerning the value of I/J.

---

† We use $-\lambda$, rather than λ, so that final equation (13.3) appears in the conventional Sturm–Liouville form.

The eigenvalues λ_i are real and will be assumed non-degenerate (for simplicity). If the corresponding normalized eigenfunctions are y_i, then ((7.18) and (7.19))

$$\int_a^b \rho y_i y_j \, dx = \delta_{ij},$$ (13.4)

and

$$\int_a^b (y_j' p y_i' + y_j q y_i) \, dx = \lambda_i \, \delta_{ij}.$$ (13.5)

In obtaining these results it was assumed that

$$y_j p \frac{dy_i}{dx} \bigg|_{x=a}^{x=b} = 0,$$ (13.6)

which can be satisfied by $y(a) = y(b) = 0$, but also by many other sets of boundary conditions (see example 13.1).

We see at once that, if the function $y(x)$ minimizes I/J, i.e. satisfies the Sturm–Liouville (S–L) equation, putting $y_i = y_j = y$ in (13.4) and (13.5) yields J and I respectively on the left-hand sides, and thus that the minimized value of I/J is just the eigenvalue λ – introduced originally as the undetermined multiplier.

▶1. For a function y satisfying the S–L equation, show directly by multiplying (13.3) through by y and integrating by parts, that provided (13.6) is satisfied, $\lambda = I/J$.

13.3   Estimation of eigenvalues and eigenfunctions

With the eigenvalues λ_i of the S–L equation being the stationary values of I/J, it follows that the absolute minimum of I/J is equal to the lowest eigenvalue λ_0. Thus *any* evaluation λ of I/J gives an upper bound for λ_0. Notice that here we have left the minimizing problem directly and made a statement about a calculation in which no actual minimization is necessary.

Further we will now show that the estimate λ obtained is a better estimate of λ_0 than the estimating [guessed] function y is of y_0 (the true eigenfunction corresponding to λ_0). The sense in which 'better' is used here will be clearer from the final result.

We first expand the estimate or trial function y in terms of the complete set y_i

$$y = y_0 + c_1 y_1 + c_2 y_2 + \cdots,$$

where if a good trial function has been guessed, the c_i will be small. Using (13.4) we have immediately that $J = 1 + \Sigma_i |c_i|^2$. The other required integral is

$$I = \int_a^b [p(y_0' + c_i y_i')^2 + q(y_0 + c_i y_i)^2] \, dx.$$

On multiplying out the squared terms, all the cross terms vanish because of (13.5) to leave

$$\lambda = \frac{I}{J} = \frac{\lambda_0 + \Sigma_i |c_i|^2 \lambda_i}{1 + \Sigma_j |c_j|^2}$$
$$= \lambda_0 + \sum_i |c_i|^2 (\lambda_i - \lambda_0) + O(c^4).$$

Hence λ differs from λ_0 by the second order in the c_i even though y differed from y_0 by the first order in c_i. We notice incidentally that, since $\lambda_0 < \lambda_i$ (all i), λ is shown to be necessarily $\geqslant \lambda_0$ with equality only if all $c_i = 0$, i.e. if $y \equiv y_0$.

The method can be extended to the second (and higher) eigenvalues by imposing, in addition to the original constraints and boundary conditions, a restriction of the trial functions to only those which are orthogonal to the eigenfunctions corresponding to lower eigenvalues. [This of course then requires complete or nearly complete knowledge of these latter eigenfunctions.] An example is given at the end of the chapter (section 13.6, example 6).

We now illustrate the method we have discussed by considering a simple example, and as on previous occasions, one for which the answer is obvious.

Example 13.1. Suppose we are required to solve

$$\frac{d^2y}{dx^2} + \lambda y = 0, \quad 0 \leqslant x \leqslant 1, \tag{13.7}$$

with boundary conditions

$$y(0) = 0, \quad y'(1) = 0. \tag{13.8}$$

In particular we wish to find the lowest value (λ_0) of λ for which (13.7) has a solution satisfying (13.8). The exact answer is of course $y = A \sin(x\pi/2)$ and $\lambda_0 = \pi^2/4 = 2.47$.

We first note that the Sturm–Liouville equation reduces to (13.7) if we take $p(x) = 1$, $q(x) = 0$ and $\rho(x) = 1$, and that the boundary conditions satisfy (13.6). Thus we are able to apply the previous theory.

We will use three trial functions so that the effect on the estimate of λ_0 of making better or worse 'guesses' can be seen. One further preliminary remark is relevant, namely that the estimate is independent of any constant multiplying factor in the functions used. This is easily verified by looking at the form of I/J. We normalize each trial function so that $y(1) = 1$, but purely in order to facilitate comparison of the various function shapes.

Figure 13.1 illustrates the trial functions used, curve (a) being the exact solution $y = \sin(\pi x/2)$. The other curves are:

(b) $y(x) = 2x - x^2$,
(c) $y(x) = x^3 - 3x^2 + 3x$,
(d) $y(x) = \sin^2(\pi x/2)$.

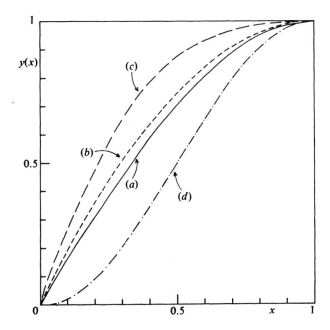

Fig. 13.1 Trial wave functions used to estimate the lowest eigenvalue λ of $y'' + \lambda y = 0$ with $y(0) = y'(1) = 0$. They are: (a) $y = \sin(\pi x/2)$, the exact result; (b) $y = 2x - x^2$; (c) $y = x^3 - 3x^2 + 3x$; (d) $y = \sin^2(\pi x/2)$.

The choice of trial wave functions is governed by the following considerations:

(i) The boundary conditions (13.8) *must* be satisfied.
(ii) A 'good' trial function ought to mimic the correct solution as well as possible, but it may not be easy to guess even the general shape of the correct solution in some cases.
(iii) The evaluation of I/J should be as simple as possible.

It is easily verified that functions (b), (c) and (d) all satisfy (13.8), but so far as mimicking the correct solution is concerned, we would expect from the figure that (b) would be superior to the other two. All three evaluations are straightforward.

▶2. $$\lambda_b = \frac{\int_0^1 (2 - 2x)^2 \, dx}{\int_0^1 (2x - x^2)^2 \, dx} = \frac{4/3}{8/15} = 2.50.$$

▶3. $$\lambda_c = \frac{\int_0^1 (3x^2 - 6x + 3)^2 \, dx}{\int_0^1 (x^3 - 3x^2 + 3x)^2 \, dx} = \frac{9/5}{9/14} = 2.80.$$

▶4. $$\lambda_d = \frac{\int_0^1 (\pi^2/4) \sin^2 (\pi x) \, dx}{\int_0^1 \sin^4 (\pi x/2) \, dx} = \frac{\pi^2/8}{3/8} = 3.29.$$

We expected all evaluations to yield estimates greater than the true lowest eigenvalue, 2.47, and this is indeed so. From these three trials alone we are (only) able to say that $\lambda_0 \leqslant 2.50$. As expected it is the best approximation (b) to the true eigenfunction which yields the lowest and therefore the best upper bound for λ_0.

▶5. Verify directly that $\lambda_a = \pi^2/4$, the true value of λ_0.

13.4   Adjustment of parameters

Instead of trying to estimate λ_0 by selecting a large number of different trial functions, we may also use trial functions which include one or more parameters which themselves may be adjusted to give the lowest value to $\lambda = I/J$ and hence the best estimate of λ_0. The justification for this method comes from the knowledge that no matter what form of function is chosen nor what values are assigned to the parameters, so long as the boundary conditions are satisfied, λ can never be less than the required λ_0.

To illustrate this method, an example from quantum mechanics will be used. The Schrodinger equation is formally written $H\psi = E\psi$, where H is a linear operator, ψ the wave function describing a quantum mechanical system and E the energy of the system. The operator H is called the Hamiltonian and, for a particle of mass m moving in a one-dimensional harmonic oscillator potential, is given by

$$H = -\frac{\hbar^2}{2m} \frac{d^2}{dx^2} + \frac{k}{2} x^2,$$ (13.9)

where $\hbar$ is Planck's constant divided by 2π.

Example 13.2. To estimate the ground state energy of a quantum mechanical harmonic oscillator.

Using (13.9) in $H\psi = E\psi$, the Schrodinger equation is

$$-\frac{\hbar^2}{2m}\frac{d^2\psi}{dx^2} + \frac{k}{2}x^2\psi = E\psi, \quad -\infty < x < \infty. \tag{13.10}$$

The boundary conditions are that ψ should vanish as $x \to \pm\infty$. Equation (13.10) is a form of the Sturm–Liouville equation in which, after changing signs throughout, $p = \hbar^2/2m$, $q = kx^2/2$, $\rho = 1$ and $\lambda = E$, and can be solved as previously.

As a trial wave function we take $\psi = \exp(-\alpha x^2)$, where α is a positive parameter whose value we will choose later. This function certainly $\to 0$ as $x \to \pm\infty$ and is convenient for calculations. Whether it approximates the true wave function is unknown, but if it doesn't our estimate will still be valid [though the upper bound will be a poor one].

With $y = \exp(-\alpha x^2)$ and therefore $y' = -2\alpha x \exp(-\alpha x^2)$, the required estimate is

►6.
$$\lambda = \frac{\int_{-\infty}^{\infty}[(\hbar^2/2m)4\alpha^2 x^2 + (k/2)x^2]e^{-2\alpha x^2}dx}{\int_{-\infty}^{\infty}e^{-2\alpha x^2}dx} = \frac{\hbar^2\alpha}{2m} + \frac{k}{8\alpha}. \tag{13.11}$$

This evaluation is easily performed using the reduction formula

►7.
$$I_n = \frac{n-1}{4\alpha}I_{n-2}, \quad \text{where } I_n = \int_{-\infty}^{\infty}x^n e^{-2\alpha x^2}\,dx. \tag{13.12}$$

So we have obtained an estimate, given by (13.11) for the ground state energy [lowest eigenvalue of H] of the oscillator, the estimate involving the parameter α. In line with our previous discussion we now minimize λ with respect to α. Putting $d\lambda/d\alpha = 0$ [clearly a minimum], gives $\alpha = (km)^{1/2}/2\hbar$ which in turn gives as the minimum value for λ

$$E = \frac{\hbar}{2}\left(\frac{k}{m}\right)^{1/2} = \frac{\hbar\omega}{2}, \tag{13.13}$$

where we have put $(k/m)^{1/2}$ equal to the classical angular frequency ω.

The method thus leads to the conclusion that the ground state energy E_0 is $\leqslant \frac{1}{2}\hbar\omega$. In fact, as is well known, the equality sign holds, $\frac{1}{2}\hbar\omega$ being just the zero-point energy of a quantum mechanical oscillator. Our estimate gives the exact value because $\exp(-\alpha x^2)$ is exactly the ground state wave function if α is as we have determined.

An alternative but equivalent approach to this and similar problems is developed in the examples at the end of this chapter, as is an extension of this particular problem to estimating the second lowest eigenvalue.

13.5   Rayleigh's method

We conclude this chapter on estimation of eigenvalues with a brief account of the related question of the eigenvalues of matrices. These topics are more fully discussed in the next chapter which is specifically concerned with matrices. Here we will develop Rayleigh's method of treating small vibrations of a stable system.

Suppose a system is described by a set of n coordinates θ_i [not necessarily angles], which are all zero at the equilibrium point. Then for *small* displacements the potential energy and kinetic energy are given by

$$V = b_{ik}\theta_i\theta_k \equiv \tilde{\theta}B\theta,$$
$$T = a_{ik}\dot{\theta}_i\dot{\theta}_k \equiv \tilde{\dot{\theta}}A\dot{\theta},$$

where the coefficients a_{ik} and b_{ik} depend upon the geometry of the system and, since the equilibrium is stable, both expressions are positive definite (or zero). The matrices A and B can be chosen real and symmetric.

Applying Lagrange's equations of motion (12.21) yields

$$a_{ik}\ddot{\theta}_k + b_{ik}\theta_k = 0, \quad i = 1,\dots,n. \tag{13.14}$$

We seek solutions of this equation which are truly periodic (normal modes) and in which the ratios of the coordinates θ_k are independent of time for a particular mode, i.e.

$$\theta_k(t) = \phi_k g(t).$$

In this case (13.14) becomes

$$a_{ik}\phi_k\ddot{g} + b_{ik}\phi_k g = 0,$$

or, since $\ddot{g}/g = -\omega^2$,

$$(b_{ik} - \omega^2 a_{ik})\phi_k = 0. \tag{13.15}$$

The problem then becomes one of finding the possible eigenvalues ω_j^2 and the corresponding eigenfunctions (eigenvectors) ϕ^j.

As is shown in chapter 14,

$$\tilde{\phi}^k A\phi^j = 0, \tag{13.16}$$

if $k \neq j$, and this can be used to establish an estimation procedure as follows.

The eigenfunctions ϕ^j form a complete set, and so any coordinate vector ϕ can be written as $\phi = c_j\phi^j$. Consider the value of

$$\lambda(\boldsymbol{\phi}) \equiv \frac{\tilde{\boldsymbol{\phi}}B\boldsymbol{\phi}}{\tilde{\boldsymbol{\phi}}A\boldsymbol{\phi}} = \frac{\tilde{\boldsymbol{\phi}}^m c_m^* B c_i \boldsymbol{\phi}^i}{\tilde{\boldsymbol{\phi}}^j c_j^* A c_k \boldsymbol{\phi}^k} = \frac{\tilde{\boldsymbol{\phi}}^m c_m^* \omega_i^2 c_i A \boldsymbol{\phi}^i}{\tilde{\boldsymbol{\phi}}^j c_j^* A c_k \boldsymbol{\phi}^k} =$$

$$\frac{\sum_i |c_i|^2 \omega_i^2 \tilde{\boldsymbol{\phi}}^i A \boldsymbol{\phi}^i}{\sum_k |c_k|^2 \tilde{\boldsymbol{\phi}}^k A \boldsymbol{\phi}^k}, \quad (13.17)$$

where we have used successively (13.15) and (13.16). Now if ω_0^2 is the lowest eigenfrequency then $\omega_i^2 \geq \omega_0^2$ for all i, and further, since $\tilde{\boldsymbol{\phi}}^i A \boldsymbol{\phi}^i \geq 0$ for all i, the numerator of (13.17) is $\geq \omega_0^2 \sum_i |c_i|^2 \tilde{\boldsymbol{\phi}}^i A \boldsymbol{\phi}^i$. Hence,

$$\lambda(\boldsymbol{\phi}) \equiv \frac{\tilde{\boldsymbol{\phi}}B\boldsymbol{\phi}}{\tilde{\boldsymbol{\phi}}A\boldsymbol{\phi}} \geq \omega_0^2 \quad (13.18)$$

for any $\boldsymbol{\phi}$ whatsoever [eigenfunction or not]. Thus we are able to estimate the lowest eigenfrequency of the system by evaluating λ for a variety of vectors $\boldsymbol{\phi}$ [the components of which, it will be recalled, give the ratios of the coordinate amplitudes]. This is sometimes a useful approach if many coordinates are involved and direct solution for the eigenvalues is intractable.

An additional result for finite matrices, not available for the Sturm–Liouville equation, is that the maximum eigenvalue ω_m^2 may also be estimated. It is obvious that if we replace the statement '$\omega_i^2 \geq \omega_0^2$ for all i' by '$\omega_i^2 \leq \omega_m^2$ for all i', then $\lambda(\boldsymbol{\phi}) \leq \omega_m^2$ for any $\boldsymbol{\phi}$.

The formal similarity between our general result and that for S–L differential equations may be noted. If we replace integration by summation over subscripts, $q(x)$ by B, $\rho(x)$ by A, and λ by ω^2, then (13.15) replaces the S–L equation with $p(x) = 0$, whilst I takes the form $\tilde{\boldsymbol{\phi}}B\boldsymbol{\phi}$ and J is $\tilde{\boldsymbol{\phi}}A\boldsymbol{\phi}$.

As an example consider a uniform rod of mass M and length l, attached by a light string of the same length to a point P, executing small oscillations in a vertical plane.

Choose as coordinates the angles θ_1 and θ_2 in fig. 13.2 (a), in which the magnitudes of the angles have been exaggerated. In terms of these, the centre of gravity of the rod has, to first order in the θ_i, a velocity in the x-direction of $l\dot{\theta}_1 + \frac{1}{2}l\dot{\theta}_2$, and in the y-direction of zero. Adding in the rotational kinetic energy of the rod about its centre of gravity we obtain

$$T = \tfrac{1}{2}Ml^2(\dot{\theta}_1^2 + \dot{\theta}_1\dot{\theta}_2 + \tfrac{1}{4}\dot{\theta}_2^2) + \tfrac{1}{24}Ml^2\dot{\theta}_2^2 + O(\theta^4),$$

and

$$A = \frac{Ml^2}{12}\begin{bmatrix} 6 & 3 \\ 3 & 2 \end{bmatrix}.$$

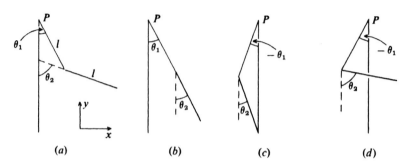

(a)   (b)   (c)   (d)

Fig. 13.2 A uniform rod of length l attached to the fixed point P by a light string of the same length: (a) the general coordinate system; (b) approximation to the normal mode with the lower eigenfrequency; (c), (d) approximations to the higher eigenfrequency mode.

The potential energy

$$V = Mlg[(1 - \cos \theta_1) + \tfrac{1}{2}(1 - \cos \theta_2)]$$
$$= \tfrac{1}{2}Mlg(\theta_1^2 + \tfrac{1}{2}\theta_2^2) + O(\theta^4),$$

and therefore

$$B = \frac{Mlg}{12} \begin{bmatrix} 6 & 0 \\ 0 & 3 \end{bmatrix}.$$

We now use (13.18) to first estimate the frequency of the slower of the two normal modes. Physical intuition suggests that the slower mode will have a configuration approximating that of a simple pendulum, fig. 13.2 (b), in which $\theta_1 = \theta_2$ and so we use this as a trial function [vector]. Taking $\boldsymbol{\phi} = (\theta, \theta)$,

$$\lambda(\boldsymbol{\phi}) = \frac{\tilde{\boldsymbol{\phi}} B \boldsymbol{\phi}}{\tilde{\boldsymbol{\phi}} A \boldsymbol{\phi}} = \frac{\tfrac{3}{4}Mlg\theta^2}{\tfrac{7}{6}Ml^2\theta^2} = \frac{9}{14}\frac{g}{l} = 0.643\,\frac{g}{l},$$

and we conclude that the lower (angular) frequency is $\leqslant (0.643g/l)^{1/2}$. The true value is $(0.641g/l)^{1/2}$ and so we have come very close to it.

Next we turn to the higher frequency. Here it is not so obvious what a typical configuration looks like but we may try those shown in fig. 13.2 (c) and (d). In (c) the angles are equal and opposite ($\theta_1 = -\theta_2$) and in (d) $\theta_2 = -2\theta_1$ [and keeps the centre of gravity of the rod under P for small values of the θ_i]. With these two forms we obtain

▶8.   (c) $\lambda = 4.5g/l$,

▶9.   (d) $\lambda = 9g/l$.

We thus conclude that the higher eigenfrequency $\geqslant (9g/l)^{1/2}$ and that the motion looks more like fig. 13.2 (d) than like (c). [The exact value is $(9.359g/l)^{1/2}$.]

For this problem it is not difficult to obtain the exact eigenvalues and eigenvectors (θ_1, θ_2) by the methods of the next chapter, but it nevertheless serves to illustrate that reasonable estimates and useful information can sometimes be obtained using a minimum amount of calculation and some physical insight.

13.6   Examples for solution

1. Show that $y'' - xy + \lambda x^2 y = 0$ has a solution for which $y(0) = y(1) = 0$ and $\lambda \leqslant 36\frac{3}{4}$.

2. A drumskin is stretched across a fixed circular rim of radius a. Small transverse vibrations of the skin have an amplitude $z(r, \phi, t)$, which satisfies

$$\nabla^2 z = \frac{1}{c^2} \frac{\partial^2 z}{\partial t^2}$$

in two-dimensional polars. For a normal mode independent of azimuth $z = Z(r) \cos \omega t$, find the differential equation satisfied by $Z(r)$. Using a trial function of the form $a^\nu - r^\nu$, obtain an estimate for the lowest normal mode frequency [the exact answer is $(5.78)^{1/2} c/a$].

3. Alternative approach to example 13.2 (page 355). Using the notation of section 13.4, $H\psi = E\psi$ and the (expectation of the) energy of the state is given by $\int \psi^* E\psi \, dv = \int \psi^* H\psi \, dv$. Denote the eigenfunctions of H by ψ_i so that $H\psi_i = E_i\psi_i$ and since H is Hermitian [self-adjoint] $\int \psi_j^* \psi_i \, dv = \delta_{ij}$.

(i) By writing any function ψ as $\sum c_i\psi_i$ and following an argument similar to (13.17), show that

$$\lambda = \frac{\int \Psi^* H\Psi \, dv}{\int \Psi^* \Psi \, dv} \geqslant E_0,$$

the energy of the lowest state. [This is the Rayleigh–Ritz principle.]
(ii) Using the same trial function as in example 13.2 show that the same result is obtained.

4. Using a trial wave function of the form $\exp(-\beta r)$ and the Rayleigh–Ritz principle, find an upper limit for the ground state energy of the hydrogen atom. Assuming that the given form is actually the correct one [which it is], find the normalized ground

state wave function and its energy in terms of the fundamental constants.

5. The vibrations of a trampoline 4 units long and 1 unit wide satisfy the equation $\nabla^2 u + k^2 u = 0$. By taking the simplest possible permissible polynomial as a trial function, show that the lowest mode of vibration has $k^2 \leqslant 10.63$ and by direct solution that the actual value is 10.49.

6. Extension to example 13.2 (page 355). With the ground state wave function as $\exp(-\alpha x^2)$, take as a trial function the orthogonal wave function $x^{2n+1} \exp(-\alpha x^2)$, using the integer n as a variable parameter. Show using Sturm–Liouville theory or the Rayleigh–Ritz principle that the energy of the second lowest state of a quantum harmonic oscillator has an energy of $\leqslant \frac{3}{2}\hbar\omega$.

7. Three particles of mass m are attached to a light string having fixed ends, the string being divided into four equal portions each of length a and being under a tension T. Show that for small transverse vibrations the amplitudes x_i of normal modes satisfy $B\mathbf{x} = (ma\omega^2/T)\mathbf{x}$, where B is the matrix

$$\begin{bmatrix} 2 & -1 & 0 \\ -1 & 2 & -1 \\ 0 & -1 & 2 \end{bmatrix}.$$

Estimate the lowest and highest eigenvalues using trial vectors $(3, 4, 3)$ and $(3, -4, 3)$. Use also the exact vectors $(1, \sqrt{2}, 1)$ and $(1, -\sqrt{2}, 1)$ and compare.

8. Use Rayleigh's method to estimate the lowest oscillation frequency of a heavy chain of n links, each of length a $(= l/n)$, which hangs freely from one end. [Try simple calculable configurations such as all links but one vertical, or all links collinear, or . . ..]

14
Matrices

At the end of the previous chapter we touched briefly on the notion of a matrix as an entity consisting of a set of quantities which are arranged in an ordered way. The ordering is usually described by the use of subscripts, the number of these being equal to the dimension of the array. Thus a_{253} stands for the element of an array which is [to use an undefined but readily visualizable description] simultaneously in the second row, in the fifth column, and in the third layer of the array. A general element of this array would be denoted by a_{ijk}.

The properties of ordered arrays of elements [usually numbers or algebraic expressions, but sometimes more complicated objects] are a large and intensively studied field and any reader who is interested in this for its own sake should consult some of the numerous books available on the subject. Our attention will be focussed on two-dimensional arrays [with elements like a_{ij}], or, as a special case, the one-dimensional vector array [with elements a_i].

Such an array is called a **matrix** A, and if i runs from 1 to M and j from 1 to N, A is called an $M \times N$ matrix. Written out in array form it is

$$A = \begin{bmatrix} a_{11} & a_{12} & \ldots & a_{1N} \\ a_{21} & a_{22} & \ldots & a_{2N} \\ \cdot & \cdot & & \cdot \\ \cdot & \cdot & & \cdot \\ \cdot & \cdot & & \cdot \\ a_{M1} & a_{M2} & \ldots & a_{MN} \end{bmatrix}, \tag{14.1}$$

and has M **rows** and N **columns**. The special case of a column (or row) vector arises when N (or M) has the value 1. In keeping with the notation used throughout the rest of this book (for 'physical' vectors) we will denote a matrix which is known to be either $M \times 1$ or $1 \times N$ by (respectively) a vector **u** or a transposed† vector $\tilde{\mathbf{v}}$ in bold type.

† Transposed vectors are introduced later when matrix manipulation is discussed.

In the present chapter we will rather narrowly pursue a path which is aimed at deriving those properties of matrices which are useful in physical applications such as the solution of simultaneous linear equations or the study of small oscillations of a mechanical system. In doing this we will have to by-pass many interesting mathematical properties of the arrays and in some cases quote results without a full proof. This will be particularly true in our treatment of determinants which will virtually consist of only those results needed to establish inverse matrices and the compatibility conditions on simultaneous equations.

14.1   Some notations

Before proceeding further, we remind the reader of the *summation convention* (section 1.9) for subscripts, since its use looms large in the work of this and the next chapter. The convention is that any *lower-case* alphabetic subscript which appears *exactly* twice in any term of an expression is to be summed over all the values that a subscript in that position can take (unless the contrary is specifically stated). This naturally implies that any such pair of repeated subscripts must only occur in subscript positions which have the same range of values. Sometimes the ranges of values have to be specified but usually they are apparent from the context.

The following simple examples illustrate what is meant:

$a_i x_i$ stands for $a_1 x_1 + a_2 x_2 + \ldots + a_M x_M$,

$a_{ij} b_{jk}$ stands for $a_{i1} b_{1k} + a_{i2} b_{2k} + \ldots + a_{iN} b_{Nk}$,

$a_{ij} b_{jk} c_k$ stands for $\sum_{j=1}^{N} \sum_{k=1}^{L} (a_{ij})(b_{jk})(c_k)$,

where of course on the right-hand side of the third example (a_{ij}) is the element in the ith row and jth column of the matrix A, etc.

Subscripts which are summed over are called **dummy subscripts** and the others **free subscripts**. It is worth remarking that when introducing a dummy subscript into an expression care should be taken not to use one that is already present, either as a free or as a dummy subscript. For example, $a_{ij} b_{jk} c_{kl}$ cannot and must not be replaced by $a_{ij} b_{jj} c_{jl}$ or by $a_{il} b_{lk} c_{kl}$, but could be replaced by $a_{im} b_{mk} c_{kl}$ or by $a_{im} b_{mn} c_{nl}$. [Naturally free subscripts should not be changed at all unless the working calls for it.]

In the study to be made of the properties of matrices we shall often have reason to consider multiple products of elements drawn from an array such as (14.1). A particular product-type of interest is one which arises from a matrix for which $M = N$ (a **square** matrix), and which contains N factors, the factors being such that there is present one and only one from each row and one and only one from each column.

Such products can be characterized as being of the general form

$$a_{1j_1}a_{2j_2}a_{3j_3}\cdots a_{Nj_N}, \tag{14.2 a}$$

where $j_1, j_2, j_3, \ldots, j_N$ is some ordering of the numbers $1, 2, 3, \ldots, N$. As an alternative, the same expression may be written so that its column subscripts occur in their natural order i.e.

$$a_{i_11}a_{i_22}a_{i_33}\cdots a_{i_NN}. \tag{14.2 b}$$

Again $i_1, i_2, i_3, \ldots, i_N$ is an ordering of the numbers $1, 2, \ldots, N$ (but in general not the same as the previous ordering of the j's).

It is a well-known result from the theory of permutations that the number of interchanges of pairs of numbers needed to obtain a specific ordering $j_1, j_2, \ldots, j_N$ starting from the natural ordering $1, 2, \ldots, N$ is either definitely odd or definitely even. That is to say that although the actual number of interchanges needed will depend on the sequence in which the changes are made, if it is odd by one method it is odd by every method (and similarly for even).

To express and use this result we define a symbol $\epsilon_{j_1j_2\ldots j_N}$ which takes only the three values ± 1 or 0, and takes these as follows,

$$\epsilon_{j_1j_2\cdots j_N} = +1 \text{ if } j_1, j_2, \ldots, j_N \text{ is an even permutation of}$$
$$1, 2, \ldots, N,$$
$$= -1 \text{ if } j_1, j_2, \ldots, j_N \text{ is an odd permutation of}$$
$$1, 2, \ldots, N,$$
$$= 0 \text{ if } j_1, j_2, \ldots, j_N \text{ is not a permutation of}$$
$$1, 2, \ldots, N. \tag{14.3}$$

▶1. Verify that, with these definitions:

(i) $\epsilon_{3412} = 1$,   (ii) $\epsilon_{321} = -1$,   (iii) $\epsilon_{2412} = 0$.

▶2. Using the ϵ notation and the summation convention show that

$$\epsilon_{ijk}a_ib_j = a_2b_3 - a_3b_2 \quad \text{for } k = 1,$$
$$= a_3b_1 - a_1b_3 \quad \text{for } k = 2,$$
$$= a_1b_2 - a_2b_1 \quad \text{for } k = 3.$$

Notice the connection between this and the kth component of $\mathbf{a} \wedge \mathbf{b}$.

14.2   Determinants

In order to prepare the way for some of our subsequent work on the properties of matrices, we will in this section explain and establish those results from the theory of determinants, which will be required. The motivation for this development work will therefore not become apparent until we come to the physical applications of later sections, but so as not to break up that work too much it will be carried out here.

The matrix A appearing in (14.1) is purely an array of numbers (elements) and as such has no particular numerical value. [One could argue, quite rightly, that it has in fact MN numerical values.] However, for a square matrix [$M = N$] it is possible to deduce a single numerical (or algebraic) expression, which it will be found later has an important role to play in the theory of matrices. This expression is known as the **determinant** of A or det A and is denoted symbolically by $|A|$. For an explicit matrix such as (14.1) with $M = N$ it is written

$$
\det A =
\begin{vmatrix}
a_{11} & a_{12} & \dots & a_{1N} \\
a_{21} & a_{22} & \dots & a_{2N} \\
\cdot & \cdot & & \cdot \\
\cdot & \cdot & & \cdot \\
\cdot & \cdot & & \cdot \\
a_{N1} & a_{N2} & \dots & a_{NN}
\end{vmatrix}.
\tag{14.4}
$$

[Notice that in our notation a determinant appears enclosed in vertical bars, whereas the corresponding matrix is enclosed in square brackets.]

The value of the determinant is given by

$$
|A| = \epsilon_{i_1 i_2 \cdots i_N} a_{i_1 1} a_{i_2 2} \dots a_{i_N N},
\tag{14.5}
$$

where the implied summation runs over all values of each i_k from 1 to N and ϵ is as defined in (14.3). Since the ϵ-factor is only non-zero when $i_1, i_2, \dots, i_N$ is some permutation of $1, 2, \dots, N$, it is apparent that the only non-zero contributions to $|A|$ in (14.5) come from terms in which the factors $a_{i_k k}$ are drawn *one and only one* from each *row* and *one and only one* from each *column*. This is just as discussed in connection with (14.2 b). Since there are N numbers $(1, 2, \dots, N)$ to fill in the values $i_1, i_2, \dots, i_N$ and all must be different, it is clear that there will be $N!$ contributing terms. [There would be N^N terms in (14.5) if ϵ did not have its special meaning.]

Let us illustrate this immediately with a simple example by evaluating det A when A is the 3 × 3 matrix

$$
A =
\begin{bmatrix}
2 & 1 & -3 \\
3 & 4 & 0 \\
1 & -2 & 1
\end{bmatrix}.
\tag{14.6}
$$

To show where each factor arises we write below it the corresponding values of i_1, i_2, i_3 in (14.5)

$$|A| = 2 \cdot 4 \cdot 1 - 2 \cdot (-2) \cdot 0 - 3 \cdot 1 \cdot 1 + 3 \cdot (-2) \cdot (-3)$$
$$\underset{i_1 i_2 i_3}{} \quad \underset{123}{} \qquad \underset{132}{} \qquad \underset{213}{} \qquad \underset{231}{}$$

$$+ 1 \cdot 1 \cdot 0 - 1 \cdot 4 \cdot (-3) = 35. \quad (14.7)$$
$$\underset{312}{} \qquad \underset{321}{}$$

As an alternative to the definition (14.5), det A can also be expressed as

$$|A| = \epsilon_{j_1 j_2 \cdots j_N} a_{1 j_1} a_{2 j_2} \ldots a_{N j_N}. \quad (14.8)$$

This definition will now be shown to be equivalent to (14.5).

It is readily apparent that the only non-zero terms in (14.8) are again those containing factors drawn one and only one from each row and column, and all that remains to be shown is that for any particular product of elements so constituted, the signs obtained from the ϵ factors are the same in (14.5) and (14.8).

To show this consider a *particular* term

$$a_{k_1 l_1} a_{k_2 l_2} \ldots a_{k_N l_N}.$$

Their occurrence in this term associates with each k-subscript a particular l-subscript, l_q with k_q. The sequence $k_1, k_2, \ldots, k_N$ is necessarily either an even or an odd permutation of $l_1, l_2, \ldots, l_N$, and both are orderings of the numbers 1 to N. Whichever it is, it is not affected by reordering the pairs $(k_1, l_1), (k_2, l_2), \ldots, (k_N, l_N)$ so as to read

$$(1, j_1), (2, j_2), \ldots, (N, j_N), \quad (14.9\ a)$$

nor by reordering them so as to read

$$(i_1, 1), (i_2, 2), \ldots, (i_N, N). \quad (14.9\ b)$$

Thus all three sets of pairs have the same even or odd permutation character; in particular (14.9 a) and (14.9 b) have the same character. But this is just the statement that $\epsilon_{j_1 j_2 \cdots j_N} = \epsilon_{i_1 i_2 \cdots i_N}$, when i and j have those values which pick out exactly the same combination of factors in (14.5) and (14.8) respectively. This completes the proof that (14.5) and (14.8) are equivalent to each other.

▶3. Use (14.8) to evaluate $|A|$, with A given in (14.6). Set out the evaluation in an analogous way to (14.7) and show that it again yields 35.

When the size of the matrix becomes much beyond 3×3, evaluation of its determinant by writing out all permitted products as in (14.5) or

(14.8) becomes very laborious and error-prone and an alternative method is preferable. This can be obtained as follows.

In the expression (14.8) for det A we group together all those terms which contain the factor a_{Ij_I} for some particular value of I. There will be N such groups corresponding to the N different values of j_I. In carrying out the development, I will be left unspecified, but it may be helpful to think of it as having a particular value (say 1). In what follows a bar over the top of a symbol, e.g. $\overline{j_I}$ or $\overline{a_{Ij_I}}$ will denote that that symbol is missing from an otherwise natural sequence.

Carrying out the grouping we can write (14.8) as

$$|A| = \sum_{j_I=1}^{N} a_{Ij_I} \sum_{j_1 j_2 \ldots \overline{j_I} \ldots j_N} \epsilon_{j_1 j_2 \ldots j_I \ldots j_N} a_{1j_1} a_{2j_2} \ldots \overline{a_{Ij_I}} \ldots a_{Nj_N}. \quad (14.10)$$

[Despite our preference for the summation convention, the summation signs have been explicitly written in here, to help to elucidate the order in which things are done.] In the second (multiple) summation, j_1, j_2, etc. range over all values from 1 to N except the value j_I.

As an intermediate aid let us denote by

$$\epsilon_{j_1 j_2 \ldots \overline{j_I} \ldots j_N},$$

the quantity analogous to $\epsilon_{j_1 j_2 \ldots j_N}$ but taking the value $+1$ or -1 according to whether $j_1, j_2, \ldots, \overline{j_I}, \ldots, j_N$ is an even or odd permutation of 1 to N excluding j_I. Now consider the sequence

$$1, 2, 3, \ldots, j_I, \ldots, N.$$

In $|j_I - I|$ successive interchanges of j_I with its successive left- (or right-) hand neighbours, j_I can be brought to the Ith position with the rest of the numbers in their natural order [omitting j_I of course]. If these interchanges are then followed by those necessary to get the remaining numbers into the order $j_1, j_2, \ldots, \overline{j_I}, \ldots, j_N$, then the total number of changes will be even or odd according to whether

$$(-1)^{j_I - I} \epsilon_{j_1 j_2 \ldots \overline{j_I} \ldots j_N}$$

is $+1$ or -1. But this is exactly the same as $\epsilon_{j_1 j_2 \ldots j_I \ldots j_N}$, since it produces the same final ordering, viz. $j_1, j_2, \ldots, j_I, \ldots, j_N$, of the numbers 1 to N starting from their natural order.

Returning now to (14.10) and inserting this result, we obtain as an alternative for $|A|$,

$$|A| = \sum_{j_1=1}^{N} a_{1j_1}(-1)^{j_1-1}$$

$$\times \left\{ \sum_{j_1 j_2 \cdots \overline{j_1} \cdots j_N} \epsilon_{j_1 j_2 \cdots \overline{j_1} \cdots j_N} a_{1j_1} a_{2j_2} \cdots \overline{a_{1j_1}} \cdots a_{Nj_N} \right\}. \quad (14.11)$$

Now the term in the curly brackets is exactly the determinant of the $(N-1) \times (N-1)$ matrix obtained by striking out the Ith row and the j_Ith column of the original matrix A, i.e. by striking out the row and column containing a_{1j_1}. This quantity is called the *minor* of a_{1j_1} and is denoted by M_{1j_1}. Taken together with the sign factor $(-1)^{j_1-1}$, it is called the *cofactor*

$$C_{1j_1} = (-1)^{j_1-1} M_{1j_1}, \quad (14.12)$$

and (14.11) can be rewritten as

$$|A| = \sum_{j_I} a_{1j_I}(-1)^{j_I-1} M_{1j_I} = a_{1j_I} C_{1j_I}, \text{ any } I \text{ in } 1 \leqslant I \leqslant N.$$
$$(14.13\,\mathrm{a})$$

By a similar argument starting from (14.5) det A can also be written as

$$|A| = \sum_{i_J} a_{i_J J}(-1)^{i_J-J} M_{i_J J} = a_{i_J J} C_{i_J J}, \text{ any } J \text{ in } 1 \leqslant J \leqslant N.$$
$$(14.13\,\mathrm{b})$$

These new expressions for det A enable it to be expressed as a weighted sum of N determinants, each with one less row and column than the original. Clearly this process can be repeated until the matrices are reduced in size to 1×1. Their values are then those of the single elements left. [When a matrix B contains only a single element, $|B|$ must not be confused with a simple modulus sign, e.g. $|-3|$ is -3 not $+3$.]

As stated earlier I was left as a fixed but arbitrary integer in the range 1 to N. But since equations (14.13) have been shown equivalent to (14.5) and (14.8), it follows that the values given by (14.13) are quite independent of which value I takes. For practical calculations, I is usually determined on the basis of which row or column has most zero entries, or is taken as $I = 1$ since this aids the visualization of the cofactor determinants.

This proof, in order to be general, has been somewhat complicated and involved subscripts to subscripts. When the form of result is applied to particular cases it is much less forbidding. To illustrate this we will again evaluate $|A|$ for A given by (14.6). In doing this we will pedantically follow out the procedure to the last single step, although in practice the evaluation of 2×2 matrices is usually carried out mentally. [In addition some properties of determinants mentioned later can greatly simplify their evaluation.]

We use (14.13 a) with $I = 1$ (and j_I running from 1 to 3) at the first stage and also evaluate the subsequent determinants by expanding in terms of the first row.

$$|A| = \begin{vmatrix} 2 & 1 & -3 \\ 3 & 4 & 0 \\ 1 & -2 & 1 \end{vmatrix}$$

$$= 2 \begin{vmatrix} 4 & 0 \\ -2 & 1 \end{vmatrix} - 1 \begin{vmatrix} 3 & 0 \\ 1 & 1 \end{vmatrix} + (-3) \begin{vmatrix} 3 & 4 \\ 1 & -2 \end{vmatrix}$$

$$= 2 [4|1| - 0|-2|] - 1[3|1| - 0|1|] - 3[3|-2| - 4|1|]$$

$$= 2 [4 \cdot 1 - 0 \cdot (-2)] - 1[3 \cdot 1 - 0 \cdot 1] - 3[3 \cdot (-2) - 4 \cdot 1]$$

$$= 8 - 3 + 30 = 35.$$

▶4. For some verification of the general results derived in this section, again evaluate the same determinant,

(i) using (14.13 a) with another value of I (say $I = 2$),
(ii) using (14.13 b) with some value of J (say $J = 3$).

Of course all methods should lead to the same value of 35.

▶5. Evaluate the determinants

$$(i) \begin{vmatrix} a & h & g \\ h & b & f \\ g & f & c \end{vmatrix}, \quad (ii) \begin{vmatrix} 1 & 0 & 2 & 3 \\ 0 & 1 & -2 & 1 \\ 3 & -3 & 4 & -2 \\ -2 & 1 & -2 & -1 \end{vmatrix}.$$

14.3   Some properties of determinants

We will now establish a series of properties of determinants which will be needed later and in addition can often be used to reduce the labour of evaluating them.

(i) The matrix B obtained from A by interchanging the rows and columns of A, i.e.

$$b_{ij} = a_{ji} \tag{14.14}$$

has the same determinant as A.

This follows immediately from the equivalence of the definitions (14.5) and (14.8). Det B evaluated using (14.5) is identical to det A evaluated using (14.8), and vice-versa.

The matrix B is known as the **transpose** of A and will be denoted by $\tilde{A}$. A matrix which is equal to its transpose, $\tilde{A} = A$, is called **symmetric.**

It also follows from the present result that any theorem established for the rows of A will apply to the columns as well, and vice-versa.

(ii) If two rows (columns) of A are interchanged, its determinant changes sign but is unaltered in magnitude.

Suppose the rth and sth $(s > r)$ rows of A are interchanged to produce a new matrix A'. Then using (14.8) to evaluate det A' we have

$$|A'| = \epsilon_{j_1 j_2 \cdots j_r \cdots j_s \cdots j_N} a_{1j_1} a_{2j_2} \cdots a_{sj_r} \cdots a_{rj_s} \cdots a_{Nj_N}.$$

Now exchanging the dummy subscripts j_r and j_s, the original products of elements which appeared in $|A|$ are recovered but the ϵ-factor is now $\epsilon_{j_1 j_2 \cdots j_s \cdots j_r \cdots j_N}$ instead of $\epsilon_{j_1 j_2 \cdots j_r \cdots j_s \cdots j_N}$. However, this reordering of the ϵ subscripts corresponds to a single interchange and so one ϵ-factor is equal to minus the other. All other factors depending on rows other than the rth and sth are unaltered and so

$$|A'| = -|A|.$$

(iii) If two rows (columns) of det A are identical, then det $A = 0$.

Interchanging the two rows clearly leaves $|A|$ unaltered. But by (ii) it changes it to $-|A|$. Thus $|A|$ must $= 0$.

(iv) (a) Multiplying each element of any one row (column) of A by a constant k multiplies $|A|$ by k.
(b) Multiplying *every* element of A by k multiplies $|A|$ by k^N.
(c) Any matrix which contains a complete row (column) of zeros has zero determinant.

(v) If two matrices A and B are identical except for the elements of one particular row (column), say the Ith, then the matrix D, having the same common elements as A and B and having its Ith row (column) constructed from the sums of the corresponding elements in A and B, has the property

$$|D| = |A| + |B|.$$

In symbols we have

$$d_{ij} = a_{ij} = b_{ij}, \quad (i \neq I), \quad j = 1, 2, \ldots, N,$$
$$d_{Ij} = a_{Ij} + b_{Ij}, \qquad\qquad j = 1, 2, \ldots, N.$$

The result is immediate if we write

$$\epsilon_{j_1 j_2 \cdots j_I \cdots j_N} d_{1j_1} d_{2j_2} \ldots d_{Ij_I} \ldots d_{Nj_N}$$
$$= \epsilon_{j_1 j_2 \cdots j_I \cdots j_N} d_{1j_1} d_{2j_2} \ldots a_{Ij_I} \ldots d_{Nj_N}$$
$$+ \epsilon_{j_1 j_2 \cdots j_I \cdots j_N} d_{1j_1} d_{2j_2} \ldots b_{Ij_I} \ldots d_{Nj_N},$$

and replace all the remaining d_{ij} in the first term by the corresponding a_{ij}, and those in the second by the corresponding b_{ij}.

(vi) The determinant of a matrix is unchanged in value by the addition to the elements of one row (column) of any fixed multiple of the elements of another row (column).

This result can be obtained from the previous one by taking the Ith row of B as the fixed multiple (λ say) of the row (the Rth say) to be added, i.e.

$$\left.\begin{array}{l} d_{ij} = a_{ij} = b_{ij}, \quad i \neq I \\[2mm] b_{Ij} = \lambda a_{Rj}, \\[2mm] d_{Ij} = a_{Ij} + \lambda a_{Rj}, \end{array}\right\} \quad j = 1, 2, \ldots, N. \qquad (14.15)$$

However, by result (iv), part (*a*), det $B = \lambda \times$ {the determinant of a matrix in which the Ith and Rth rows are identical}; thus by result (iv), part (*c*), this must $= 0$. Putting this into result (v) gives that $|D| = |A|$, which equations (14.15) show is the stated result.

Example 14.1. Use properties (ii) to (vi) to evaluate the matrix (ii) of ▶5.

There is no explicit procedure for using the results (ii)–(vi), and judging the quickest method is a matter of experience. A general guide is to try to reduce all terms but one in a row or column to zero and hence effectively obtain a determinant of smaller size. The steps taken below are certainly not the fastest series but have been chosen in order to illustrate the use of most of the properties. To save space the various stages are written sequentially and the commentary given afterwards.

$$\begin{vmatrix} 1 & 0 & 2 & 3 \\ 0 & 1 & -2 & 1 \\ 3 & -3 & 4 & -2 \\ -2 & 1 & -2 & -1 \end{vmatrix} \quad (a) \quad = 2 \begin{vmatrix} 1 & 0 & 1 & 3 \\ 0 & 1 & -1 & 1 \\ 3 & -3 & 2 & -2 \\ -2 & 1 & -1 & -1 \end{vmatrix}$$

$$(b) \quad = 2 \begin{vmatrix} 1 & 0 & 1 & 3 \\ 0 & 1 & 0 & 1 \\ 3 & -3 & -1 & -2 \\ -2 & 1 & 0 & -1 \end{vmatrix} \quad (c) \quad = 2 \begin{vmatrix} 1 & 0 & 1 & 3 \\ 0 & 1 & 0 & 0 \\ 3 & -3 & -1 & 1 \\ -2 & 1 & 0 & -2 \end{vmatrix}$$

(d)

$$= 2 \cdot 1 \cdot (-1)^{2-2} \begin{vmatrix} 1 & 1 & 3 \\ 3 & -1 & 1 \\ -2 & 0 & -2 \end{vmatrix}$$

(e)

$$= 2 \begin{vmatrix} 4 & 0 & 4 \\ 3 & -1 & 1 \\ -2 & 0 & -2 \end{vmatrix}$$

(f)

$$= 2(-2) \begin{vmatrix} -2 & 0 & -2 \\ 3 & -1 & 1 \\ -2 & 0 & -2 \end{vmatrix}$$

(g)

$$= 0.$$

(a) property (iv), part (a).
(b) add column 2 to column 3.
(c) add $(-1) \times$ column 2 to column 4.
(d) expand using (14.13 a) with $I = 2$.
(e) add row 2 to row 1.
(f) property (iv), part (a).
(g) property (iii).

▶6. Evaluate

$$\begin{vmatrix} gc & ge & a+ge & gb+ge \\ 0 & b & b & b \\ c & e & e & b+e \\ a & b & b+f & b+d \end{vmatrix}.$$

Before leaving determinants we may note one further property which will be useful later.

Consider the matrix A' obtained from an original matrix A by replacing the Ith row by one of the other rows (say the Rth) of A. Then A' is a matrix with two identical rows and thus has zero determinant. However replacing the Ith row by something else does not change the cofactors C_{Ij} of the elements in the Ith row which are therefore the same in A and A'. Thus applying (14.13 a) to the new matrix A' we obtain

$$0 = |A'| = a'_{Ij}C'_{Ij} = a_{Rj}C_{Ij}, \quad I \neq R,$$

and we may now combine this result more compactly with (14.13 a) to read

$$a_{Rj}C_{Ij} = \delta_{RI}|A|. \tag{14.16 a}$$

An analogous argument applies to column expansions to enable (14.13 b) to be generalized to

$$a_{IS}C_{Ij} = \delta_{SJ}|A|. \tag{14.16 b}$$

These two results show that there is a special relationship between the elements of a matrix and their cofactors. That such a relationship exists is not surprising since each element itself appears in the cofactors of other elements. In vector language, equations (14.16) state that any row (column) vector in the matrix is orthogonal to the vector formed by the cofactors of any other row (column).

14.4   Simultaneous homogeneous linear equations

If we consider the N simultaneous homogeneous linear equations

$$\left.\begin{array}{l} a_{11}x_1 + a_{12}x_2 + \cdots + a_{1N}x_N = 0, \\ a_{21}x_1 + a_{22}x_2 + \cdots + a_{2N}x_N = 0, \\ \quad\vdots \qquad\qquad\qquad \vdots \\ a_{N1}x_1 + a_{N2}x_2 + \cdots + a_{NN}x_N = 0, \end{array}\right\} \tag{14.17}$$

in N unknowns $x_1, x_2, \ldots, x_N$, we would expect, in view of every right-hand side being zero, that in general the only solution is that all x_j are themselves zero.

However, this is clearly not *necessarily* so, since, for example, the three simultaneous equations

$$\left.\begin{array}{l} 3x_1 - x_2 + 2x_3 = 0, \\ x_1 - 2x_2 + 2x_3 = 0, \\ x_1 + 3x_2 - 2x_3 = 0, \end{array}\right\} \tag{14.18}$$

have the non-zero solution $x_1 = -2$, $x_2 = 4$, $x_3 = 5$ [and also any multiple of this set]. We may thus consider what conditions are necessary for the general equations (14.17) to have a non-zero solution.

The general question of whether M simultaneous homogeneous linear equations in N unknowns have no solution, a unique solution, or infinities of solutions, is lengthy to consider in full generality, and if pursued would take us well beyond what is usually required for physical problems. We will restrict our attention to the case $M = N$ (as in (14.17)) and derive only a necessary condition that at least one non-zero solution exists.

Consider the constants a_{ij} $(i, j = 1, 2, \ldots, N)$, in (14.17) as the elements of a matrix A and suppose there exists a non-zero solution to the equations, i.e. a solution with at least one $x_i \neq 0$. For definiteness suppose $x_1 \neq 0$.

Next consider det A. In view of the results of the previous section the following operation can be carried out on A without changing the value of det A. To the first column of A add

$$\frac{x_2}{x_1} \times (\text{2nd column of } A) + \frac{x_3}{x_1} \times (\text{3rd column of } A) + \cdots$$

$$+ \frac{x_N}{x_1} \times (N\text{th column of } A).$$

Now every element of the first column of the new matrix has the form

$$a_{i1} + \frac{a_{i2}x_2}{x_1} + \frac{a_{i3}x_3}{x_1} + \cdots + \frac{a_{iN}x_N}{x_1}, \quad i = 1, 2, \ldots, N. \quad (14.19)$$

But because the x_j have been selected to be solutions of (14.17), each of these elements is individually zero. Hence the new matrix contains a column of zeros and so its determinant is zero. Thus necessarily det A is zero also.

This result (which will later form an important step in the discussion of eigenvalues of matrices) may be stated as:

> If equations (14.17) have a non-zero solution, then $|A| = 0$ where A is the matrix whose elements are the constants in (14.17).

In the matrix and vector notation of later sections, this reads:

> The equation $A\mathbf{x} = \mathbf{0}$, can have a solution with $\mathbf{x} \neq \mathbf{0}$ only if $|A| = 0$. (14.20)

▶7. Verify that the matrix appropriate to equations (14.18) has zero determinant.

▶8. Do the following sets of equations have non-zero solutions? If so find them.

(i) $3x + 2y + z = 0$, $x - 3y + 2z = 0$, $2x + y + 3z = 0$;
(ii) $2x = b(y + z)$, $x = 2a(y - z)$, $x = (6a - b)y - (6a + b)z$.

14.5   Matrix algebra

So far in our development, matrices have been nothing more than arrays of numbers (or algebraic expressions), and it is natural to enquire as to whether it is possible or useful to define operations by means of which two or more matrices may be combined to give further (related) matrices or other quantities. The kind of operations on ordinary numbers, for which analogues may be sought for matrices, are addition, subtraction, multiplication and division.

1. *Addition and subtraction.* At one point in the derivation of property (v) of section 14.3 an equation appears which has something of the appearance of the addition of two matrices. The actual context there is one of determinants of matrices rather than the matrices themselves. However, it seems natural to define the sum $D = A + B$ of two matrices A and B

[we avoid the symbol C which we employ in connection with cofactors] as that matrix whose elements are given by

$$d_{ij} = a_{ij} + b_{ij} \tag{14.21}$$

for every pair of subscripts (i, j), with $i = 1, 2, \ldots, M$ and $j = 1, 2, \ldots, N$. For this to have any meaning, the matrices A and B must both be $M \times N$ and their sum D will be the same.

From definition (14.21) it follows that $A + B = B + A$ and that the sum of a number of matrices can be written unambiguously without bracketing, i.e. the *commutative* and *associative* laws hold.

It is clear that for consistency the matrix $E = A - B$ must be defined by

$$e_{ij} = a_{ij} - b_{ij}, \quad i = 1, 2, \ldots, M, \quad j = 1, 2, \ldots, N. \tag{14.22}$$

2. *Multiplication.* A useful definition of multiplication of matrices is less obvious, but one is suggested by a study of *linear transformations*. For discussion purposes and to save space, we will limit ourselves to matrices of small size until the general formalism has been established. The results to be obtained will be applicable to matrices of arbitrary size and in general can be compactly stated using the summation convention.

Consider a transformation from one set of variables x_1, x_2 to a new set of variables y_1, y_2, y_3 [we suppose there are a different number of them so that the matrix properties can be developed for general matrices]. If the transformation is *linear* it will take the form

$$\begin{aligned} x_1 &= a_{11}y_1 + a_{12}y_2 + a_{13}y_3, \\ x_2 &= a_{21}y_1 + a_{22}y_2 + a_{23}y_3. \end{aligned} \tag{14.23}$$

These equations can be written more compactly using the summation convention as

$$x_i = a_{ij}y_j \quad (i = 1, 2), (j = 1, 2, 3). \tag{14.24}$$

Alternatively if we view the entities

$$X \equiv \begin{bmatrix} x_1 \\ x_2 \end{bmatrix} \quad \text{and} \quad Y \equiv \begin{bmatrix} y_1 \\ y_2 \\ y_3 \end{bmatrix} \tag{14.25}$$

as a 2×1 matrix and a 3×1 matrix respectively, we see that we could conveniently define multiplication of matrices by saying that

$$X = AY \tag{14.26}$$

means that each element of the *product matrix* X is given by (14.24) for a particular value of the subscript i. For this to be a meaningful definition it is clearly necessary that the number of columns in A equals the number of rows in Y [in this case both $= 3$].

As mentioned earlier, it is usual when a matrix has only a single column (row) to call it a column (row) vector and, to be in line with the vector notation of chapters 2–4, to write it in bold type, e.g. $\mathbf{x}(\tilde{\mathbf{x}})$. Thus (14.26) would normally be written

$$\mathbf{x} = A\mathbf{y}, \tag{14.27}$$

and we will follow this practice for column or row vectors.

Consideration of a *further* transformation, say

$$y_k = b_{k1}z_1 + b_{k2}z_2 + b_{k3}z_3 = b_{kl}z_l, \quad \text{for } l = 1, 2, 3, \tag{14.28}$$

will show how the definition of multiplication of matrices can be extended from (14.24) and (14.26) to more general cases. Substituting (14.28) into (14.24) [with $j = k$] gives

$$x_i = a_{ij}b_{jl}z_l \tag{14.29}$$

and so gives the connection between the variables (x_1, x_2) and the variables (z_1, z_2, z_3) directly. Written as a single transformation it would have the form

$$x_i = d_{il}z_l, \tag{14.30}$$

and comparison between this and (14.29) shows that we should define the product of the matrices A and B as

$$D = AB \tag{14.31}$$

with each component d_{il} of D defined by

$$d_{il} \equiv a_{ij}b_{jl}. \tag{14.32}$$

Again the number of columns in A must match the number of rows in B.

Multiplication of more than two matrices follows naturally and we simply state the defining relationship for the individual elements

$$(H)_{ij} = (ABC\ldots G)_{ij} = A_{ik}B_{kl}C_{lm}\ldots G_{qj}. \tag{14.33}$$

The expression (14.33) is independent of the order in which the sums over $k, l, m, \ldots, q$ are carried out and so the *associative* law holds in multiplication.

Returning to (14.32), it is apparent that if A is $M \times N$ and B is an $N \times M$ matrix, then two product matrices are possible, given by

$$\text{(i) } d_{il} = a_{ij}b_{jl}, \quad \text{and} \quad \text{(ii) } e_{il} = b_{ij}a_{jl}.$$

They are clearly not the same since D is an $M \times M$ matrix whilst E is an $N \times N$ matrix. Care must thus be taken to write matrix products in the intended order; $D = AB$ but $E = BA$.

Even if both A and B were square (i.e. both $N \times N$) the product AB is *not in general* the same as the product BA. If they are the same the two matrices are said to *commute*.

For the rest of this chapter we will consider only square matrices. These will be of arbitrary size $N \times N$ unless otherwise stated although most specific examples will be 2×2 or 3×3. All summations over dummy subscripts must be understood to run from 1 to N.

To give a concrete example of the multiplication procedure and also to illustrate the points of the last paragraph but one, we now carry out the following example.

Example 14.2. Evaluate AB and BA when A and B are the 3×3 matrices given by

$$A = \begin{bmatrix} 3 & 2 & -1 \\ 0 & 3 & 2 \\ 1 & -3 & 4 \end{bmatrix}, \quad B = \begin{bmatrix} 2 & -2 & 3 \\ 1 & 1 & 0 \\ 3 & 2 & 1 \end{bmatrix}.$$

The layout to be used for the working is that usually found most convenient. The elements (d_{ij} say) of the resultant matrices are (for simple elements) found by mentally taking the 'scalar product' of the ith row of the first matrix with the jth column of the second one. For example, $d_{11} = 3 \cdot 2 + 2 \cdot 1 + (-1) \cdot 3 = 5$, $d_{12} = 3 \cdot (-2) + 2 \cdot 1 + (-1) \cdot 2 = -6$, etc.

$$D = AB = \begin{bmatrix} 3 & 2 & -1 \\ 0 & 3 & 2 \\ 1 & -3 & 4 \end{bmatrix} \begin{bmatrix} 2 & -2 & 3 \\ 1 & 1 & 0 \\ 3 & 2 & 1 \end{bmatrix} = \begin{bmatrix} 5 & -6 & 8 \\ 9 & 7 & 2 \\ 11 & 3 & 7 \end{bmatrix}.$$

$$E = BA = \begin{bmatrix} 2 & -2 & 3 \\ 1 & 1 & 0 \\ 3 & 2 & 1 \end{bmatrix} \begin{bmatrix} 3 & 2 & -1 \\ 0 & 3 & 2 \\ 1 & -3 & 4 \end{bmatrix} = \begin{bmatrix} 9 & -11 & 6 \\ 3 & 5 & 1 \\ 10 & 9 & 5 \end{bmatrix}.$$

It is apparent from these results that $AB \neq BA$.

▶9. Check the remainder of the evaluation of D and E in the above example.

Further properties of matrix multiplication which can be deduced directly from its definition include the *distributive laws*

$$(A + B)F = AF + BF,$$
$$F(A + B) = FA + FB. \tag{14.34}$$

Although $AB \neq BA$ in general, it is the case for any square matrices A and B that

$$\det (AB) = \det A \cdot \det B = \det (BA). \tag{14.35}$$

A proof of this can be found in any textbook on matrix algebra and will not be given here.

▶10. Show that this is true for the particular matrices of example 14.2.

▶11. (i) By considering matrices

$$A = \begin{bmatrix} 1 & 0 \\ 0 & 0 \end{bmatrix}, \qquad B = \begin{bmatrix} 0 & 0 \\ 3 & 4 \end{bmatrix},$$

show that $AB = O$, where O is the *zero matrix* with all entries equal to zero, does not imply that either A or B is the zero matrix,
(ii) Verify that result (14.35) is still true however.

3. *Division.* If we were dealing with ordinary numbers we would consider equation (14.31), $D = AB$, as equivalent to $B = D/A$ provided that $A \neq 0$. However, if A, B, and D are matrices this notation does not have an obvious meaning, since we have elected to define multiplication by

$$d_{il} = a_{ij}b_{jl} \tag{14.32 bis}$$

rather than, say, $d_{il} = a_{il} \times b_{il}$ (no summation), and we would wish division to be in some way the inverse of multiplication. The question becomes: 'Can an explicit formula for B be obtained in terms of A and D?'

It will be shown that this is possible for those cases where $\det A \neq 0$. A square matrix whose determinant $= 0$ is called a **singular** matrix, otherwise it is described as *non-singular*. What will be shown is that if A is nonsingular, we can define a matrix, denoted by A^{-1} and called the **inverse** of A, which has the property that,

if $\quad AB = D,$
then $B = A^{-1}D.$ \tag{14.36 a}

In words, B can be obtained by multiplying D 'on the left' by A^{-1}. Analogously, if B is non-singular, then by multiplication 'on the right'

$$A = DB^{-1}. \tag{14.36 b}$$

The construction of the inverse of A is based upon the results (14.16) involving the cofactors C_{ij} of the elements of A, namely

$$a_{Rj}C_{ij} = \delta_{RI}|A|, \qquad\qquad\qquad\qquad (14.16\text{ a bis})$$

$$a_{iS}C_{iJ} = \delta_{SJ}|A|. \qquad\qquad\qquad\qquad (14.16\text{ b bis})$$

From these equations it is clear that if we take as the (i, j) elements of A^{-1} the C_{ji} (note the inversion of the order of the subscripts) divided by the non-zero number $|A|$, then the k, l element of $A^{-1}A$ will be

$$(A^{-1}A)_{kl} = (A^{-1})_{km}(A)_{ml} = \frac{C_{mk}}{|A|} a_{ml} = \frac{\delta_{kl}}{|A|}|A| = \delta_{kl}. \quad (14.37)$$

In the next to last step, (14.16 b) was used.

Thus, with this definition of A^{-1}, the matrix $A^{-1}A$ has elements $= 1$ on the leading diagonal (i.e. when $k = l$) and $= 0$ elsewhere. Such a matrix is called a **unit** matrix and denoted by I [it too is $N \times N$];

$$I = \begin{bmatrix} 1 & 0 & 0 & \ldots & 0 \\ 0 & 1 & 0 & & \cdot \\ 0 & 0 & 1 & & \cdot \\ \cdot & & & \ddots & \cdot \\ \cdot & & & 1 & 0 \\ 0 & & \ldots & 0 & 1 \end{bmatrix}. \qquad (14.38)$$

In this notation (14.37) can be written as

$$A^{-1}A = I. \qquad\qquad\qquad\qquad (14.39)$$

▶12. Show by a similar argument to (14.37) that $AA^{-1} = I$, i.e. A and its inverse commute.

▶13. Verify formally the following intuitively obvious properties of the unit matrix.

(i) For any matrix A, $AI = A$.
(ii) I commutes with any other matrix.
(iii) I is its own inverse.

We may now return to our original equation $AB = D$ and solve it for B as follows.

Because A is non-singular ($|A| \neq 0$) we can define the matrix A^{-1}. Multiply both sides of $AB = D$ on the left by A^{-1} to obtain

$$A^{-1}AB = A^{-1}D.$$

Using (14.39) and the results of ▶13, this becomes

$$A^{-1}D = A^{-1}AB = IB = BI = B,$$

which is the stated result (14.36 a).

It is clear that A^{-1} cannot be singular [e.g. by using (14.39) and (14.35) and noticing that det $I = 1$] and so it too has an inverse. Applying the result (14.36) to (14.39) shows that its inverse is A itself; in symbols,

$$A^{-1}A = I,$$

thus $\quad A = (A^{-1})^{-1}I = (A^{-1})^{-1}.$ \hfill (14.40)

By similar arguments, other properties of the inverse may be found; these are left as exercises for the student.

▶14. Show that for non-singular matrices $A, B, \ldots,$

(i) $(\tilde{A})^{-1} = (\widetilde{A^{-1}}),$
(ii) $(AB)^{-1} = B^{-1}A^{-1},$
(iii) $(AB\ldots G)^{-1} = G^{-1}\ldots B^{-1}A^{-1}.$

14.6 Simultaneous linear equations

As a particular case of matrix division we may consider a matrix equation

$$A\mathbf{x} = \mathbf{b}, \tag{14.41}$$

where A is $N \times N$ and non-singular and $\mathbf{x}$ and $\mathbf{b}$ are $N \times 1$ column vectors. Written out in full this represents the N simultaneous linear equations

$$\left.\begin{array}{l} a_{11}x_1 + a_{12}x_2 + \cdots + a_{1N}x_N = b_1, \\ a_{21}x_1 + a_{22}x_2 + \cdots + a_{2N}x_N = b_2, \\ \vdots \\ a_{N1}x_1 + a_{N2}x_2 + \cdots + a_{NN}x_N = b_N. \end{array}\right\} \tag{14.42}$$

Consideration has already been given in section 14.4 to the case where $\mathbf{b} = 0$ and so we will not consider that case further.

A solution for the unknowns $x_1, x_2, \ldots, x_N$ can be immediately obtained in terms of the elements of A and $\mathbf{b}$ by using (14.36 a), namely

$$\mathbf{x} = A^{-1}\mathbf{b}. \tag{14.43}$$

To illustrate this method and demonstrate the construction of an inverse matrix we will solve the following.

Example 14.3. Solve the simultaneous equations

$$2x_1 + 4x_2 + 3x_3 = 4,$$
$$x_1 - 2x_2 - 2x_3 = 0,$$
$$-3x_1 + 3x_2 + 2x_3 = -7.$$

Referring to the equations as $Ax = b$, we first determine $|A|$ and find it $= 11$. This is non-zero and so an inverse matrix A^{-1} can be constructed.

To do this we need the cofactors of all the elements of A. The cofactor of a_{21}, for example, is $(-1)^{2-1}[(4 \times 2) - (3 \times 3)] = 1$ and this is to be placed as the 1, 2 [note the order] element of A^{-1}. The matrix A^{-1} is in this way built up as

▶15. $$A^{-1} = \frac{1}{11} \begin{bmatrix} 2 & 1 & -2 \\ 4 & 13 & 7 \\ -3 & -18 & -8 \end{bmatrix}.$$

The solution of the equations is then easily obtained as

$$\mathbf{x} = A^{-1}\mathbf{b} = \frac{1}{11} \begin{bmatrix} 2 & 1 & -2 \\ 4 & 13 & 7 \\ -3 & -18 & -8 \end{bmatrix} \begin{bmatrix} 4 \\ 0 \\ -7 \end{bmatrix} = \begin{bmatrix} 2 \\ -3 \\ 4 \end{bmatrix}.$$

Thus the solution is $x_1 = 2$, $x_2 = -3$, $x_3 = 4$.

The solution of an equation of the form (14.41) can also be considered as inverting a transformation of variables such as is described by

$$\mathbf{x} = A\mathbf{y}. \tag{14.27 bis}$$

The solution

$$\mathbf{y} = A^{-1}\mathbf{x} \tag{14.44}$$

gives the equations which express the y_i in terms of the x_i, whereas the original equation (14.27) expressed the x_i in terms of the y_i. [It should again be noted that for a unique inverse transformation to exist there must be equal numbers of x_i and $y_i - A$ is square – and A must be non-singular.]

14.7   Derived matrices and their properties

In addition to the inverse matrix A^{-1}, there are a number of other matrices which can be derived from an $N \times N$ matrix A. In describing them we will begin by assuming all quantities involved are real.

The transposed matrix $\tilde{A}$ has already been defined in equation (14.14) of section 14.3, where it was stated that if $\tilde{A} = A$ then A is said to be symmetric. A matrix which has a transpose equal to minus itself, i.e.

$$a_{ij} = -a_{ji}, \tag{14.45}$$

is called **skew-** (or **anti-**) **symmetric**. Its diagonal elements $a_{11}, a_{22}, \ldots, a_{NN}$, are necessarily zero.

The transpose of a product of two matrices $\widetilde{AB}$ is given by the product of their transposes taken in reverse order,

$$\widetilde{AB} = \tilde{B}\tilde{A}. \tag{14.46}$$

This is easily shown by the following chain of equalities,

$$(\widetilde{AB})_{ij} = (AB)_{ji} = A_{jk}B_{ki} = (\tilde{A})_{kj}(\tilde{B})_{ik} = (\tilde{B})_{ik}(\tilde{A})_{kj} = (\tilde{B}\tilde{A})_{ij}.$$

▶16. Extend this to a proof that

$$(\widetilde{ABC\ldots G}) = \tilde{G}\ldots\tilde{C}\tilde{B}\tilde{A}.$$

A matrix with the particular property that its transpose is also its inverse,

$$\tilde{A} = A^{-1}, \tag{14.47}$$

is called an **orthogonal** matrix. This clearly requires that A is non-singular; but further, since

$$\tilde{A}A = I \tag{14.48}$$

for such a matrix, $|A|^2 = |\tilde{A}||A| = |I| = 1$ and $|A| = \pm 1$.

If an orthogonal matrix describes a change of variable, as in (14.27)

$$\mathbf{x} = A\mathbf{y}, \tag{14.27 bis}$$

then, using (14.46), the row vector $\tilde{\mathbf{x}}$ is given by $\tilde{\mathbf{x}} = \tilde{\mathbf{y}}A^{-1}$ and the scalar product [1 × 1 matrix] $\tilde{\mathbf{x}}\mathbf{x}$, when expressed in terms of the new variables y_i, becomes

$$\tilde{\mathbf{x}}\mathbf{x} = \tilde{\mathbf{y}}A^{-1}A\mathbf{y} = \tilde{\mathbf{y}}I\mathbf{y} = \tilde{\mathbf{y}}\mathbf{y}. \tag{14.49}$$

Thus a transformation described by an orthogonal matrix (an orthogonal transformation) is one which preserves the lengths of vectors.

As an alternative to the matrix equation (14.48), we may write a relationship between the elements themselves of an orthogonal matrix,

$$a_{ki}a_{kj} = (\tilde{A})_{ik}(A)_{kj} = \delta_{ij}. \tag{14.50}$$

As a simple example of an orthogonal matrix we may consider

$$A = \begin{bmatrix} \cos\theta & -\sin\theta & 0 \\ \sin\theta & \cos\theta & 0 \\ 0 & 0 & 1 \end{bmatrix}.$$

The transpose of A is also its inverse as is shown by

$$\tilde{A}A = \begin{bmatrix} \cos\theta & \sin\theta & 0 \\ -\sin\theta & \cos\theta & 0 \\ 0 & 0 & 1 \end{bmatrix} \begin{bmatrix} \cos\theta & -\sin\theta & 0 \\ \sin\theta & \cos\theta & 0 \\ 0 & 0 & 1 \end{bmatrix}$$

$$= \begin{bmatrix} 1 & 0 & 0 \\ 0 & 1 & 0 \\ 0 & 0 & 1 \end{bmatrix} = I.$$

This particular matrix A represents the transformation obtained by a rotation of the axes through an angle θ about the x_3-axis.

▶17. Calculate A^{-1} directly as in section 14.5 and show that it is the same as $\tilde{A}$.

When the elements of matrices are allowed to be complex, and not restricted to real values, correspondingly more general derived matrices can be suitably defined. Roughly speaking the additional operation of complex conjugation is added to each of those defined for real quantities.

Thus the **Hermitian conjugate** of the matrix A is given by $(\tilde{A})^*$ (clearly identical with $\widetilde{(A^*)}$), where * indicates the operation of taking the complex conjugate of each element of the array. Some authors use A^H or $A^\dagger$ instead of $\tilde{A}^*$. A matrix for which $\tilde{A}^* = A$ is said to be **Hermitian**, a real symmetric matrix being a special case of this when all the elements of the matrix are real. Similarly if $\tilde{A}^* = -A$, then A is called **anti-Hermitian**.

The 'complex analogue' of an orthogonal matrix is a **unitary** one, which is defined as a matrix A for which

$$\tilde{A}^* = A^{-1}. \tag{14.51}$$

A unitary transformation maintains the value of the modulus of a (complex) vector $\mathbf{x}$, which is given by $(\bar{\mathbf{x}}^*\mathbf{x})^{1/2}$. On transforming, $\bar{\mathbf{x}}^*\mathbf{x}$ becomes

$$\bar{\mathbf{x}}^*\mathbf{x} = (\widetilde{A\mathbf{y}})^*A\mathbf{y} = \bar{\mathbf{y}}^*\tilde{A}^*A\mathbf{y} = \bar{\mathbf{y}}^*A^{-1}A\mathbf{y} = \bar{\mathbf{y}}^*\mathbf{y},$$

which is the modulus squared of the vector in the new coordinates.

▶18. Show that a Hermitian matrix H can be written as $H = R + iS$, where R is a real symmetric matrix and S is real and skew-symmetric.

▶19. Consider the matrices

$$
\text{(i)} \ B = \begin{bmatrix} 0 & -i & i \\ i & 0 & -i \\ -i & i & 0 \end{bmatrix}, \ \text{(ii)} \ C = \frac{1}{\sqrt{8}} \begin{bmatrix} \sqrt{3} & -\sqrt{2} & -\sqrt{3} \\ 1 & \sqrt{6} & -1 \\ 2 & 0 & 2 \end{bmatrix}.
$$

Are they (a) real, (b) symmetric, (c) skew-symmetric, (d) singular, (e) orthogonal, (f) Hermitian, (g) anti-Hermitian, (h) unitary?

14.8 Eigenvalues and eigenvectors

We have seen that a non-singular transformation $\mathbf{x} = A\mathbf{y}$ and its inverse $\mathbf{y} = A^{-1}\mathbf{x}$ describe a 'one-to-one' correspondence between the vectors with components x_i and those with components y_i. Putting it another way, we may say that given one vector $\mathbf{x}$, say, it determines a second one $\mathbf{y}$ through the matrix A. The vectors $\mathbf{x}$ and $\mathbf{y}$ may describe different types of quantities [such as voltage gradient and current] if A has physical dimensions [such as those of electrical resistivity], or they may both be purely mathematical entities and to all intents and purposes dimensionless.

However, particularly interesting cases arise when the two vectors $\mathbf{x}$ and $\mathbf{y}$ are 'parallel' to each other, i.e. expressed in their own units if necessary, one is just a multiple of the other. As a formula, $\mathbf{x} = \lambda\mathbf{y}$ and

$$
A\mathbf{y} = \mathbf{x} = \lambda\mathbf{y}. \tag{14.52}
$$

A non-zero vector $\mathbf{y}$ which satisfies (14.52) for some value of λ is called an **eigenvector** of A, and λ is called the corresponding **eigenvalue**.

As will be seen in the remaining sections of this chapter and much of the next, eigenvectors and eigenvalues of matrices play an important role in many physical situations. The term eigenvalue has already been used extensively in earlier chapters in connection with the solution of differential equations, and maintains a corresponding position here. The eigenvector is the discrete analogue, appropriate to a matrix, of the continuous eigenfunction of a differential operator.

In the remainder of this section we will prove some general results concerning eigenvectors and eigenvalues in preparation for their use in later sections. The results will be established for matrices whose elements may be complex, the corresponding properties for real matrices being obtained as special cases.

The ith eigenvector will be noted by $\mathbf{y}^i$ and the corresponding eigenvalue by λ_i, for $i = 1, 2, \dots$ [for an $N \times N$ matrix we will find that there are in general N distinct eigenvalues], so that

$$
A\mathbf{y}^i = \lambda_i\mathbf{y}^i \ \text{(no summation)}. \tag{14.53}
$$

1. The eigenvalues of a Hermitian matrix A are real.

For any particular value of i take the Hermitian conjugate (complex conjugate transposed) of (14.53),

$$\bar{\mathbf{y}}^{i*}\tilde{A}^* = \lambda_i^*\bar{\mathbf{y}}^{i*}. \tag{14.54}$$

Using $\tilde{A}^* = A$, since A is Hermitian, and multiplying on the right by $\mathbf{y}^i$, we obtain

$$\bar{\mathbf{y}}^{i*}A\mathbf{y}^i = \lambda_i^*\bar{\mathbf{y}}^{i*}\mathbf{y}^i. \tag{14.55}$$

But multiplying (14.53) through on the left by $\bar{\mathbf{y}}^{i*}$ gives

$$\bar{\mathbf{y}}^{i*}A\mathbf{y}^i = \lambda_i\bar{\mathbf{y}}^{i*}\mathbf{y}^i.$$

Subtracting this from (14.55) yields

$$0 = (\lambda_i^* - \lambda_i)\bar{\mathbf{y}}^{i*}\mathbf{y}^i.$$

But $\bar{\mathbf{y}}^{i*}\mathbf{y}^i$ is the modulus squared of the non-zero vector $\mathbf{y}^i$ and is thus non-zero. Hence λ_i^* must equal λ_i and thus be real.

2. The eigenvectors corresponding to different eigenvalues of a Hermitian matrix are orthogonal.

Consider two unequal eigenvalues λ_i and λ_j $(\neq\lambda_i)$. Multiply (14.54) through on the right by $\mathbf{y}^j$ to obtain

$$\bar{\mathbf{y}}^{i*}\tilde{A}^*\mathbf{y}^j = \lambda_i^*\bar{\mathbf{y}}^{i*}\mathbf{y}^j,$$

and (14.53), for the eigenvalue λ_j, through on the left by $\bar{\mathbf{y}}^{i*}$ to obtain

$$\bar{\mathbf{y}}^{i*}A\mathbf{y}^j = \lambda_j\bar{\mathbf{y}}^{i*}\mathbf{y}^j.$$

Then using the fact that $\tilde{A}^* = A$, the two left-hand sides are equal and, since the λ_i are real, on subtraction we obtain

$$0 = (\lambda_i - \lambda_j)\bar{\mathbf{y}}^{i*}\mathbf{y}^j.$$

Finally $\lambda_j \neq \lambda_i$ and so we must have $\bar{\mathbf{y}}^{i*}\mathbf{y}^j = 0$, which is the stated result.

3. The eigenvectors of a Hermitian matrix are mutually orthogonal or can be chosen to be so.

This follows immediately from result 2 if all the eigenvalues are unequal. However if some are the same, further justification is needed. An eigenvalue corresponding to two (or more) different eigenvectors (one not simply a multiple of the other) is said to be *degenerate*.

Suppose, for the sake of our proof, λ_1 is K-fold degenerate, i.e.

$$A\mathbf{y}^i = \lambda_1\mathbf{y}^i \quad \text{for } i = 1, 2, \ldots, K, \tag{14.56}$$

but that λ_1 is different from any of λ_{K+1}, λ_{K+2}, etc. Then any linear combination of these $\mathbf{y}^i$ is also an eigenvector with eigenvalue λ_1 since

$$A\mathbf{z} \equiv A \sum_1^K c_i\mathbf{y}^i = \sum_1^K c_i A\mathbf{y}^i = \sum_1^K c_i\lambda_1\mathbf{y}^i = \lambda_1\mathbf{z}, \tag{14.57}$$

for arbitrary values of the coefficients c_i.

If the $\mathbf{y}^i$ defined in (14.56) are not already mutually orthogonal, consider the new eigenvectors $\mathbf{z}^i$ constructed by the following procedure, in which each of the new vectors is to be normalized before proceeding to the construction of the next one. [The normalization can be carried out by dividing each element of the vector $\mathbf{z}^i$ by $(\tilde{\mathbf{z}}^{i*}\mathbf{z}^i)^{1/2}$.]

(i) $\mathbf{z}^1 = \mathbf{y}^1$,

(ii) $\mathbf{z}^2 = \mathbf{y}^2 - (\tilde{\mathbf{z}}^{1*}\mathbf{y}^2)\mathbf{z}^1$,

(iii) $\mathbf{z}^3 = \mathbf{y}^3 - (\tilde{\mathbf{z}}^{2*}\mathbf{y}^3)\mathbf{z}^2 - (\tilde{\mathbf{z}}^{1*}\mathbf{y}^3)\mathbf{z}^1$,

         . . .

(K) $\mathbf{z}^K = \mathbf{y}^K - (\tilde{\mathbf{z}}^{K-1*}\mathbf{y}^K)\mathbf{z}^{K-1} - \cdots - (\tilde{\mathbf{z}}^{1*}\mathbf{y}^K)\mathbf{z}^1. \tag{14.58}$

Each of the factors in brackets, $(\tilde{\mathbf{z}}^{m*}\mathbf{y}^n)$, is a scalar product and thus only a number; hence each new vector $\mathbf{z}^i$ is, as shown in (14.57), an eigenvector of A with eigenvalue λ_1, and will remain so on normalization. It is straightforward to check that, provided the previous new eigenfunctions have been normalized as prescribed, each $\mathbf{z}^i$ is orthogonal to all of its predecessors.

Thus, by this explicit construction procedure, the truth of result 3 is established.

As special cases of these three results, applicable when the matrices under consideration contain only real elements, we may conclude that the eigenvalues of a real symmetric matrix are real and that its eigenfunctions are (or can be made) mutually orthogonal.

14.9   Determination of eigenvalues and eigenvectors

The next step is to show how the eigenvalues and eigenvectors of a given matrix are found. To do this we refer to the definition (14.52) and rewrite it as

$$A\mathbf{y} - \lambda I\mathbf{y} = (A - \lambda I)\mathbf{y} = \mathbf{0}. \tag{14.59}$$

The slight rearrangement used here is to write $I\mathbf{y}$ instead of $\mathbf{y}$ where I is the unit matrix of the same dimension (size) as A. The point of doing this is immediate since (14.59) now has the form developed in (14.20) of section 14.4. What was proved there was that the equation $B\mathbf{x} = 0$ only has a non-trivial solution $\mathbf{x}$ if $|B| = 0$. Correspondingly, therefore, we must have in the present case that

$$|A - \lambda I| = 0, \tag{14.60}$$

if there are to be non-zero solutions $\mathbf{y}$ to (14.59).

Equation (14.60) is known as the **characteristic equation** for A and its left-hand side as the **characteristic** or **secular determinant** of A. The equation is a polynomial one (of degree N) in the quantity λ. The roots of this equation λ_i ($i = 1, 2, \ldots, N$) give the possible eigenvalues of A; there will be N of them. Corresponding to each λ_i there will be a vector $\mathbf{y}^i$ which is the ith eigenvector of A.

Before proceeding further with the use of the eigenvectors and values so found, we will work through a specific example.

Example 14.4. Verify the preceding results using the particular real symmetric matrix

$$A = \begin{bmatrix} 2 & 0 & 1 \\ 0 & 1 & 0 \\ 1 & 0 & 0 \end{bmatrix}. \tag{14.61}$$

We are seeking non-zero solutions $\mathbf{y}^i$ of the equation $A\mathbf{y}^i = \lambda_i \mathbf{y}^i$, and (14.60) shows that the possible values of λ_i are given by $|A - \lambda I| = 0$, i.e.

$$\begin{vmatrix} 2 - \lambda & 0 & 1 \\ 0 & 1 - \lambda & 0 \\ 1 & 0 & -\lambda \end{vmatrix} = 0. \tag{14.62}$$

Expanding out this determinant gives

$$(2 - \lambda)(-\lambda + \lambda^2) - (1 - \lambda) = 0,$$
$$(\lambda - 1)(\lambda^2 - 2\lambda - 1) = 0,$$

which has the roots $\lambda_1 = 1$, $\lambda_2 = 1 + \sqrt{2}$, $\lambda_3 = 1 - \sqrt{2}$.

For the first root, $\lambda_1 = 1$, a suitable eigenvector with elements y_1, y_2, y_3 must satisfy

$$\left. \begin{array}{l} 2y_1 \qquad + y_3 = 1 \cdot y_1, \\ \qquad y_2 \qquad = 1 \cdot y_2, \\ \quad y_1 \qquad \quad = 1 \cdot y_3. \end{array} \right\} \tag{14.63}$$

These three equations are consistent [that was the purpose in finding the particular values of λ] and yield $y_1 = y_3 = 0$, $y_2 =$ anything non-zero. A suitable (normalized) vector would thus be

$$\mathbf{y}^1 = \begin{bmatrix} 0 \\ 1 \\ 0 \end{bmatrix}. \tag{14.64}$$

Repeating the last paragraph, but with the 1 on the right-hand side of (14.63) replaced successively by $\lambda_2 = 1 + \sqrt{2}$ and $\lambda_3 = 1 - \sqrt{2}$, gives two further (normalized) eigenvectors

▶ 20. $$\mathbf{y}^2 = (4 - 2^{3/2})^{-1/2} \begin{bmatrix} 1 \\ 0 \\ \sqrt{2} - 1 \end{bmatrix},$$

$$\mathbf{y}^3 = (4 + 2^{3/2})^{-1/2} \begin{bmatrix} 1 \\ 0 \\ -\sqrt{2} - 1 \end{bmatrix}.$$

The three values of λ are all different and A is a real symmetric matrix. We thus expect that the three eigenvectors are mutually orthogonal. This is easily checked,

▶21. $$\mathbf{y}^1 \cdot \mathbf{y}^2 = \bar{\mathbf{y}}^{1*} \mathbf{y}^2 = 0 = \mathbf{y}^1 \cdot \mathbf{y}^3 = \mathbf{y}^2 \cdot \mathbf{y}^3.$$

The *, which must be present in general, here has no effect since the eigenvectors are real. It will also be apparent that the normalization of the vectors has no effect on their orthogonality, which is as would be expected.

▶22. Find the eigenvalues and a set of eigenvectors of the symmetric matrix

$$\begin{bmatrix} 1 & 3 & -1 \\ 3 & 4 & -2 \\ -1 & -2 & 2 \end{bmatrix}.$$

Verify that its eigenvectors are mutually orthogonal.

It will be observed that when (14.60) is written out as a polynomial equation in λ, the coefficient of $-\lambda^{N-1}$ in the equation will be simply $a_{11} + a_{22} + \cdots + a_{NN}$ relative to the coefficient of λ^N being unity. The quantity $\sum_{i=1}^{N} a_{ii}$ is called the **trace** or **spur** of A and from the ordinary theory of polynomial equations the sum of the roots of (14.60), $\sum_{i=1}^{N} \lambda_i$, must therefore be equal to the trace of the original matrix A. This can be used as some check that a computation of the eigenvalues has been done correctly. In example 14.4 the trace of $A = 2 + 1 + 0 = 3$, and the sum of the computed eigenvalues is $1 + (1 + \sqrt{2}) + (1 - \sqrt{2}) = 3$; this is thus consistent with the general property.

14.10   Diagonalization of matrices

Having gained a little practical experience of dealing with eigenvalues and vectors, we now turn to the rather more complex question of '*diagonalizing*' a real symmetric matrix.†

We may illustrate what is meant by considering a homogeneous expression such as

$$Q \equiv x_1^2 + x_2^2 - 3x_3^2 + 2x_1x_2 + 6x_1x_3 - 6x_2x_3. \tag{14.65}$$

Q is called a *quadratic form* and can be represented as a product of matrices by

$$\tilde{x}Ax = [x_1 \quad x_2 \quad x_3] \begin{bmatrix} 1 & 1 & 3 \\ 1 & 1 & -3 \\ 3 & -3 & -3 \end{bmatrix} \begin{bmatrix} x_1 \\ x_2 \\ x_3 \end{bmatrix}. \tag{14.66}$$

It may now be asked whether a transformation of variables $x = Uy$, along the lines of (14.27), can be found which changes this form to one in the new variables y_1, y_2, y_3 containing no cross-terms. In other words Q takes the form $\mu_1 y_1^2 + \mu_2 y_2^2 + \mu_3 y_3^2 = \tilde{y}\Lambda y$, where Λ is a diagonal matrix i.e. one with all elements not on the principal diagonal equal to zero. [The diagonal elements are μ_1, μ_2, μ_3.]

If such a transformation were made $\tilde{x}Ax$ would be transformed as follows,

$$\tilde{x}Ax = (\widetilde{Uy})AUy = \tilde{y}\tilde{U}AUy. \tag{14.67}$$

---

† To keep the principles of the method as clear as possible, we will work only with real symmetric matrices, which have been shown to have real eigenvalues and whose eigenvectors can be chosen real. The same ideas apply to the more general Hermitian matrix, but 'Hermitian conjugate' and 'unitary' must be used instead of 'transpose' and 'orthogonal' respectively.

So, what is required is to find a matrix U such that $\tilde{U}AU$ is a diagonal matrix Λ. We will now find such a matrix.

Consider the matrix U, whose columns are the *normalized* eigenvectors of A, i.e.

$$U_{ij} = (\mathbf{y}^j)_i = y^j_i. \qquad (14.68\,a)$$

Its transposed matrix $\tilde{U}$ will then have elements

$$\tilde{U}_{ij} = U_{ji} = y^i_j. \qquad (14.68\,b)$$

Now, bearing in mind that the $\mathbf{y}^j$ are normalized and are (or can be made) orthogonal to each other, we may examine the matrix $\tilde{U}U$;

$$(\tilde{U}U)_{ij} = \tilde{U}_{ik}U_{kj} = y^i_k y^j_k = \mathbf{y}^i \cdot \mathbf{y}^j = \delta_{ij},$$

or $\qquad \tilde{U}U = I. \qquad (14.69)$

Thus $\tilde{U}$ is the inverse of U, and U is therefore an orthogonal matrix.

Using this result, we now examine the form of $\tilde{U}AU$ appearing in (14.67)

$$\begin{aligned}
(\tilde{U}AU)_{ij} &= (U^{-1}AU)_{ij} \\
&= U^{-1}_{ik}A_{kl}U_{lj} \\
&= U^{-1}_{ik}A_{kl}y^j_l \\
&= U^{-1}_{ik}\lambda_j y^j_k \\
&= \lambda_j U^{-1}_{ik}U_{kj} \text{ (not summed over } j) \\
&= \lambda_j I_{ij} \text{ (not summed over } j) \\
&= \Lambda_{ij}.
\end{aligned} \qquad (14.70)$$

In this sequence, successive use has been made of the orthogonality and construction of U, and then of the eigenvalue property of A. It is seen that the matrix $\Lambda = \tilde{U}AU$ is diagonal and has as its elements $\mu_i = \lambda_i$ ($i = 1$, 2, 3).

To summarize these results, the orthogonal (and thus non-singular) matrix U constructed according to equations (14.68) has the property that a change of variables $\mathbf{x} = U\mathbf{y}$ transforms the quadratic form $\tilde{\mathbf{x}}A\mathbf{x}$ into the 'diagonal' form $\tilde{\mathbf{y}}\Lambda\mathbf{y}$, with the diagonal elements of Λ being the eigenvalues of A.

Example 14.5. Find an orthogonal transformation (i.e. one for which the matrix is orthogonal) which takes the quadratic form (14.66) into the form $\mu_1 y^2_1 + \mu_2 y^2_2 + \mu_3 y^2_3$.

We first find the eigenvalues of the matrix A of (14.66) by means of its characteristic determinant

$$\begin{vmatrix} 1 - \lambda & 1 & 3 \\ 1 & 1 - \lambda & -3 \\ 3 & -3 & -3 - \lambda \end{vmatrix} = 0,$$

of which the solutions are

▶23. $\lambda_1 = 2, \qquad \lambda_2 = 3, \qquad \lambda_3 = -6.$

Corresponding normalized eigenvectors are

▶24. $\mathbf{x}^1 = 2^{-1/2} \begin{bmatrix} 1 \\ 1 \\ 0 \end{bmatrix}, \quad \mathbf{x}^2 = 3^{-1/2} \begin{bmatrix} 1 \\ -1 \\ 1 \end{bmatrix}, \quad \mathbf{x}^3 = 6^{-1/2} \begin{bmatrix} 1 \\ -1 \\ -2 \end{bmatrix}.$

Following the previous construction we now form the matrix

$$U = 6^{-1/2} \begin{bmatrix} \sqrt{3} & \sqrt{2} & 1 \\ \sqrt{3} & -\sqrt{2} & -1 \\ 0 & \sqrt{2} & -2 \end{bmatrix}. \tag{14.71}$$

The general result already proved now shows that the transformation $\mathbf{x} = U\mathbf{y}$ will carry the form (14.66) into the form $2y_1^2 + 3y_2^2 - 6y_3^2$.

This may be verified most easily by writing out the inverse transform $\mathbf{y} = U^{-1}\mathbf{x} = \tilde{U}\mathbf{x}$ and substituting. The inverse equations will be

$$y_1 = 2^{-1/2}(x_1 + x_2),$$
$$y_2 = 3^{-1/2}(x_1 - x_2 + x_3),$$
$$y_3 = 6^{-1/2}(x_1 - x_2 - 2x_3).$$

If these are now substituted into the form $Q = 2y_1^2 + 3y_2^2 - 6y_3^2$ the original expression

▶25. $Q = x_1^2 + x_2^2 - 3x_3^2 + 2x_1x_2 + 6x_1x_3 - 6x_2x_3$

is recovered.

▶26. (i) Verify directly that U given by (14.71) is orthogonal.
(ii) Verify directly that $\tilde{U}AU$ has the appropriate diagonal form.

▶27. Find an orthogonal transformation which takes the quadratic form $Q \equiv -x_1^2 - 2x_2^2 - x_3^2 + 8x_2x_3 + 6x_1x_3 + 8x_1x_2$ into the form $\mu_1 y_1^2 +$

$\mu_2 y_2^2 - 4y_3^2$, where μ_1 and μ_2 are to be determined. Express the y_i in terms of the x_i.

The change of variables by an orthogonal transformation is a special case of a **similarity** transformation on the matrix A. In a similarity transformation the matrix A is transformed to the matrix B given by

$$B = S^{-1}AS, \tag{14.72}$$

where S is a non-singular matrix.

Under a similarity transformation by an orthogonal matrix U several properties of A (if it has them originally) are preserved. Without explanation of the individual steps (all of which are simple, but the reader should justify each one to himself) we will prove some of them.

(a) A is the zero (obvious) or unit matrix, $A = I$.

$$B = U^{-1}IU = U^{-1}U = I.$$

(b) The determinant of A.

$$|B| = |U^{-1}AU| = |U^{-1}||A||U| = |A||U^{-1}||U|$$
$$= |A||U^{-1}U| = |A|.$$

(c) Symmetry or antisymmetry of A, $\tilde{A} = \pm A$.

$$\tilde{B} = \widetilde{(U^{-1}AU)} = \tilde{U}\tilde{A}U = \pm U^{-1}AU = \pm B.$$

(d) Orthogonality of A, $\tilde{A}A = I$.

$$\tilde{B}B = \widetilde{(U^{-1}AU)}(U^{-1}AU) = \tilde{U}\tilde{A}UU^{-1}AU = U^{-1}\tilde{A}AU$$
$$= U^{-1}IU = I.$$

(e) Trace A.

$$\text{tr } B = b_{ii} = u_{ij}^{-1}a_{jk}u_{ki} = u_{ki}u_{ij}^{-1}a_{jk} = \delta_{kj}a_{jk} = a_{kk} = \text{tr } A.$$

(f) Eigenvalues of A.

$$|B - \lambda I| = |U^{-1}AU - \lambda I| = |U^{-1}(A - \lambda I)U|$$
$$= |U^{-1}||U||A - \lambda I| = |A - \lambda I|;$$

thus A and B have the same characteristic determinant and hence the same eigenvalues.

▶28. Which of the above properties would be preserved in a general similarity transformation, in which S is non-singular but not necessarily orthogonal?

14.11   Quadratic and Hermitian forms

The notion of a **quadratic form** was introduced in the previous section (equations (14.65) and (14.66)). The example given there involved a real symmetric 3×3 matrix. For the purposes of illustration we will continue to work in a system with three components for which the column (vector) matrices have a ready interpretation; however, all our results will apply to general $N \times N$ matrices.

It is clear that if the components x_1, x_2, x_3 are real, then the quadratic form

$$Q \equiv \tilde{\mathbf{x}} A \mathbf{x} \equiv [x_1 \quad x_2 \quad x_3] \begin{bmatrix} 1 & 1 & 3 \\ 1 & 1 & -3 \\ 3 & -3 & -3 \end{bmatrix} \begin{bmatrix} x_1 \\ x_2 \\ x_3 \end{bmatrix} \qquad (14.66\ bis)$$

is real also.

Another, rather more general, expression which is also real is a **Hermitian form**

$$H \equiv \tilde{\mathbf{x}}^* A \mathbf{x}, \qquad (14.73)$$

where A is Hermitian, $\tilde{A}^* = A$, but the components of $\mathbf{x}$ may now be complex. It is straightforward to show that H is real, since

$$H^* = \tilde{H}^* = \tilde{\mathbf{x}}^* \tilde{A}^* \mathbf{x} = \tilde{\mathbf{x}}^* A \mathbf{x} = H.$$

With suitable generalization, the properties of quadratic forms apply also to Hermitian forms, but to keep the presentation simple we will restrict ourselves to the former, which is a special case of the latter.

The stationary property of the eigenvectors. Consider a quadratic form such as (14.66). As the vector $\mathbf{x}$ is varied, through its three components x_1, x_2 and x_3, the value of the quantity Q also varies, so that $Q = Q(\mathbf{x})$. Because of the homogeneous form of Q we may restrict any investigation of these variations to vectors of unit length [since multiplying any vector $\mathbf{x}$ by any scalar k simply multiplies the value of Q by a factor k^2].†

Of particular interest are any vectors $\mathbf{x}$ which make the value of the quadratic form a maximum or minimum. A necessary (but not sufficient) condition for this is that Q is stationary with respect to small variations in $\mathbf{x}$, whilst $\tilde{\mathbf{x}} \mathbf{x}$ is maintained at a constant value (unity).

Using Lagrange undetermined multipliers to deal with the conditional variations, we are led to seek solutions of

$$\delta[\tilde{\mathbf{x}} A \mathbf{x} - \lambda(\tilde{\mathbf{x}} \mathbf{x} - 1)] = 0. \qquad (14.74)$$

---

† An equivalent alternative is to consider variations in the quantity $(\tilde{\mathbf{x}} A \mathbf{x})/(\tilde{\mathbf{x}} \mathbf{x})$ with no restrictions placed on $\mathbf{x}$.

This may be used directly, together with the fact that $(\delta\bar{\mathbf{x}})A\mathbf{x} = \bar{\mathbf{x}}A(\delta\mathbf{x})$ (since A is symmetric), to obtain the necessary condition

$$A\mathbf{x} = \lambda\mathbf{x}, \tag{14.75 a}$$

that $\mathbf{x}$ must satisfy.

However, the use of variations of a vector may be avoided by expressing Q in its subscript form $Q = x_i a_{ij} x_j$ and employing differentiation with respect to x_k ($k = 1, 2, \ldots, N$), subject to $x_l x_l = 1$. It is then required that

$$0 = \frac{\partial}{\partial x_k}(x_i a_{ij} x_j - \lambda x_l x_l + \lambda)$$

$$= \delta_{ki} a_{ij} x_j + x_i a_{ij} \delta_{jk} - \lambda \delta_{kl} x_l - \lambda x_l \delta_{lk}$$

$$= a_{kj} x_j + a_{ik} x_i - 2\lambda x_k$$

$$= 2a_{kj} x_j - 2\lambda x_k, \text{ since } a_{ik} = a_{ki}.$$

Thus

$$a_{kj} x_j = \lambda x_k, \tag{14.75 b}$$

which is just the eigenvalue equation (14.75 a) in its subscript form.

Furthermore if equations (14.75) are satisfied for some eigenvector $\mathbf{x}$, the value of $Q(\mathbf{x})$ is then just

$$Q = \bar{\mathbf{x}}A\mathbf{x} = \bar{\mathbf{x}}\lambda\mathbf{x} = \lambda. \tag{14.76 a}$$

On the other hand if $\mathbf{x}$ and $\mathbf{y}$ are eigenvectors corresponding to different eigenvalues they are (or can be chosen) orthogonal. Consequently the expression $\bar{\mathbf{y}}A\mathbf{x}$ is necessarily zero, since

$$\bar{\mathbf{y}}A\mathbf{x} = \bar{\mathbf{y}}\lambda\mathbf{x} = \lambda\bar{\mathbf{y}}\mathbf{x} = 0. \tag{14.76 b}$$

Summarizing, those vectors $\mathbf{x}$ of given magnitude which make the quadratic form Q stationary are eigenvalues of A, and the stationary value of Q is then equal to the corresponding eigenvalue. It is straightforward to see from the proofs of (14.75) that conversely any eigenvector of A makes Q stationary.

The eigenvectors of A thus lie along those directions in space for which the quadratic form Q has stationary values, given a fixed magnitude for the vector $\mathbf{x}$. This last sentence may be turned round to state that the surface given by

$$\bar{\mathbf{x}}A\mathbf{x} = \text{constant} = 1 \text{ (say)} \tag{14.77}$$

has stationary values of its radius (origin-surface distance) in those directions which are along the eigenvectors of A. More specifically in three dimensions the quadric $\tilde{\mathbf{x}} A \mathbf{x} = 1$ has its principal axes along the three mutually perpendicular eigenvectors of A, and the squares of the corresponding principal radii are given by λ_i^{-1} ($i = 1, 2, 3$). If any of the eigenvalues are degenerate, then the quadric has rotational symmetry about some axis and the choice of a pair of axes perpendicular to that axis is not uniquely defined.

▶29. Obtain this last result by taking the surface as $\phi = x_i a_{ij} x_j$, determining the direction of its normal $\nabla\phi$, and then expressing the condition that the normal at a point on the surface be parallel to the radius vector there if the point lies on a principal axis.

As an illustration [involving no further calculation] we may interpret the result of example 14.5 geometrically. The quadric to which the result refers is

$$x_1^2 + x_2^2 - 3x_3^2 - 6x_2 x_3 + 6x_1 x_3 + 2x_1 x_2 = 1. \tag{14.78}$$

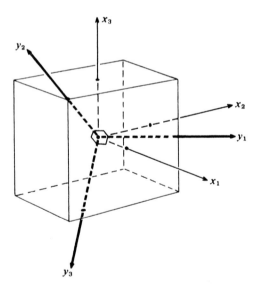

Fig. 14.1 The three mutually perpendicular eigenvector directions of the quadric given by equation (14.78). Referred to the (x_1, x_2, x_3)-axes they are $(1, 1, 0)$, $(1, -1, 1)$ and $(1, -1, -2)$.

If, instead of expressing the quadric in terms of x_1, x_2, x_3, we were to use new variables y_1, y_2, y_3, with axes along the three mutually perpendicular eigenvector directions $(1, 1, 0)$, $(1, -1, 1)$ and $(1, -1, -2)$

[referred to the x_i-axes] as in fig. 14.1, then the quadric would take the form

$$\frac{y_1^2}{(2^{-1/2})^2} + \frac{y_2^2}{(3^{-1/2})^2} - \frac{y_3^2}{(6^{-1/2})^2} = 1. \qquad (14.79)$$

Thus, for example, a section of the quadric in the plane $y_3 = 0$, i.e. $x_1 - x_2 - 2x_3 = 0$ is an ellipse with semi-axes $2^{-1/2}$ and $3^{-1/2}$. Similarly a section in the plane $y_1 = x_1 + x_2 = 0$ is a hyperbola.

Clearly the simplest situation to visualize is that in which all the eigenvalues are positive, since then the quadric is an ellipsoid.

Several examples of the use of quadratic forms and quadric surfaces in physics and engineering problems appear in the next chapter in connection with Cartesian tensors and so further consideration will not be given to them here.

14.12   Mechanical oscillations and normal modes

The subject of oscillations and normal modes was discussed briefly in the previous chapter. There, the equations for the **normal modes** (motions which are truly periodic) were derived using Lagrange's equations which were in turn derived from Euler's variational equations. As an alternative, simpler, but less rigorous approach, we will here derive them [or at least make them plausible] from the more physical consideration of energy conservation.

So far as mechanical systems are concerned we consider only those for which a potential exists. That is the potential energy of the system in any particular configuration depends upon the coordinates of the configuration [which need not be restricted to spatial positions] but must not depend upon the time derivative (general velocity) of these coordinates. A further restriction which we place is that the potential has a strict (local) minimum at the equilibrium point; this is a physically-obvious necessary and sufficient condition for stable equilibrium. By suitably defining the origin of the potential, we may take its value at the equilibrium point as zero and we will do this.

We denote the coordinates chosen to describe a configuration of the system by q_i ($i = 1, 2, \ldots, N$). [In chapter 13 coordinates θ_i were used to avoid confusion with the function $q(x)$ also appearing in that chapter.] The q_i need not be distances; some could be angles, for example, as in the swinging rod on a string of section 13.5. For convenience we can define the q_i so that they are all zero at the equilibrium point.

The instantaneous velocities of various parts of the system will depend on the time derivatives of the q_i, denoted by $\dot{q}_i$. For small oscillations the velocities will be linear in the $\dot{q}_i$ and consequently the total kinetic energy

T will be quadratic in them – including cross terms of the form $\dot{q}_i\dot{q}_j$ with $i \neq j$. The general expression can be written as a quadratic form

$$T = a_{ij}\dot{q}_i\dot{q}_j. \tag{14.80}$$

The matrix A can be taken as symmetric, since if it is not so already, it may be replaced by another A' whose elements are given by $a'_{ij} = a'_{ji} = \frac{1}{2}(a_{ij} + a_{ji})$ and is therefore symmetric, without affecting the value of T. The coefficients a_{ij} depend upon the geometry of the system and are therefore real. Thus A is a real symmetric $N \times N$ matrix.

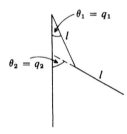

Fig. 14.2 The relationship between the coordinates θ_i of section 13.5 and the coordinates q_i of section 14.12.

As an example, the kinetic energy of the rod of section 13.5 (resketched in fig. 14.2) was found in that section to be given to second order in $\dot{q}_i$ ($= \dot{\theta}_i$ in that problem) by

$$T = \tfrac{1}{2}Ml^2(\dot{q}_1^2 + \dot{q}_1\dot{q}_2 + \tfrac{1}{4}\dot{q}_2^2) + \tfrac{1}{24}Ml^2\dot{q}_2^2,$$

and therefore represented by the real symmetric matrix

$$A = \frac{Ml^2}{12}\begin{bmatrix} 6 & 3 \\ 3 & 2 \end{bmatrix}. \tag{14.81}$$

This, like any matrix corresponding to a kinetic energy, is positive definite (more strictly non-negative definite); that is, whatever real values the $\dot{q}_i$ take, the quadratic form (14.80) has a value $\geqslant 0$.

Turning now to the potential energy, we may write its value for a configuration $\mathbf{q} = q_1, q_2, \ldots, q_N$ by means of a Taylor expansion about the origin $\mathbf{q} = \mathbf{0}$,

$$V(\mathbf{q}) = V(\mathbf{0}) + \frac{\partial V(\mathbf{0})}{\partial q_i}q_i + \frac{1}{2}\frac{\partial^2 V(\mathbf{0})}{\partial q_i\partial q_j}q_iq_j + O(q_i^3). \tag{14.82}$$

However we have chosen $V(0) = 0$ and, since the origin is an equilibrium point, there is no force there and so $\partial V(0)/\partial q_i = 0$. Consequently, to second order in the q_i we also have a quadratic form for the potential energy – although this time a quadratic in the coordinates rather than in their time derivatives,

$$V = b_{ij}q_iq_j. \tag{14.83}$$

In the example already mentioned, V was shown to be the quadratic form

$$V = \tfrac{1}{4}Mlg(2q_1^2 + q_2^2),$$

to second order in the q_i, with corresponding real symmetric matrix

$$B = \frac{Mlg}{12} \begin{bmatrix} 6 & 0 \\ 0 & 3 \end{bmatrix}. \tag{14.84}$$

In this case and in general, the requirement that the potential is a strict minimum means that the potential matrix, like the kinetic energy one, is real and positive definite. It too may be made symmetric if not initially so.

The development via (14.82) is not intended to imply that the matrix B is in general to be calculated as $\tfrac{1}{2}\,\partial^2 V(0)/\partial q_i\,\partial q_j$ [which would automatically make it symmetric]; in practice it is usually found by direct calculation of V to second order in the small coordinates.

With these expressions for T and V we now proceed with our simplified approach to obtaining the normal mode equations, based upon the fact that

$$\frac{d}{dt}(T + V) = 0, \tag{14.85}$$

i.e. there are no external forces.

Substituting from (14.80) and (14.83) into (14.85) and recalling that matrices A and B are symmetric we obtain

$$a_{ij}\dot{q}_i\ddot{q}_j + a_{ij}\ddot{q}_i\dot{q}_j + b_{ij}q_i\dot{q}_j + b_{ij}\dot{q}_iq_j = 0,$$
$$2(a_{ij}\ddot{q}_j + b_{ij}q_j)\dot{q}_i = 0. \tag{14.86}$$

Now it is not clear from this that it is the only solution possible, but we will assume that this implies that the coefficient of each $\dot{q}_i$ in this summation is separately zero, i.e.

$$a_{ij}\ddot{q}_j + b_{ij}q_j = 0, \quad i = 1, 2, \ldots, N. \tag{14.87}$$

For a rigorous derivation Lagrange's equations should be used as in chapter 13. The present treatment cannot hope to produce this result of N

separate equations in a rigorous way, since the starting equation (14.85) expresses only a single overall conservation condition.

Now we search for sets of coordinates q_i which *all* go through their oscillations with the same period, i.e. the total motion repeats itself *exactly* after a definite *finite* interval.† We do this by seeking solutions of the form

$$q_i = x_i \cos \omega t, \tag{14.88}$$

where ω does not depend on i, but of course the x_i do, and the relative values of the x_i [assuming we can find such a solution] will indicate how each coordinate is involved in this special motion.

Putting (14.88) into (14.87) yields directly

$$-\omega^2 a_{ij}x_j + b_{ij}x_j = 0, \quad i = 1, 2, \ldots, N,$$

which may be written as the simultaneous homogeneous linear equations

$$(b_{ij} - \omega^2 a_{ij})x_j = 0. \tag{14.89}$$

Our previous work of section 14.4 shows that this can only have non-trivial solutions if

$$|B - \omega^2 A| = 0. \tag{14.90}$$

This is a form of characteristic equation for B except that A has replaced the unit matrix I. It has the more familiar form if a choice of coordinates is made in which the kinetic energy T is a simple sum of squared terms in the $\dot{q}_i$, i.e. $T = \sum_i \dot{q}_i^2$.

However even in the present case, (14.90) is an Nth degree polynomial equation in ω^2 whose roots (which can all be shown to be positive since A and B are positive definite) may be labelled as ω_k^2 ($k = 1, 2, \ldots, N$). Substituting each in turn into (14.89) will enable the corresponding set of values x_j^k to be established and the initial (stationary) physical configuration, which on release will execute motion with period $2\pi/\omega_k$, to be found.

To illustrate this we complete the solution of the problem of the suspended rod of length l swinging in one plane at the end of a light string also of length l.

Equations (14.81), (14.84) and (14.90) show that we have to solve

$$\left| \frac{Mlg}{12} \begin{bmatrix} 6 & 0 \\ 0 & 3 \end{bmatrix} - \frac{\omega^2 Ml^2}{12} \begin{bmatrix} 6 & 3 \\ 3 & 2 \end{bmatrix} \right| = 0.$$

† From this point on, the analysis applies equally well to many coupled electrical circuit problems with no resistive damping, since they may often be stated in the form of coupled differential equations like those in (14.87).

Writing $\omega^2 l/g = \lambda$, this becomes

$$\begin{vmatrix} 6 - 6\lambda & -3\lambda \\ -3\lambda & 3 - 2\lambda \end{vmatrix} = 0,$$

$$\lambda^2 - 10\lambda + 6 = 0,$$

$$\lambda = 5 \pm \sqrt{19}.$$

Thus the two frequencies, known as **normal frequencies** or **eigenfrequencies**, of the system are $\omega_1 = (0.641g/l)^{1/2}$ and $\omega_2 = (9.359g/l)^{1/2}$ [the two values quoted in section 13.5].

Putting the lower of the two values for ω^2, namely $(5 - \sqrt{19})g/l$ into (14.89) shows that for this mode

▶30.　　$x_1:x_2 = 3(5 - \sqrt{19}):6(\sqrt{19} - 4) = 1.923:2.154.$

This is close to, but not exactly the same as, the 'straight-out' position used in section 13.5 to estimate the lowest frequency. [Hence the good estimate obtained there.]

▶31. Find the ratio of the initial inclinations of the string and rod if, when released from rest, they are to execute simple harmonic motion at the higher of the two normal frequencies.

In connection with quadratic forms it has been shown how to make a change of coordinates so that the matrix for a particular form becomes diagonal. In example 17 at the end of this chapter a method is developed for simultaneously diagonalizing two quadratic forms (but not with an orthogonal transformation in general). If this process is carried out for A and B of a general system undergoing stable oscillations, the kinetic and potential energies in the new variables η_i take the forms

$$T = \sum_i \mu_i \dot{\eta}_i^2, \tag{14.91 a}$$

$$V = \sum_i \nu_i \eta_i^2, \tag{14.91 b}$$

and the equations of motion are *uncoupled* equations

$$\mu_i \ddot{\eta}_i + \nu_i \eta_i = 0 \text{ (no summation)}, \quad i = 1, 2, \ldots, N. \tag{14.92}$$

Clearly a simple renormalization of the η_i can be made to reduce all the μ_i in (14.91 a) to unity. When this is done the variables so formed are called **normal coordinates**, and equations (14.92) the **normal equations**.

When a system is executing one of these truly periodic motions it is said to be in a **normal mode**, and once started in such a mode it will repeat its motion exactly after each interval of $2\pi/\omega_i$. Any arbitrary motion may be written as a superposition of the normal modes and each component will execute harmonic motion with the corresponding eigenfrequency; however, unless by chance the eigenfrequencies are integrally related, the system will never return to its initial configuration after any finite time interval.

As a second example we will consider a number of masses coupled together by springs. For this type of situation the potential and kinetic energies are automatically quadratic functions of the coordinates and their derivatives (provided the elastic limits of the springs are not exceeded), and the oscillations do not have to be vanishingly small for the analysis to be valid.

Example 14.6. Find the normal frequencies and modes of oscillation of three particles of masses m, μm, m connected in that order in a straight line to two equal light springs of force constant k. (This arrangement could serve as a model for some linear molecules.)

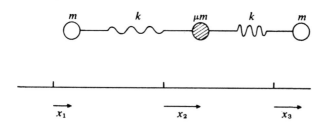

Fig. 14.3 The coordinate system of example 14.6. The three masses m, μm and m are connected by two equal light springs of force constant k.

The situation is shown in fig. 14.3 in which the coordinates of the particles x_1, x_2, x_3, are measured from their equilibrium positions at which the springs are neither extended nor compressed.

The kinetic energy of the system is simply

$$T = \tfrac{1}{2}m(\dot{x}_1^2 + \mu\dot{x}_2^2 + \dot{x}_3^2),$$

whilst the potential energy stored in the springs is

$$V = \tfrac{1}{2}k[(x_2 - x_1)^2 + (x_3 - x_2)^2].$$

The kinetic and potential energy symmetric matrices are thus

$$A = \frac{m}{2}\begin{bmatrix} 1 & 0 & 0 \\ 0 & \mu & 0 \\ 0 & 0 & 1 \end{bmatrix}, \quad B = \frac{k}{2}\begin{bmatrix} 1 & -1 & 0 \\ -1 & 2 & -1 \\ 0 & -1 & 1 \end{bmatrix}.$$

To find the normal frequencies we have, following (14.90), to solve $|B - \omega^2 A| = 0$. Thus writing $m\omega^2/k = \lambda$ we have

$$\begin{vmatrix} 1 - \lambda & -1 & 0 \\ -1 & 2 - \mu\lambda & -1 \\ 0 & -1 & 1 - \lambda \end{vmatrix} = 0,$$

which leads to

▶32. $\lambda = 0, 1,$ or $1 + (2/\mu)$.

The corresponding eigenvectors are (respectively)

▶ 33. $\mathbf{x}^1 = 3^{-1/2} \begin{bmatrix} 1 \\ 1 \\ 1 \end{bmatrix},$ $\mathbf{x}^2 = 2^{-1/2} \begin{bmatrix} 1 \\ 0 \\ -1 \end{bmatrix},$

$$\mathbf{x}^3 = [2 + (4/\mu^2)]^{-1/2} \begin{bmatrix} 1 \\ -2/\mu \\ 1 \end{bmatrix}. \quad (14.93)$$

The physical motions associated with these solutions are illustrated in fig. 14.4. The first, $\lambda = \omega = 0$ and all the x_i equal, merely describes the bodily translation of the whole system, with no (i.e. zero frequency) internal oscillations.

In the second solution the central particle remains stationary, $x_2^2 = 0$, whilst the other two oscillate with equal amplitudes in antiphase with each other. This motion of frequency $\omega = (k/m)^{1/2}$ is illustrated in the middle of fig. 14.4.

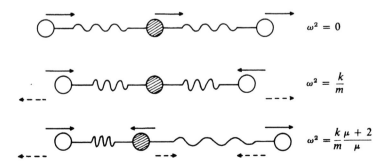

$$\omega^2 = 0$$

$$\omega^2 = \frac{k}{m}$$

$$\omega^2 = \frac{k}{m} \frac{\mu + 2}{\mu}$$

Fig. 14.4 The normal modes of the masses and springs of example 14.6.

The final and most complicated of the three normal modes has a frequency $\omega = (k(\mu + 2)/m\mu)^{1/2}$, and involves a motion of the central particles which is in antiphase with that of the two outer ones and has an amplitude which is $2/\mu$ times as great. In this motion the two springs are compressed and extended in turn.

The eigenvectors $\mathbf{x}^k$ obtained by solving $(B - \omega^2 A)\mathbf{x} = 0$ are not mutually orthogonal unless A is a multiple of the unit matrix $[\mu = 1]$, but it is shown in example 17 of section 14.13 that they do satisfy

$$\tilde{\mathbf{x}}^i A \mathbf{x}^j = 0 \quad \text{and} \quad \tilde{\mathbf{x}}^i B \mathbf{x}^j = 0 \quad \text{for } i \neq j. \tag{14.94}$$

▶34. Verify that this is so for the particular case of solutions (14.93).

The general property (14.94) can be used [as in the development of (13.17) and (13.18)] to justify the general result that no matter what trial vector $\mathbf{x}$ may be used, the quantity $Q = [(\tilde{\mathbf{x}}B\mathbf{x})/(\tilde{\mathbf{x}}A\mathbf{x})]^{1/2}$ always lies between the lowest and the highest eigenfrequencies of the system, for which upper and lower bounds respectively may thus be found. Furthermore, as with the quadratic forms considered earlier, Q has a stationary value of ω_k^2 when $\mathbf{x}$ is the kth eigenvector.

14.13   Examples for solution

1. Evaluate the following: (a) ϵ_{53214}, (b) $\epsilon_{ijk}\epsilon_{ijk}$, (c) $\epsilon_{ijk}\epsilon_{jik}$, (d) $\epsilon_{ijk}a_{ij}$, where a_{ij} are the elements of a symmetric matrix.

2. In the following suppose all subscripts run from 1 to 3 and that the a_i are the components of a vector $\mathbf{a}$ etc.

(a) Verify that $\delta_{ij}a_i b_j = \mathbf{a} \cdot \mathbf{b}$ [$\delta_{ij} = 1$ if $i = j$ and equals zero otherwise].
(b) Verify that $\epsilon_{ijk}a_i b_j = (\mathbf{a} \wedge \mathbf{b})_k$.
(c) Express $\epsilon_{ijk}\epsilon_{klm}a_i b_j c_l d_m$ in vector form, then use result (a) to show that, since $\mathbf{a}$, $\mathbf{b}$, $\mathbf{c}$ and $\mathbf{d}$ are arbitrary vectors, $\epsilon_{ijk}\epsilon_{klm} = \delta_{il}\delta_{jm} - \delta_{im}\delta_{jl}$.

3. Find the characteristic equation [det $(A - \lambda I) = 0$] of the matrix A and prove that the matrices B and C have the same characteristic equation.

$$A = \begin{bmatrix} b & c & a \\ c & a & b \\ a & b & c \end{bmatrix}, \quad B = \begin{bmatrix} c & a & b \\ a & b & c \\ b & c & a \end{bmatrix}, \quad C = \begin{bmatrix} a & b & c \\ b & c & a \\ c & a & b \end{bmatrix}.$$

Show that, if $BC = CB$, then two of the roots of the characteristic equation are zero.

4. Use the properties of determinants to solve the following for x with a minimum of actual calculation.

(i) $\begin{vmatrix} x & a & a & 1 \\ a & x & b & 1 \\ a & b & x & 1 \\ a & b & c & 1 \end{vmatrix} = 0;$ (ii) $\begin{vmatrix} x+2 & x+4 & x-3 \\ x+3 & x & x+5 \\ x-2 & x-1 & x+1 \end{vmatrix} = 0.$

5. Show that the following equations have solutions only if $\eta = 1$ or 2,

$$x + y + z = 1, \quad x + 2y + 4z = \eta, \quad x + 4y + 10z = \eta^2,$$

and find them in these cases.

6. Solve the following equations for x_1, x_2, x_3 using matrix methods:

$x_1 + 2x_2 + 3x_3 = 1,$
$3x_1 + 4x_2 + 5x_3 = 2,$
$x_1 + 3x_2 + 4x_3 = 3.$

7. Solve the simultaneous equations:

$2x + 3y + z = 11,$
$x + y + z = 6,$
$5x - y + 10z = 34.$

8. (i) Show that if A is Hermitian and U is unitary, then $U^{-1}AU$ is Hermitian.
(ii) Show that if A is anti-Hermitian, then iA is Hermitian.
(iii) Prove that the product of two Hermitian matrices A and B is Hermitian if and only if A and B commute.
(iv) Prove that if S is a real skew matrix, then the matrix $A = (I - S)(I + S)^{-1}$ is orthogonal and express the matrix

$$A = \begin{bmatrix} \cos\theta & \sin\theta \\ -\sin\theta & \cos\theta \end{bmatrix}$$

in this form.

9. For the matrix

$$A = \begin{bmatrix} 1 & \alpha & 0 \\ \beta & 1 & 0 \\ 0 & 0 & 1 \end{bmatrix},$$

where α and β are non-zero complex numbers, find the eigenvalues and eigenvectors. Find the respective conditions (a) for the eigenvalues to be real, (b) for the eigenvectors to be orthogonal. Show that the conditions are jointly satisfied only if A is Hermitian.

10. Following the procedure of (14.58) construct an orthonormal set of vectors from the following $\bar{x}^1 = (0, 0, 1, 1)$, $\bar{x}^2 = (1, 0, -1, 0)$, $\bar{x}^3 = (1, 2, 0, 2)$, $\bar{x}^4 = (2, 1, 1, 1)$.

11. Find three real orthogonal vectors each of which is a simultaneous eigenvector of

$$A = \begin{bmatrix} 0 & 0 & 1 \\ 0 & 1 & 0 \\ 1 & 0 & 0 \end{bmatrix} \quad \text{and} \quad B = \begin{bmatrix} 0 & 1 & 1 \\ 1 & 0 & 1 \\ 1 & 1 & 0 \end{bmatrix}.$$

12. What are the maximum and minimum values taken by the expression

$$Q = 5x^2 + 4y^2 + 4z^2 + 2xz + 2xy$$

on the unit sphere $x^2 + y^2 + z^2 = 1$? For what values of (x, y, z) do they occur?

13. [The following problem is effectively the inverse of the preceding question. It uses the properties of eigenvectors and eigenvalues to construct a matrix with given eigenvectors and values. The required answer is the given data of the previous question, but an understanding of the method of procedure is the purpose of the problem.]

Find a quadratic form in $x = (x, y, z)$ on the unit sphere $x^2 + y^2 + z^2 = 1$, which takes a maximum value of 6 at the points $\pm 6^{-1/2} (2, 1, 1)$, a minimum value of 3 at $\pm 3^{-1/2} (1, -1, -1)$ and the value 4 at its other pair of stationary points. Proceed as follows:

(i) Find the other pair of stationary points.
(ii) Consider the expression $Q = \sum_{i=1}^3 \lambda_i (\bar{x}^i x)^2$ relating the terms in it to the components of $U^{-1}x$, where U is the orthogonal matrix whose columns are the given eigenvectors x^i.

14. Find the lengths of the semi-axes of the ellipse

$$73x^2 + 72xy + 52y^2 = 100,$$

and determine its orientation.

15. (i) Show that the quadric

$$5x^2 + 11y^2 + 5z^2 - 10yz + 2xz - 10xy = 4$$

is an ellipsoid and has semi-axes of 2, 1, and $\frac{1}{2}$. Find the direction of the longest axis.
(ii) Find the direction of the axis of symmetry of the quadric

$$7x^2 + 7y^2 + 7z^2 - 20yz - 20xz + 20xy = 3.$$

16. [The following problem is constructed directly through Newton's laws and not by the methods of section 14.12.]

A system of three coupled pendulums satisfies the following equations of motion:

$$-m\ddot{x}_1 = cmx_1 + d(x_1 - x_2),$$
$$-M\ddot{x}_2 = cMx_2 + d(x_2 - x_1) + d(x_2 - x_3),$$
$$-m\ddot{x}_3 = cmx_3 + d(x_3 - x_2),$$

where x_1, x_2 and x_3 are measured from the equilibrium point, m and M are masses, and c and d are positive constants. Find the normal frequencies of the system and sketch the patterns of oscillation. What happens as $d \to 0$, or $d \to \infty$?

17. *Simultaneous reduction to diagonal form of two real, symmetric quadratic forms.* [This is a more advanced piece of development, based upon the methods used in the body of the chapter. Its results have previously been referred to in section 14.12.]

Consider two real symmetric forms $\tilde{u}Au$ and $\tilde{u}Bu$, where $\tilde{u}$ stands for (x, y, z), and denote by u^n those vectors which satisfy

$$Bu^n = \lambda^n Au^n \text{ (no summation)}, \tag{a}$$

in which n is a label and the λ^n are real, non-zero and all different.

(i) By multiplying equation (a) on the left by $\tilde{u}^m$ and the transpose of the corresponding equation for u^m on the right by u^n, show that $\tilde{u}^m Au^n = 0$ for $n \neq m$.

(ii) By noting $Au^n = (\lambda^n)^{-1}Bu^n$, deduce $\tilde{u}^m Bu^n = 0$ for $m \neq n$.

It can be shown that the u^n are linearly independent and the next step is to construct a matrix P whose columns are the vectors u^n, i.e. $P_{jn} = u_j^n$, e.g. P_{23} is the y element of u^3.

(iii) Make a change of variable $u = Pv$ so that $\tilde{u}Au$ becomes $\tilde{v}Cv$, and $\tilde{u}Bu$ becomes $\tilde{v}Dv$. Show that C and D are diagonal by showing $c_{ij} = 0$ if $i \neq j$ and similarly for d_{ij}.

Thus $u = Pv$ or $v = P^{-1}u$ reduces both quadratics to diagonal form.
To summarize, the method is:

(*A*) find the λ^n which allow $Bu = \lambda Au$ a non-zero solution, by solving $|B - \lambda A| = 0$,
(*B*) for each λ^n construct u^n,
(*C*) construct the non-singular matrix P whose columns are the vectors u^n,
(*D*) make the change of variable $u = Pv$.

Notes: (*a*) the method can be extended to the case of equal roots of $|B - \lambda A| = 0$,
(*b*) if A is taken as I, the analysis is that of the main text and P is orthogonal.

18. [Do not attempt this question until question 17 has been studied.]
Find a real linear transformation which simultaneously reduces the quadratic forms

$$x^2 + 4y^2 + 2z^2 - 4yz - 2zx + 6xy$$

and

$$2x^2 + y^2 + 2zx,$$

to the forms $a_1\xi^2 + a_2\eta^2 + a_3\chi^2$ and $b_1\xi^2 + b_2\eta^2 + b_3\chi^2$.

19. Find a real linear transformation which simultaneously reduces the quadratic forms

$$3x^2 + 5y^2 + 5z^2 + 2yz + 6zx - 2xy$$

and

$$5x^2 + 12y^2 + 8yz + 4zx$$

to diagonal form. [Take the first quadratic as $\tilde{u}Au$ in example 17.]

20. This problem is an alternative [and, as it happens, more complicated] solution to the situation analysed in example 7 of section 13.6.

Three particles of mass m are attached to a light string with fixed ends so that they divide it into four equal portions of length a. Find the normal frequencies and modes of the system by calculating the matrices A and B in $T = \dot{x}_i a_{ij} \dot{x}_j$ and $V = x_i b_{ij} x_j$. Calculate the coefficients b_{ij} as follows, starting from the equilibrium configuration:

(i) Calculate the work done in moving particle 1 to position $x_1 (\ll a)$ showing it is equal to $2Sx_1^2/3a$, where S is the tension in the string.
(ii) Keeping x_1 fixed, calculate the work done in moving particle 2 from its present position $[2x_1/3]$ to its final position x_2. $[(\frac{3}{4}x_2^2 - x_1 x_2 + \frac{1}{4}x_1^2)S/a]$
(iii) Do the same for particle 3. $[(x_3^2 - x_2 x_3 + \frac{1}{4}x_2^2)S/a]$

21. (i) The base vectors of the unit cell of a crystal, with O at one corner, are denoted by e_1, e_2, e_3. G is the matrix with elements g_{ij} where

$$g_{ij} = e_i \cdot e_j,$$

and h_{ij} are the elements of a matrix $H \equiv G^{-1}$. Show that the vectors $f_i = h_{ij} e_j$ are the reciprocal vectors and that $h_{ij} = f_i \cdot f_j$.

(ii) If the vectors $\mathbf{u}$ and $\mathbf{v}$ are given by

$$\mathbf{u} = u_i\mathbf{e}_i, \qquad \mathbf{v} = v_i\mathbf{f}_i,$$

obtain expressions for $|\mathbf{u}|$, $|\mathbf{v}|$, and $\mathbf{u}\cdot\mathbf{v}$.

(iii) If the base vectors are each of length a and the angle between each pair is $\pi/3$, write down G and hence obtain H.

(iv) Calculate (a) the length of the normal from O onto the plane containing the points $p^{-1}\mathbf{e}_1, q^{-1}\mathbf{e}_2, r^{-1}\mathbf{e}_3$, and ($b$) the angle between this normal and $\mathbf{e}_1$.

15
Cartesian tensors

It is often said by physicists and engineers when approaching the quantitative description of a physical process that 'of course it doesn't matter how I choose my axes, the physical result must always be the same'. Having thus justified doing so they proceed quite naturally to choose a coordinate system in the way that seems the most convenient for the particular investigation. However, we may turn the argument around and ask, 'As the physical results are independent of the choice of coordinate axes, what must this imply about the nature of the quantities involved in the description of the physical processes?' The study of these implications and classification of physical quantities by means of them, for a particular type of coordinate change – namely rotations – form the content of the present chapter.

Some attention was given to rotations of coordinate axes in chapter 14, where it was shown that under such changes different types of quantities behaved in different ways. For example, a real vector $\mathbf{x}$ was transformed into the vector $\mathbf{y} = A^{-1}\mathbf{x}$, or more specifically for rotations, into $\mathbf{y} = \tilde{A}\mathbf{x}$ since A was then an orthogonal matrix. On the other hand a scalar product $\tilde{\mathbf{x}}^{(1)}\mathbf{x}^{(2)}$ was transformed into a scalar product $\tilde{\mathbf{y}}^{(1)}\mathbf{y}^{(2)}$ which had the same numerical value as $\tilde{\mathbf{x}}^{(1)}\mathbf{x}^{(2)}$ since

$$\tilde{\mathbf{y}}^{(1)}\mathbf{y}^{(2)} = \tilde{\mathbf{x}}^{(1)}\widetilde{(A^{-1})}A^{-1}\mathbf{x}^{(2)} = \tilde{\mathbf{x}}^{(1)}AA^{-1}\mathbf{x}^{(2)} = \tilde{\mathbf{x}}^{(1)}\mathbf{x}^{(2)}. \qquad (15.1)$$

Different again was the behaviour of a matrix B, since it transformed into a new matrix given by $\tilde{A}BA$.

In this chapter we develop a general formulation to describe and classify these transformations. In the development, the generic name **tensor** is introduced, and certain scalars, vectors and matrices are described respectively as tensors of zeroth, first and second order. [The *order* – or *rank*† – corresponding to the number of subscripts needed to specify a particular element of the tensor.] Tensors of third and fourth order will also occupy some of our attention.

---

† 'Order' and 'rank' are used equivalently in respect of general tensors. For matrices 'rank' has a particular meaning, and may have values other than 2.

The general study of tensors of arbitrary order, in many dimensions, and allowing oblique axes, is a difficult one and is far beyond the scope of this book. We will confine our attention strictly to orthogonal coordinate systems in two or three dimensions, but allow tensors of any finite order (in principle). But even this restricted area is too wide for a reasonably concise treatment and a further restriction to *Cartesian coordinates* will be assumed.

Our object then is to study the properties of various types of mathematical quantities and their associated physical interpretations, when they are described in terms of Cartesian coordinates and the axes of coordinates are rigidly rotated whilst keeping the origin fixed. Naturally our ultimate interest is in physical quantities and their interrelationships, but as an introduction we will develop mathematical properties and illustrate them with physical examples, rather than the other way round. The mathematical quantities are called **Cartesian tensors**.

Before the presentation of the main development it should be pointed out that, as in the previous chapter, the summation convention is assumed except where the contrary is specifically stated. On the other hand, it is perhaps also worth remarking again that, in the belief that excessively condensed notations hinder rather than help many students, many equations which could be written more compactly (e.g. by suppressing subscripts or using a suffix notation for partial differentiation) have been left in their lengthier forms so as to reduce the 'mental unpacking' required.

15.1 First- and zeroth-order Cartesian tensors

The whole concept of a tensor is dependent upon the behaviour of Cartesian coordinates x_i ($i = 1, 2, 3$) when the axes are rotated about the origin, and so we examine this first. If we denote coordinates with respect to the new rotated axes by x'_i ($i = 1, 2, 3$), then the connecting relationships are

$$x_i = a_{ij}x'_j, \qquad (15.2\,a)$$

$$x'_i = a_{ji}x_j, \qquad (15.2\,b)$$

the second equation following because the matrix A, of which a_{ij} is an element, is orthogonal ($A^{-1} = \tilde{A}$). The orthogonality also implies relationships among the elements of A, expressing the fact that $A\tilde{A} = \tilde{A}A = I$, and given in subscript notation by

$$a_{ik}a_{jk} = \delta_{ij}, \qquad (15.3\,a)$$

and $\qquad a_{ki}a_{kj} = \delta_{ij}. \qquad (15.3\,b)$

These results have been obtained previously in section 14.7.

It is natural to think of the coordinates x_i as the components of a (position) vector [the most obvious vector], and taking this as a guide, to consider any set of (three) quantities u_i which are directly or indirectly functions of the x_i together with possible constants, and ask how their values are changed by any rotation of the Cartesian axes. The specific question to be answered is whether or not the values of the specific forms u_i', in terms of the new variables, can or cannot be obtained from the old ones u_i, in terms of the old variables, simply by replacing x by u in equations (15.2). If they can, the u_i are said to form the components of a **vector** or **first-order Cartesian tensor**. By definition the position co-ordinates themselves are the components of such a tensor.

Thus our *definition* is that the expressions u_i form the components of a first-order Cartesian tensor if, for all rotations of the axes of coordinates, given by (15.2 a, b), subject to (15.3 a, b), the same expressions using the new coordinate variables are u_i', and these are the same as the u_i' given by

$$u_i = a_{ij}u_j',$$ (15.4 a)

and $u_i' = a_{ji}u_j.$ (15.4 b)

The two conditions (15.4 a) and (15.4 b) are not independent, the second being obtainable from the first by multiplying through by a_{ik} and using (15.3). Equally the second implies the first.

That the question is not a redundant one and really does pick out sets of two or three quantities which have particular properties under rotations will now be shown by considering explicit examples. In order to keep the equations to reasonable proportions the examples will be restricted to two-dimensional ones. Three-dimensional cases are no different in principle – only much longer to write out.

In two dimensions the most general rotation of coordinates from the original set Ox_1x_2 to the new set $Ox_1'x_2'$ through an angle θ (fig. 15.1) is given by

$$x_1 = x_1' \cos \theta - x_2' \sin \theta,$$
$$x_2 = x_1' \sin \theta + x_2' \cos \theta.$$ (15.5 a)

Comparing this with (15.2 a) gives for the orthogonal matrix

$$A(\theta) = \begin{bmatrix} \cos \theta & -\sin \theta \\ \sin \theta & \cos \theta \end{bmatrix}.$$ (15.5 b)

The inverse equations are

$$x_1' = x_1 \cos \theta + x_2 \sin \theta,$$
$$x_2' = -x_1 \sin \theta + x_2 \cos \theta,$$ (15.6)

in line with (15.2 b).

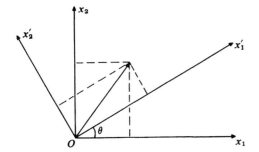

Fig. 15.1 Rotation of Cartesian axes by an angle θ about the x_3-direction.

Example 15.1. Which of the following pairs of quantities (u_1, u_2) are the components of a first-order Cartesian tensor in two dimensions;

$$\text{(i) } (x_2, -x_1), \qquad \text{(ii) } (x_2, x_1), \qquad \text{(iii) } (x_1^2, x_2^2)?$$

To save space we denote $\cos\theta$ by c and $\sin\theta$ by s in the working.

(i) Here $u_1 = x_2$ and $u_2 = -x_1$ referred to the old axes. In terms of the new coordinates they will be $u_1' = x_2'$ and $u_2' = -x_1'$, i.e.

$$\begin{aligned} u_1' &= x_2' = -sx_1 + cx_2, \\ u_2' &= -x_1' = -cx_1 - sx_2. \end{aligned} \tag{15.7}$$

Now if we start again and evaluate u_1' and u_2' as given by (15.4 b) we find that

$$\begin{aligned} u_1' &= a_{11}u_1 + a_{21}u_2 = cx_2 + s(-x_1), \\ u_2' &= a_{12}u_1 + a_{22}u_2 = (-s)x_2 + c(-x_1). \end{aligned} \tag{15.8}$$

Expressions (15.7) and (15.8) for u_1' and u_2' are the same whatever the values of θ [i.e. for *all* rotations] and thus by definition (15.4) the pair $(x_2, -x_1)$ *are* components of a first-order tensor.

(ii) Here $u_1 = x_2$ and $u_2 = x_1$. Following the same procedure,

$$\begin{aligned} u_1' &= x_2' = -sx_1 + cx_2, \\ u_2' &= x_1' = cx_1 + sx_2. \end{aligned}$$

But by (15.4 b), for a Cartesian tensor we must have,

$$\begin{aligned} u_1' &= cu_1 + su_2 = cx_2 + sx_1, \\ u_2' &= (-s)u_1 + cu_2 = -sx_2 + cx_1. \end{aligned}$$

These two sets of expressions do not agree and thus the pair (x_2, x_1) do *not* form the components of a first-order tensor.

(iii) $u_1 = x_1^2$ and $u_2 = x_2^2$. The first component alone [as in (ii)] is sufficient to show that these do *not* form a first-order tensor, since directly

$$u_1' = x_1'^2 = c^2 x_1^2 + 2cs x_1 x_2 + s^2 x_2^2,$$

whilst (15.4 b) requires that

$$u_1' = cu_1 + su_2 = cx_1^2 + sx_2^2,$$

which is quite different.

▶1. (i) Show for any general (but fixed) ϕ that $(u_1, u_2) = (x_1 \cos \phi - x_2 \sin \phi, x_1 \sin \phi + x_2 \cos \phi)$ are the components of a first-order Cartesian tensor in two dimensions.
(ii) Identify example 15.1 (i) as a particular case of this and show that (u_1, u_2) of example 15.1 (ii) cannot be so represented.

There are many physical examples of first-order tensors which will be familiar to the reader. As a straightforward one, we may take the Cartesian components of momentum for a particle, $(m\dot{x}_1, m\dot{x}_2, m\dot{x}_3)$. These transform in all essentials as do (x_1, x_2, x_3) themselves since the other operations involved, multiplication by a number and differentiation with respect to time, are quite unaffected by any orthogonal transformation of axes. Similarly acceleration and force are represented by the components of first-order tensors.

Other more complicated vectors involving the position coordinates 'more than once', such as angular momentum $\mathbf{J} = \mathbf{r} \wedge \mathbf{p} = m(\mathbf{r} \wedge \dot{\mathbf{r}})$, are also first-order tensors. That this is so is less obvious than for the above examples, but it may be verified by writing out the components of $\mathbf{J}$ explicitly or by appealing to the quotient law of section 15.4 and the use of ϵ_{ijk} from section 15.5.

Having considered the effects of rotations on 'vector-like' sets of quantities we may consider quantities which are unchanged by a rotation of axes. In our previous nomenclature these have been called **scalars** but we may also describe them as **tensors of zero order**. They contain only one element [formally they need zero subscripts to identify a particular element] and the most obvious non-trivial example associated with a rotation of the axes is the square of the distance of a point from the origin, $u = x_1^2 + x_2^2 + x_3^2$. In the new coordinate system it will have the form $u' = x_1'^2 + x_2'^2 + x_3'^2$, which for any rotation has the same value as $x_1^2 + x_2^2 + x_3^2$, as is shown by (15.1) in the particular case when $\mathbf{x}^{(1)} = \mathbf{x}^{(2)}$.

Any 'scalar-product' of two first-order tensors (vectors) is in fact a zero-order tensor (scalar), since in the original system it is $u_i v_i$ (summed over i) and in the rotated system is given by

$$u_i' v_i' = a_{ji} u_j a_{ki} v_k = a_{ji} a_{ki} u_j v_k = \delta_{jk} u_j v_k = u_j v_j, \tag{15.9}$$

which is exactly the same.

This result leads directly to the identification of many physically important quantities as zero-order tensors. Perhaps the most immediate of these is energy, either as a potential, or as an energy density (e.g. $\mathbf{F} \cdot d\mathbf{r}$, $e\mathbf{E} \cdot d\mathbf{r}$, $\mathbf{D} \cdot \mathbf{E}$, $\mathbf{B} \cdot \mathbf{H}$, $\boldsymbol{\mu} \cdot \mathbf{B}$), but others, such as the angle between two directed quantities, are important.

It is the fact that in most analyses of physical situations it is a scalar quantity (such as energy) that is required to be found, that leads to the situation described in the first paragraph of this chapter. Such quantities are *invariant* under a rotation of the axes and so it is possible to work with the most convenient set of axes and still have confidence in the result.

Complementing the way in which a zero-order tensor was obtained from two first-order tensors, so a first-order tensor can be obtained from a scalar. We show this by taking a specific example, that of the electric field $\mathbf{E}$ derived from an electrostatic scalar potential ϕ by

$$E_i = -\frac{\partial \phi}{\partial x_i}. \tag{15.10}$$

Under a rotation of the axes (equation (15.2)), ϕ is invariant ($\phi' = \phi$) but the components of the electric field E_i' are given by

$$E_i' = \left(-\frac{\partial \phi}{\partial x_i}\right)' = -\frac{\partial \phi'}{\partial x_i'} = -\frac{\partial x_j}{\partial x_i'}\frac{\partial \phi}{\partial x_j} = a_{ji} E_j, \tag{15.11}$$

where (15.2 a) has been used to evaluate $\partial x_j/\partial x_i'$. Now (15.11) is in the form (15.4 b) thus showing that the components of the electric field do behave as the components of a first-order tensor.

▶2. If u ($\equiv u_i$) is a first-order Cartesian tensor, show that $\nabla \cdot \mathbf{u} = \partial u_i/\partial x_i$ is a zero-order tensor. [Model the argument on (15.11) and use (15.3).]

15.2   Second- and higher-order Cartesian tensors

Following on from scalars with no subscripts and vectors with one subscript, we turn next to sets of quantities which require two subscripts to identify a particular element of the set. Let these quantities be denoted by u_{ij}.

Taking (15.4) as a guide we define a **second-order Cartesian tensor** by saying that the u_{ij} form the components of such a tensor if, under the same conditions as for (15.4),

$$u_{ij} = a_{ik}a_{jl}u'_{kl},$$ (15.12 a)

$$u'_{ij} = a_{ki}a_{lj}u_{kl}.$$ (15.12 b)

We may at the same time define a Cartesian tensor of general order as follows. The set of expressions $u_{ij...k}$ form the components of a Cartesian tensor if, for all rotations of the axes of coordinates given by (15.2 a, b) subject to (15.3 a, b), the expressions using the new coordinates are $u'_{ij...k}$, and these are the same as the $u'_{ij...k}$ given by

$$u_{ij...k} = a_{ip}a_{jq}...a_{kr}u'_{pq...r},$$ (15.13 a)

$$u'_{ij...k} = a_{pi}a_{qj}...a_{rk}u_{pq...r}.$$ (15.13 b)

It is apparent that in three dimensions, an Nth-order Cartesian tensor has 3^N components.

Although a second-order tensor has two subscripts and it is natural to display its components in matrix form, a tensor and a matrix are not identical. A matrix may in a particular case be a representation of a tensor with respect to a particular coordinate system, but equally the elements of a matrix may have no connection at all with any set of axes. Indeed the algebra of matrices can be developed purely by defining the basic operations (addition, multiplication, etc.) and proceeding from there. Under orthogonal transformations, $A \rightarrow U^{-1}AU$, the elements of matrices do behave like the components of a second-order tensor, but their behaviour under these particular transformations is only one of many properties, whilst for tensors it is almost the only one that matters. [We may consider, for example, the behaviour of matrices under general similarity transformations as mentioned at the end of section 14.10. It may also be noted that although the elements a_{ij} of the transformation are written with two subscripts, they cannot be the components of a tensor since the two subscripts refer one each to two different coordinate systems.]

As examples of sets of quantities which are readily shown to be second-order tensors we consider the following.
1. *The outer-product of two vectors.* Let u_i and v_i be two vectors ($i = 1, 2, 3$) and consider the set of quantities w_{ij} defined by

$$w_{ij} = u_i v_j.$$ (15.14)

The set w_{ij} is called the *outer product* of u_i and v_i. Under rotation it becomes

$$w'_{ij} = u'_i v'_j = a_{ki}u_k a_{lj}v_l = a_{ki}a_{lj}w_{kl},$$ (15.15)

which shows that it does transform as the components of a second-order tensor. Use has been made in (15.15) of the fact that u_i and v_i are first-order tensors.

2. *The gradient of the components of a vector.* Suppose u_i is a vector and consider the quantities generated by forming the derivatives of each u_i $(i = 1, 2, 3)$ with respect to each x_j $(j = 1, 2, 3)$, i.e.

$$w_{ij} = \frac{\partial u_i}{\partial x_j}. \tag{15.16}$$

These nine quantities form a second-order tensor as can be seen from the following argument:

$$w'_{ij} = \frac{\partial u'_i}{\partial x'_j} = \frac{\partial(a_{ki}u_k)}{\partial x_l}\frac{\partial x_l}{\partial x'_j}$$

$$= a_{ki}\frac{\partial u_k}{\partial x_l}a_{lj}$$

$$= a_{ki}a_{lj}w_{kl}, \tag{15.17}$$

as required by (15.13 b).

A test as to whether any given set of quantities form a second-order tensor can always be made by direct substitution for the x'_i into the transformed set and comparison of this with the right-hand side of (15.12 b). This procedure is extremely laborious and it is almost always better to try to recognize the set as being of one of the forms just developed, or to make alternative tests based on the quotient law of section 15.4.

However, one example of direct substitution will now be carried out, but, as in example 15.1, we will work only in two dimensions and again abbreviate $\cos\theta$ and $\sin\theta$ of matrix (15.5 b) to c and s respectively.

Example 15.2. Show that the u_{ij} given by

$$U = \begin{bmatrix} x_2^2 & -x_1x_2 \\ -x_1x_2 & x_1^2 \end{bmatrix} \tag{15.18}$$

are components of a second-order tensor – or, more briefly, that U is a second-order tensor.

Carrying out first the direct evaluation we obtain (see (15.6))

i	j	u'_{ij}
1	1	$u'_{11} = x_2'^2 = s^2x_1^2 - 2scx_1x_2 + c^2x_2^2,$
1	2	$u'_{12} = -x_1'x_2' = scx_1^2 + (s^2 - c^2)x_1x_2 - scx_2^2,$
2	1	$u'_{21} = -x_1'x_2' = scx_1^2 + (s^2 - c^2)x_1x_2 - scx_2^2,$
2	2	$u'_{22} = x_1'^2 = c^2x_1^2 + 2scx_1x_2 + s^2x_2^2.$

Now evaluating the right-hand side of (15.12 b)

$$
\begin{array}{cccc}
& u_{11} & u_{12} & u_{21} & u_{22} \\
\end{array}
$$

$$
\begin{aligned}
u'_{11} &= \text{cc}x_2^2 &+& \text{cs}(-x_1x_2) &+& \text{sc}(-x_1x_2) &+& \text{ss}x_1^2, \\
u'_{12} &= \text{c}(-\text{s})x_2^2 &+& \text{cc}(-x_1x_2) &+& \text{s}(-\text{s})(-x_1x_2) &+& \text{sc}x_1^2, \\
u'_{21} &= (-\text{s})\text{c}x_2^2 &+& (-\text{s})\text{s}(-x_1x_2) &+& \text{cc}(-x_1x_2) &+& \text{cs}x_1^2, \\
u'_{22} &= (-\text{s})(-\text{s})x_2^2 &+& (-\text{s})\text{c}(-x_1x_2) &+& \text{c}(-\text{s})(-x_1x_2) &+& \text{cc}x_1^2.
\end{aligned}
$$

The corresponding expressions are seen to be the same, showing (as required) that U is a second-order tensor.

The same result could be inferred much more easily by noting that U in (15.18) is in fact the outer product of the vector $(x_2, -x_1)$ with itself, and using results (15.14) and (15.15). That $(x_2, -x_1)$ is a vector was established in example 15.1 (i).

▶3. Show that

$$
C = \begin{bmatrix} x_2^2 & x_1x_2 \\ x_1x_2 & x_1^2 \end{bmatrix}
$$

is *not* a Cartesian tensor of order 2. [To establish a single element that does not transform correctly is sufficient.]

Physical examples involving second-order tensors will be discussed in the later sections of this chapter, but we might for example note here that the susceptibility and conductivity of materials are given by second-order tensors and that the form (15.16) has close connections with the theory of elastic strain and fluid flow.

15.3   The algebra of tensors

Because of the similarities between first- and second-order tensors, and vectors and matrices, it would be expected that similar types of algebraic operations can be carried out with them.

The addition and subtraction of tensors follows an obvious definition; namely that if $u_{ij...k}$ and $v_{ij...k}$ are tensors (of the same order) then their sum and difference, $w_{ij...k}$ and $y_{ij...k}$ respectively, are given by

$$
w_{ij...k} = u_{ij...k} + v_{ij...k}, \tag{15.19 a}
$$

$$
y_{ij...k} = u_{ij...k} - v_{ij...k}, \tag{15.19 b}
$$

for each set of values $i, j, \ldots, k$. That w and v are tensors follows immediately from the linearity of a rotation of coordinates.

It is equally straightforward to show that if $u_{ij...k}$ is a tensor, then so is the set of quantities formed by interchanging the order of (a pair of) indices, e.g. $u_{ji...k}$.

►4. Write out the formal demonstration of this last property.

If $u_{ji...k}$ is found to be identical with $u_{ij...k}$, then $u_{ij...k}$ is said to be a *symmetric* tensor with respect to its first two subscripts [or simply 'symmetric' for second-order tensors]. If $u_{ji...k} = -u_{ij...k}$ (every element) then u is an *antisymmetric* tensor. An arbitrary tensor is neither symmetric nor antisymmetric but can always be written as the sum of a symmetric and an antisymmetric tensor:†

$$u_{ij...k} = \tfrac{1}{2}(u_{ij...k} + u_{ji...k}) + \tfrac{1}{2}(u_{ij...k} - u_{ji...k})$$
$$\equiv s_{ij...k} \qquad\qquad + t_{ij...k}. \qquad (15.20)$$

Of course these properties are valid for any pair of subscripts.

In equation (15.14) of the previous section we had an example of a kind of 'multiplication' of two tensors, thereby producing a tensor of higher order – in that case two first-order tensors multiplied to give a second-order tensor. Inspection of line (15.15) shows that there is nothing particular about the actual orders involved and thus in general that the outer product of an Nth-order tensor with an Mth-order one will produce an $(M + N)$th-order tensor.

An operation which produces the opposite effect – namely generates a tensor of smaller order, rather than larger – is known as **contraction** and consists of making two of the subscripts equal and summing over all values of the equalized subscripts. That this produces another tensor, but with order reduced by 2, is shown by the following argument. Let $u_{ij...l...m...k}$ be an Nth-order tensor, then

$$u'_{ij...l...m...k} = \underbrace{a_{pi}a_{qj}...a_{rl}...a_{sm}...a_{nk}}_{N \text{ factors}}\, u_{pq...r...s...n}. \qquad (15.21)$$

Thus

$$u'_{ij...l...l...k} = a_{pi}a_{qj}...a_{rl}...a_{sl}...a_{nk}u_{pq...r...s...n}$$
$$= a_{pi}a_{qj}...\delta_{rs}...a_{nk}u_{pq...r...s...n}$$
$$= \underbrace{a_{pi}a_{qj}...a_{nk}}_{N-2 \text{ factors}}u_{pq...r...r...n}, \qquad (15.22)$$

showing that $u_{ij...l...l...k}$ is a Cartesian tensor of order $N - 2$.

For a second-rank tensor, the process of contraction is the same as taking the trace of the corresponding matrix (section 14.9). The trace u_{ii}

---

† The connection between an antisymmetric second-order tensor and a vector is considered in example 3 of section 15.10. This work requires results from sections 15.4 and 15.5.

itself is thus a zero-order tensor or scalar and hence invariant under rotations, as has been noted in the previous chapter.

The process of taking the scalar product of two vectors can be recast into tensor language as forming the outer product $w_{ij} = u_i v_j$ of two first-order tensors u and v and then contracting the second-order tensor W so formed, to give $w_{ii} = u_i v_i$, a scalar (invariant under a rotational change of axes).

As yet another example of a familiar operation which is a particular case of a contraction, we may note that the multiplication of a vector u_i by a matrix B (with elements b_{ij}) to produce another vector v_i,

$$b_{ij}u_j = v_i, \tag{15.23}$$

can be looked on as the contraction w_{ijj} of the third-order tensor w_{ijk} formed from the outer product of b_{ij} and u_k.

15.4   The quotient law

The previous paragraph appears to be a heavy-handed way of describing a familiar operation, but it leads us to ask whether it has a converse. To put the question in more general terms – if we know that u and v are tensors and also that

$$w_{pq...k...m}u_{ij...k...n} = v_{pq...mij...n},$$

does this imply that w is also a tensor? Here w, u and v are respectively of Mth, Nth, and $(M + N - 2)$th order and it should be noted that the subscript k which has been contracted may be any of the indices in w and u independently.

The answer to the question is 'yes' provided that the components of u can be varied independently [the components of v are then not arbitrary but depend upon u]. This result is called the **quotient law** for tensors. To prove it for general M and N is no more difficult in the ideas involved than to show it for specific M and N, but does involve the introduction of a large number of subscript symbols. We will therefore take the case $M = N = 2$, but it will be readily apparent that the principle of the proof holds for general M and N.

We thus start with (say)

$$w_{pk}u_{ik} = v_{pi}, \tag{15.24}$$

with u_{ik} an arbitrary second-order tensor. Under a rotation of the coordinates the set w_{pk} (tensor or not) transforms into a new set of quantities which we will denote by w'_{pk}. We thus obtain in succession the following steps

$$w'_{pk}u'_{ik} = v'_{pi} \qquad \text{(transforming (15.24))}$$
$$= a_{qp}a_{ji}v_{qj} \qquad \text{(v is a tensor)}$$
$$= a_{qp}a_{ji}w_{ql}u_{jl} \qquad \text{(from (15.24))}$$
$$= a_{qp}a_{ji}w_{ql}a_{jm}a_{ln}u'_{mn} \qquad \text{(u is a tensor)}$$
$$= a_{qp}a_{ln}w_{ql}u'_{in}. \qquad (a_{ji}a_{jm} = \delta_{im})$$

Now k on the left and n on the right are dummy subscripts and thus this is equivalent to

$$(w'_{pk} - a_{qp}a_{lk}w_{ql})u'_{ik} = 0. \tag{15.25}$$

Since u_{ik} and hence u'_{ik} is an arbitrary tensor, we must have

$$w'_{pk} = a_{qp}a_{lk}w_{ql},$$

showing that the w'_{pk} are given by the general formula (15.13 b) and hence that the set w_{pk} are the components of a second-order tensor.

▶5. (i) Show, by following an analogous argument, that the same result (15.25) and deduction could be obtained if (15.24) were

$$w'_{pk}u_{ki} = v_{pi},$$

i.e. with the contraction with respect to a different pair of indices.
(ii) Verify directly that $w_{kk}u_{pi} = v_{pi}$ implies (as given by the quotient law) that w_{kk} is a scalar.

Use of the quotient law to test whether or not a given set of quantities is a tensor is generally much more convenient than making a direct substitution along the lines of example 15.2. A particular way in which it is applied is to contract the given set of quantities (with N subscripts) with an arbitrary (independently variable components) Nth-order tensor and determine whether the result is a scalar.

▶6. Use the quotient law to show that, (a) the coefficients of a quadratic form in the position coordinates, and (b) U of equation (15.18), are second-order tensors, but that (c) C in ▶3 is not.

15.5 The tensors δ_{ij} and ϵ_{ijk}

Throughout this book and particularly in chapter 14 we have made frequent use of the two quantities δ_{ij} and ϵ_{ijk} defined by

$$\delta_{ij} = 1 \quad \text{if } i = j,$$
$$= 0 \quad \text{otherwise}, \tag{15.26}$$

and

$$\begin{aligned}
\epsilon_{ijk} &= +1 \quad \text{if } i, j, k \text{ is an even permutation of 1, 2, 3,} \\
&= -1 \quad \text{if } i, j, k \text{ is an odd permutation of 1, 2, 3,} \\
&= 0 \quad\;\; \text{otherwise.}
\end{aligned} \tag{15.27}$$

We will now show that these are respectively second- and third-order Cartesian tensors. Notice that the coordinates x_i do not appear explicitly in the components of these tensors, their components consisting entirely of 0 and 1.

Treating first δ_{ij}, the proof is straightforward, since if we consider the quantities given by $a_{ik}a_{jl}\delta_{ij}$ they are

$$\begin{aligned}
a_{ik}a_{jl}\delta_{ij} &= a_{ik}a_{il} \\
&= \delta_{kl}, \quad \text{from (15.3 b),}
\end{aligned} \tag{15.28}$$

which is exactly the same expression in the new coordinates. Thus δ_{ij} is a second-order tensor.

Turning now to ϵ_{ijk}, we have to consider the quantity ϵ'_{lmn} given by $a_{il}a_{jm}a_{kn}\epsilon_{ijk}$. For any particular set of values l, m, n, this is just the determinant of a matrix. If any two or more of l, m, n are equal, then two columns of the matrix are equal and the determinant is zero. Thus we need to consider further only the case of l, m, n, all different. For $l = 1, m = 2$ and $n = 3$, or any even permutation of this, the determinant is simply that of the matrix A describing the rotation. Since A is orthogonal this has value 1.† If l, m, n, is an odd permutation of 1, 2, 3 [the only remaining possibility], then the matrix has determinant -1.

Collecting together the various cases, we see that ϵ'_{lmn} has exactly the properties of (15.27) but with i, j, k replaced by l, m, n, i.e. it is the same as the expression ϵ_{ijk} written using the new coordinates. This shows that ϵ_{ijk} is a third-order Cartesian tensor.

Many of the familiar expressions of vector algebra and calculus studied in chapters 2–4 can be written as contracted tensors involving δ_{ij} and ϵ_{ijk}, e.g. the vector product $\mathbf{b} \wedge \mathbf{c}$ has as its ith component $\epsilon_{ijk}b_j c_k$. Others may be found by the reader in the following exercise.

▶7. Write the following as contracted Cartesian tensors (i) $\mathbf{a} \cdot \mathbf{b}$; (ii) $\nabla^2\phi$; (iii) curl $\mathbf{u}$; (iv) grad (div $\mathbf{u}$); (v) curl (curl $\mathbf{u}$); (vi) $(\mathbf{a} \wedge \mathbf{b}) \cdot \mathbf{c}$.

An important relationship between the ϵ- and δ-tensors is expressed by the identity

$$\epsilon_{ijk}\epsilon_{klm} = \delta_{il}\delta_{jm} - \delta_{im}\delta_{jl}. \tag{15.29}$$

† But see section 15.7 later.

To establish the validity of this identity between two fourth-order tensors [the left-hand side is a once-contracted sixth-order tensor] we consider the various possibilities which arise.

The right-hand side has the values

$$+1 \text{ if } i = l \text{ and } j = m \neq i, \qquad (15.30\,a)$$

$$-1 \text{ if } i = m \text{ and } j = l \neq i, \qquad (15.30\,b)$$

$$0 \text{ for any other set of subscript values } i, j, l, m. \qquad (15.30\,c)$$

In each product on the left-hand side, k has the same value in both ϵ factors and for a non-zero contribution none of i, l, j, m, can have the same value as k. Since there are only three values [1, 2 and 3] that any of the subscripts may take, the only non-zero possibilities are $i = l$ and $j = m$ or vice versa, but not with all four subscripts equal [since then each ϵ factor is zero, as it would be if $i = j$ or $l = m$].

This reproduces (15.30 c) for the left-hand side of (15.29) and also the conditions of (15.30 a) and (15.30 b). The values in (15.30 a and b) are also reproduced in the left-hand side of (15.29) since,

(a) if $i = l$ and $j = m$, $\epsilon_{ijk} = \epsilon_{lmk} = \epsilon_{klm}$, and whether ϵ_{ijk} is $+1$ or -1, the product of the two factors is $+1$, and,

(b) if $i = m$ and $j = l$, $\epsilon_{ijk} = \epsilon_{mlk} = -\epsilon_{klm}$, and thus the product $\epsilon_{ijk}\epsilon_{klm}$ (no summation) has value -1.

This concludes the establishment of identity (15.29).

A useful application of (15.29) is to obtain an alternative expression for curl curl of a vector. As in ▶7 (v), curl curl $\mathbf{u}$ expressed in tensor form is

$$(\text{curl curl } \mathbf{u})_i = \epsilon_{ijk}\epsilon_{klm}\frac{\partial^2 u_m}{\partial x_j\,\partial x_l}. \qquad (15.31)$$

Using identity (15.29) this becomes

$$(\text{curl curl } \mathbf{u})_i = (\delta_{il}\delta_{jm} - \delta_{im}\delta_{jl})\,\frac{\partial^2 u_m}{\partial x_j\,\partial x_l}$$

$$= \frac{\partial}{\partial x_i}\left(\frac{\partial u_j}{\partial x_j}\right) - \frac{\partial^2 u_i}{\partial x_j^2}$$

$$= [\text{grad}\,(\text{div } \mathbf{u})]_i - \nabla^2 u_i. \qquad (15.32)$$

This result has already been mentioned in section 4.5 and the reader is referred there for a discussion of its applicability.

▶8. By examining the various possibilities satisfy yourself that

$$
\epsilon_{ijk}\epsilon_{pqr} = \begin{vmatrix} \delta_{ip} & \delta_{iq} & \delta_{ir} \\ \delta_{jp} & \delta_{jq} & \delta_{jr} \\ \delta_{kp} & \delta_{kq} & \delta_{kr} \end{vmatrix}.
$$

Equation (15.29) is a special case of this more general result.

15.6   Isotropic tensors

It will have been noticed that, unlike most of the tensors discussed [except for scalars], δ_{ij} and ϵ_{ijk} have the property that all their components have values [as opposed to expressions] which are the same whatever rotation of axes is made, i.e. the component values are independent of the a_{ij} of the transformation. Specifically δ_{11} has the value 1 in all coordinate frames, whereas for a general second-order tensor U all we know is that if $u_{11} = f_{11}(x_1, x_2, x_3)$ then $u'_{11} = f_{11}(x'_1, x'_2, x'_3)$. Tensors with this particular property are called **isotropic** (or **invariant**) **tensors.**

It is important to know how general a tensor can be and still be iso-tropic, since the description of the physical properties (e.g. conductivity, magnetic susceptibility or tensile strength) of an isotropic medium [i.e. one which has the same properties whichever way it is orientated] will involve an isotropic tensor. In the previous section it was shown that δ_{ij} and ϵ_{ijk} are second- and third-order isotropic tensors; we will now show that, to within a scalar multiple, they are the only such isotropic tensors.

$\lambda\delta_{ij}$ *is the only isotropic second-order tensor.* Suppose u_{ij} is an isotropic tensor; then for *any* rotation of the axes we must, by definition, have that

$$
u_{ij} = u'_{ij} = a_{ki}a_{lj}u_{kl} \tag{15.33}
$$

for each of the 9 components.

First consider a rotation of the axes about the $(1, 1, 1)$ direction taking Ox_1, Ox_2, Ox_3 into Ox'_2, Ox'_3, Ox'_1 respectively. This requires that $u_{11} = u'_{11} = u_{33}$. Similarly $u_{12} = u'_{12} = u_{31}$. Continuing in this way,

▶9.   (a) $u_{11} = u_{22} = u_{33}$,
    (b) $u_{12} = u_{23} = u_{31}$,
    (c) $u_{21} = u_{32} = u_{13}$. $\tag{15.34}$

Next consider a rotation of the axes (from their original position) by $\pi/2$ about the Ox_3-axis. For this rotation $a_{12} = 1$, $a_{21} = -1$, $a_{33} = 1$ and all other $a_{ij} = 0$. Amongst other relationships, we must have from

(15.33), that

$$u_{13} = (-1) \cdot 1 \cdot u_{23},$$
and $$u_{23} = 1 \cdot 1 \cdot u_{13}.$$

Hence $u_{13} = u_{23}$ equals 0 and so therefore, by parts (b) and (c) of (15.34) do all elements except u_{11}, u_{22} and u_{33}, which are all the same. This shows that $u_{ij} = \lambda \delta_{ij}$.

$\lambda \epsilon_{ijk}$ is the only isotropic third-order tensor. The general line of attack is as in the previous example and so only a minimum of explanation will be given.

$$u_{ijk} = u'_{ijk} = a_{li} a_{mj} a_{nk} u_{lmn} \quad \text{(all 27 elements)}.$$

Rotate about the (1, 1, 1) direction; $1 \to 2 \to 3 \to 1$.

(a) $u_{111} = u_{222} = u_{333}$.
(b) $u_{112} = u_{223} = u_{331}$ (and 2 similar sets involving repeated subscripts).
(c) $u_{123} = u_{231} = u_{312}$ (and a set involving odd permutations of 1, 2, 3).

Rotate by $\pi/2$ about the Ox_3-axis; $a_{12} = 1$, $a_{21} = -1$, $a_{33} = 1$, others $= 0$.

(d) $u_{111} = (-1) \cdot (-1) \cdot (-1) \cdot u_{222} = -u_{222}$.
(e) $u_{112} = (-1) \cdot (-1) \cdot 1 \cdot u_{221}$.
(f) $u_{221} = 1 \cdot 1 \cdot (-1) \cdot u_{112}$.
(g) $u_{123} = (-1) \cdot 1 \cdot 1 \cdot u_{213}$.

(a) and (d) show that elements with all subscripts the same are zero.
(e), (f) and (b) show that all elements with repeated subscripts are zero.
(g) and (c) show that $u_{123} = u_{231} = u_{312} = -u_{213} = -u_{321} = -u_{132}$.
    In total, u_{ijk} is at most a multiple of ϵ_{ijk}, but since ϵ_{ijk} (and hence $\lambda \epsilon_{ijk}$) has already been explicitly shown to be an isotropic tensor, it must be the most general third-order one.

▶10. Show that the only isotropic first-order tensor is the trivial one with all its elements zero.

15.7 Polar and axial vectors

It is something of an aside in the context of the rest of this chapter, but sufficiently important from a physical point of view that the behaviour of mathematical entities (and the physical quantities they represent) under a different kind of axis change, namely *reflection* in the origin, should receive some mention. This kind of transformation is represented in all its essentials by the equation

$$x'_i = -x_i, \tag{15.35}$$

but may in practice also involve a 'rigid axes' rotation and hence appear more complicated. The change may alternatively be looked upon as one which changes from an initial right-handed coordinate system to a left-handed one; any prior or subsequent rotation will not change this state of affairs.

It is obvious that a reflection plus rotation preserves the length of the position vector and is thus represented by an orthogonal matrix A as in (15.2) and (15.3). However, in this case det A is equal to -1 and not $+1$ [the matrix corresponding to (15.35) itself being the most obvious example].

Vectors (first-order tensors) which, like the position coordinates x_i, reverse sign on a change to a new coordinate system, obtained from the original by a reflection in the origin, are called **polar vectors**. Vectors whose sign remains unaltered under such a change are called **axial vectors**.

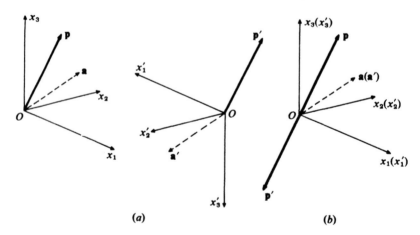

(a)   (b)

Fig. 15.2 (a) The behaviour of polar (p) and axial (a) vectors under reflection in the origin of the coordinate system (x_1, x_2, x_3) giving the new system (x'_1, x'_2, x'_3). (b) In practice the axes are drawn unchanged and the vectors reversed (p') or not (a').

For drawing purposes it is, in practice, usually much more convenient to imagine that the axes are unchanged, but that the vectors are reversed (or not), i.e. to use fig. 15.2 (b) rather than the two diagrams of fig. 15.2 (a) which show the behaviour of a polar vector **p** and an axial vector **a**.

Since the position vector of a particle $\mathbf{r}$ ($=x_i$) is a polar vector, so is its momentum $m\dot{\mathbf{r}}$, but its angular momentum $\mathbf{J}$ (as can be seen from fig. 15.3) does not change sign on a reflection of the coordinate system and is therefore an axial vector.

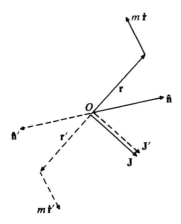

Fig. 15.3 Angular momentum $\mathbf{J} = \mathbf{r} \wedge m\dot{\mathbf{r}}$ is an axial vector. Vector $\mathbf{J}$ is perpendicular to the plane containing $\mathbf{r}$ and $m\dot{\mathbf{r}}$.

Vectors which are obtained as the gradient of scalars, such as the components E_i of an electrostatic field, will also be polar vectors since

$$\left(\frac{\partial \phi}{\partial x_i}\right)' = \frac{\partial \phi'}{\partial x_i'} = -\frac{\partial \phi}{\partial x_i}, \tag{15.36}$$

if x_i' is given by (15.35) and ϕ is a scalar, but vectors obtained as the curl of another vector may be polar or axial depending upon the character of this second vector.

Corresponding to polar and axial vectors for first-order tensors, zeroth-order tensors may be divided into scalars and **pseudo-scalars** – the latter being invariant under rotations but changing sign on reflection. As an example of a pseudo-scalar we may consider the triple scalar product of three polar vectors such as the component of angular momentum about a particular axis given by $(\mathbf{r} \wedge m\dot{\mathbf{r}}) \cdot \hat{\mathbf{n}}$ (see fig. 15.3).

This last expression also provides an example of a form of quotient law relevant in this connection, which can be used to test the characteristic of an unknown vector.

> The contraction of two vectors with each other yields a scalar or pseudo-scalar according as the vectors are both of the same kind or one is polar and the other axial.

From the physics point of view the importance of the distinction between polar and axial vectors is principally that:

(i) Any physically possible equation must have all its terms with the same characteristics,† i.e. all terms vector, or all pseudo-scalar, etc.
(ii) Any system which is symmetric (antisymmetric) with respect to reflection in the origin *cannot* have quantities describing it which are pseudo-scalar (scalar).

▶11. Determine the behaviour of the following under reflection.

(*a*) Electrical current **I**, (*b*) magnetic field **H** [consider that due to a current-carrying wire], (*c*) magnetic flux **B**, (*d*) magnetic moment **μ**, (*e*) the couple on a magnetic dipole in magnetic flux **B**.

▶12. Verify that the following physical equations and expressions are invariant under reflection:

(*a*) the force on a current-carrying wire in a magnetic field: $\mathbf{F} = \mathbf{I} \wedge \mathbf{B}$,
(*b*) Poynting vector for energy flow in an electromagnetic wave: $\mathbf{S} = \mathbf{E} \wedge \mathbf{H}$,
(*c*) rate of ohmic heating per unit volume: $w = \mathbf{E} \cdot \mathbf{j}$,
(*d*) acceleration in rotating coordinate systems:

$$\ddot{\mathbf{R}} = \ddot{\mathbf{r}} + \dot{\boldsymbol{\omega}} \wedge \mathbf{r} + 2\boldsymbol{\omega} \wedge \dot{\mathbf{r}} + \boldsymbol{\omega} \wedge (\boldsymbol{\omega} \wedge \mathbf{r}).$$

[This may alternatively be looked on as a proof that **ω** must be axial.]

15.8   Physical tensors

In this section some physical applications of tensors will be illustrated. First-order tensors are fairly familiar as vectors, and so attention will be concentrated on second-order tensors, starting with a mechanical example.

Consider a collection of rigidly connected point particles‡ of which the αth with mass $m^{(\alpha)}$ is a typical one, positioned at $\mathbf{r}^{(\alpha)}$ with respect to an origin O. Suppose that the rigid assembly is rotating about an axis through O with angular velocity **ω** (fig. 15.4).

---

† Ignoring here the non-conservation of parity which appears in some nuclear physics reactions.
‡ In this paragraph and all that follow on this topic, a more realistic situation obtains if a continuous rigid body is considered. In this case $m^{(\alpha)}$ must be replaced everywhere by $\rho(\mathbf{r})\, dx\, dy\, dz$, and all summations by integrations over the volume of the body.

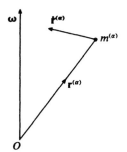

Fig. 15.4 The velocity $\dot{\mathbf{r}}^{(\alpha)}$ of the αth particle is $\boldsymbol{\omega} \wedge \mathbf{r}^{(\alpha)}$ and is perpendicular to the plane containing $\boldsymbol{\omega}$ and $\mathbf{r}^{(\alpha)}$.

The angular momentum $\mathbf{J}$ about O of the assembly is given by

$$\mathbf{J} = \sum_{\alpha} (\mathbf{r}^{(\alpha)} \wedge \mathbf{p}^{(\alpha)}). \qquad (15.37)$$

But $\mathbf{p}^{(\alpha)} = m^{(\alpha)}\dot{\mathbf{r}}^{(\alpha)}$ and $\dot{\mathbf{r}}^{(\alpha)}$ itself is given by $\boldsymbol{\omega} \wedge \mathbf{r}^{(\alpha)}$ [for any α] and so in subscript form,

$$\begin{aligned}
J_i &= \sum_{\alpha} m^{(\alpha)} \epsilon_{ijk} x_j^{(\alpha)} \dot{x}_k^{(\alpha)} \\
&= \sum_{\alpha} m^{(\alpha)} \epsilon_{ijk} x_j^{(\alpha)} \epsilon_{klm} \omega_l x_m^{(\alpha)} \, . \\
&= \sum_{\alpha} m^{(\alpha)} (\delta_{il}\delta_{jm} - \delta_{im}\delta_{jl}) x_j^{(\alpha)} x_m^{(\alpha)} \omega_l \qquad \text{(by (15.29))} \\
&= \sum_{\alpha} m^{(\alpha)} (r^{(\alpha)2} \delta_{il} - x_i^{(\alpha)} x_l^{(\alpha)}) \omega_l \\
&\equiv I_{il}\omega_l, \qquad\qquad\qquad\qquad\qquad\qquad\qquad (15.38)
\end{aligned}$$

where I_{il} is a symmetric second-order Cartesian tensor [by the quotient rule, since $\mathbf{J}$ and $\boldsymbol{\omega}$ are vectors] which depends only on the distribution of masses in the assembly and not upon the direction or magnitude of $\boldsymbol{\omega}$. The tensor is called the **inertia tensor** at O of the assembly.

Written out in full (in ordinary Cartesians) for (say) a continuous body (see footnote) it would be

$$\mathsf{I} = \begin{bmatrix} \int (y^2 + z^2)\rho \, d\tau & -\int xy\rho \, d\tau & -\int xz\rho \, d\tau \\ -\int xy\rho \, d\tau & \int (z^2 + x^2)\rho \, d\tau & -\int yz\rho \, d\tau \\ -\int xz\rho \, d\tau & -\int yz\rho \, d\tau & \int (x^2 + y^2)\rho \, d\tau \end{bmatrix},$$

$$(15.39)$$

where $\rho = \rho(x, y, z)$ is the mass distribution and $d\tau$ stands for $dx\, dy\, dz$, the integrals being over the whole body. The diagonal elements of this tensor are called the *moments of inertia* and the off-diagonal elements without the minus signs are known as the *products of inertia*.

▶13. Verify that, for any one α, both terms in the (15.38) definition of I_{tl} are separately tensors, and hence that I_{tl} must be.

By a parallel argument to that already made, the kinetic energy of the rotating system is given by

▶14. $$T = \tfrac{1}{2} \sum_\alpha m^{(\alpha)}(\dot{\mathbf{r}}^{(\alpha)} \cdot \dot{\mathbf{r}}^{(\alpha)}) = \tfrac{1}{2} I_{jl} \omega_j \omega_l = \tfrac{1}{2} J_j \omega_j, \qquad (15.40)$$

and so can be expressed as the scalar obtained by twice contracting $\boldsymbol{\omega}$ with the inertia tensor. This also shows that the moment of inertia of the body about any line given by unit vector $\hat{\mathbf{n}}$ is

$$I_{jl}\hat{n}_j\hat{n}_l \qquad (15.41)$$

(or $\tilde{\mathbf{n}}\mathit{I}\hat{\mathbf{n}}$ in matrix and vector form).

Since $\mathit{I}\,(\equiv I_{tl})$ is a real symmetric second-order tensor, it has associated with it three mutually perpendicular directions which are its *principal axes* and have the properties [proofs in the previous chapter]:

(i) with each axis is associated a principal moment of inertia λ_μ ($\mu = 1, 2, 3$),
(ii) when the rotation of the body is about one of these axes, the angular velocity and the angular momentum are parallel and given by

$$\lambda_\mu \boldsymbol{\omega} = \mathbf{J} = \mathit{I}\boldsymbol{\omega}, \qquad (15.42)$$

i.e. $\boldsymbol{\omega}$ is an eigenvector of I with eigenvalue λ_μ,
(iii) referred to these axes as coordinate axes, the inertia tensor is diagonal with diagonal entries $\lambda_1, \lambda_2, \lambda_3$.

Two further examples of physical quantities represented by second-order tensors are magnetic susceptibility and electrical conductivity. In the first case we have (in standard notation)

$$M_i = \chi_{ij}H_j, \qquad (15.43)$$

and in the second

$$j_i = \sigma_{ij}E_j. \qquad (15.44)$$

M is the magnetic moment per unit volume and **j** the current density (per unit area). In both cases we have on the left-hand side a vector and on the right-hand side the contraction of a set of quantities with another vector. The sets of quantities must therefore form the components of a second-order tensor.

For isotropic media $\mathbf{M} \propto \mathbf{H}$ and $\mathbf{j} \propto \mathbf{E}$, but for anisotropic materials such as crystals the susceptibility and conductivity may be different along different crystal axes, making χ_{ij} and σ_{ij} general second-order tensors, although they are usually symmetric.

The susceptibility and conductivity tensors have analogous properties to those of the inertia tensor. They can be deduced from the analogies between (15.38) and equations (15.43) and (15.44) and by noting that the expressions for energy are

$$-\tfrac{1}{2}\mu_0 \mathbf{M} \cdot \mathbf{H} \text{ per unit volume} \tag{15.45}$$

in the magnetic case, and

$$\mathbf{E} \cdot \mathbf{j} \text{ per unit volume per sec} \tag{15.46}$$

in the electrical one.

15.9   Stress and strain tensors

The theory of *small elastic deformations* of solid bodies provides further examples of tensors. We consider first the description of *strain*, i.e. the displacement of particles of the body from the positions they occupied before the body was deformed. It will be found that in the neighbourhood of a point P the displacement vector for the strained body of any other point Q, relative to that of P, is given in terms of the position vector of Q relative to P for the unstrained body by a second-order tensor.

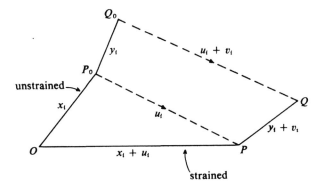

Fig. 15.5 An element at $P_0 Q_0$ when the body is unstrained, moves to PQ under the strain. For an explanation of the other symbols see the text.

As illustrated in fig. 15.5, consider a particle at P_0 ($\equiv x_i$) which is displaced under the strain to P ($\equiv x_i + u_i$) and another one in its neighbourhood which is initially at Q_0 ($\equiv x_i + y_i$) and under the strain moves to Q ($\equiv x_i + u_i + y_i + v_i$). All of y_i, u_i and v_i ($i = 1, 2, 3$ in all cases) are taken as small, so that a first-order theory may be developed. Now, assuming that $u_i(x_1, x_2, x_3)$ is differentiable up to any needed order, we note that the total displacement at Q is both $u_i + v_i$ and also $u_i(x_k + y_k)$ and thus write

$$u_i + v_i = u_i(x_k + y_k)$$

$$= u_i(x_k) + y_j \left(\frac{\partial u_i}{\partial x_j}\right)_{P_0} + \text{second order}.$$

Hence

$$v_i = \left(\frac{\partial u_i}{\partial x_j}\right)_{P_0} y_j, \tag{15.47}$$

which is the symbolic form of the statement made in the first paragraph of this section.

That $(\partial u_i/\partial x_j)$ is a Cartesian tensor of second order is apparent either from the quotient law (section 15.4) since v_i and y_j are vectors, or by direct demonstration as follows.

Under a rotation of axes

$$x_i = a_{ij}x'_j, \qquad x'_i = a_{ji}x_j, \tag{15.2 bis}$$

$(\partial u_i/\partial x_j)$ becomes

$$\left(\frac{\partial u_i}{\partial x_m}\right)' = \frac{\partial u'_i}{\partial x'_m} = \frac{\partial(a_{ji}u_j)}{\partial x_k} \frac{\partial x_k}{\partial x'_m} = a_{ji}a_{km} \frac{\partial u_j}{\partial x_k},$$

showing that $(\partial u_i/\partial x_j)$ is indeed a second-order tensor.

As has been shown previously for a general second-order tensor, $(\partial u_i/\partial x_j)$ can be written as the sum of a symmetric tensor

$$e_{ij} = \frac{1}{2}\left(\frac{\partial u_i}{\partial x_j} + \frac{\partial u_j}{\partial x_i}\right) \tag{15.48}$$

and an antisymmetric one

$$\omega_{ij} = \frac{1}{2}\left(\frac{\partial u_i}{\partial x_j} - \frac{\partial u_j}{\partial x_i}\right). \tag{15.49}$$

Also, as proved in example 3 of section 15.10, the contribution $\omega_{ij}y_j$ to v_i represents a small rotation $-\boldsymbol{\Omega} \wedge \mathbf{y}$ with $\boldsymbol{\Omega} = (\omega_{23}, \omega_{31}, \omega_{12})$.

The symmetric tensor e_{ij} is called the **strain tensor** and describes the local *deformation* $e_{ij}y_j$ at the point Q which had [when Q_0] initial coordinates y_i with respect to P_0.

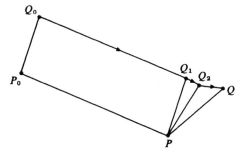

Fig. 15.6 The total displacement of P_0Q_0 consists of a rigid translation to PQ_1, a rotation to PQ_2 and a deformation to PQ.

The full displacement of the point Q (still confined to be in the neighbourhood of P) from its unstrained position is thus made up of three parts (see fig. 15.6),

(i) a rigid body translation, u_i $P_0Q_0 \rightarrow PQ_1$,
(ii) a local rigid body rotation, $\omega_{ij}y_j$ $PQ_1 \rightarrow PQ_2$,
(iii) a local distortion, $e_{ij}y_j$ $PQ_2 \rightarrow PQ$.

The main interest from the point of view of the present chapter is in the symmetric strain tensor e_{ij} describing a pure deformation.

The simplest deformation is a uniform stretching of the material of the body in one direction, say the Ox_1 direction,

$$(u_1, u_2, u_3) = (kx_1, 0, 0). \tag{15.50}$$

In this case $e_{11} = k$ and all other elements of the tensor are zero.

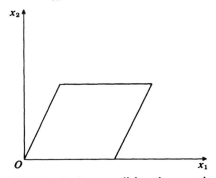

Fig. 15.7 A simple shear parallel to the x_1x_3-plane.

A simple shear (see fig. 15.7) in the x_1-direction and parallel to the x_1x_3-plane is described by

$$(u_1, u_2, u_3) = (kx_2, 0, 0).$$

This displacement is not a pure deformation but consists partially of a rotation, as is confirmed by the fact that only $\partial u_1/\partial x_2$ is non-zero and thus $\partial u_i/\partial x_j$ is not symmetric.

▶15. A *pure shear* is described by a tensor in which $\omega_{ij} = 0$, $e_{12} = e_{21} = k$, and all other elements of e_{ij} are zero. Show that this corresponds to the superposition of two simple shears.

Associated with the strain tensor we may define a **strain quadric** such that the direction of the displacement at any point Q on a radius vector PP' is parallel to the normal to the quadric at the point where the radius vector meets the quadric (fig. 15.8). In addition, the fractional extension of an element in any direction from P is inversely proportional to the square of the quadric radius (r in fig. 15.8) in that direction.

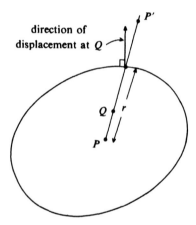

direction of displacement at Q

Fig. 15.8 The strain quadric at P.

To derive the quadric suppose that, although we are working to first order in both y_i and v_i, $|v|$ is small compared with $|y|$. Then the *fractional extension e* due to the deformation described by e_{ij} is

$$\frac{PQ - P_0Q_0}{P_0Q_0} = \frac{|y_i + v_i| - |y_i|}{|y_i|}$$

$$= \frac{(|y|^2 + 2v_iy_i + |v|^2)^{1/2} - |y|}{|y|}$$

$$= \frac{v_iy_i}{|y|^2} + O\left(\frac{v^2}{y^2}\right). \tag{15.51}$$

Now replacing v_i by $e_{ij}y_j$ gives for e the expression

$$e = \frac{e_{ij}y_iy_j}{|y|^2} . \qquad (15.52)$$

Thus if we take a quadric with centre P and given by

$$e_{ij}y_iy_j = c, \qquad (15.53)$$

where c is chosen to make the quadric real, it will have the properties stated.

▶16. Verify that this is so.

In addition, just like the quadrics discussed in section 14.11, it will possess three mutually perpendicular principal axes. An element of the body in any of these directions is extended in the direction of its own radius vector (y_i).

▶17. Show that the strain quadric for a simple extension along the direction l, m, n, is a pair of parallel planes.

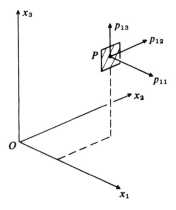

Fig. 15.9 Definition of the stress tensor p_{ij} at the point P for the case $i = 1$. The plane across which the components of stress give p_{1j} has its normal parallel to the positive x_1-axis.

As well as the tensor describing the strain of a deformed solid a **stress tensor** at a point P may be specified; this is denoted by p_{ij}. The quantity p_{ij} is the x_j-component of the stress vector acting across a plane through P whose normal lies in the x_i-direction. The sense of the stress vector is that of the action of the region of greater x_i on that of lesser x_i. Fig. 15.9 illustrates the case in which $i = 1$ and j has the values 1, 2 and 3; the other six components correspond [in threes] to two other planes through P parallel to the x_1x_3- and x_1x_2-planes.

If an element of area containing P (inside or on the surface of the body) has unit normal $\mathbf{n}$ then the stress $\mathbf{p}^{(n)}$ across it can be shown to be given by

$$p_j^{(n)} = p_{ij}n_i, \tag{15.54}$$

where $p_{ij} = p_{ji}$. That p_{ij} is a second-order Cartesian tensor follows either from the quotient law or by direct verification.

Just as for strain, a stress quadric

$$p_{ij}y_iy_j = c, \quad \text{with } y_i = rn_i \tag{15.55}$$

and analogous properties, can be defined. The normal to the quadric at the point where any radius vector meets it gives the direction of stress across any plane to which the radius vector is a normal.

Example 15.3. A generalization of Hooke's Law relates the stress and strain tensors by

$$p_{ij} = c_{ijkl}e_{kl}, \tag{15.56}$$

where c_{ijkl} is a fourth-order Cartesian tensor. Assuming that the most general fourth-order isotropic tensor is†

$$c_{ijkl} = \lambda\delta_{ij}\delta_{kl} + \eta\delta_{ik}\delta_{jl} + \nu\delta_{il}\delta_{jk}, \tag{15.57}$$

find the form of (15.56) for an *isotropic* medium of Young's modulus E and Poisson's ratio σ.

For an isotropic medium we must have an isotropic tensor for c_{ijkl} and so assume the form (15.57). Substituting this into (15.56) yields

$$p_{ij} = \lambda\delta_{ij}e_{kk} + \eta e_{ij} + \nu e_{ji}.$$

But e_{ij} is symmetric, and if we write $\eta + \nu = 2\mu$ this takes the form

$$p_{ij} = \lambda e_{kk}\delta_{ij} + 2\mu e_{ij}. \tag{15.58}$$

[λ and μ are called *Lamé constants*. It will be noted that if $e_{ij} = 0$ for $i \neq j$, then so does p_{ij}, i.e. the principal axes of stress and strain coincide.]

Now consider a simple tension in the x_1-direction, i.e. $p_{11} = S$, all other $p_{ij} = 0$. Then denoting e_{kk} [summed over k] by θ we have (in addition to $e_{ij} = 0$ for $i \neq j$) the three equations

$$S = \lambda\theta + 2\mu e_{11},$$
$$0 = \lambda\theta + 2\mu e_{22},$$
$$0 = \lambda\theta + 2\mu e_{33}.$$

† This may be shown by methods similar to, but lengthier than, those of section 15.6.

Adding them gives

$$S = \theta(3\lambda + 2\mu).$$

Substituting for θ from this into the first of the three, and recalling that Young's modulus is defined by $S = Ee_{11}$, gives E as

►18.
$$E = \frac{\mu(3\lambda + 2\mu)}{\lambda + \mu}.$$
(15.59 a)

Further Poisson's ratio is defined as $\sigma = -e_{22}/e_{11}$ which is thus

$$\sigma = \frac{1}{e_{11}} \frac{\lambda\theta}{2\mu} = \frac{1}{e_{11}} \frac{\lambda}{2\mu} \frac{Ee_{11}}{3\lambda + 2\mu} = \frac{\lambda}{2(\lambda + \mu)}.$$
(15.59 b)

Solving (15.59) for λ and μ gives finally

►19.
$$p_{ij} = \frac{\sigma E}{(1 + \sigma)(1 - 2\sigma)} e_{kk}\delta_{ij} + \frac{E}{(1 + \sigma)} e_{ij}.$$
(15.60)

15.10 Examples for solution

1. Show how to decompose the tensor T_{ij} into three tensors

$$T_{ij} = U_{ij} + V_{ij} + S_{ij},$$

where U_{ij} is symmetric and traceless ($U_{ii} = 0$), V_{ij} is isotropic, and S_{ij} has only three independent components.

2. Use the quotient law to show that

$$\begin{bmatrix} y^2 + z^2 - x^2 & -2xy & -2xz \\ -2yx & x^2 + z^2 - y^2 & -2yz \\ -2xz & -2yz & x^2 + y^2 - z^2 \end{bmatrix}$$

is a second-order tensor.

3. *Antisymmetric tensor.* Suppose w_{ij} is an antisymmetric second-order Cartesian tensor in three-dimensions.

(i) Show it has only three independent components.
(ii) By writing the tensor in terms of w_{23}, w_{31}, and w_{12} and considering $v_i = \frac{1}{2}\epsilon_{ijk}w_{jk}$, show that the three independent components of W form the components of a vector **v**.
(iii) Verify $w_{ij} = \epsilon_{ijk}v_k$.
(iv) For a general vector **b** evaluate $(\mathbf{v} \wedge \mathbf{b})_i$ and hence show that vector multiplication by **v** is equivalent to tensor contraction with $-w_{ij}$.

4. Use tensor methods to establish the following vector identities.

(i) $(\mathbf{u} \wedge \mathbf{v}) \wedge \mathbf{w} = (\mathbf{u} \cdot \mathbf{w})\mathbf{v} - (\mathbf{v} \cdot \mathbf{w})\mathbf{u}$.

(ii) $\operatorname{curl}(\phi\mathbf{u}) = \phi \operatorname{curl} \mathbf{u} + (\operatorname{grad} \phi) \wedge \mathbf{u}$.

(iii) $\operatorname{div}(\mathbf{u} \wedge \mathbf{v}) = \mathbf{v} \cdot \operatorname{curl} \mathbf{u} - \mathbf{u} \cdot \operatorname{curl} \mathbf{v}$.

(iv) $\operatorname{curl}(\mathbf{u} \wedge \mathbf{v}) = (\mathbf{v} \cdot \operatorname{grad})\mathbf{u} - (\mathbf{u} \cdot \operatorname{grad})\mathbf{v} + \mathbf{u} \operatorname{div} \mathbf{v} - \mathbf{v} \operatorname{div} \mathbf{u}$.

(v) $\operatorname{grad}(\tfrac{1}{2}\mathbf{u} \cdot \mathbf{u}) = \mathbf{u} \wedge \operatorname{curl} \mathbf{u} + (\mathbf{u} \cdot \operatorname{grad})\mathbf{u}$.

5. In four dimensions define second-order antisymmetric tensors F_{ij} and Q_{ij} and a vector S_i as follows:

(a) $F_{23} = H_1$, $Q_{23} = B_1$ and their cyclic permutations,

(b) $F_{i4} = -D_i$, $\qquad Q_{i4} = E_i$ for $i = 1, 2, 3$,

(c) $S_i = J_i$ for $i = 1, 2, 3$; $\quad S_4 = \rho$.

Then taking x_4 as t and the other symbols to have their usual meaning in electromagnetic theory show that the equations $\partial(F_{ij})/\partial x_j = S_i$, and $\partial(Q_{jk})/\partial x_i + \partial(Q_{ki})/\partial x_j + \partial(Q_{ij})/\partial x_k = 0$, where i, j, k are any set chosen from 1, 2, 3, 4 (all different), reproduce Maxwell's equations.

6. (i) Evaluate $(\mathbf{A} \wedge \operatorname{curl} \mathbf{A})_i$ for any vector function $\mathbf{A}$.

(ii) Assuming that the divergence theorem holds for any tensor field

$$\int_V \frac{\partial u_{ij\ldots k\ldots n}}{\partial x_k} \, dV = \int_S u_{ij\ldots k\ldots n} \, dS_k,$$

show that

$$\int_S [\mathbf{A}(\mathbf{A} \cdot d\mathbf{S}) - \tfrac{1}{2}A^2 \, d\mathbf{S}] = \int_V [\mathbf{A} \operatorname{div} \mathbf{A} - \mathbf{A} \wedge \operatorname{curl} \mathbf{A}] \, dV.$$

(iii) How does this general result simplify (a) if $\mathbf{A}$ is the electric field $\mathbf{E}$ in a time-independent situation, and (b) if $\mathbf{A}$ is the magnetic flux $\mathbf{B}$?

7. Ohm's law says that the current density $\mathbf{j}$ is linearly related to the electric field $\mathbf{E}$ in a conductor. Show that

$$j_i = s_{ik}E_k + (\mathbf{E} \wedge \mathbf{a})_i,$$

where s_{ik} is a symmetric tensor and $\mathbf{a}$ is an axial vector.

8. A rigid body consists of four particles of masses $m, 2m, 3m, 4m$ respectively situated at the points (a, a, a), $(a, -a, -a)$, $(-a, a, -a)$, $(-a, -a, a)$ and connected together by a light framework.

(i) Find the inertia tensor at the origin and show that the principal moments of inertia are $20ma^2$, $(20 + 2\sqrt{5})ma^2$, and $(20 - 2\sqrt{5})ma^2$.

(ii) Find the principal axes and verify that they are mutually orthogonal.

9. A rigid body comprises 8 particles each of mass m, held together by light rods. In a certain coordinate frame the particles are at

$$\pm a(3, 1, -1), \quad \pm a(1, -1, 3), \quad \pm a(1, 3, -1), \quad \pm a(-1, 1, 3).$$

Show that when the body rotates about an axis through the origin, if the angular velocity and angular momentum vectors are parallel then they are in one of the ratios $40ma^2$, $64ma^2$ or $72ma^2$.

10. The paramagnetic tensor χ_{ij} of a body placed in a magnetic field is

$$\begin{bmatrix} 2k & 0 & 0 \\ 0 & 3k & k \\ 0 & k & 3k \end{bmatrix}.$$

Assuming depolarizing effects are negligible, find how the body will orientate itself if the field is horizontal and the body (i) can rotate freely, (ii) is suspended with the axis $(1, 0, 0)$ vertical, (iii) is suspended with the axis $(0, 1, 0)$ vertical.

11. A block of wood contains a number of thin soft iron nails. A unit magnetic field directed eastwards induces a magnetic moment with components $(3, 1, -2)$ in the block, and fields of unit strength directed northwards and vertically upwards induce moments $(1, 3, -2)$ and $(-2, 2, 2)$ respectively. Show that the nails all lie parallel to a certain plane. [Assume soft iron has a constant permeability.]

12. For tin the conductivity tensor has the form

$$\sigma = \begin{bmatrix} a & 0 & 0 \\ 0 & a & 0 \\ 0 & 0 & b \end{bmatrix}$$

when referred to its crystal axes. A single crystal is grown in the shape of a long wire of length l and radius r, the axis of the wire making polar angle θ with respect to the crystal third axis. Show that the resistance of the wire is

$$\frac{l}{\pi r^2} \frac{1}{ab} (a \cos^2 \theta + b \sin^2 \theta).$$

13. By considering an isotropic body subjected to a uniform 'hydrostatic' pressure (no shearing stress) show that the bulk modulus, defined by

$$k = \frac{\text{pressure}}{\text{fractional decrease in volume}},$$

is given by $k = E/[3(1 - 2\sigma)]$.

14. In a certain isotropic elastic medium the displacement vector u_i is given by

$$u_1 = K(x_1 + x_3), \qquad u_2 = K(x_1 + x_2), \qquad u_3 = K(x_2 + x_3).$$

Find the magnitudes of the principal stresses in terms of Young's modulus and Poisson's ratio.

15. An elastic cylinder of finite length and initially unstressed, fits accurately into a rigid smooth cylindrical tube open at both ends. Equal and opposite compressive normal stresses are applied uniformly over the ends of the cylinder. Show that the modulus of elasticity is

$$\frac{E(1 - \sigma)}{(1 + \sigma)(1 - 2\sigma)}.$$

16. (More difficult vector analysis.) For an isotropic elastic medium under stress, the displacement u_i satisfies

$$\frac{\partial p_{ij}}{\partial x_j} = \rho \frac{\partial^2 u_i}{\partial t^2}, \qquad p_{ij} = c_{ijkl} \left(\frac{\partial u_k}{\partial x_l} + \frac{\partial u_l}{\partial x_k} \right),$$

[p_{ij} is the stress tensor] where c_{ijkl} is the isotropic tensor

$$c_{ijkl} = a\delta_{ij}\delta_{kl} + b(\delta_{ik}\delta_{jl} + \delta_{il}\delta_{jk}) + c(\delta_{ik}\delta_{jl} - \delta_{il}\delta_{jk}),$$

and ρ is a constant. Show that div **u** and curl **u** both satisfy wave equations.

16
Complex variables

Throughout this book references have been made to results 'which are derived from the theory of complex variables'. This theory thus becomes an integral part of the mathematics appropriate to physical applications. The difficulty with it, from the point of view of a book such as the present one, is that, although the applications for which it is needed are very real and applied, the underlying basis of complex variable theory has a distinctly pure mathematics flavour.

To adopt this more rigorous approach correctly would involve developing a large amount of groundwork in analysis, for example, precise definitions of continuity and differentiability, the theory of sets and a detailed study of boundedness. It has been decided not to do so here, but rather to pursue only those parts of the formal theory which are needed to establish the results used elsewhere in this book and some others of general utility. Specifically, the subjects treated are,

(i) complex potentials for two-dimensional potential problems,
(ii) location of zeros of a function, in particular a polynomial,
(iii) summation of series and evaluation of integrals,
(iv) the inverse Laplace transform integral.

In this spirit, the proofs that are adopted for some of the standard results of complex variable theory have been chosen with an eye to simplicity rather than sophistication. This means that in some cases the imposed conditions are more stringent than would be strictly necessary if more sophisticated proofs were used; where this happens the less restrictive results are usually stated as well. Some proofs have been omitted altogether or merely sketched in exercises for the student to do. The reader who is interested in a fuller treatment of the fascinating subject of complex variable theory should consult one of the many excellent textbooks on the subject.†

† For example, Knopp, *Theory of functions*, Part I (Dover, 1945); Phillips, *Functions of a complex variable* (Oliver & Boyd, 1954); Titchmarsh, *The theory of functions* (Oxford, 1952).

One further concession to 'hand-waving' has been made in the interests of keeping the treatment to a moderate length. In several places phrases such as '... can be made as small as we like ...' are used, rather than a careful treatment in terms of '... given $\epsilon > 0$, there exists a $\delta > 0$ such that ...'. In the author's experience, some students are more at ease with the former type of statement despite its lack of precision, whilst others, who would only contemplate the latter, are usually well able to supply it for themselves.

16.1   Functions of a complex variable

The quantity $f(z)$ is said to be a function of the complex variable z if to every value of z in a certain domain (region of the Argand diagram) there corresponds one or more values of $f(z)$. Stated like this $f(z)$ could be any function consisting of a real and an imaginary part, each of which is itself a function of x and y. However, we will only be concerned ultimately with functions which are differentiable in a particular sense, and so we proceed immediately to this restricted class of functions.

A function $f(z)$ which is single-valued in a domain is **differentiable** at the point z_0 if the *derivative*

$$f'(z_0) \equiv \lim_{z \to z_0} \frac{f(z) - f(z_0)}{z - z_0} \tag{16.1}$$

exists and is unique (in that it does not depend upon the direction in the Argand diagram from which z tends to z_0).

To illustrate that this notion is restrictive, consider the following two examples which are both complex functions of a complex variable; one is differentiable, the other is not. We denote the real and imaginary parts of $f(z)$ by u and v respectively; for a general function both will depend on x and y,

$$f(z) = u(x, y) + iv(x, y). \tag{16.2}$$

(i) Take $f(z) = x^2 - y^2 + i\,2xy$.

Consider the definition (16.1) when $z = z_0 + \Delta z = x_0 + \Delta x + i(y_0 + \Delta y)$; then

$$\frac{f(z) - f(z_0)}{z - z_0} = \frac{(x_0 + \Delta x)^2 - (y_0 + \Delta y)^2 + i2(x_0 + \Delta x)(y_0 + \Delta y) - x_0^2 + y_0^2 - i2x_0y_0}{\Delta x + i\,\Delta y}$$

$$= \frac{2x_0 \, \Delta x + (\Delta x)^2 - 2y_0 \, \Delta y - (\Delta y)^2 + i2(x_0 \, \Delta y + y_0 \, \Delta x + \Delta x \, \Delta y)}{\Delta x + i \, \Delta y}$$

$$= 2x_0 + i2y_0 + \frac{(\Delta x)^2 - (\Delta y)^2 + i2 \, \Delta x \, \Delta y}{\Delta x + i \, \Delta y},$$

$$(\text{using } -1 = i^2).$$

Now, in whatever way Δx and Δy are allowed to tend to zero, the last term on the right will tend to zero and the unique limit $2x_0 + i \, 2y_0$ will be obtained. Thus $f(z)$ with $u = x^2 - y^2$ and $v = 2xy$ is differentiable at $z_0 = x_0 + iy_0$.

(ii) Take $f(z) = 2y + ix$. Following the same procedure we have

$$\frac{f(z) - f(z_0)}{z - z_0} = \frac{2y_0 + 2 \, \Delta y + ix_0 + i \, \Delta x - 2y_0 - ix_0}{\Delta x + i \, \Delta y}$$

$$= \frac{2 \, \Delta y + i \, \Delta x}{\Delta x + i \, \Delta y}.$$

Now suppose $z \to z_0$ along a line through z_0 of slope m, so that $\Delta y = m \, \Delta x$. Then

$$\lim_{z \to z_0} \frac{f(z) - f(z_0)}{z - z_0} = \lim_{\Delta x, \Delta y \to 0} \frac{2 \, \Delta y + i \, \Delta x}{\Delta x + i \, \Delta y} = \frac{2m + i}{1 + im}.$$

This limit is dependent on m and hence on the direction from which z tends to z_0. Thus $f(z) = 2y + ix$ is not a differentiable function at $x_0 + iy_0$.

A function which is one-valued and differentiable at all points of a domain D is said to be **analytic** (or **regular**) in D. A function may be analytic in a domain except at a finite (or infinite if the domain is infinite) number of points; in this case it is said to be analytic except at these points, which are called the **singularities** of $f(z)$. [In our treatment we will not consider cases in which an infinite number of singularities occur in a finite domain.]

From examining the two previous examples, it is apparent that for a function $f(z)$ to be differentiable and hence analytic there must be some particular connection between its real and imaginary parts u and v. We next establish what this connection must be by repeating the procedures of the two examples, but for a general function.

If the limit

$$L = \lim_{z \to z_0} \frac{f(z) - f(z_0)}{z - z_0} \tag{16.3}$$

is to exist and be unique, in the way required for differentiability, then two particular ways of letting z tend to z_0 (any point in D), parallel to the real axis and parallel to the imaginary axis, must produce the same limit. This is certainly a necessary result although it may not be sufficient.

So first suppose $z - z_0$ is purely real and equal to Δx, then L of (16.3) is given by

$$L = \lim_{\Delta x \to 0} \frac{u(x_0 + \Delta x, y_0) + iv(x_0 + \Delta x, y_0) - u(x_0, y_0) - iv(x_0, y_0)}{\Delta x}$$

$$= \lim_{\Delta x \to 0} \left[\frac{u(x_0 + \Delta x, y_0) - u(x_0, y_0)}{\Delta x} + i \frac{v(x_0 + \Delta x, y_0) - v(x_0, y_0)}{\Delta x} \right].$$

For this limit to exist we must have that $\partial u/\partial x$ and $\partial v/\partial x$ exist at $x_0 + iy_0$ and that L has the value

$$\frac{\partial u}{\partial x} + i \frac{\partial v}{\partial x}. \tag{16.4 a}$$

Similarly if $z - z_0$ is purely imaginary and equal to $i \Delta y$, then the partial derivatives $\partial u/\partial y$ and $\partial v/\partial y$ must exist at $x_0 + iy_0$ and L must have the value

▶ 1.
$$\frac{\partial v}{\partial y} - i \frac{\partial u}{\partial y}. \tag{16.4 b}$$

For f to be differentiable, expressions (16.4 a) and (16.4 b) for L must be identical, and thus equating real and imaginary parts we must have, as a *necessary* condition, that

$$\frac{\partial u}{\partial x} = \frac{\partial v}{\partial y}, \qquad \frac{\partial v}{\partial x} = -\frac{\partial u}{\partial y}, \qquad \text{at } z = x_0 + iy_0. \tag{16.5}$$

The two equations are known as the **Cauchy–Riemann equations**.

We can now see why example (i) $f(z) = x^2 - y^2 + i\, 2xy$ was differentiable, or at least why example (ii) $f(z) = 2y + ix$ was not.

(i) $u = x^2 - y^2$, $v = 2xy$:

$$\frac{\partial u}{\partial x} = 2x = \frac{\partial v}{\partial y} \qquad \text{and} \qquad \frac{\partial v}{\partial x} = 2y = -\frac{\partial u}{\partial y},$$

(ii) $u = 2y$, $v = x$:

$$\frac{\partial u}{\partial x} = 0 = \frac{\partial v}{\partial y} \qquad \text{but} \qquad \frac{\partial v}{\partial x} = 1 \neq -2 = -\frac{\partial u}{\partial y}.$$

It is apparent that for $f(z)$ to be analytic something more than the existence of the partial derivatives of u and v with respect to x and y is required; this something is that they satisfy the Cauchy–Riemann equations.

We may also enquire as to the *sufficient* conditions for $f(z)$ to be analytic in D. It can be shown† that a sufficient condition is that the four partial derivatives exist, *are continuous* and satisfy the Cauchy–Riemann equations. It is the additional requirement of continuity which makes the difference between the necessary conditions and the sufficient conditions.

▶2. Which of the following complex functions are analytic?

 (a) $3x + 2y + i(3y - 2x)$,

 (b) $\sin x \cosh y + i \cos x \sinh y$,

 (c) $x^2 + y^2$,

 (d) $x + \dfrac{x}{x^2 + y^2} + i\left(y - \dfrac{y}{x^2 + y^2} \right).$

▶3. In which domain(s) of the Argand diagram is

$$f = |x| - i|y|$$

an analytic function?

Since

$$x = \frac{1}{2}(z + z^*) \quad \text{and} \quad y = \frac{1}{2i}(z - z^*), \tag{16.6}$$

we may formally regard any function $f = u + iv$ as a function of z and its conjugate z^*, rather than x and y. If we do this and examine $\partial f/\partial z^*$ we obtain

$$\frac{\partial f}{\partial z^*} = \frac{\partial f}{\partial x}\frac{\partial x}{\partial z^*} + \frac{\partial f}{\partial y}\frac{\partial y}{\partial z^*}$$

$$= \left(\frac{\partial u}{\partial x} + i\frac{\partial v}{\partial x}\right) \times \frac{1}{2} + \left(\frac{\partial u}{\partial y} + i\frac{\partial v}{\partial y}\right)\left(-\frac{1}{2i}\right)$$

$$= \frac{1}{2}\left(\frac{\partial u}{\partial x} - \frac{\partial v}{\partial y}\right) + \frac{i}{2}\left(\frac{\partial v}{\partial x} + \frac{\partial u}{\partial y}\right). \tag{16.7}$$

Now if f is analytic, the Cauchy–Riemann equations (16.5) must be satisfied, and these immediately give that $\partial f/\partial z^*$ is identically zero. Thus we conclude that if f is analytic, then f cannot be a function of z^* and any expression representing an analytic function of z can contain x and y only in the combination $x + iy$ [*not* in the combination $x - iy$].

 † See for example any of the references given earlier.

►4. Write the functions in ►2 in terms of z and z^* and hence verify the results obtained there.

One further result of great practical importance in the theoretical physics can be obtained simply from the satisfying of the Cauchy–Riemann equations by the real and imaginary parts of an analytic function. Differentiating one equation again with respect to one independent variable, and the other with respect to the other, we obtain

$$\frac{\partial}{\partial x}\left(\frac{\partial u}{\partial x}\right) = \frac{\partial}{\partial x}\left(\frac{\partial v}{\partial y}\right) = \frac{\partial}{\partial y}\left(\frac{\partial v}{\partial x}\right) = -\frac{\partial}{\partial y}\left(\frac{\partial u}{\partial y}\right),$$

and $$\frac{\partial}{\partial x}\left(\frac{\partial v}{\partial x}\right) = -\frac{\partial}{\partial x}\left(\frac{\partial u}{\partial y}\right) = -\frac{\partial}{\partial y}\left(\frac{\partial u}{\partial x}\right) = -\frac{\partial}{\partial y}\left(\frac{\partial v}{\partial y}\right).$$

Thus *both* u and v are separately solutions of Laplace's equation in two dimensions

$$\nabla^2\phi = \frac{\partial^2\phi}{\partial x^2} + \frac{\partial^2\phi}{\partial y^2} = 0. \tag{16.8}$$

Further use of this will be made in section 16.4.

16.2   Power series in a complex variable

The theory of power series in a real variable was discussed in section 3 of chapter 1. A natural extension of this in the present context is to consider a series such as

$$f(z) = \sum_{n=0}^{\infty} a_n z^n, \tag{16.9}$$

where z is the complex variable and the a_n are in general complex. Expression (16.9) is a power series about the origin and may be used for general discussion; a power series about any other point z_0 is obtained by a change of variable from z to $z - z_0$.

If z were written in its modulus and argument form $z = r \exp(i\theta)$, expression (16.9) would be

$$f(z) = \sum_{n=0}^{\infty} a_n r^n \exp(in\theta). \tag{16.10}$$

This series is absolutely convergent if the same can be said of

$$\sum_{n=0}^{\infty} |a_n| r^n, \tag{16.11}$$

which is a series of positive real terms. Tests for the absolute convergence of real series can thus be used, and of these the most appropriate form is based on the Cauchy root test. A **radius of convergence** R is defined by

$$1/R = \overline{\lim_{n \to \infty}} |a_n|^{1/n}. \tag{16.12}$$

The series (16.9) is absolutely convergent if $|z| < R$ and divergent if $|z| > R$; if $|z| = R$ no particular conclusion may be drawn. [$\overline{\lim}$ is not strictly identical with lim, but for our purposes the distinction may be ignored.]

A circle of radius R and centred on the origin is called the **circle of convergence** of the series $\sum a_n z^n$. $R = 0$ and $R = \infty$ correspond respectively to convergence at the origin only, and convergence everywhere. For R finite the convergence occurs in a restricted part of the z-plane (Argand diagram). For a power series about a general point z_0, the circle of convergence is of course centred on that point.

Example 16.1. Find the parts of the z-plane for which the following series are convergent

$$\text{(i)} \sum_0^\infty \frac{z^n}{n!}, \quad \text{(ii)} \sum_0^\infty n! z^n, \quad \text{(iii)} \sum_1^\infty \frac{z^n}{n}.$$

(i) $(n!)^{1/n}$ behaves like n as $n \to \infty$. Thus $\overline{\lim} (1/n!)^{1/n} = 0$. Hence $R = \infty$ and the series is convergent for all z.
(ii) Correspondingly, $\overline{\lim} (n!)^{1/n} = \infty$. Thus $R = 0$ and the series converges only at $z = 0$.
(iii) As $n \to \infty$, $(n)^{1/n}$ has a lower limit of 1 and hence $\overline{\lim} (1/n)^{1/n} = 1/1 = 1$. Thus the series is absolutely convergent if $|z| < 1$.

▶5. By taking $z = 1$ and $z = -1$ in case (iii) above, demonstrate that a power series may or may not converge on its circle of convergence.

The ratio test may also be employed to investigate the absolute convergence of a complex power series. The series is absolutely convergent if

$$1 > \overline{\lim} \frac{|a_{n+1}| |z|^{n+1}}{|a_n| |z|^n} = \overline{\lim} \frac{|a_{n+1}| |z|}{|a_n|}. \tag{16.13}$$

For example in case (i) of example 16.1,

$$\frac{|a_{n+1}| |z|}{|a_n|} = \frac{n! |z|}{(n+1)!} = \frac{|z|}{n+1},$$

which is < 1 for all $n > n_0$ if n_0 is sufficiently large [n_0 will depend upon $|z|$]. Thus the series is absolutely convergent for all (finite) z, confirming the previous result.

▶6. Apply the ratio test to cases (ii) and (iii) of example 16.1.

Before turning to particular power series, we next prove the following important theorem.

> The power series $\sum_0^\infty a_n z^n$ has a sum which is an analytic
> function of z inside its circle of convergence.   (16.14)

To prove this write $f(z)$ for $\sum a_n z^n$ with $|z| = r < R$ and consider

$$\frac{f(z + h) - f(z)}{h} = \sum_0^\infty a_n \left[\frac{(z + h)^n - z^n}{h} \right],$$

where $\rho = |h|$ is small enough that $r + \rho$ is also $< R$. [This is always possible since z is *inside* the circle.] Now

$$\left| \frac{(z + h)^n - z^n}{h} - nz^{n-1} \right| = \rho \left| \binom{n}{2} z^{n-2} + \binom{n}{3} z^{n-3} h + \cdots \right.$$

$$\left. + \binom{n}{n} h^{n-2} \right|$$

$$\leqslant \rho \binom{n}{2} \left| r^{n-2} + \binom{n-2}{1} r^{n-3} \rho + \cdots \right.$$

$$\left. + \binom{n-2}{n-2} \rho^{n-2} \right|$$

$$= \rho \binom{n}{2} (r + \rho)^{n-2}.$$

The inequality used here is justified by the observation that

$$\binom{n}{m} = \frac{n!}{m!(n-m)!} \leqslant \frac{n!}{2!(m-2)!(n-m)!}$$

$$= \binom{n}{2}\binom{n-2}{m-2} \quad \text{for } m \geqslant 2.$$

Thus

$$\left| \frac{f(z + h) - f(z)}{h} - \sum_0^\infty na_n z^{n-1} \right| \leqslant \rho \left[\sum_0^\infty |a_n| \binom{n}{2} (r + \rho)^{n-2} \right].$$

$$(16.15)$$

Now since $\sum a_n z^n$ has radius of convergence R, $\overline{\lim} (a_n)^{1/n} = R^{-1}$. Thus, since $n^{1/n} \to 1$ as $n \to \infty$, $\overline{\lim} (\frac{1}{2}n(n-1)a_n)^{1/n}$ must also $= R^{-1}$. Therefore the series in brackets on the right-hand side of (16.15) converges and consequently is bounded, $< M$ say, where M does not depend on ρ.

So finally in the limit $h \to 0$, the right-hand side of (16.15) tends to zero and shows that

$$\lim_{h \to 0} \frac{f(z + h) - f(z)}{h} = \sum_{0}^{\infty} n a_n z^{n-1}. \tag{16.16}$$

This establishes statement (16.14) and also shows (by repeated application) that any power series can be differentiated any number of times inside its circle of convergence.

16.3 Some elementary functions

In example 16.1 it was shown that the function exp (z) *defined* by

$$\exp(z) = \sum_{0}^{\infty} \frac{z^n}{n!} \tag{16.17}$$

is convergent for all z of finite modulus and is thus by the previous section an analytic function over the whole z-plane.† Like its real variable counterpart it is called the **exponential function**; also like its real counterpart it is equal to its own derivative.

The multiplication of two exponential functions results in a further exponential function, in accordance with the corresponding result for real variables.

▶7. By considering the coefficients of $z_1^r z_2^s$ in exp $(z_1) \cdot$ exp (z_2) and exp $(z_1 + z_2)$, and noting that all series involved are absolutely convergent for all z, show that

$$\exp(z_1) \cdot \exp(z_2) = \exp(z_1 + z_2). \tag{16.18}$$

As an extension of (16.17) we may also define the complex exponent of a real number $a > 0$ by the equation

$$a^z = \exp(z \ln a), \tag{16.19}$$

where $\ln a$ is the natural logarithm of a. The particular case $a = e$ and the fact that $\ln e = 1$, enables us to write exp (z) interchangeably with e^z. If z is real, the definition agrees with the familiar one.

The result that when $z = iy$,

$$\exp(iy) = \cos y + i \sin y, \tag{16.20}$$

has already been met in equation (1.15 a) of chapter 1. Its immediate extension is that

$$\exp(z) = \exp(x)(\cos y + i \sin y). \tag{16.21}$$

† Functions which are analytic in the *whole* z-plane are usually called *integral* functions.

As z varies over the complex plane the modulus of exp (z) takes all real positive values, except that of 0. However, two values of z which differ by $2\pi n$i, for any integral n, produce the same value of exp (z), as given by (16.21), and so exp (z) is periodic with period 2πi. If we denote exp (z) by t, then the strip in the z-plane, $-\pi < y \leqslant \pi$ corresponds to the whole of the t-plane, except for the point $t = 0$.

▶8. To which regions of the t-plane do the following in the z-plane correspond? (i) $-\infty < x < 0$, $-\pi < y \leqslant \pi$, (ii) $0 < x < \infty$, $-\pi < y \leqslant \pi$, (iii) the half-ray arg $z = \theta$ with $0 < \theta < \pi/2$, (iv) the half-ray arg $z = \theta$ with $\pi/2 < \theta < \pi$.

The **sine**, **cosine**, **sinh** and **cosh** functions of a complex variable are defined from the exponential function exactly as are those for real variables. The functions derived from them (e.g. tan and tanh), the identities they satisfy, and their derivative properties, are also just as for real variables. In view of this we will not give them further attention here.

The 'inverse function' of exp (z) is given by w the solution of

$$\exp(w) = z. \tag{16.22}$$

By virtue of the discussion following (16.21), w is not uniquely defined, since it is indeterminate to the extent of any integral multiple of 2πi. If we denote w by

$$w = \text{Log } z = \log |z| + \text{i arg } z, \tag{16.23}$$

where $\log |z| \equiv \ln |z|$ is the natural logarithm (to base e) of the real positive quantity $|z|$, then Log z is an infinitely many-valued function of z. Its **principal value** is obtained by giving arg z its principal value $(-\pi < \text{arg } z \leqslant \pi)$, and is denoted by log z. Thus

$$\log z = \log |z| + \text{i}\theta, \quad \text{with } -\pi < \theta \equiv \text{arg } z \leqslant \pi. \tag{16.24}$$

Now that a logarithm of a complex variable has been defined, definition (16.19) of a general power can be extended to cases other than those in which a is real and positive. If t $(\neq 0)$ and z are both complex, the zth power of t is defined by

$$t^z = \exp(z \text{ Log } t). \tag{16.25}$$

Since Log t is multiple valued, so is this definition. Its principal value is obtained by giving Log t its principal value, log t.

If t $(\neq 0)$ is complex but z is real and equal to $1/n$, then (16.25) provides a definition of the **nth root** of t. Because of the multiple-valuedness of Log t, there will be more than one nth root of any given t.

►9. Show that there are exactly n distinct nth roots of t.

In the definition of an analytic function, one of the conditions imposed was that the function was single valued. Now the logarithmic function, a complex power and a root, are all multiple valued. However it happens that the properties of analytic functions can still be applied to these and other multiple-valued functions of a complex variable provided suitable care is taken. This care amounts to recognizing that if z is varied in such a way that its path in the Argand diagram forms a closed curve enclosing particular points (dependent upon the function in question) of the diagram, then the function will not return to its original value. A general discussion of how this is done, using a 'cut' plane, is beyond the scope of this book and is not needed in a general form for the particular applications of later sections. Again the reader is referred to more complete texts, e.g. the references given at the beginning of this chapter.

►10. Evaluate the following, (a) Re exp $(2iz)$, (b) Im $\cosh^2 z$, (c) $(-1 + \sqrt{3}i)^{1/2}$, (d) $|\exp(i^{1/2})|$, (e) exp (i^3), (f) Im 2^{i+3}, (g) i^i, (h) Log $[(\sqrt{3} + i)^3]$.

►11. Verify that d $(\exp(z))/dz = \exp(z)$ and deduce that

$$\frac{d}{dz}(\text{Log } z) = \frac{1}{z}.$$

16.4   Complex potentials and conformal transformations

At the end of section 16.1 it was shown that both the real and the imaginary parts of an analytic function of z are separately solutions of Laplace's equation in two dimensions. Analytic functions thus offer a possible way of solving some two-dimensional physical problems describable by a potential satisfying $\nabla^2\phi = 0$. The general method is known as that of **complex potentials**. As preliminaries to showing that they can be so used, some further properties of functions of a complex variable and their behaviour under 'transformations' will be established.

If, as previously, we denote an analytic function $f(z)$ by $u + iv$, then:

The contours of constant u and constant v are orthogonal.

On a curve $u =$ constant in the z-plane

$$0 = \frac{du}{dx} = \frac{\partial u}{\partial x} + \frac{\partial u}{\partial y} \times \frac{dy}{dx}\bigg|_u, \tag{16.26}$$

and similarly on a curve $v = $ constant,

$$\left.\frac{dy}{dx}\right|_v = -\left(\frac{\partial v}{\partial x}\right)\Big/\left(\frac{\partial v}{\partial y}\right). \tag{16.27}$$

Thus

$$\left.\frac{dy}{dx}\right|_u \times \left.\frac{dy}{dx}\right|_v = \left[-\left(\frac{\partial u}{\partial x}\right)\Big/\left(\frac{\partial u}{\partial y}\right)\right] \times \left[-\left(\frac{\partial v}{\partial x}\right)\Big/\left(\frac{\partial v}{\partial y}\right)\right]. \tag{16.28}$$

But, because f is analytic, the Cauchy–Riemann equations imply that $\partial u/\partial x = \partial v/\partial y$ and $\partial u/\partial y = -\partial v/\partial x$, and hence that the right-hand side of (16.28) is equal to -1. However, this is just the condition that the tangents to the two curves $u = $ constant and $v = $ constant [whose slopes are given by dy/dx] meet at right angles. This establishes the result.

In the context of solutions of Laplace's equation, this result implies that the real and imaginary parts of $f(z)$ have an additional connection between them, for if the set of contours on which one of them is a constant represents the equipotentials of a system, the contours on which the other is constant, being orthogonal to each of the first set, must represent the *corresponding* field lines [or stream lines, depending on the context].

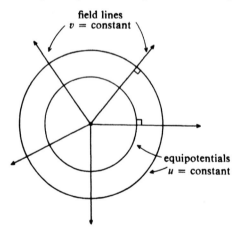

Fig. 16.1 The equipotentials and field lines for a line charge perpendicular to the plane of the paper.

As an example consider the function

$$f(z) = \frac{q}{2\pi\epsilon_0}\,\log z, \tag{16.29}$$

in connection with the physical situation of a line charge of strength q per unit length passing through the origin, perpendicular to the z-plane (fig. 16.1). Its real and imaginary parts are

$$u = \frac{q}{2\pi\epsilon_0} \log |z|, \qquad v = \frac{q}{2\pi\epsilon_0} \arg z. \tag{16.30}$$

The contours in the z-plane of $u = $ constant are concentric circles and of $v = $ constant are radial lines. As expected these are orthogonal sets, but also they are respectively the equipotentials and field lines appropriate to the field produced by the line charge.

Suppose we make the choice that the real part u of the analytic function f shall give the conventional potential function [v could equally well be selected]. Then we may consider how the field direction and magnitude is related to f.

Because $u = $ constant is an equipotential, the field has components [we use the electrostatic case for definiteness]

$$E_x = -\frac{\partial u}{\partial x} \quad \text{and} \quad E_y = -\frac{\partial u}{\partial y}. \tag{16.31}$$

Since f is analytic, (i) we may use the Cauchy–Riemann equations to change the second of these

$$E_x = -\frac{\partial u}{\partial x} \quad \text{and} \quad E_y = \frac{\partial v}{\partial x}, \tag{16.32}$$

and (ii) the direction of differentiation at a point is immaterial and so

$$\frac{df}{dz} = \frac{\partial f}{\partial x} = \frac{\partial u}{\partial x} + i\frac{\partial v}{\partial x} = -E_x + iE_y. \tag{16.33}$$

From these it can be seen that the field at a point is given in magnitude by

$$E = \left|\frac{df}{dz}\right|, \tag{16.34 a}$$

and that it makes an angle with the x-axis given by

$$\pi - \arg\frac{df}{dz}. \tag{16.34 b}$$

It is apparent that much of physical interest can be calculated by working in terms of f and z directly.

Next, we turn our attention to the subject of 'transformations' by which we mean a change of coordinates from the complex variable z to another one $t \equiv r + is$ by means of a prescribed formula

$$z = g(t). \tag{16.35}$$

Under such a **transformation**, or **mapping**, the Argand diagram for the z-variable is transformed into one for the t-variable, although the complete z-plane may be 'mapped' onto only a part of the t-plane, or onto the whole of the t-plane, or onto some or all of the t-plane covered more than once. An example appropriate to the particular transformation $t = \exp(z)$ was discussed briefly in section 16.3 and ▶8.

We consider only those mappings for which z and t are related by an analytic function g; such mappings are called **conformal**. The important points about them are that, *except* at points at which $g'(t) = 0$ or is infinite:

1. Continuous lines in the z-plane transform into continuous lines in the t-plane.
2. Any analytic function of z transforms to an analytic function of t.
3. The magnification, as between the z- and t-plane, of a small line element in the neighbourhood of any particular point is independent of the direction of the element.
4. The angle between two intersecting curves in the z-plane equals the angle between the corresponding curves in the t-plane.

Result 1 is immediate and result 2 is almost self-evident, since if $f(z)$ is analytic in z (i.e. an analytic function of z) and $z = g(t)$ is analytic in t, then $F(t) = f(g(t))$ is analytic in t. Its importance lies in the fact that the real and imaginary parts of $F(t)$ are, since F is analytic, necessarily solutions of

$$\frac{\partial^2 \phi}{\partial r^2} + \frac{\partial^2 \phi}{\partial s^2} = 0. \tag{16.36}$$

Further, suppose (say) $\operatorname{Re} f(z)$ is constant over a boundary B in the z-plane. Then $\operatorname{Re} F(t)$ is constant over B in the z-plane. But this is the same as saying that $\operatorname{Re} F(t)$ is constant over the boundary B' in the t-plane, B' being the curve into which B is transformed by (16.35).

Taking these results together shows that if a solution of Laplace's equation, constant over a particular boundary, can be found as the real or imaginary part of an analytic function† of z in the xy-plane, then the same expression put in terms of r and s will be a solution of Laplace's equation in the rs-plane and in addition will be constant over the corresponding boundary curve expressed in terms of r and s. Thus from any two-dimensional solution of Laplace's equation derived from an analytic function, for a particular geometry, further solutions for some other geometries can be obtained by making conformal transformations. Naturally the initial geometry is usually simple and the final one more

---

† In fact the original potential function need not be explicitly given as the real or imaginary part of an analytic function. Any solution of $\nabla^2 \phi = 0$ is carried over into another solution of $\nabla^2 \phi = 0$ in the new variables by a conformal transformation. A lengthier proof is needed in this case.

complicated. From the physical point of view the problem is the converse, since the geometry is usually given and the solution sought. However, experience working from simpler to more complicated situations is necessary before the reverse procedure can be successfully tackled.

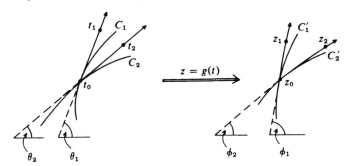

Fig. 16.2 Under the transformation $z = g(t)$ two curves C_1 and C_2 passing through t_0 become two curves passing through $z_0 = g(t_0)$. The angle between the curves is unchanged by the transformation.

The final two results, 3 and 4, can be justified by the following argument. Let fig. 16.2 show two curves C_1 and C_2 passing through the point t_0 in the t-plane and t_1 and t_2 be two points on their respective tangents at t_0, each distance ρ from t_0. The same prescription with z replacing t describes the transformed situation; however, the transformed tangents may not be straight lines and the distances of z_1 and z_2 from z_0 have not yet been shown to be equal.

In the t-plane t_1 and t_2 are given by

$$t_1 - t_0 = \rho \exp(i\theta_1) \quad \text{and} \quad t_2 - t_0 = \rho \exp(i\theta_2).$$

The corresponding descriptions in the z-plane are

$$z_1 - z_0 = \rho_1 \exp(i\phi_1) \quad \text{and} \quad z_2 - z_0 = \rho_2 \exp(i\phi_2).$$

The angles θ_i and ϕ_i are clear from fig. 16.2.

Now since $z = g(t)$ and is analytic

$$\lim_{t_1 \to t_0} = \frac{z_1 - z_0}{t_1 - t_0} = \lim_{t_2 \to t_0} \frac{z_2 - z_0}{t_2 - t_0} = \frac{dg}{dt},$$

i.e.

$$\lim_{\rho \to 0} \frac{\rho_1}{\rho} \exp[i(\phi_1 - \theta_1)] =$$

$$\lim_{\rho \to 0} \frac{\rho_2}{\rho} \exp[i(\phi_2 - \theta_2)] = g'(t). \quad (16.37)$$

Comparing magnitudes and phases (arguments) in the equalities (16.37) gives the stated results 3 and 4 and adds quantitative information to them.

Namely that for *small* line elements

$$(\rho_1/\rho) \simeq (\rho_2/\rho) \simeq |g'(t)|, \tag{16.38 a}$$
$$\phi_1 - \theta_1 \simeq \phi_2 - \theta_2 \simeq \arg g'(t). \tag{16.38 b}$$

For strict comparison with result 4, (16.38 b) must be written as $\theta_1 - \theta_2 = \phi_1 - \phi_2$, with an ordinary equality sign, since the angles are only defined in the limit $\rho \to 0$ when (16.38 b) becomes a true identity.

Since in the neighbourhoods of corresponding points in a transformation, angles are preserved and magnifications are independent of direction, it follows that small plane figures are transformed into figures of the same shape, but in general with magnification and rotation [but no distortion].

16.5   Examples of complex potentials

Since one of the best ways of explaining a mathematical method is to use it, this section consists of worked examples and exercises using complex potentials in electrostatics.

Example 16.2. Find (i) the complex electrostatic potential associated with an infinite charged conducting plate ($s = 0$), and thus obtain those associated with:
(ii) a semi-infinite charged conducting plate ($x > 0, y = 0$),
(iii) the inside of a right-angled charged conducting wedge ($x > 0, y = 0$ and $x = 0, y > 0$).

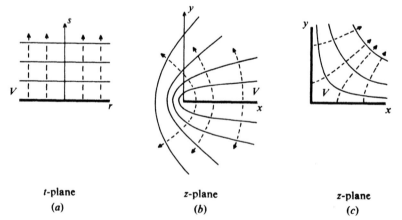

t-plane   z-plane   z-plane
(a)   (b)   (c)

Fig. 16.3 (a) The equipotentials (solid lines) and field lines (broken) for an infinite charged conducting plane at $s = 0$, where $t = r + is$; (b) after the transformation $z = t^2$; (c) after the transformation $z = t^{1/2}$ of the situation shown in (a).

(i) Figure 16.3 (a) shows the equipotentials (solid lines) and field lines (broken) for an infinite charged conducting plane $s = 0$. Suppose we elect to make the real part of the complex potential coincide with the conventional electrostatic potential. This latter is clearly

$$\phi(r, s) = V - ks, \tag{16.39}$$

where k is related to the charge density σ by $k = \sigma/\epsilon_0$, since physically $\mathbf{E} = (0, \sigma/\epsilon_0)$ and $\mathbf{E} = -\nabla\phi$.

Thus what is needed is an analytic function of t of which the real part is $V - ks$. This can be obtained by inspection, but we may proceed formally and use the Cauchy–Riemann equations to obtain the imaginary part $\psi(r, s)$ thus:

$$\frac{\partial\psi}{\partial s} = \frac{\partial\phi}{\partial r} = 0 \quad \text{and} \quad \frac{\partial\psi}{\partial r} = -\frac{\partial\phi}{\partial s} = k.$$

Hence $\psi = kr + c$, and, absorbing c into V,

$$f(t) = V - ks + ikr = V + ikt. \tag{16.40}$$

This is the required complex potential.

(ii) Now consider the transformation

$$z = g(t) = t^2. \tag{16.41}$$

This satisfies the criteria for a conformal mapping (except at $t = 0$) and carries the upper half of the t-plane into the entire z-plane, with the equipotential line (plane) $s = 0$ going into the half line $x > 0$, $y = 0$.

By the general results proved, $f(t)$ when expressed in terms of x and y will give a complex potential of which the real part will be constant on the half line in question;

$$F(z) = f(t) = V + ikt = V + ikz^{1/2} \tag{16.42}$$

is thus the required potential. Expressed in terms of x, y and $\rho = (x^2 + y^2)^{1/2}$, $z^{1/2}$ is

▶12. $$z^{1/2} = \rho^{1/2}\left[\left(\frac{\rho + x}{2\rho}\right)^{1/2} + i\left(\frac{\rho - x}{2\rho}\right)^{1/2}\right], \tag{16.43}$$

and so, for example, the electrostatic potential is given by

$$\text{Re } F(z) = V - \frac{k}{\sqrt{2}}\,[(x^2 + y^2)^{1/2} - x]^{1/2}. \tag{16.44}$$

The corresponding equipotentials and field lines are roughly sketched in fig. 16.3 (b).

(iii) A 'converse' transformation to that used in (ii),

$$z = g(t) = t^{1/2},$$

has the effect of mapping the upper half t-plane into the first quadrant of the z-plane and the conducting plane $s = 0$ into the wedge $x > 0, y = 0$ and $x = 0, y > 0$.

The complex potential now becomes

$$F(z) = V + ikz^2$$
$$= V + ik[(x^2 - y^2) + i2xy], \qquad (16.45)$$

showing that the electrostatic potential is

$$V - 2kxy, \qquad (16.46)$$

and the electric field has components

$$\mathbf{E} = (2ky, 2kx). \qquad (16.47)$$

Fig. 16.3 (c) indicates the approximate equipotentials and field lines. [Note that in both transformations no violation at the origin of result 4 of the previous section occurs, since $g'(t)$ is either 0 or ∞ there and so the conditions for result 4 are not satisfied.]

▶13. Use the results expressed in (16.34 a, b) to show directly

(a) in (iii) that $|\mathbf{E}| = 2k(x^2 + y^2)^{1/2}$,
(b) in (ii) that the charge density on the plate $\propto x^{-1/2}$,
(c) in (ii) that $|\mathbf{E}| = \frac{1}{2}k(x^2 + y^2)^{-1/4}$,
(d) in (iii) that the direction of $\mathbf{E}$ is at any point complementary to that of the radius vector to that point. [Their sum is $\pi/2$.]

▶14. Carry out (a), (b) and (c) of ▶13 in terms of x and y.

▶15. (i) Verify that $f(t) = E(t - a^2t^{-1})$ is the complex potential appropriate to a conducting circular cylinder of radius a placed perpendicular to a uniform electric field E.
(ii) Show that the transformation $z = t + \frac{1}{2}a^2t^{-1}$ transforms the circular cylinder into an elliptical one of semi-axes $\frac{3}{2}a$ and $\frac{1}{2}a$.
(iii) Show that the only 'singular' (non-conformal) points of the transformation are located inside the cylinder.
(iv) Deduce that the potential appropriate to the elliptic cylinder is

$$F(z) = \frac{3}{2}E(z^2 - 2a^2)^{1/2} - \frac{1}{2}Ez.$$

(v) Verify that on the elliptic cylinder, $\mathrm{Re}\, F(z) = 0$, as it should be.

16.6   Complex integrals

Corresponding to the integral of a real variable, the integral of a complex variable between two (complex) limits can be defined. Since the z-plane is two dimensional there is clearly greater freedom and hence ambiguity in what is meant by a complex integral. For example, in fig. 16.4, we might ask, 'Does the integral of some function $f(z)$ between A and B mean one involving the values of f at points along the straight line C_1, or along the curved lines C_2 or C_3, or does it not make any difference anyway?' What will be found is that in general they will have different values, i.e. in general, the value of the integral depends upon the path adopted in

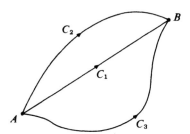

Fig. 16.4 Alternative paths for an integral of a function $f(z)$ between A and B.

the complex plane. However, it will also be found that for different paths bearing a particular relationship to each other the value of the integral does *not* depend upon which of the paths is adopted.

Let a particular path C be described by a continuous parameter q ($\alpha \leqslant q \leqslant \beta$) which gives successive positions on C by means of the equations

$$x = x(q), \qquad y = y(q), \tag{16.48}$$

with $q = \alpha$ and $q = \beta$ corresponding to the points A and B respectively. Then the integral along path C of a continuous function $f(z)$ is written

$$\int_C f(z)\,\mathrm{d}z, \tag{16.49}$$

and is given more explicitly as the sum of the four real integrals obtained as follows:

$$\int_C f(z)\,\mathrm{d}z = \int_C (u + iv)(\mathrm{d}x + i\,\mathrm{d}y)$$

$$= \int_C u\,\mathrm{d}x - \int_C v\,\mathrm{d}y + i\int_C u\,\mathrm{d}y + i\int_C v\,\mathrm{d}x$$

$$= \int_\alpha^\beta u\,\frac{\mathrm{d}x}{\mathrm{d}q}\,\mathrm{d}q - \int_\alpha^\beta v\,\frac{\mathrm{d}y}{\mathrm{d}q}\,\mathrm{d}q + i\int_\alpha^\beta u\,\frac{\mathrm{d}y}{\mathrm{d}q}\,\mathrm{d}q$$

$$+ i\int_\alpha^\beta v\,\frac{\mathrm{d}x}{\mathrm{d}q}\,\mathrm{d}q. \tag{16.50}$$

The question of when such an integral exists will not be gone into, except to state that a sufficient condition is that dx/dq and dy/dq are continuous.

To illustrate this definition and also the final sentences of the first paragraph, we will now consider some simple examples.

Example 16.3. (i) $f(z) = z^{-1}$ and C_1 is the circle $|z| = R$, starting and finishing at $z = R$.

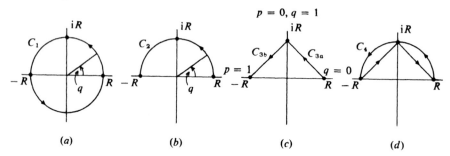

(a)   (b)   (c)   (d)

Fig. 16.5 The paths and parameterizations of example 16.3. See text for details.

The path C_1 is parameterized as (fig. 16.5 (a))

$$z(q) = R \cos q + iR \sin q, \quad 0 \leqslant q \leqslant 2\pi,$$

whilst $f(z)$ is given by

$$f(z) = \frac{1}{x + iy} = \frac{x - iy}{x^2 + y^2}.$$

Thus

$$u = \frac{x}{x^2 + y^2} = \frac{R \cos q}{R^2}$$

and

$$v = \frac{-y}{x^2 + y^2} = -\frac{R \sin q}{R^2}.$$

Hence using expression (16.50),

$$\int_{C_1} \frac{1}{z}\, dz = \int_0^{2\pi} \frac{\cos q}{R}(-R \sin q)\, dq - \int_0^{2\pi} \frac{-\sin q}{R} R \cos q\, dq$$

$$+ i \int_0^{2\pi} \frac{\cos q}{R} R \cos q\, dq$$

$$+ i \int_0^{2\pi} \frac{-\sin q}{R}(-R \sin q)\, dq \quad (16.51)$$

$$= 0 + 0 + i\pi + i\pi = 2\pi i. \tag{16.52}$$

This very important result will be used many times later, and the following should be carefully noted,

(a) its value,
(b) that its value is independent of R.

With a bit of experience, integrals like this one can sometimes be evaluated directly without writing them as four separate real integrals (see also ▶16). In the present case:

$$\int_{C_1} \frac{dz}{z} = \int_0^{2\pi} \frac{-R \sin q + iR \cos q}{R \cos q + iR \sin q} \, dq = \int_0^{2\pi} i \, dq = 2\pi i. \tag{16.53}$$

(ii) $f(z) = z^{-1}$ and C_2 is the semicircle $|z| = R$ in the half plane $y \geqslant 0$, joining $z = R$ to $z = -R$.
   This is just as in (i) except that now $0 \leqslant q \leqslant \pi$. With this change we have from line (16.51) or line (16.53) that

$$\int_{C_2} \frac{dz}{z} = \pi i.$$

(iii) Yet again take $f(z) = z^{-1}$, but this time with a contour C_3 made up of the two straight lines C_{3a} and C_{3b} (fig. 16.5 (c)). These may be parameterized as

$$C_{3a}: z = (1 - q)R + iqR \quad (0 \leqslant q \leqslant 1),$$
$$C_{3b}: z = -pR + i(1 - p)R \quad (0 \leqslant p \leqslant 1).$$

With these parameterizations the required integrals may be written

$$\int_{C_3} \frac{dz}{z} = \int_0^1 \frac{-R + iR}{R + q(-R + iR)} \, dq + \int_0^1 \frac{-R - iR}{iR + p(-R - iR)} \, dp. \tag{16.54}$$

If we could accept from real variable theory that, for real q, $\int (a + bq)^{-1} \, dq = b^{-1} \log (a + bq)$, even if a and b are complex, then these integrals could be evaluated immediately. However to do this would be presuming to some extent what we wish to show, and so the evaluation must be made in terms of entirely real integrals. For example, the first is

$$\int_0^1 \frac{-R + iR}{R(1 - q) + iqR} \, dq = \int_0^1 \frac{(-1 + i)(1 - q - iq)}{(1 - q)^2 + q^2} \, dq$$

$$= \int_0^1 \frac{2q - 1}{1 - 2q + 2q^2} \, dq$$

$$+ i \int_0^1 \frac{1}{1 - 2q + 2q^2} \, dq$$

$$= \tfrac{1}{2} [\ln (1 - 2q + 2q^2)]_0^1$$

$$+ \frac{i}{2} \left[2 \arctan \left(\frac{q - 1/2}{1/2} \right) \right]_0^1$$

$$= 0 + \frac{i}{2} \left[\frac{\pi}{2} - \left(-\frac{\pi}{2} \right) \right] = \tfrac{1}{2}\pi i.$$

The second integral on the right of (16.54) can also be shown to have value $\tfrac{1}{2}\pi i$. Thus

$$\int_{C_3} \frac{dz}{z} = \pi i.$$

(iv) Take $f(z) = \text{Re} (z)$ and the same contour as in (i). Then

$$\int_{C_1} \text{Re} (z) \, dz = \int_0^{2\pi} R \cos q(-R \sin q + iR \cos q) \, dq = i\pi R^2.$$

(v) As (iv) but using C_2 as the contour.

$$\int_{C_2} \text{Re} (z) \, dz = \int_0^{\pi} R \cos q(-R \sin q + iR \cos q) \, dq = \tfrac{1}{2} i\pi R^2.$$

(vi) As (iv) but using $C_3 = C_{3a} + C_{3b}$ as the contour.

$$\int_{C_3} \text{Re} (z) \, dz = \int_0^1 (1 - q)R(-R + iR) \, dq$$

$$+ \int_0^1 (-pR)(-R - iR) \, dp$$

$$= \tfrac{1}{2}R^2(-1 + i) + \tfrac{1}{2}R^2(1 + i) = iR^2.$$

Considering results (i)–(vi) together and recalling that (i)–(iii) and (iv)–(vi) have common integrands, some interesting observations are possible. Firstly the two integrals (ii) and (iii) from $z = R$ to $z = -R$ have the same value even though the paths taken (C_2 and C_3) are different, but this is not true for (v) and (vi). It also follows that if we took a closed path C_4 (fig. 16.5 (*d*)) given by C_2 from R to $-R$ and C_3 traversed backwards from $-R$ to R, then the integral round C_4 of z^{-1} would be zero [both parts contributing equal and opposite amounts]. This is to be compared with result (i), in which closed path C_1, beginning and ending at the same place as C_4, yields a value $2\pi i$.

These results demonstrate that the value of an integral between the same two points may depend upon the path that is taken between, but at the same time suggests that in some circumstances it is independent of the path. The general result is embodied in the results of the next section – namely Cauchy's theorem which is the corner-stone of the integral calculus of complex variables.

▶16. Relate line (16.53) to ▶11 and the multivalued nature of Log z.

16.7   Cauchy's theorem and integral

Cauchy's theorem. If $f(z)$ is an analytic function, and $f'(z)$ is continuous at each point within and on a closed contour C, then

$$\int_C f(z) \, dz = 0. \tag{16.55}$$

To prove this we will need a two-dimensional form of the divergence theorem (section 4.2) (also known as **Green's lemma**). This says that if u and v are two functions with continuous first derivatives within and on a closed contour C (bounding a domain D) in the xy-plane, then

$$\int\int_D \left(\frac{\partial u}{\partial x} + \frac{\partial v}{\partial y} \right) dx \, dy = \int_C (lu + mv) \, ds = \int_C (u \, dy - v \, dx). \tag{16.56}$$

Here l, m are the direction cosines of the outward normal to C at the point (x, y).

If the three-dimensional divergence theorem is applied to a 'vector' whose three components are $(u, v, 0)$ over a volume which is a uniform cylinder with axis perpendicular to the xy-plane and whose intersection with that plane is the contour C, then the first equality of (16.56) is apparent. The first expression is the volume integral and the second the surface integral, both being in terms of 'per unit length of the cylinder'.

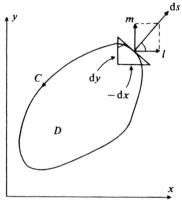

Fig. 16.6 The relationship used in Green's lemma between the outward normal to domain D and an element of its enclosing contour C. The two marked angles are equal.

The second equality follows from the geometric relationship between the outward normal to the cylinder (or C) and the line element in traversing C in the positive sense, the tangent to C at (x, y) having direction cosines $(-m, l)$ if the outward normal is (l, m). [See fig. 16.6.]

This can now be applied to

$$I = \int_C f(z)\,dz = \int_C (u\,dx - v\,dy) + i \int_C (v\,dx + u\,dy)$$

to give

$$I = \int\int_D \left[\frac{\partial(-u)}{\partial y} + \frac{\partial(-v)}{\partial x} \right] dx\,dy + i \int\int_D \left[\frac{\partial(-v)}{\partial y} + \frac{\partial u}{\partial x} \right] dx\,dy.$$

$$(16.57)$$

Now recalling that $f(z)$ is analytic and therefore the Cauchy–Riemann equations (16.5) apply, we see that each integrand is identically zero and therefore I is also. This proves the theorem.

In actual fact the conditions of the above proof are more stringent than are needed. The continuity of $f'(z)$ is not necessary for the proof of Cauchy's theorem, analyticity of $f(z)$ within and on C being sufficient. However, the proof then becomes more complicated and is too long to be given here.†

A sort of converse of Cauchy's theorem is known as **Morera's theorem.** We state it without proof.

> If $f(z)$ is a continuous function of z in a closed domain D bounded by a curve C and further $\int_C f(z)\,dz = 0$, then $f(z)$ is analytic in D.

† The reader may refer to almost any book devoted to complex variables and the theory of functions.

▶17. The function $f(z) = z^2$ is analytic in any region of the z-plane. Verify by a direct evaluation that $\int_C z^2 \, dz$ around a closed contour C has zero value,

(i) when C is the circle $|z| = R$ starting at $z = R$,

(ii) when C is the square whose four corners are $x = \pm L$, $y = \pm L$ starting at any corner.

The connection between Cauchy's theorem and the zero value of the integral of z^{-1} around the composite path C_4 discussed towards the end of the previous section, is apparent – the function z^{-1} being analytic in the two regions of the z-plane enclosed by contours (C_2 and C_{3a}) and (C_2 and C_{3b}).

> **Cauchy's integral.** If $f(z)$ is analytic within and on a closed contour C and ξ is a point within C, then
>
> $$f(\xi) = \frac{1}{2\pi i} \int_C \frac{f(z)}{z - \xi} \, dz .$$
> (16.58)

This is saying that for an analytic function its value anywhere inside a closed contour is uniquely determined by its values on the contour† and that the specific expression (16.58) can be given for the value at the interior point.

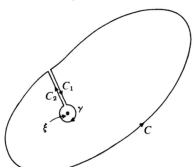

Fig. 16.7 The contour used to prove Cauchy's integral formula.

To prove the validity of Cauchy's integral consider a contour as shown in fig. 16.7. In this γ is a circle centred on the point $z = \xi$ and of small enough radius ρ that it all lies inside C. The two close parallel lines C_1 and C_2 join γ and C, which are 'cut' to accommodate them. The new contour Γ so formed consists of C, C_1, γ and C_2.

† The similarity between this and the Uniqueness theorem for Dirichlet boundary conditions of chapter 10 will be noticed.

Within the area bounded by Γ [which does not include the point $z = \xi$] the function $f(z)/(z - \xi)$ is analytic and therefore by Cauchy's theorem (16.55),

$$\int_\Gamma \frac{f(z)}{z - \xi} \, dz = 0. \tag{16.59}$$

Now the parts C_1 and C_2 of Γ are traversed in opposite directions and lie (in the limit) on top of each other and so their contributions to (16.59) cancel. Thus

$$\int_C \frac{f(z)}{z - \xi} \, dz + \int_\gamma \frac{f(z)}{z - \xi} \, dz = 0. \tag{16.60}$$

The sense of the integral round γ is opposite to the conventional (anti-clockwise) one and its value is given by the following argument, which uses the fact that any point z on γ is given by $z = \xi + \rho \exp(i\theta)$,

$$\int_\gamma \frac{f(z)}{z - \xi} \, dz = \int_\gamma \frac{f(\xi)}{z - \xi} \, dz + \int_\gamma \frac{f(z) - f(\xi)}{z - \xi} \, dz$$

$$= -f(\xi) \int_0^{2\pi} \frac{\rho i \exp(i\theta)}{\rho \exp(i\theta)} \, d\theta + I_1 \tag{†}$$

$$= -2\pi i f(\xi) + I_1. \tag{16.61}$$

But $|I_1|$ cannot be greater than

$$\frac{1}{\rho} \max_{z \text{ on } \gamma} \{|f(z) - f(\xi)|\} 2\pi\rho.$$

Since $f(z)$ is continuous the maximum of $|f(z) - f(\xi)|$ on γ can be made as small as we wish by taking ρ small enough. Hence $I_1 = 0$ and (16.61) can be rearranged in the form (16.58) thus establishing that result.

An extension to Cauchy's integral formula can be made to yield an integral expression for $f'(\xi)$ as

$$f'(\xi) = \frac{1}{2\pi i} \int_C \frac{f(z)}{(z - \xi)^2} \, dz, \tag{16.62}$$

under the same conditions as previously.

To show this, the definition of a derivative and (16.58) itself are used to evaluate

$$\frac{f(\xi + h) - f(\xi)}{h} = \frac{1}{2\pi i} \int_C \frac{f(z)}{h} \left(\frac{1}{z - \xi - h} - \frac{1}{z - \xi} \right) dz. \tag{16.63}$$

† Compare with example 16.3 (i) after a change of origin to $z - \xi$.

The right-hand side of (16.63) can be rearranged as the right-hand side of (16.62) together with an integral I_2 given by

$$I_2 = \frac{1}{2\pi i} \int_C \frac{hf(z)\,dz}{(z - \xi)^2(z - \xi - h)}.$$ (16.64)

All that remains to be shown is that $|I_2| \to 0$ as $|h| \to 0$. But since ξ is inside C and z is on C, for sufficiently small h the denominator of (16.64) will be finite [perhaps small – but definitely non-zero]. However, since f is analytic and therefore bounded in the domain, the numerator $\to 0$ as $|h| \to 0$. Thus in the limit $|h| \to 0$, I_2 must tend to 0 and (16.62) is established.

▶18. Prove by induction that the nth derivative of $f(z)$ is given by a Cauchy integral,

$$f^{(n)}(\xi) = \frac{n!}{2\pi i} \int_C \frac{f(z)\,dz}{(z - \xi)^{n+1}}.$$ (16.65)

Following on from (16.65), **Taylor's theorem** may be established although we will only indicate how.

If $f(z)$ is analytic in the region $|z - a| \leq R$ and z is a point inside that region, then

$$f(z) = \sum_{n=0}^{\infty} a_n(z - a)^n,$$ (16.66)

where a_n is given by $f^{(n)}(a)/n!$

It can be proved by expanding $(\xi - z)^{-1}$ as a geometric series in $(z - a)/(\xi - a)$, multiplying through by $f(\xi)$, integrating $d\xi$ around a contour of radius $\rho < R$ centred on a, and using Cauchy's formula and result (16.65).

▶19. Carry through these procedures.

The Taylor expansion is valid inside the region of analyticity and, for any particular a, can be shown to be unique.

16.8   Zeros and singularities

So far we have considered functions only in domains where they are analytic. We now examine their behaviour at points where they cease to be so.

Suppose $z = a$ is a point in the z-plane and that a power series expansion of a function about that point can be made for all z inside some region $0 < |z - a| < R$,

$$f(z) = \sum_{n = -\infty}^{+\infty} a_n(z - a)^n. \tag{16.67}$$

This series, which is an extension of the Taylor expansion, is called a **Laurent expansion**. The previously stated Taylor's theorem indicates that if $f(z)$ is analytic at $z = a$, then all a_n for $n < 0$ must be zero.

It may happen that not only are all a_n zero for $n < 0$, but that a_0, $a_1, \ldots, a_{m-1}$, are all zero as well. In this case the first non-vanishing term in (16.67) is $a_m(z - a)^m$ with $m > 0$, and $f(z)$ is then said to have a **zero** of **order** m at $z = a$.

If $f(z)$ is not analytic at $z = a$ then two cases arise (m is here taken as positive):

(i) it is possible to find an integer m such that $a_{-m} \neq 0$ but $a_{-(m+k)} = 0$ for all integral $k > 0$,

(ii) it is not possible to find such a lowest value of $-m$.

In case (i), $f(z)$ has the form

$$\frac{a_{-m}}{(z - a)^m} + \frac{a_{-m+1}}{(z - a)^{m-1}} + \cdots + \frac{a_{-1}}{z - a}$$

$$+ \sum_{0}^{\infty} a_n(z - a)^n, \quad a_{-m} \neq 0, \tag{16.68}$$

and is described as having a **pole** of **order** m at $z = a$; the value of a_{-1} [not a_{-m}] is called the **residue** of $f(z)$ at the pole $z = a$, and will play an important part in later applications.

It can be shown that:

(a) zeros of $f(z)$ are isolated, i.e. around each one there exists a neighbourhood which contains no other zero of $f(z)$,

(b) poles are isolated,

(c) if $f(z)$ has a pole at $z = a$ then $|f(z)| \to \infty$ as $z \to a$, from whatever direction.

These three results are perhaps intuitively reasonable although they really require formal demonstration by analysis.

For case (ii) in which the negatively decreasing powers of $(z - a)$ do not terminate, $f(z)$ is said to have an **essential singularity**. Such cases and other more complicated singularities, whilst interesting in their own right, will not be of importance to the physical applications considered and will not be pursued.

An expression common in mathematics and which we have so far avoided using explicitly, is 'z tends to infinity'. For a real variable such as

$|z|$ or R, 'tending to infinity' has a reasonably well-defined meaning. For a complex variable needing a two-dimensional plane to represent it, the meaning is not well defined. However, it is convenient to have a unique meaning and this is provided by the following *definition*.

The behaviour of $f(z)$ **at infinity** is given by that of $f(1/\xi)$ at $\xi = 0$. $\hfill$ (16.69)

Different functions have different behaviours at $z = \infty$; as examples:

(a) $f(z) = a + bz^{-2}$; on putting $z = 1/\xi$, $f(1/\xi) = a + b\xi^2$ which is analytic at $\xi = 0$; thus f is analytic at $z = \infty$.
(b) $f(z) = z(1 + z^2)$: $f(1/\xi) = 1/\xi + 1/\xi^3$, thus f has a pole of order 3 at $z = \infty$.
(c) $f(z) = \exp(z)$: $f(1/\xi) = \sum_0^\infty (n!)^{-1}\xi^{-n}$, thus f has an essential singularity at $z = \infty$.

16.9   Residue theorem

Having seen that the value of an integral round a closed contour C is zero if the integrand is analytic inside the contour, the next question is 'what is the value when it is not analytic inside C?' The answer to this is contained in the residue theorem which will now be proved in a number of stages.

1. Suppose the point $z = a$ is a pole of $f(z)$ of order m, then $f(z)$ can be written as

$$f(z) = \phi(z) + \sum_{n=1}^{m} \frac{a_{-n}}{(z-a)^n},$$ $\hfill$ (16.70)

where $\phi(z)$ is analytic within some neighbourhood surrounding a. Now consider the integral I of $f(z)$ along an arc C of a circle

$$|z - a| = \rho, \qquad \theta_1 \leqslant \arg(z - a) \leqslant \theta_2,$$ $\hfill$ (16.71)

where ρ is chosen small enough so that no singularity of f, other than $z = a$, lies within the circle.

$$I = \int_C \phi(z)\,dz + \sum_{n=1}^{m} \int_C \frac{a_{-n}\,dz}{(z-a)^n}$$

$$= \int_C \phi(z)\,dz + \sum_{1}^{m} \int_{\theta_1}^{\theta_2} \frac{a_{-n}i\rho \exp(i\theta)\,d\theta}{\rho^n \exp(in\theta)}$$

$$= \int_C \phi(z)\,dz + i\sum_{1}^{m} a_{-n}\rho^{1-n} \int_{\theta_1}^{\theta_2} \exp[(1-n)i\theta]\,d\theta.$$ $\hfill$ (16.72)

Two particular cases now arise.

2. C is a complete circle, i.e. $\theta_2 = \theta_1 + 2\pi$.

In this case all the angular integrals in the sum, except that for which $1 - n = 0$, vanish. In addition, by Cauchy's theorem, so does the first integral. Hence for a closed circular contour C

$$\int_C f(z)\,dz = I = ia_{-1}\rho^0 \int_{\theta_1}^{\theta_1 + 2\pi} d\theta = 2\pi ia_{-1}. \tag{16.73}$$

Notice that this result does not depend on the value of ρ and compare it with that of example 16.3(i) (p. 459).

3. $m = 1$ ($z = a$ is then called a **simple pole**) and $\rho \to 0$.

As $\rho \to 0$ the first integral tends to zero, since the path becomes of zero length and ϕ is analytic and therefore continuous along it. The *only* term in the sum, and hence the value of I, is thus given by

$$\lim_{\rho \to 0} \int_C f(z)\,dz = I = \lim_{\rho \to 0} ia_{-1}\rho^0 \int_{\theta_1}^{\theta_2} d\theta = ia_{-1}(\theta_2 - \theta_1). \tag{16.74}$$

A similar, but slightly more useful result than this can be obtained by replacing a_{-1} in (16.74) by

$$\lim_{z \to a} [(z - a)f(z)], \tag{16.75}$$

if such a limit exists.

> **Residue theorem.** If $f(z)$ is continuous within and on a closed contour C and analytic, except for a finite number of poles, within C, then
>
> $$\int_C f(z)\,dz = 2\pi i \sum_j R_j, \tag{16.76}$$
>
> where $\sum R_j$ is the sum of the residues of $f(z)$ at its poles within C.

The method of proof is indicated by fig. 16.8 in which (a) shows the original contour C referred to in (16.76), and (b) shows a contour C' giving the same value to the integral – because f is analytic between C and C'. Now the contribution to the C' integral from the polygon (a triangle for the case illustrated) joining the small circles is nothing, since f is also analytic inside C'. Hence the whole value of the integral comes from the circles and, by result (16.73), each of these contributes $2\pi i$ {residue at the pole it encloses}. All the circles are traversed in their positive sense if C is and so the residue theorem follows.

Formally Cauchy's theorem (16.55) is a particular case of this in which C encloses no poles.

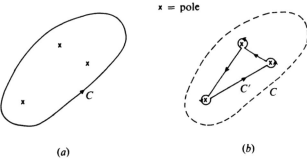

(a)   (b)

Fig. 16.8 The contours to prove the residue theorem: (a) the original contour; (b) the contracted one encircling each of the poles.

▶20. Obtain results (i)–(iii) of example 16.3 (p. 458) by inspection and use of the residue theorem.

▶21. What is the value of the integral of $f(z) = z^{-2}(2 - z)^{-1}$ around the contours (a) the unit circle, (b) a square of unit side centred on the origin, and (c) a square of side 6 centred on the origin?

16.10   Location of zeros

As the basis of a method of locating the zeros of functions of a complex variable we next prove three theorems.

1. If $f(z)$ has poles as its only singularities inside a contour C and is not zero at any point on C, then

$$\int_C \frac{f'(z)}{f(z)} \, dz = 2\pi i \sum_j (N_j - P_j). \tag{16.77}$$

Here $N_j(P_j)$ is the order of the zero (pole) of $f(z)$ at the jth of the points a_j enclosed by C at which f has a zero or pole. [Obviously only one case at each point.]

To prove this we note that at each position a_j, $f(z)$ can be written as

$$f(z) = (z - a_j)^{m_j}\phi(z), \tag{16.78}$$

where $\phi(z)$ is analytic and non-zero at $z = a_j$ and m_j is positive for a zero and negative for a pole. Then the integrand $f'(z)/f(z)$ takes the form

$$\frac{f'(z)}{f(z)} = \frac{m_j}{z - a_j} + \frac{\phi'(z)}{\phi(z)}. \tag{16.79}$$

Since $\phi(a_j) \neq 0$, the second term on the right is analytic; thus the integrand has a simple pole at $z = a_j$ with residue m_j. For zeros $m_j = N_j$, and for poles $m_j = -P_j$, and thus by the residue theorem (16.77) follows.

2. If $f(z)$ is analytic inside C and not zero at any point on it, then

$$2\pi \sum_j N_j = \Delta_C[\arg f(z)],\tag{16.80}$$

where $\Delta_C[x]$ denotes the variation in x around the contour C.

Since f is analytic there are no P_j, and further since

$$\frac{f'(z)}{f(z)} = \frac{d}{dz}[\operatorname{Log} f(z)],\tag{16.81}$$

equation (16.77) can be written

$$\Delta_C[\operatorname{Log} f(z)] = \int_C \frac{f'(z)}{f(z)}\,dz = 2\pi i \sum N_j.\tag{16.82}$$

But

$$\Delta_C[\operatorname{Log} f(z)] = \Delta_C[\log |f(z)|] + i\Delta_C[\arg f(z)],\tag{16.83}$$

and since C is a closed contour, $\log |f(z)|$ cannot change around it and the real term on the right is zero. Comparison of (16.82) and (16.83) then establishes (16.80), which is known as the *principle of the argument*.

3. (**Rouché's theorem.**) If $f(z)$ and $g(z)$ are analytic within and on a closed contour C and $|g(z)| < |f(z)|$ on C, then $f(z)$ and $f(z) + g(z)$ have the same number of zeros inside C.

With the conditions given, neither $f(z)$ nor $f(z) + g(z)$ can have a zero on C. So applying theorem 2 with an obvious notation,

$$\begin{aligned}2\pi \sum_j N_j(f + g) &= \Delta_C[\arg (f + g)]\\ &= \Delta_C[\arg f] + \Delta_C[\arg (1 + g/f)]\\ &= 2\pi \sum_k N_k(f) + \Delta_C[\arg (1 + g/f)].\end{aligned}\tag{16.84}$$

Further since $|g| < |f|$ on C, $1 + g/f$ always lies *within* a unit circle centred on $z = 1$, thus its argument *always* lies in the range $-\pi/2 < \arg (1 + g/f) < \pi/2$ and cannot change by any multiple of 2π. It must therefore return to its original value when z returns to its starting point having traversed C. Hence the second term on the right of (16.84) is zero and the theorem is established.

These three theorems are of value in locating the zeros of functions of a complex variable. The location of such zeros has particular application in electrical network and general oscillations theory, since the complex zeros of certain functions give the system parameters (usually frequencies) at which system instabilities occur.

The importance of Rouché's theorem is that for some functions, in particular, polynomials, only the behaviour of a single term in the function need be considered if the contour is chosen appropriately. For example, for a polynomial [treated as $f(z) + g(z)$] only the properties of its leading (smallest) power [treated as $f(z)$] need be investigated, if a circular contour of large (small) enough radius R is chosen so that |leading (smallest) power term| $> \sum$ |all other terms| on the contour. Further if the zeros of $f(z) + g(z) = \sum_0^N b_n z^n$ are considered as the roots of $f(z) + g(z) = 0$ written in the form

$$1 + \frac{g(z)}{f(z)} = 0, \tag{16.85}$$

then it is apparent that no roots can lie outside (inside) $|z| = R$ and also that $f(z) = b_N z^N$ (or b_0) has N (or 0) zeros inside $|z| = R$; $f + g$ consequently has the same number of zeros inside the same circle.

A weak form of the **maximum-modulus theorem** may also be deduced. This states that

> If $f(z)$ is analytic within and on a simple closed contour C, then $|f(z)|$ attains its maximum value on the boundary of C.   (16.86 a)

Let $|f(z)| \leqslant M$ on C with equality at at least one point of C. Now suppose there is a point $z = a$ inside C such that $|f(a)| > M$. Then the function $h(z) \equiv f(a)$ is such that $|h(z)| > |-f(z)|$ on C, and thus $h(z)$ and $h(z) - f(z)$ have the same number of zeros inside C. But $h(z)$ ($\equiv f(a)$) has no zeros inside C and by Rouché's theorem this would imply $f(a) - f(z)$ has no zeros in C. However $f(a) - f(z)$ clearly has a zero at $z = a$, and so we have a contradiction; the assumption of a point $z = a$ inside C such that $|f(a)| > M$ must be invalid. This establishes the theorem.

The stronger form of the theorem, which we do not prove, allows the following to be added to (16.86 a).

> The maximum value is not attained at any interior point except in the case in which $f(z)$ is a constant.   (16.86 b)

To illustrate some uses of the theorems, some information about the zeros of $h(z) = z^4 + z + 1$ is now deduced.

(a) Putting $z = x$ and $z = iy$ shows that no zeros occur on the real or imaginary axes. They must therefore occur in conjugate pairs [as taking the complex conjugate of $h(z) = 0$ shows].

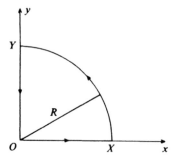

Fig. 16.9 A contour for locating zeros of a polynomial occurring in the first quadrant.

(b) Take C as the contour $OXYO$ as shown in fig. 16.9 and consider the change in the argument of $h(z)$.

OX: along OX, $\arg h$ is everywhere zero (since h is real) and thus $\Delta_{OX}[\arg h] = 0$.
XY: $z = R\exp(i\theta)$.

$$\begin{aligned}
\Delta_{XY}[\arg h] &= \Delta_{XY}[\arg z^4] + \Delta_{XY}[\arg(1 + z^{-3} + z^{-4})] \\
&= \Delta_{XY}[\arg R^4 e^{4i\theta}] + \Delta_{XY}[\arg(1 + O(R^{-3}))] \\
&= 2\pi + O(R^{-3}). \qquad\qquad (16.87)
\end{aligned}$$

YO: $\arg h = y/(y^4 + 1)$, which starts at $O(R^{-3})$ and finishes at 0 as y goes from R (large) to 0. It never reaches $\pi/2$ because $y^4 + 1 = 0$ has no real positive root. Thus $\Delta_{YO}[\arg h] = 0$. Hence for the complete contour $\Delta_C[\arg h] = 0 + 2\pi + 0 + O(R^{-3})$ and if R is allowed to tend to infinity we deduce from (16.80) that $h(z)$ has one zero in the first quadrant.

Since the roots occur in conjugate pairs, a second root must be in the fourth quadrant and the other pair in the second and third quadrants.

(c) Apply Rouché's theorem,

(i) with C as $|z| = 3/2$, $f = z^4$, $g = z + 1$.

Now $|f| = 81/16$ on C and $|g| \leqslant 1 + |z| \leqslant 5/2 < 81/16$. Thus since $z^4 = 0$ has four roots inside $|z| = 3/2$, so also does $z^4 + z + 1 = 0$.

(ii) with C as $|z| = 2/3$, $f = 1$, $g = z^4 + z$.

Now $f = 1$ on C and $|g| \leqslant |z^4| + |z| = 16/81 + 2/3 = 70/81 < 1$. Thus since $1 = 0$ has no roots inside $|z| = 2/3$, neither does $1 + z + z^4 = 0$.

Hence to summarize at this point, the four zeros of $h(z) = z^4 + z + 1$ occur one in each quadrant and all lie between the circles $|z| = 2/3$ and $|z| = 3/2$.

▶22. Use a quadrant contour $0 \to 1 \to i \to 0$ to show that the zero in the first quadrant has $|z| > 1$. [Use de Moivre's theorem to obtain an expression for the argument.]

A further technique useful in locating function zeros is explained in example 11 of section 16.16.

16.11   Integrals of sinusoidal functions

The remainder of this chapter is devoted to methods of applying contour integration and the residue theorem to various types of definite integrals. In each case not much preamble is given since, for this material, the simplest explanation is felt to be a series of worked examples which can be used as models.

Suppose an integral of the form

$$\int_0^{2\pi} F(\cos\theta, \sin\theta)\, d\theta \tag{16.88}$$

is to be evaluated. It can be made into a contour integral around the unit circle C by writing $z = \exp(i\theta)$ and hence

$$\cos\theta = \tfrac{1}{2}(z + z^{-1}), \qquad \sin\theta = -\frac{i}{2}(z - z^{-1}),$$

$$d\theta = -iz^{-1}\, dz. \tag{16.89}$$

This contour integral can then be evaluated using the residue theorem, provided the transformed integrand has only a finite number of poles inside the unit circle and none on it.

Example 16.4. Evaluate

$$I = \int_0^{2\pi} \frac{\cos 2\theta}{a^2 + b^2 - 2ab\cos\theta}\, d\theta, \qquad b > a > 0. \tag{16.90}$$

By de Moivre's theorem

$$\cos n\theta = \tfrac{1}{2}(z^n + z^{-n}). \tag{16.91}$$

Using $n = 2$ in (16.91) and straightforward substitution for the other functions of θ in (16.90) gives

$$I = \frac{i}{2ab} \int_C \frac{z^4 + 1}{z^2[z - (a/b)][z - (b/a)]}\, dz.$$

There are 2 poles inside C,

(i) at $z = 0$,
(ii) at $z = a/b$ [recall $b > a$].

(i) Rearrange the integrand as

$$\frac{1}{z^2}(1 + z^4)\left(1 - \frac{bz}{a}\right)^{-1}\left(1 - \frac{az}{b}\right)^{-1}$$

and expand. The coefficient of z^{-1}, i.e. the residue, is

▶ 23. $\dfrac{a}{b} + \dfrac{b}{a}$.

(ii) The residue at $z = a/b$ (a simple pole) is

$$\lim_{z \to (a/b)}\left[\left(z - \frac{a}{b}\right) \times \text{integrand}\right] = \frac{(a/b)^4 + 1}{(a/b)^2(a/b - b/a)}$$

$$= -\frac{a^4 + b^4}{ab(b^2 - a^2)}.$$

And so by the residue theorem

$$I = 2\pi i \times \frac{i}{2ab}\left(\frac{a^2 + b^2}{ab} - \frac{a^4 + b^4}{ab(b^2 - a^2)}\right) = \frac{2\pi a^2}{b^2(b^2 - a^2)}.$$

16.12   Some infinite integrals

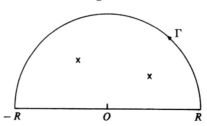

Fig. 16.10 A semicircular contour in the upper half-plane.

If (i) $f(z)$ is analytic in the upper half-plane, Im $z \geqslant 0$ except for a finite number of singularities, but with none on the real axis,
(ii) on the semicircle Γ (fig. 16.10) of radius R, $\{R \times \text{maximum of } |f| \text{ on } \Gamma\}$ tends to zero as $R \to \infty$,
(iii) $\int_{-\infty}^{0} f(x)\,dx$ and $\int_{0}^{\infty} f(x)\,dx$ both exist, then

$$\int_{-\infty}^{\infty} f(x)\,dx = 2\pi i\{\text{sum of the residues at poles with Im } z \geqslant 0\}.$$

$$(16.92)$$

Condition (ii) ensures that

$$\left|\int_{\Gamma} f(z)\,dz\right| \leqslant 2\pi R\{\text{maximum of } |f| \text{ on } \Gamma\},$$

which tends to 0 as $R \rightarrow \infty$, after which (16.92) is obvious from the residue theorem.

Example 16.5. Evaluate

$$I = \int_0^\infty \frac{dx}{(x^2 + a^2)^4}, \qquad a \text{ real.}$$

The complex function $(z^2 + a^2)^{-4}$ has poles only at $z = \pm ai$ of which only $z = ai$ is in the upper half-plane. Conditions (ii) and (iii) are clearly satisfied. If we put $z = ai + \xi$ and expand for small ξ we obtain

$$\frac{1}{(z^2 + a^2)^4} = \frac{1}{(2ai\xi + \xi^2)^4} = \frac{1}{(2ai\xi)^4} \left(1 - \frac{i\xi}{2a}\right)^{-4}.$$

The coefficient of ξ^{-1} is

$$\frac{1}{(2a)^4} \frac{(-4)(-5)(-6)}{3!} \left(\frac{-i}{2a}\right)^3 = \frac{-5i}{32a^7}.$$

Hence

$$\int_{-\infty}^\infty \frac{dx}{(x^2 + a^2)^4} = \frac{10\pi}{32a^7}$$

and $I = 5\pi/32a^7$.

Condition (i) of the previous method required no poles of the integrand on the real axis, but *simple* poles on the real axis can be accommodated by indenting the contour as shown in fig. 16.11. The indentation at the pole $z = a$ is in the form of a semicircle γ of radius ρ in the upper half-plane, thus excluding the pole from the interior of the contour.

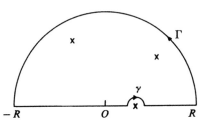

Fig. 16.11 An indented semicircular contour used when the integrand has a simple pole on the real axis.

What is then obtained from a contour integration is, apart from the contributions for Γ and γ,

$$\mathbf{P} \int_{-R}^R f(x)\, dx \equiv \int_{-R}^{a-\rho} f(x)\, dx + \int_{a+\rho}^R f(x)\, dx, \qquad (16.93)$$

and is called the **principal value of the integral** as $\rho \rightarrow 0$.

The remainder of the calculation goes through as before, but the contribution from the semicircle γ must be included. Result 3 of section 16.9 shows that, since only a simple pole is involved, its contribution is

$$-ia_{-1}\pi,\tag{16.94}$$

where a_{-1} is the residue at the pole and the minus sign arises because γ is traversed in the clockwise (negative) sense.

We defer giving an example of an indented contour until we have established Jordan's lemma and then work through an example illustrating both. **Jordan's lemma** enables infinite integrals involving sinusoidal functions to be evaluated. It states that if,

(i) $f(z)$ is analytic in the upper half-plane except for a finite number of poles in Im $z > 0$,
(ii) the maximum of $|f| \to 0$ as $|z| \to \infty$ in the upper half-plane,
(iii) $m > 0$, then

$$I_\Gamma \equiv \int_\Gamma e^{imz} f(z)\, dz \to 0 \quad \text{as} \quad R \to \infty,\tag{16.95}$$

where Γ is the same semicircular contour as in fig. 16.10. Notice that condition (ii) is less stringent than the earlier condition (ii), since we now only require $M(R) \to 0$ and not $RM(R) \to 0$, where M is the maximum† of $|f(z)|$ on $|z| = R$.

The proof of the lemma is straightforward, once it has been observed that for $0 \leqslant \theta \leqslant \frac{1}{2}\pi$

$$1 \geqslant \frac{\sin \theta}{\theta} \geqslant \frac{2}{\pi}.\tag{16.96}$$

It then proceeds as follows

$$I_\Gamma \leqslant \int_\Gamma |e^{imz} f(z)|\, |dz| \leqslant \int_0^\pi M e^{-mv} R\, d\theta$$

$$= 2MR \int_0^{\pi/2} e^{-mR\sin\theta}\, d\theta.$$

Thus

$$I_\Gamma < 2MR \int_0^{\pi/2} e^{-mR(2\theta/\pi)}\, d\theta$$

$$= \frac{\pi M}{m} (1 - e^{-mR}) < \frac{\pi M}{m},$$

and hence tends to zero since M does, as $R \to \infty$.

† More strictly the least upper bound.

Example 16.6. Find the principal value of

$$\int_{-\infty}^{\infty} \frac{\cos mx}{x - a}\, dx, \qquad\qquad a \text{ real}, m > 0.$$

Consider the function $(z - a)^{-1} \exp(imz)$; it has no poles in the upper half-plane and a simple pole at $z = a$, and further $|(z - a)^{-1}| \to 0$ as $|z| \to \infty$. We use a contour like that shown in fig. 16.11 and apply the residue theorem. Symbolically,

$$\int_{-R}^{a-\rho} + \int_{\gamma} + \int_{a+\rho}^{R} + \int_{\Gamma} = 0. \qquad (16.97)$$

Now as $R \to \infty$ and $\rho \to 0$, $\int_{\Gamma} \to 0$ by Jordan's lemma and as in (16.93) and (16.94) we obtain

$$\mathbf{P}\int_{-\infty}^{\infty} \frac{e^{imx}}{x - a}\, dx - i\pi a_{-1} = 0, \qquad (16.98)$$

where a_{-1} is the residue of $(z - a)^{-1} \exp(imz)$ at $z = a$, which is $\exp(ima)$. Then taking real and imaginary parts of (16.98) gives

$$\mathbf{P}\int_{-\infty}^{\infty} \frac{\cos mx}{x - a}\, dx = -\pi \sin ma, \text{ as required},$$

$$\mathbf{P}\int_{-\infty}^{\infty} \frac{\sin mx}{x - a}\, dx = \pi \cos ma, \text{ as a bonus}.$$

16.13 Integrals of many-valued functions

We have discussed briefly some of the properties and difficulties associated with several many-valued functions such as $z^{1/2}$ or Log z. It was mentioned, but not discussed, that one method of managing such functions is by means of a 'cut plane'. A similar technique can be used with advantage to evaluate some kinds of infinite integrals involving real functions for which the corresponding complex functions are many-valued. A typical contour employed is shown in fig. 16.12. Here Γ is a large circle of radius R and γ a small one of radius ρ, both centred on the origin. Eventually we will let $R \to \infty$ and $\rho \to 0$.

The value of the method comes from the fact that because the integrand is multivalued, its value along the two lines AB and CD joining $z = \rho$ to $z = R$ are *not* equal and opposite although both are related to the corresponding real integral. Again an example is the best explanation.

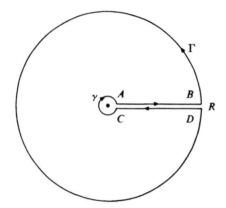

Fig. 16.12 A cut-plane contour for use with multivalued integrands.

Example 16.7. Evaluate

$$I = \int_0^\infty \frac{dx}{(x + a)^3 x^{1/2}}, \qquad\qquad a > 0.$$

We consider the integrand $f(z) = (z + a)^{-3} z^{-1/2}$ and note that $|zf(z)| \to 0$ on the two circles as $\rho \to 0$ and $R \to \infty$. Thus the two circles make no contributions to the contour integral.

The only pole of the integrand inside the contour is at $z = -a$. To determine its residue put $z = -a + \xi$ and expand [note $z^{1/2}$ is $a^{1/2} \exp(\tfrac{1}{2}i\pi) = ia^{1/2}$];

$$\frac{1}{(z + a)^3 z^{1/2}} = \frac{1}{\xi^3 ia^{1/2}(1 - \xi/a)^{1/2}}$$

$$= \frac{1}{i\xi^3 a^{1/2}} \left(1 + \frac{1}{2}\frac{\xi}{a} + \frac{3}{8}\frac{\xi^2}{a^2} + \cdots \right).$$

The residue is thus $-3i/8a^{5/2}$.

The residue theorem now gives

$$\int_{AB} + \int_\Gamma + \int_{DC} + \int_\gamma = 2\pi i \left(\frac{-3i}{8a^{5/2}} \right).$$

We have seen that $\int_\Gamma$ and $\int_\gamma$ vanish, and if we denote z by x along the line AB then it has the value $z = x \exp(2\pi i)$ along the line DC [note $\exp(2\pi i)$ must not be set equal to 1, until after the substitution for z has been made in $\int_{DC}$]. Putting in these forms,

$$\int_0^\infty \frac{dx}{(x + a)^3 x^{1/2}}$$

$$+ \int_\infty^0 \frac{dx}{[x \exp(2\pi i) + a]^3 x^{1/2} \exp(\tfrac{1}{2}2\pi i)} = \frac{3\pi}{4a^{5/2}}.$$

Thus

$$\left[1 - \frac{1}{\exp(\pi i)}\right] \int_0^\infty \frac{dx}{(x+a)^3 x^{1/2}} = \frac{3\pi}{4a^{5/2}},$$

and

$$I = \frac{1}{2} \times \frac{3\pi}{4a^{5/2}}.$$

▶24. Use the same contour and technique to evaluate

$$\int_0^\infty \frac{dx}{(x+a)^2 x^\beta}, \quad a > 0, \, 0 < \beta < 1.$$

16.14 Summation of series

Sometimes a real infinite series may be summed if a suitable complex function can be found which has poles on the real axis at the positions corresponding to the values of the dummy variable in the summation, and whose residues at these poles are equal to the values of the terms of the series there.

Example 16.8. By considering

$$\int_C \frac{\pi \cot \pi z}{(a+z)^2} \, dz, \qquad\qquad a \text{ non-integral,}$$

where C is a circle of large radius evaluate

$$\sum_{n=-\infty}^\infty \frac{1}{(a+n)^2}.$$

The integrand has (i) simple poles at $z = $ integral n, $-\infty < n < \infty$, (ii) a double pole at $z = -a$.

(i) To find the residue of $\cot \pi z$, put $z = n + \xi$ for small ξ;

$$\cot \pi z = \frac{\cos(n\pi + \xi\pi)}{\sin(n\pi + \xi\pi)} \simeq \frac{\cos(n\pi)}{\cos(n\pi)\xi\pi} = \frac{1}{\xi\pi}.$$

The residue of the integrand is thus $\pi(a+n)^{-2} \cdot \pi^{-1}$.

(ii) Putting $z = -a + \xi$ for small ξ and determining the coefficient of ξ^{-1},

$$\frac{\pi \cot \pi z}{(a+z)^2} = \frac{\pi}{\xi^2} \cot(-a\pi + \xi\pi)$$

$$= \frac{\pi}{\xi^2} \left\{ \cot(-a\pi) + \xi \left[\frac{d}{dz}(\cot \pi z)\right]_{z=-a} + \cdots \right\}. \quad (\dagger)$$

† This illustrates a useful technique for determining residues.

Thus the residue is

$$\pi[-\pi \cosec^2 \pi z]_{z=-a} = -\pi^2 \cosec^2 (\pi a).$$

Collecting together these results to express the residue theorem gives

$$I = \int_C \frac{\pi \cot \pi z}{(a+z)^2} \, dz = 2\pi i \left[\sum_{n=-N}^{N} \frac{1}{(a+n)^2} - \pi^2 \cosec^2 (\pi a) \right]$$

(16.99)

with N = integral part of R. But as the radius R of C tends to ∞, $|\cot \pi z| \rightarrow \mp 1$ (depending on whether $\mathrm{Im}\, z \gtrless 0$). Thus

$$I < k \int \frac{dz}{(a+z)^2},$$

which tends to 0 as $R \rightarrow \infty$. Thus $I \rightarrow 0$ as $R \rightarrow \infty$ and (16.99) establishes the result

$$\sum_{n=-\infty}^{\infty} \frac{1}{(a+n)^2} = \frac{\pi^2}{\sin^2 (\pi a)}.$$

Series with alternating signs in the terms, i.e. $(-1)^n$, can also be attempted this way, but using $\cosec \pi z$ instead of $\cot \pi z$ since this has residue $(-1)^n \pi^{-1}$ at $z = n$. See section 16.16 for an example.

16.15   Inverse Laplace transform

As a final example of contour integration we mention a method whereby the process of Laplace transformation, discussed in section 5.11, can be inverted.

It will be recalled (equation (5.60)) that the Laplace transform $F(s)$ of a function $f(x)$, $x \geqslant 0$, is given by

$$F(s) = \int_0^{\infty} e^{-sx} f(x) \, dx, \quad \mathrm{Re}\, s > 0. \tag{16.100}$$

In chapter 5, functions were deduced from the transforms by means of a prepared dictionary. However, an explicit formula for an unknown inverse may be written in the form of an integral. It is known as the **Bromwich integral**, and is given by

$$f(x) = \frac{1}{2\pi i} \int_{\lambda - i\infty}^{\lambda + i\infty} e^{sx} F(s) \, ds, \quad \lambda > 0, \tag{16.101}$$

where s is treated as a complex variable and the integration is along the line L indicated in fig. 16.13. The position of the line is dictated by the requirements that λ is positive and that all singularities of $F(s)$ lie to the left of the line.

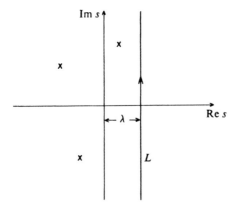

Fig. 16.13 The integration path of the inverse Laplace transform is along the infinite line L. The quantity λ must be positive and large enough for all poles of the integrand to lie to the left of L.

That (16.101) really is the unique inverse of (16.100) is difficult to show for general functions and transforms, but the following verification at least makes it plausible,

$$f(x) = \frac{1}{2\pi i} \int_{\lambda - i\infty}^{\lambda + i\infty} ds \, e^{sx} \int_0^\infty e^{-sx'} f(x') \, dx',$$

$$\text{Re} \, (s) > 0, \quad \text{i.e. } \lambda > 0,$$

$$= \frac{1}{2\pi i} \int_0^\infty dx' f(x') \int_{\lambda - i\infty}^{\lambda + i\infty} e^{s(x - x')} \, ds$$

$$= \frac{1}{2\pi i} \int_0^\infty dx' f(x') \int_{-\infty}^\infty e^{\lambda(x - x')} e^{ip(x - x')} i \, dp,$$

$$\text{putting } s = \lambda + ip,$$

$$= \frac{1}{2\pi} \int_0^\infty f(x') \, e^{\lambda(x - x')} 2\pi \, \delta(x - x') \, dx'$$

$$\left. \begin{array}{l} = f(x), \; x \geqslant 0, \\ = 0, \; x < 0. \end{array} \right\} \tag{16.102}$$

Our main interest here is in the use of contour integration. To employ it to evaluate the line integral in (16.101), the path L must be made into a closed contour in such a way that the contribution from the completion either vanishes or is simply calculable.

A typical completion is shown in fig. 16.14(a) and would be appropriate if $F(s)$ had a finite number of poles. For more complicated cases in which F has an infinite sequence of poles (but all to the left of L, as in fig. 16.14(b))

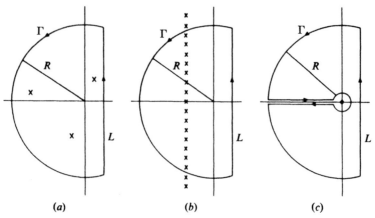

Fig. 16.14 Some contour completions of the integration path L of the inverse Laplace transform. For details of when each is appropriate see the main text.

a sequence of circular arc completions which pass between the poles must be used and $f(x)$ is obtained as a series. If $F(s)$ is a multivalued function, a cut plane is needed and a contour such as that shown in fig. 16.14(c) may be appropriate.

We consider here only one simple example [for which the answer is already given in chapter 5] and refer the reader to the examples of section 16.16 for others.

Example 16.9. Find the function $f(x)$ whose Laplace transform is

$$F(s) = s^{-1}(1 - e^{-sX}),$$

where X is fixed and positive.

From (16.101) we have the integral

$$f(x) = \frac{1}{2\pi i} \int_{\lambda - i\infty}^{\lambda + i\infty} \frac{e^{sx} - e^{s(x - X)}}{s} \, ds, \qquad (16.103)$$

whose integrand has a singularity only at the origin. Thus any positive value of λ will suffice.

Using the general result (16.102),

$$f(x) = 0 \quad \text{for } x < 0. \qquad (16.104\,a)$$

[This could be obtained formally by closing L with a circular arc Γ in the right half-plane, thus enclosing no poles, and observing that the integrand $\to 0$ everywhere on Γ since $\text{Re } s > 0$ and $x < 0$. With no poles enclosed and no contribution from Γ, the integral along L must also be zero.] So we have to consider $x > 0$, and here two cases arise.

(i) $x > X > 0$. Close L as in fig. 16.14(a). On Γ, $\text{Re } s < 0$ [λ can be as small as desired] and $s \cdot s^{-1}$ {exp (sx) − exp $[s(x − X)]$} tends to zero everywhere on Γ as $R \to \infty$. As in section 16.12 the integral round Γ vanishes for infinite R and so by the residue theorem,

$$\frac{1}{2\pi i} \int_L \frac{e^{sx} - e^{s(x-X)}}{s} \, ds = \frac{2\pi i}{2\pi i} \{\text{residue at } s = 0\} = 0.$$

Thus

$$f(x) = 0 \quad \text{for } x > X > 0. \tag{16.104 b}$$

(ii) $X > x > 0$. Here the two parts of the integrand behave in different ways and have to be treated separately,

$$2\pi i(I_1 - I_2) \equiv \int_L \frac{e^{sx}}{s} \, ds - \int_L \frac{e^{s(x-X)}}{s} \, ds.$$

The integrand of I_1 then vanishes in the far left-hand half-plane and I_1 is evaluated as in (i) to yield

$$2\pi i I_1 = 2\pi i\{\text{residue of } s^{-1} e^{sx} \text{ at } s = 0\} = 2\pi i. \tag{16.104 c}$$

The integrand of I_2 vanishes in the far right-hand half-plane and is evaluated by a circular arc completion in that half-plane. Such a contour encloses no poles and, as explained immediately following result (16.104 a), leads to $I_2 = 0$.

Thus collecting together results (16.104 a–c) we obtain:

$$\begin{aligned} f(x) &= 0, & x &< 0, \\ &= 1, & 0 &< x < X, \\ &= 0, & x &> X, \end{aligned}$$

as shown in fig. 16.15. This is the same result as given by entries 1 and 13 of table 5.1.

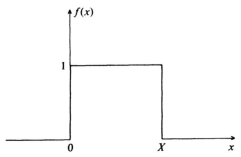

Fig. 16.15 The result of the Laplace inversion carried out in example 16.9.

16.16   Examples for solution

1. Find an analytic function of $z = x + iy$ whose imaginary part is $(y \cos y + x \sin y) \exp (x)$.

2. Find a function $f(z)$, analytic in a suitable part of the Argand diagram, for which

$$\text{Re} f = \frac{\sin 2x}{\cosh 2y - \cos 2x}.$$

Where are the singularities of $f(z)$?

3. Find the radii of convergence of the following series

(a) $\sum_{1}^{\infty} \frac{z^n}{n^n}$;   (b) $\sum_{2}^{\infty} \frac{z^n}{\ln n}$;   (c) $\sum_{1}^{\infty} \frac{n! z^n}{n^n}$;

(d) $\sum_{1}^{\infty} z^n n^{\ln n}$;   (e) $\sum_{1}^{\infty} \left(\frac{n + p}{n}\right)^{n^2} z^n$ (p real in (e)).

4. For the function

$$f(z) = \log\left(\frac{z + c}{z - c}\right),$$

where c is real, show that the real part u of f is constant on a circle of radius c cosech u centred on the point $z = c \coth u$.

Use this result to show that the (electrical) capacity/unit length of two parallel cylinders of radii a, placed with their axes $2d$ apart, is proportional to $(\text{arcosh} (d/a))^{-1}$.

5. Find a complex potential in the t-plane appropriate to the physical situation in which the half-plate $r > 0$, $s = 0$ has zero potential and the half-plate $r < 0$, $s = 0$ has potential V.

By making the transformation $z = \frac{1}{2}a(t + t^{-1})$ with a real and positive, find the electrostatic potential associated with the two half-plates $x > a$, $y = 0$ and $x < -a$, $y = 0$ at potentials 0 and V respectively.

6. By considering in turn the transformations

$$z/c = \frac{1}{2}(w + w^{-1}), \quad w = \exp \zeta,$$

where $z = x + iy$, $w = r \exp (i\theta)$, $\zeta = \xi + i\eta$, and c is a real positive constant, show that $z/c = \cosh \zeta$ maps the strip $\xi \geq 0$, $0 \leq \eta \leq 2\pi$ onto the whole z-plane. What curves in the z-plane correspond to the lines $\xi = $ constant, $\eta = $ constant? Identify those corresponding to $\xi = 0$, $\eta = 0$, $\eta = 2\pi$.

The electric potential ϕ of a charged conducting strip $-c \leq x \leq c$, $y = 0$ satisfies

$$\phi \sim -k \ln (x^2 + y^2)^{1/2} \quad \text{for large } (x^2 + y^2)^{1/2},$$

with ϕ constant on the strip. Show that $\phi = \text{Re}(-k \operatorname{arcosh} (z/c))$ and that the magnitude of the electric field near the strip is $k(c^2 - x^2)^{-1/2}$.

7. Find the Taylor series expansion about the origin for the function $f(z)$ defined by

$$f(z) = \sum_{r=1}^{\infty} (-1)^{r+1} \sin\left(\frac{pz}{r}\right),$$

where p is a constant. Hence verify formally that the definition of $f(z)$ is a convergent series for all z.

8. Identify the zeros, poles and essential singularities of the function $w = \tan z$.

9. For the equation $8z^3 + z + 1 = 0$,

(i) show that all three roots lie between the circles $|z| = \frac{3}{8}$ and $|z| = \frac{5}{8}$;
(ii) approximately locate the real root, and hence deduce that the complex ones lie in the first and fourth quadrants and have moduli greater than 0.5.

10. (i) Prove that $z^8 + 3z^3 + 7z + 5$ has two zeros in the first quadrant.
(ii) Find in which quadrants the zeros of $2z^3 + 7z^2 + 10z + 6$ lie. Try to locate them.

11. *Zeros inside* $|z| = R$. The following is a method of determining the number of zeros of an nth degree polynomial $f(z)$ inside the contour C given by $|z| = R$;

(i) put $z = \dfrac{R(1 + it)}{(1 - it)}$ with $t\,(= \tan \theta/2)$ in $-\infty \leqslant t \leqslant \infty$,

(ii) obtain $f(z)$ as $\dfrac{A(t) + iB(t)}{(1 - it)^n} \dfrac{(1 + it)^n}{(1 + it)^n}$,

(iii) $\arg f(z) = \arctan (B/A) + n \arctan t$,

(iv) $\Delta_C [\arg f(z)] = \Delta_C[\arctan (B/A)] + n\pi$.

(v) By inspection or sketch graph, determine $\Delta_C[\arctan (B/A)]$, by finding the discontinuities in B/A and evaluating $\arctan (B/A)$ at $t = \pm\infty$.

Use this method, together with ▶22 (page 472) to show that the zeros of $z^4 + z + 1$ in the second and third quadrants have $|z| < 1$.

12. By considering the real part of

$$\int \frac{-i z^{n-1}\, dz}{1 - a(z + z^{-1}) + a^2}, \quad n \text{ integral and} \geqslant 0,$$

where $z = \exp(i\theta)$, evaluate

$$\int_0^\pi \frac{\cos n\theta\, d\theta}{1 - 2a \cos \theta + a^2}, \quad a \text{ real and} > 1.$$

13. Evaluate $\displaystyle\int_{-\pi}^{\pi} \frac{\sin\theta}{a - \sin\theta}\, d\theta$, where a is real and > 1.

[Prove first that if $f(z)$ has a simple zero at z_0, then $1/f(z)$ has residue $1/f'(z_0)$ there.]

14. Evaluate $\displaystyle\int_0^\infty \frac{t\sin\alpha t}{1 + t^2}\, dt$.

15. Prove $\displaystyle\int_0^\infty \frac{\cos mx\, dx}{4x^4 + 5x^2 + 1} = \frac{\pi}{6}(4e^{-m/2} - e^{-m}), \quad m > 0$.

16. Show that the principal value of the integral

$$\int_{-\infty}^\infty \frac{\cos(x/a)}{x^2 - a^2}\, dx$$

is $-(\pi/a)\sin 1$.

17. [Harder problem] (i) Evaluate the integral of $\exp(i\pi z^2)\, \mathrm{cosec}(\pi z)$ around the parallelogram with corners $\pm\frac{1}{2} \pm R\exp(i\pi/4)$.
(ii) Show that the parts of the contour parallel to the real axis give no contribution when $R \to \infty$.
(iii) Evaluate the integrals along the other two sides by putting $z' = r\exp(i\pi/4)$ and working in terms of $z' + \frac{1}{2}$ and $z' - \frac{1}{2}$. Hence by letting $R \to \infty$ show

$$\int_{-\infty}^\infty e^{-\pi r^2}dr = 1.$$

18. Evaluate $\displaystyle\int_0^\infty \frac{x^{-\alpha}}{x + 1}\, dx$, α real, $0 < \alpha < 1$.

19. Show $\displaystyle\int_0^\infty \frac{\ln x}{x^{3/4}(1 + x)}\, dx = -\sqrt{2}\pi^2$ and $\displaystyle\int_0^\infty \frac{dx}{x^{3/4}(1 + x)}$

$$= \sqrt{2}\pi.$$

20. Prove that $\displaystyle\sum_{-\infty}^\infty (n^2 + \tfrac{3}{4}n + \tfrac{1}{8})^{-1} = 4\pi$.

[It may prove instructive to carry out the summation numerically, say between -4 and 4, and note how much of the sum comes from values near the poles of the contour integration.]

21. By considering the integral of

$$\left(\frac{\sin z\alpha}{z\alpha}\right)^2 \frac{\pi}{\sin \pi z}$$

around a circle of large radius, evaluate, for $\alpha < \pi/2$, the sum

$$\sum_{m=1}^\infty (-1)^{m-1} \frac{\sin^2(m\alpha)}{(m\alpha)^2}.$$

22. Use the Bromwich inversion and contours like fig. 16.14(a) to find the functions of which the following are the Laplace transforms,

(i) $s(s^2 + b^2)^{-1}$.
(ii) $n!(s - a)^{-(n+1)}$, n integral and > 0, $s > a$ [change variable to $t = s - a$].
(iii) $a(s^2 - a^2)^{-1}$, $s > |a|$ [change to $t = s - |a|$].
Compare your answers with table 5.1.

23. Use the contour in fig. 16.14(c) to show that the function whose Laplace transform is $F(s) = s^{-1/2}$, is $(\pi x)^{-1/2}$. [For an integrand of the form $r^{-1/2} \exp(-rx)$ change to a variable $t = r^{1/2}$.]

24. In example 18(iii) of section 9.14 an expression was given for the transform of a function $v(x, t)$,

$$V(x, s) = s^{-1} \exp(-kxs^{1/2}).$$

Show that its Bromwich integral inversion can be converted into a real integral as follows:

(i) Consider a completion such as shown in fig. 16.14(c) and verify that it encloses no poles of the integrand,
(ii) on the upper (side of the) cut take s as $r \exp(i\pi)$ and on the lower as $r \exp(-i\pi)$. Write $k^2r = u^2$ and show that the contributions from the two cuts are together

$$\frac{2}{\pi} \int_0^\infty \exp\left(-\frac{u^2t}{k^2}\right) \frac{\sin ux}{u} \, du,$$

(iii) show that, in the limit of its radius ρ tending to zero, the small circle γ contributes -1 to the integral to give for the complete expression for $v(x, t)$ the *real* integral

$$v(x, t) = 1 - \frac{2}{\pi} \int_0^\infty \exp\left(-\frac{u^2t}{k^2}\right) \frac{\sin ux}{u} \, du.$$

[Assume, as can be shown, that the contribution from Γ vanishes as $R \to \infty$.]

Solutions and hints for exercises and examples

CHAPTER 1

▶4. Square each of the equations (1.23) and subtract.

▶5. Use (1.16).

▶7. $a(1 - x)^{-1}(1 - x^{N+1})$, convergent for $|x| < 1$, divergent for $x \geqslant 1$, oscillates finitely for $x = -1$, oscillates infinitely for $x < -1$.

▶8. (a) $Na + \frac{1}{2}dN(N - 1)$, divergent; (b) $\ln (N + 1)$, divergent; (c) $\frac{1}{3}(1 - (-2)^N)$, oscillates infinitely; (d) $1 - (N + 1)^{-1}$, convergent to $S = 1$; (e) add $S_N/3$ to the S_N series,

$$(3/16)(1 - (-3)^{-N}) + (3N/4)(-3)^{-(N+1)},$$

convergent to $3/16$.

▶9. (i) Convergent, compare with ▶8(d); (ii) divergent, compare with $\sum n^{-1}$; (iii) convergent, compare with ▶8(d); (iv) divergent, $a_n \nrightarrow 0$; (v) convergent, ratio test; (vi) convergent, root test; (vii) divergent, ratio $\rightarrow$ e, or $a_n \nrightarrow 0$.

▶10. (a) Absolutely convergent, compare with ▶9(i); (b) oscillates finitely; (c) convergent, alternate signs; (d) absolutely convergent for all x; (e) absolutely convergent, use partial fractions; (f) oscillates infinitely.

▶11. (a) 3; (b) $2x + 1$, use the binomial expansion of $(A + B)^2$; (c) 2; (d) 0; (e) $\cos x$, use the formula for $\sin (A + B)$ and that $\sin B \simeq B$ and $\cos B \simeq 1$ in the limit $B \rightarrow 0$.

▶12. (a) $8x^7$; (b) $-3/x^4$; (c) $x(1 + x^2)^{-1/2}$; (d) $-2 \sin (2x)$; (e) $a \sec^2 (ax)$; (f) $a(1 + a^2x^2)^{-1}$; (g) $\exp (x)$; (h) $\sinh (x)$, see definition (1.23 a) and result (g); (i) $(1 + x^2)^{-1/2}$; (j) x^{-1}; (k) $6 \sin 3x \cos 3x = 3 \sin 6x$.

▶13. Write $\dfrac{1}{\Delta x}\left(\dfrac{1}{f(x + \Delta x)} - \dfrac{1}{f(x)}\right)$

as $-\dfrac{1}{f(x + \Delta x)f(x)} \dfrac{f(x + \Delta x) - f(x)}{\Delta x}$,

(a) $- 6(2x + 3)^{-4}$; (b) $- 2 \exp (- 2x)$; (c) $2 \sec^2 x \tan x$; (d) $- 9 \operatorname{cosech}^3 (3x) \coth (3x)$.

▶15. (a) $(x^2 + 2x) \exp (x)$;
(b) $\cos x \cosh x + \sin x \sinh x$;
(c) $\ln x$;
(d) $3x^2 \tan^2 x(\tan x + x \sec^2 x)$;
(e) $[\sin (ax) + ax \cos (ax) + \lambda x \sin (ax)] \exp (\lambda x)$;

(f) $2x \cos(1 - x^2) + 2x^3 \sin(1 - x^2)$;

(g) $-2x(a^2 + x^2)^{-1}(b^2 + x^2)^{-2}(c^2 + x^2)^{-3}[(a^2 + x^2)^{-1} + 2(b^2 + x^2)^{-1}$
$+ 3(c^2 + x^2)^{-1}]$.

▶16. (a) $\dfrac{a - x}{(a + x)^3}$; (b) $\dfrac{1 - \frac{1}{2}x}{(1 - x)^{3/2}}$;

  (c) $\dfrac{1}{\cos^2 x} = \sec^2 x$; (d) $\dfrac{-7x^2 - x + 2}{(4x^2 - 2x + 1)^2}$.

▶19. (a) $(2 \sec^2 x - 4) \tan x \sin 2x + 4 \sec^2 x \cos 2x$;

  (b) $\sin x(2x^{-3} - 3x^{-1}) - \cos x(3x^{-2} + \ln x)$;

  (c) $8(4x^3 + 30x^2 + 62x + 38) \exp(2x)$.

▶20. Show $xy^{(1)} = (x + 2)y$ and then use Leibniz.

▶21. (a) $\displaystyle\sum_{n=0}^{\infty} \dfrac{(-1)^n(2x)^{2n+1}}{(2n + 1)!}$; (b) $x + \dfrac{x^3}{3} + \dfrac{2x^5}{15}$;

  (c) $-\displaystyle\sum_{n=1}^{\infty} \dfrac{(1 - x)^n}{n}$; (d) $x + x^2 + \dfrac{x^3}{3} - \dfrac{x^5}{30}$.

▶22. (a) $x = -2$, max., $x = 3$, min.; (b) $x = (n + \frac{1}{2})(\pi/a)$, max./min. for n even/odd; (c) no real max. or min., inflection at $x = 0$; (d) max. at $x = -(3/5)^{1/2}$, min. at $x = (3/5)^{1/2}$, inflection at $x = 0$.

▶23. -26 at $x = -2$, other stationary values are 6 at $x = 0$ and 1 at $x = 1$.

▶24. (i)

	(a)	(b)	(c)	(d)	(e)
$\dfrac{\partial f}{\partial x}$	$2xy$	$2x$	$\dfrac{1}{y}\cos\left(\dfrac{x}{y}\right)$	$\dfrac{-y}{x^2 + y^2}$	$\dfrac{x}{r}$
$\dfrac{\partial f}{\partial y}$	x^2	$2y$	$-\dfrac{x}{y^2}\cos\left(\dfrac{x}{y}\right)$	$\dfrac{x}{x^2 + y^2}$	$\dfrac{y}{r}$

for (e) $\partial f/\partial z = z/r$;

(ii) (a) $2y$, 0, $2x$, (b) 2, 2, 0, (c) $(y^2 + z^2)r^{-3}$, $(x^2 + z^2)r^{-3}$, $-xyr^{-3}$;

(iii) both $= (y^2 - x^2)(x^2 + y^2)^{-2}$.

▶26. $3x + (2x + 3y)(1 - x^2)^{1/2} = 3 \sin y + (2 \sin y + 3y) \cos y$.

▶27. $\dfrac{\partial f}{\partial x} = \dfrac{1}{(x^2 + y^2 + z^2)^{1/2}}\left(x + \dfrac{y \cos^2 y \cot y}{2} + \dfrac{z^3 + 2xz}{1 - 2xz}\right)$.

▶28. $2^{-1/2}[1 + x - \pi/4 + y - \frac{1}{2}(x - \pi/4)^2 - \frac{1}{2}y^2 - (x - \pi/4)y + y(z - 1)]$, $\partial f/\partial z = \partial^2 f/\partial z^2 = \partial^2 f/\partial x\,\partial z = 0$, but all the other partial derivatives contribute.

▶29. $\dfrac{dr}{dt} = \dfrac{8t^3 + \sin 4t - e^{-2t}}{(4t^4 + \sin^2 2t + e^{-2t})^{1/2}}$.

▶31. $\dfrac{\sin 2\phi}{2}\left(\dfrac{\partial^2 f}{\partial r^2} - \dfrac{1}{r^2}\dfrac{\partial^2 f}{\partial \phi^2} - \dfrac{1}{r}\dfrac{\partial f}{\partial r}\right) + \cos 2\phi\left(\dfrac{1}{r}\dfrac{\partial^2 f}{\partial \phi\,\partial r} - \dfrac{1}{r^2}\dfrac{\partial f}{\partial \phi}\right)$.

▶32. $4 \dfrac{\partial^2 f}{\partial \xi \, \partial \eta} + 2 \dfrac{\partial f}{\partial \xi} = 0$.

▶33. Use (1.89) twice on (1.91), or once on (1.90).

▶34. (i) $\dfrac{x}{z} \times \dfrac{-y}{z} \times \dfrac{z}{y} = -1$;

(ii) $\dfrac{y}{y^2 + x^2} \times \dfrac{x}{y} \times (-x)\left(1 + \dfrac{y^2}{x^2}\right) = -1$.

▶35. Take x, y as (i) S, p; (ii) T, p; (iii) T, V, in (1.95).

▶37. The second partial derivatives are (apart from a common factor of $\exp(-x^2 - y^2)$), $4x^5 - 14x^3 + 6x$, $y(4x^4 - 6x^2)$, $x^3(4y^2 - 2)$.

▶38. Min. at $(2, 3, 1)$. In this example all second derivatives are independent of x, y, z which is not the case in general.

▶39. Max. value of 15 at $\pm 5^{-1/2} (2, 1)$, min. value of 5 at $\pm 5^{-1/2}(1, -2)$.

▶40. (i) Yes, for $\lambda > 0$, value λ^{-1}; (ii) yes, value 0; (iii) no, $\ln(1 + R) \to \infty$ as $R \to \infty$; (iv) no, $\epsilon^{-1} \to \infty$ as $\epsilon \to 0$; (v) no, $\ln(\sin \theta) \to -\infty$ as $\theta \to 0$; (vi) yes, value 1.

▶41. (i) $(2 - y^2) \cos y + 2y \sin y - 2$;
(ii) $\frac{1}{2}y^2 \ln y + \frac{1}{4}(1 - y^2)$;
(iii) $y \arcsin y + (1 - y^2)^{1/2}$;
(iv) $-y^{-1} \ln(a^2 + y^2) + (2/a) \arctan(y/a)$;
(v) and (vi) 3/5.

▶42. $n!$.

▶43. $J_n = \frac{1}{2}(n - 1)J_{n-2}$.

▶44. (i) 1/60.

▶45. $(a^2 b/6)(ab + 3)$.

▶46. $\frac{8}{3}abc(a^2 + b^2 + c^2)$.

▶47. If e.g. the order of integration used is z, y, x, the corresponding limits are $[0, 1 - x - y]$, $[0, 1 - x]$, and $[0, 1]$; 1/360.

▶48. $2\pi(1 + \lambda^2)^{-1}$.

▶50. $\displaystyle\int_0^1 \left(\int_{-\eta}^{\eta} \dfrac{\xi^2}{2} \, d\xi \right) d\eta = \dfrac{1}{12}$.

▶51. $\displaystyle\int_0^a \int_{-b}^{b} \int_0^{2\pi} \rho(r^2 + z^2) r \, d\phi \, dz \, dr$.

Section 1.20

1. $x = x' \cosh \phi + ct' \sinh \phi$, $ct = x' \sinh \phi + ct' \cosh \phi$, use (1.24).
2. (a) (i) 1 for $N \neq 1$, 2 for $N = 1$, (ii) 2 for $N = 1$, 3 for $N = 2$, 1 otherwise,
(iii) 2 for $x = 1$ and $N = 1$, $(1 - \exp(2\pi i/N)x^{N+1})/(1 - \exp(2\pi i/N)x)$ otherwise;
(b) (i) -2 for $N = 2$, 0 otherwise, (ii) $-\surd(3)$, consider $\operatorname{Im} \sum \omega_m 2^m$.

3. To sum numerator, subtract $\exp(-x)$ times the series from the original series, $\bar{E} = h\nu[\exp(h\nu/kT) - 1]^{-1}$.

4. $r = 0$ divergent, $a_n \nrightarrow 0$; $r = 1$ divergent, see (1.33); $r = 2$ convergent, resolve into partial fractions; $r > 2$ convergent, by comparison with $r = 2$ case.

5. $x < e^2$ by the root test (1.36).

6. Divide the series into two series n odd and n even. For $r = 2$ both are absolutely convergent by comparison with $\sum n^{-2}$. For $r = 1$ neither series is convergent by comparison with $\sum n^{-1}$. However, the sum of the two is by the alternating sign test, or by showing that the terms cancel in pairs.

7. Use Leibniz; $x = 2a$, $x = 6a$, $x = \infty$; $f^{(8)}$ has values $28a^{-6}e^{-2}$, $-4a^{-6}e^{-6}$, 0.

8. Use Leibniz; (i) and (ii) differentiate the definition of g_n and substitute; (iii) multiply the $y^{(n)}$ recurrence relation through by $\exp(\tfrac{1}{2}x^2)$; use the g_n recurrence relation, $g_0 = 1$, $g_1 = -x$, $g_2 = x^2 - 1$, $g_3 = 3x - x^3$.

9. $x - \dfrac{x^3}{3!} + \dfrac{9x^5}{5!}$.

10. (a) 3; (b) -4, repeat the procedure until 0/0 is not obtained; (c) 0.

11. (i) 0; (ii) $t(s^2 + t^2)$; (iii) $-\sin 2t$.

12. Stationary points at (a) $x = 0$, $y = 0$; (b) $x = 0$, $y = \pm 1$; (c) $x = \pm 1$, $y = 0$. The second-order determinants are

$$(a) \begin{vmatrix} 2 & 0 \\ 0 & -2 \end{vmatrix}; \quad (b) \begin{vmatrix} 4e^{-1} & 0 \\ 0 & 4e^{-1} \end{vmatrix}; \quad (c) \begin{vmatrix} -4e^{-1} & 0 \\ 0 & -4e^{-1} \end{vmatrix};$$

showing saddle point in case (a), minima in cases (b), maxima in cases (c).

13. Minimize $y_1 + y_2 + \tfrac{1}{2}a[(x_1 - 2a)^2 + (y_1 - y_2)^2]$ subject to $x_1 = -(a^2 + y_1^2)^{1/2}$.

14. Consider the real and imaginary parts, $I = \alpha/(\alpha^2 + \omega^2)$, $J = \omega/(\alpha^2 + \omega^2)$.

15. (i) Draw a sketch or investigate the monotonic behaviour of $\sin x$ in the range to give $(2/\pi)x \leqslant \sin x \leqslant x$. This leads to

$$\tfrac{2}{3}[(1 + \pi/2)^{3/2} - 1] > I > (\pi/3)(2^{3/2} - 1),$$

i.e. $2.08 > I > 1.91$; (ii) $I = 2$.

16. Form dI/dy showing that it equals

$$-\frac{\operatorname{sech}^2 y}{\tanh y}\left[\operatorname{arsinh}\left(\frac{\cos x}{\operatorname{cosech} y}\right)\right]_0^{\pi/2}.$$

$K = \pi^2/4$.

17. $\tfrac{7}{6}$.

18. (i) Use integration by parts, $\tfrac{1}{2}\nu \ln(1 + \mu^2/\nu^2) + \mu \arctan(\nu/\mu)$;
    (ii) $4[(\nu + \mu)^{1/2} - \nu^{1/2} - \mu^{1/2}]$.

19. Jacobian $= (u/v)^{1/2} + (v/u)^{1/2}$. Area in uv-plane is triangle bounded by $v = 0$, $u = v$, $u = a$. Integral has value a^2.

CHAPTER 2

▶2. Note $\hat{\mathbf{n}} \cdot \mathbf{a} = a \cos AOP = p$, since OPA is a right angle.

▶4. $(\mathbf{a} \wedge \mathbf{b}) \wedge \mathbf{c} = [(a_z b_x - a_x b_z)c_z - (a_x b_y - a_y b_x)c_y]\mathbf{i} + \text{cyclic}$
$= [(a_x c_x + a_y c_y + a_z c_z)b_x - (b_x c_x + b_y c_y + b_z c_z)a_x]\mathbf{i} + \text{cyclic}$
$= (\mathbf{a} \cdot \mathbf{c})\mathbf{b} - (\mathbf{b} \cdot \mathbf{c})\mathbf{a}$.

▶5. $(\mathbf{a} \wedge \mathbf{b}) \wedge \mathbf{c}$ lies in the plane containing $\mathbf{a}$ and $\mathbf{b}$; $\mathbf{a} \wedge (\mathbf{b} \wedge \mathbf{c})$ lies in that containing $\mathbf{b}$ and $\mathbf{c}$.

▶7. Use (2.26) to evaluate $\mathbf{a}' \wedge \mathbf{b}'$ and then (2.28) to simplify the result and to evaluate $\lambda' = (\mathbf{a}' \wedge \mathbf{b}') \cdot \mathbf{c}'$. Hence conclude that $\mathbf{c}'' \equiv (\lambda')^{-1}(\mathbf{a}' \wedge \mathbf{b}') = \mathbf{c}$.

Section 2.10

1. (a) False; (b) false, see ▶5; (c) true, note $\mathbf{a} \wedge (\mathbf{b} \wedge \mathbf{c}) = -(\mathbf{b} \wedge \mathbf{c}) \wedge \mathbf{a} = (\mathbf{c} \wedge \mathbf{b}) \wedge \mathbf{a}$ and then use (2.26); (d) true; (e) true, $\mathbf{c}$ is parallel to $\mathbf{a} - \mathbf{b}$; (f) false, left side $= -$ right side, from (2.26).

2. $R \arccos (\cos \theta_1 \cos \theta_2 + \sin \theta_1 \sin \theta_2 \cos (\phi_2 - \phi_1))$.

3. Note $(\mathbf{a} \wedge \mathbf{b}) \cdot (\mathbf{c} \wedge \mathbf{d}) = \mathbf{d} \cdot ((\mathbf{a} \wedge \mathbf{b}) \wedge \mathbf{c})$.

4. Common perpendicular is parallel to $\mathbf{b}' \wedge \mathbf{b}$. Required distance is the projection of $\mathbf{a}' - \mathbf{a}$ in this direction, leading to

$$|(\mathbf{a}' - \mathbf{a}) \cdot (\mathbf{b}' \wedge \mathbf{b})|/|\mathbf{b}' \wedge \mathbf{b}|.$$

5. Note that the plane contains the vectors $h^{-1}\mathbf{a} - k^{-1}\mathbf{b}$ and $h^{-1}\mathbf{a} - l^{-1}\mathbf{c}$.

6. (a) $\pm 2a^2(1, 0, -1)$ m^2; (b) $\arccos (2/3)^{1/2}$; (c) $-x + z = 2a$; (d) (i) $a/\sqrt{2}$ m, (ii) $a^3/3$ m^3; (e) (i) $\mathbf{M} = (20\sqrt{6})a(1, -1, 0)$ N m, using $\overrightarrow{AH} \wedge$ force, (ii) moment about line $= F \times \sin$ (angle between directions HJ and AF) $\times$ length of common perpendicular (see example 4), which can be manipulated to give $\mathbf{M} \cdot \widehat{\mathbf{AF}} = -(10/\sqrt{3})a$ N m, (iii) $5\sqrt{6}a$ Joules; (f) total $\boldsymbol{\omega} = (1, 1, 3)$ rad s^{-1}, (i) $\sqrt{2a}/3$ m, (ii) $\sqrt{(24)}a$ m s^{-1}.

CHAPTER 3

($\dot{x}$ denotes dx/dt.)

▶1. Note $\mathbf{a} \wedge \ddot{\mathbf{b}} = \mathrm{d}(\mathbf{a} \wedge \dot{\mathbf{b}})/\mathrm{d}t - \dot{\mathbf{a}} \wedge \dot{\mathbf{b}}$.

▶3. $-\int_1^2 c\{c^3/t^3\,\mathbf{i} + 2/t^2\,\mathbf{j} + c/t\,\mathbf{k}\}\,\mathrm{d}t$ is the required value.

▶4. $\int_c^{2c} (\tfrac{1}{4}x(3c - x)^2 - 1 + 0)\,\mathrm{d}x$ is the required value.

▶5. $\mathrm{d}\mathbf{H} = \dfrac{I}{4\pi r^3}\,(\mathrm{d}\mathbf{s} \wedge \mathbf{r})$;

$$\mathbf{F}_1 = \int_{C_1} \int_{C_2} \frac{\mu\mu_0 I_1 I_2}{4\pi r^3}\,(\mathrm{d}\mathbf{s}_1 \wedge (\mathrm{d}\mathbf{s}_2 \wedge \mathbf{r})),$$

where $\mathbf{r} = \mathbf{s}_1 - \mathbf{s}_2$.

▶6. $H_i = \omega_i \int r^2 \rho\,\mathrm{d}V - \int x_i x_j \omega_j \rho\,\mathrm{d}V$.

Section 3.8

1.   (a) $\mathbf{F} = m\ddot{\mathbf{r}} + \dot{m}\dot{\mathbf{r}}$; (b) $m\ddot{\mathbf{r}} + k\mathbf{r} = \mathbf{0}$; (c) $\mathbf{E} = \dot{\mathbf{r}} \wedge \mathbf{B}$; (d) $m\ddot{\mathbf{r}} = e(\dot{\mathbf{r}} \wedge \mathbf{B})$
     $+ e\mathbf{E}$; (e) $\sum_i \mathbf{r}_i \wedge \mathbf{F}_i = \mathrm{d}(\sum_i m_i(\mathbf{r}_i \wedge \dot{\mathbf{r}}_i))/\mathrm{d}t$.

2. $= \int \dfrac{\mathrm{d}}{\mathrm{d}t}\, ((\mathbf{b}\cdot\mathbf{a})\mathbf{a} - \mathbf{a}^2\mathbf{b})\,\mathrm{d}t = \mathbf{a} \wedge (\mathbf{a} \wedge \mathbf{b}) + \mathbf{h}$.

3.   Proceed as in example 3.3 and ▶4. This is an example of the type of
     function discussed there.

4. $\mathbf{M} = I\int \mathbf{r} \wedge (\mathrm{d}\mathbf{r} \wedge \mathbf{B})$; consider integrals along the four sides separately,
     $\mathbf{B} = (B, 0, 0)$ and, e.g. on one vertical side $\mathbf{r} = (a\cos\phi, a\sin\phi, z)$,
     $\mathrm{d}\mathbf{r} = (0, 0, \mathrm{d}z)$, the first integral contributes $(0, 0, 2abBI\cos\phi)$ and the
     second $(0, 0, 0)$.

5.   Equation (3.17) leads to $\iint c(x^2 + y^2 + c^2)^{-3/2}\,\mathrm{d}x\,\mathrm{d}y$, then put $x = (y^2 + c^2)^{1/2}\tan\theta$.

6.   Note that for a band element of surface of width $\mathrm{d}l$, $\mathrm{d}l\,\hat{\mathbf{n}}\cdot\hat{\mathbf{n}}_z = -\mathrm{d}r$, and
     that $2zr\,\mathrm{d}r$ can be integrated by parts as $z\,\mathrm{d}(r^2)$.

7.   (a) $z = u_1 u_2$, $\rho^2 = u_1^2 + u_2^2 - u_1^2 u_2^2 - 1$;
     (b) $u_1(1 - u_2^2)\cos u_3/\rho$, $u_1(1 - u_2^2)\sin u_3/\rho$, u_2; $u_2(1 - u_1^2)\cos u_3/\rho$,
     $u_2(1 - u_1^2)\sin u_3/\rho$, u_1; $-\rho\sin u_3$, $\rho\cos u_3$, 0;

     (c) $\left(\dfrac{u_1^2 - u_2^2}{u_1^2 - 1}\right)^{1/2}$, $\left(\dfrac{u_2^2 - u_1^2}{u_2^2 - 1}\right)^{1/2}$, ρ; $|u_1^2 - u_2^2|\,\mathrm{d}u_1\,\mathrm{d}u_2\,\mathrm{d}u_3$;

     (d) confocal ellipsoids, hyperboloids, planes containing the z-axis.

CHAPTER 4

▶2.   Use the chain rule of differentiation. $\partial\phi/\partial u$ is common.

▶6.   Use $\partial r/\partial x = x/r$ etc.

▶8. $\nabla \wedge \mathbf{a} = (\mathbf{i} \wedge \mathbf{j})\,\partial a_y/\partial x + (\mathbf{j} \wedge \mathbf{i})\,\partial a_x/\partial y + $ similar terms.

▶9. $\mathbf{a} = (0, r\omega, 0)$.

▶10. The terms of $\mathbf{a} \wedge \mathbf{b}$ are products and so ∇ produces 2 terms from each.

▶11. Faraday, $\displaystyle\int \mathbf{E}\cdot\mathrm{d}\mathbf{l} = -\dfrac{\partial}{\partial t}\int \mathbf{B}\cdot\mathrm{d}\mathbf{S}$.

▶12. $\nabla\cdot\mathbf{a} = \dfrac{\partial}{\partial x}\left(\dfrac{\partial b_z}{\partial y} - \dfrac{\partial b_y}{\partial z}\right) + $ cyclic.

     The six terms on the right cancel in pairs.

▶13. $\nabla \wedge \mathbf{a} = (0, -1, -2xy)$.

▶15–18. Substitute for h_1, h_2, h_3 from table 3.1.

Section 4.8

1.   (a) = (d) = $\left[\phi\left(\dfrac{\partial a_z}{\partial y} - \dfrac{\partial a_y}{\partial z}\right), \ldots, \ldots\right]$;

$(b) = \left(a_z \dfrac{\partial \phi}{\partial y} - a_y \dfrac{\partial \phi}{\partial z}, \cdots, \cdots \right);$

$(c) = \left(\phi \dfrac{\partial a_z}{\partial y} + a_z \dfrac{\partial \phi}{\partial y} - \phi \dfrac{\partial a_y}{\partial z} - a_y \dfrac{\partial \phi}{\partial z}, \cdots, \cdots \right);$

$(e) = \left(a_y \dfrac{\partial \phi}{\partial z} - a_z \dfrac{\partial \phi}{\partial y}, \cdots, \cdots \right).$

2. $\mathbf{H} = -(1/4\pi)\nabla[(\mathbf{M} \cdot \mathbf{r})r^{-3}]$ and use (4.6); $H_A = (1/4\pi)2Mr^{-3}$ along direction of $\mathbf{M}$, $H_B = (1/4\pi)Mr^{-3}$ opposed to direction of $\mathbf{M}$.

3. (a) The surface element is $a^2 \sin\theta\, d\theta\, d\phi(\sin\theta\cos\phi\,\mathbf{i} + \sin\theta\sin\phi\,\mathbf{j} + \cos\theta\,\mathbf{k})$;
    (b) div $\mathbf{F} = 0$. Close volume with plane surface $x^2 + y^2 \leqslant a^2, z = 0$; common answer $\pi a^4/4$.

4. See $\blacktriangleright 6$, $\nabla \cdot (r^n \mathbf{r}) = (3 + n)r^n$; $\mathbf{a} = f(r)\mathbf{r}$ for some f, and use $\blacktriangleright 6$ again.

5. $\phi(x, y, z) = zr^{-1} +$ constant.

6. (a) Yes, show $\nabla \wedge \mathbf{F} = 0$; $(x - y)\exp(-r^2)$;
    (b) Yes; $-\frac{1}{2}(x^2 + y^2)\exp(-r^2)$; (c) No, $\nabla \wedge \mathbf{F} \neq 0$.

7. $(-cr^{-2}, 0, 0) \neq (0, 0, 0)$.

8. $df/du_1 = Ch_1/(h_2 h_3) = C(u_1^2 - 1)^{-1}; f = B \ln[(u_1 - 1)/(u_1 + 1)]$, B and C constants.

CHAPTER 5

$\blacktriangleright 1$. Equation is $\sum_2^\infty i(i - 1)a_i x^{i-2} + \sum_0^\infty a_i x^i = 0$ implying $a_0 = (-1)^n (2n)! a_{2n}$ and $a_1 = (-1)^n (2n + 1)! a_{2n+1}$.

$\blacktriangleright 2$. $x + B = \arcsin(y/A^{1/2})$.

$\blacktriangleright 3$. (a) $4\cos x - \sin x$; (b) $4\cos x + 2\sin x$;
    (c) $a_1 \cos x + (\sqrt{8} - a_1)\sin x$; (d) $\cos x + a_2 \sin x$;
    (e) $2\cos x + \sin x$; (f) no solution.

$\blacktriangleright 4$. (a) $\pm(c - x^2)^{-1/2}$; (b) $c \arctan x$; (c) $(\ln x + 4x^{-1} - c)^{-1}$.

$\blacktriangleright 5$. (a) exact, $x^2 y^4 + x^2 + y^2 = c$; (b) IF $= x^{-1/2}$, $x^{1/2}(x + y) = c$; (c) IF $= \cos^{-2} x$, $y^2 \tan x + y = c$.

$\blacktriangleright 6$. (a) IF $= (1 - x^2)^{-2}$, $y = (1 - x^2)\arcsin x$; (b) IF $= \operatorname{cosec} x$, $y = k\sin x + \cos x$; (c) exact equation is $y^{-1}(dx/dy) - y^{-2}x = y^2$, $x = ky + \frac{1}{4}y^4$.

$\blacktriangleright 8$. (iii) $(z^2 - z)(dy/dz) + 2y = z$, leading to (v) $(1 - z)^2 y = z^2(\ln z + c') + z$; (vi) same form for y with $c - 2 = c'$.

$\blacktriangleright 10$. (i) $2x + 3$; (ii) try $a\cos 2x + b\sin 2x$ finding $a = 2$, $b = 1$; (iii) try $a\exp(2x)$ finding $a = 1$.

$\blacktriangleright 12$. $y_1' = zy_2' + z'y_2$, $y_1'' = zy_2'' + 2z'y_2' + z''y_2$.

$\blacktriangleright 14$. IF $= \sin x$, $z = \ln(\sin x) - x\cot x - c'\cot x + k'$.

$\blacktriangleright 15$. Repeated integration by parts.

$\blacktriangleright 16$. Integrate by parts twice and rearrange.

$\blacktriangleright 18$. Replace $f(x)$ in (5.65) by $g(x) = \int_0^x f(z)\, dz$, clearly $g(0) = 0$.

$\blacktriangleright 19$. Write $\sinh(ax) = \frac{1}{2}[\exp(ax) - \exp(-ax)]$.

$\blacktriangleright 22$. Use $a^2 - b^2 = (a - b)(a + b)$.

Section 5.12

1. (a) $y = \frac{1}{2}x \tan (\ln kx^{1/2})$; (b) $y = 2x \arctan (cx)$.

2. IF $= (x + 1)^2$, $y = \frac{1}{4}(x + 1)^2 + c(x + 1)^{-2}$.

3. Put $(dx/dy) = z^{-1}$ to obtain $y = 2z + \ln (z - 1) + c$; $x = 0$ when $z(z - 1) = 1$, i.e. $z = 1.618$, giving $c = -2.755$; $x = \ln z(z - 1)$.

4. (a) $(A + Bx) \exp (2x) + C \exp (-4x) + 2x + 1$;
   (b) $A \cos x + B \sin x + (C \cos x + D \sin x) \exp (3x) + \frac{1}{10}$.

5. As in example 5.5 (iv), giving $y(x) = \frac{4}{3} \exp (2x) + \frac{5}{3} \exp (-x) - 3 \cos x - \sin x$.

6. Use equation (5.43) with $\Lambda(D) = D + 1 + \sqrt{2}i$. Solution,
$$y = \exp (-x)[A \cos (\sqrt{2}x) + B \sin (\sqrt{2}x) + 8^{-1/2}x \sin (\sqrt{2}x)].$$

7. λ^2 is negative, since L is physically necessarily larger than M. Hence λ is purely imaginary, $i_2 = \omega^3 ME \cos (\omega t)[(L^2 - M^2)\omega^{4\cdot} - 2LG\omega^2 + G^2]^{-1}$.

8. $m\ddot{x} = Ee + Be\dot{y}$, $m\ddot{y} = -Be\dot{x}$, $x = A(1 - \cos \omega t)$, $y = A(\sin \omega t - \omega t)$, where $A = Ee/m\omega^2$ and $\omega = Be/m$.

9. (a) $z = x^3(\exp (x) + A)$,
   $$y = \frac{1}{4}x^4 \exp (x) + B \exp (x) - A(x^3 + 3x^2 + 6x + 6);$$
   (b) $z = A \exp (2x) - 8x - 10$ for form $(D - 2)(D - 4)y = 16x + 12$,
   $\quad = B \exp (4x) - 4x - 4$ for form $(D - 4)(D - 2)y = 16x + 12$.
   $y = C \exp (2x) + D \exp (4x) + 2x + 3$ for both forms.

10. $y = A(x) \exp (4x) + B(x) \exp (2x)$ with $A(x) = \frac{1}{20} (\sin x - 3 \cos x)$ $\times \exp (-3x)$ and $B(x) = \frac{1}{4}(\cos x - \sin x) \exp (-x)$. General solution $a \exp (4x) + b \exp (2x) + \frac{1}{10}(\cos x - 2 \sin x) \exp (x)$.

11. $y = [(m - n)(n + m + 1)]^{-1}x^m$.

12. (i) $y = x$ is a CF solution; putting $y = zx$ leads to $y = Ax + Bx \sin x - x^2 - \frac{1}{2}x \sin 2x$;
   (ii) $y = \exp (x)(A + Bx - \frac{1}{3}Bx^3)$;
   (iii) $y = \exp (x)$ is a CF solution;
   $$y = \exp (x)[B \int^x \exp (-u - \frac{1}{4}u^3) du + A].$$

13. (i) put $y' = z$, $y = a^{-1} \text{artanh} (x/a)$;
   (ii) put $y' = z$, $z = \frac{1}{2}y + Ay^{-1}$, $y^2 = B \exp (x) - 2A$.

14. Use $d /dx = \exp (-t)(d /dt)$. Equation reduces to $\ddot{y} + \dot{y} + \frac{1}{4}y = \exp (-t/2)$. General solution is $\exp (-t/2)(A + Bt + \frac{1}{4}t^2)$. Including boundary conditions, $y(x) = \frac{1}{2}x^{-1/2}(\ln x) \ln (x/e)$.

15. Make successive substitutions $y = zx$ and $x = \exp (t)$ as in section 5.10. This yields an equation in $z = z(t)$ which is solved by putting $dz/dt = u$. $y = Ax \tan (\frac{1}{2}A \ln x + AB)$.

16. $Y(s) = (s - 2)^{-1}(s - 3)^{-2}(3s^2 - 15s + 19)$ giving
$$y(x) = (2 + x) \exp (3x) + \exp (2x).$$

17. Transients $\dfrac{C + Ds}{(s + b)^2 + k^2}$, persistent $\dfrac{A + Bs}{s^2 + \omega^2}$, where $b = R/2L$ and $k^2 = (G/L) - b^2$.
   (b) damped oscillatory, exponential decay, $\propto \exp (-bt)(F + Ht)$;
   (c) $A = L(G - L\omega^2)X^{-1}$, $B = -RLX^{-1}$, where $X = (G - L\omega^2)^2 + R^2\omega^2$, $\tan \phi = -A/(B\omega)$.

18. L (amount z of C present) $= x_0(s + 8)s^{-1}(s + 2)^{-1}(s + 4)^{-1}$, yielding
$z = x_0[1 + \frac{1}{4}\exp(-4t) - \frac{3}{2}\exp(-2t)]$.

19. Write $s(s + n)^{-4}$ as $(s + n)^{-3} - n(s + n)^{-4}$.

CHAPTER 6

▶1. Bessel $x^2y'' + xy' + (x^2 - m^2)y = 0$; Legendre $X^2y'' + [2(X + 1)/(2 + X)]Xy' - [l(l + 1)X/(2 + X)]y = 0$, where $X = x - 1$; similarly at $x = -1$.

▶2. Recall $p_0 = q_0 = 0$ and hence $\sigma(\sigma - 1) = 0$.

▶3. $p_0 = -\frac{3}{2}, q_0 = q_1 = 1$, all other p_i and q_i are zero.
(i) $\sigma^2 - \frac{5}{2}\sigma + 1 = 0$; (ii) $a_n = -n^{-1}(n + \frac{3}{2})^{-1}a_{n-1}$; (iii) the first term of the quoted final answer; (iv) $a_n = -n^{-1}(n - \frac{3}{2})^{-1}a_{n-1}$; (v) by the ratio test, or since $t_n \to 0$ and the signs alternate.

▶4. Differentiate as a product of n factors like $(\sigma + 2r)^{-2}$.

▶5. $a_{2n} = -a_{2n-2}[2n(2n + 1)]^{-1}$.

▶7. $a_{2n} = -a_{2n-2}[2n(2n - 1)]^{-1}$.

▶8. $a_{2n} = -a_{2n-2}[2n(2n + 2)]^{-1}$.

▶9. Differentiate as a product of $n - 1$ factors like $(\sigma + 2r + 1)^{-2}$ and one other $(\sigma + 2n + 1)^{-1}$.

▶10. Multiplier $= -k/(2a_0)$.

▶11. $y = a_0x \sum_0^\infty \dfrac{x^{2n}\, 2^n}{(2n + 1)!!}$, $\qquad y = b_0 \sum_0^\infty \dfrac{x^{2n}}{n!} = b_0 \exp(x^2)$.

▶13. $(\sigma + 2)(\sigma + 1)a_2 = -[l(l + 1) - 2\sigma]a_0$,
$(\sigma + 3)(\sigma + 4)a_4 = -[l(l + 1) - 2\sigma]a_0 - [l(l + 1) - 2(\sigma + 2)]a_2$.

▶14.

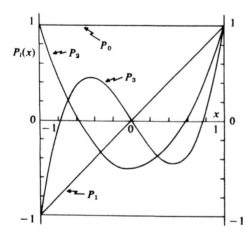

Fig. S.1   The Legendre polynomials $P_l(x)$ plotted against x for $l = 0$ to 3.

▶17. Write $2l/x^2$ as $-2l(1 - x^2) + 2l$.

▶19. (a) Take the coefficient of h^l in

$$(x - h) \sum P_l h^l = (1 - 2xh + h^2) \sum l P_l h^{l-1}.$$

Section 6.10

1. $y = a_0 \sum_0^\infty \frac{1}{n!} \left(\frac{x}{2}\right)^n = a_0 \exp\left(\frac{x}{2}\right), \qquad y = b_0 x^{1/2} \sum_0^\infty \frac{x^n}{(2n + 1)!!}.$

2. Only the larger root series, $\sigma = m$, is needed because of the finiteness requirement:

$$a_{2n} = \frac{(-1)^n m! a_0}{2^{2n} n! (m + n)!} \quad \text{and} \quad J_m(x) = \sum_{n=0}^\infty \frac{(-1)^n}{n!(m + n)!} \left(\frac{x}{2}\right)^{2n+m}.$$

3. $|x^2 + 5x| < 4$; range including $x = 0$, $-2 < 2x < \sqrt{(41)} - 5$.

4. Repeated root of the indicial equation, $\sigma = 2$. Following section 6.3 the solution is

$$y(x) = ax^2 + \sum_{n=1}^\infty \frac{(-2x)^{n+2}(n + 1)}{n!}$$

$$\times \left[\tfrac{1}{4}a + b\left(\ln x + \frac{1}{n + 1} - \frac{1}{n} - \frac{1}{n - 1} - \cdots - \frac{1}{2} - 2\right)\right].$$

5. Putting $y(x) = f(x) \exp(x^2)$ gives $f(x) = A \int_0^x \exp(-u^2)\, du + B$.

6. Transformed equation is $\xi y'' + 2y' + y = 0$, where $y = y(\xi)$. $a_n = (-1)^n (n + 1)^{-1} (n!)^{-2} a_0$. Convergent for all $x \neq 0$. $y^2(x)\, (df/dx) = A = $ constant, giving

$$\frac{df}{dx} = \frac{A}{a_0^2}\left(1 + \frac{1}{x} + \frac{\alpha}{x^2} + \cdots\right).$$

7. For even n,

$$y = a_0\left[1 + \sum_{r=1}^{n/2} \frac{(-2)^r n(n - 2)\ldots(n + 2 - 2r)}{(2r)!} x^{2r}\right].$$

For odd n, $y = b_0\left[x + \right.$

$$\left. \sum_{r=1}^{\frac{1}{2}(n-1)} \frac{(-2)^r (n - 1)(n - 3)\ldots(n + 1 - 2r)}{(2r + 1)!} x^{2r+1}\right].$$

8. Equate coefficients of h^n in the partial differential equation; (i) use $\partial\Phi/\partial x = 2h\Phi$; (ii) use $\partial\Phi/\partial h = 2x\Phi - 2h\Phi$.

9. $\Phi = \exp(x^2) \exp[-(x - h)^2]$. This gives $H_n(x) = \exp(x^2)(-1)^n \times \partial^n[\exp(-(x - h)^2)]/\partial x^n$, which at $h = 0$ yields the quoted result.

10. At step (iv) $\dfrac{1}{h} \ln \dfrac{1 + h}{1 - h} = \sum_{n=0}^{\infty} h^{2n} \int_{-1}^{1} P_n^2(x)\, dx$.

11. The distances from the two charges $-e$ to the point (r, θ, ϕ) are $r(1 \pm 2hx + h^2)^{1/2}$, where $x = \cos\theta$ and $h = (a/r)$. Use also $P_l(-x) = (-1)^l P_l(x)$.

CHAPTER 7

▶2. $\alpha_l = [(2l + 1)/2]^{1/2}$.

▶3. $2l(l + 1)/(2l + 1)$.

▶5. (7.26) $F = \exp(-x^2)$; $p = \rho = \exp(-x^2)$, $q = 0$, $\lambda = 2\alpha$.
    (7.27) $F = \exp(-x)$; $p = x\exp(-x)$, $q = 0$, $\rho = \exp(-x)$, $\lambda = \alpha$.

▶6. $F = (1 - x^2)^{-1/2}$; $p = (1 - x^2)^{1/2}$, $q = 0$, $\rho = (1 - x^2)^{-1/2}$, $\lambda = n^2$.

▶7. $\int_0^{\pi/2} \sin(nz)\, dz - \int_{\pi/2}^{\pi} \sin(nz)\, dz$.

▶8. Show $L(G) = \sum_j y_j(x)y_j^*(z)$, using G from (7.33).

▶9. $y' = \frac{1}{2}A \cos(x/2) - \frac{1}{2}B \sin(x/2)$, hence y'', and substitute in original equation.

▶11. Make the indefinite integrals for A and B start from π and 0 respectively.

▶12. $A = (p_+ - p_-)^{-1} \exp(-p_+ z)$, $B = -(p_+ - p_-)^{-1} \exp(-p_- z)$,
    $p_+ - p_- = 2(\beta^2 - \omega^2)^{1/2}$.

▶13. Substitute y given by (7.52) and (7.53) into the left-hand side of (7.50) and use (7.51 a) and (7.51 c) in turn.

Section 7.7

1. (a) All u_i are *linear* combinations of y_i; (b) assume all u_i are normalized and mutually orthogonal for $i = 1, \ldots, n$, and consider $\int u_k^* \rho u_{n+1}\, dx$ for any k, $1 \leqslant k \leqslant n$.

2. Integrating factor $= \exp\{\int^x [(\alpha + \beta - 1)z + (1 - \gamma)]/[z(z - 1)]\}\, dz = x^{\gamma - 1}(1 - x)^{\alpha + \beta - \gamma}$, $p = \mp x^\gamma(1 - x)^{\alpha + \beta - \gamma + 1}$, $\lambda = \pm\alpha\beta$, $q = 0$, $\rho = x^{\gamma - 1}(1 - x)^{\alpha + \beta - \gamma}$.

3. If multiplied through by x, equation is of Sturm–Liouville form with weight function x.

4. $y_n = (2/\pi)^{1/2} \sin[(n + \frac{1}{2})x]$; $f(x) = -(8/\pi)^{1/2} \sum_0^\infty (n + \frac{1}{2})^{-3} y_n$; $y(x) = (4/\pi) \sum_0^\infty (n^2 + n - \frac{3}{4})^{-1}(n + \frac{1}{2})^{-3} \sin[(n + \frac{1}{2})x]$.

5. (a) If $y(x) = \sum_0^\infty a_n P_n(x)$, $a_n = (n + \frac{1}{2})[b - n(n + 1)]^{-1} \int_{-1}^1 f(z)P_n(z)\, dz$.
    (b) Note $5x^3 = 2P_3(x) + 3P_1(x)$; $a_1 = \frac{1}{4}$, $a_3 = 1$ all other $a_n = 0$; $y(x) = (5/4)(2x^3 - x)$.

6. Putting $x = \exp(t)$, $(d/dx) = \exp(-t)(d/dt)$ gives $d^2y/dt^2 + dy/dt + (\frac{1}{4} + \lambda)y = 0$ with solution $y(t) = [A \exp(i\lambda^{1/2}t) + B \exp(-i\lambda^{1/2}t)] \times \exp(-\frac{1}{2}t)$. Boundary conditions give $y_n(x) = B_n x^{-1/2} \sin(n\pi \ln x)$ with $\lambda_n = n^2\pi^2$. Normalization gives $B_n = \sqrt{2}$.

$a_n = -(n\pi)^{-2} \int_1^e \sqrt{2}\, x^{-1} \sin(n\pi \ln x)\, dx = -\sqrt{8}\,(n\pi)^{-3}$ [n odd], $= 0$
[n even].

7. $G(r, r') = (4\pi|r - r'|)^{-1}$.

8. The main steps are:

$$x(t) = n^{-1}F \exp(-\beta t)I,$$

where

$$I = \tfrac{1}{2}J(p + n, -n) - \tfrac{1}{2}J(p - n, n),$$

where

$$J(\lambda, \mu)(\beta^2 + \lambda^2) = \{\beta \cos[(\lambda + \mu)t] + \lambda \sin[(\lambda + \mu)t]\}$$
$$\times \exp(\beta t) - \{\beta \cos(\mu t) + \lambda \sin(\mu t)\}.$$

Transients:

$$FX^{-1}p \exp(-\beta t)[n^{-1}(p^2 - n^2) \sin(nt) + 2\beta \cos(nt)].$$

Steady state:

$$FX^{-1}[(\omega^2 - p^2) \sin(pt) - 2\beta p \cos(pt)],$$

where

$$X = (\omega^2 - p^2)^2 + 4\beta^2 p^2.$$

9. $$G_K(r, r') = \frac{1}{(2\pi)^3} \int \frac{\exp(ik \cdot r) \exp(-ik \cdot r')}{k^2 - K^2}\, dk,$$

as in section 7.6. Here the 'index' k replaces the i of that section.

$$G_K(r, r') = \frac{1}{(2\pi)^3} \int_0^\infty dk \int_0^\pi d\theta \int_0^{2\pi} d\phi \, \frac{\exp(ik\rho \cos\theta)k^2 \sin\theta}{k^2 - K^2}$$

which, when integrated successively with respect to ϕ and θ, gives the required form. Here $\rho = |r - r'|$.

CHAPTER 8

▶1. Consider an even function, like $\sin^2(\omega t)$ for example.

▶4. Use (8.11 a–b) in (8.9).

▶5. Equations (8.11 a–b) show A_m and B_m are real, then use (8.9).

▶7. (i) A_0 and odd cosines; (ii) all, there is no symmetry about $T/4$ for the periodic function; (iii) odd cosines.

▶8. The coefficient of the $1/n$ term vanishes;
$$a_n = (32y_0/3n^2k^2l^2)[\cos(n\pi/2) - 1] \text{ for } n \neq 0.$$

▶9. See ▶8; $b_n = (32y_0/3n^2k^2l^2) \sin(n\pi/2)$.

▶10. See ▶8 concerning the $1/n$ term.

▶11. See ▶8 concerning the $1/n$ term.

▶13. Evaluate $\sum (\cos (2m + 1)x/(2m + 1)^2)$ at the suggested values; true values $\pm(1.221, 0.823, 0.411, 0)$, values after 3 terms $\pm(1.151, 0.831, 0.409, 0)$.

▶14. Differentiate (8.24) and put $x = \pi$.

▶15. (i) $y(\theta) = (\pi/4)(\pi/2 - |\theta|)$, see (8.23);
(ii) $y(\theta) = (\pi\theta/4)(\pi/2 - |\theta|/2)$, use integration result on (i);
(iii) put $x = 0$ in (8.23) to obtain $S_o = \pi^2/8$.

If $S_e = \sum_{m=1}^{\infty} (2m)^{-2}$, then $S_e = \frac{1}{4}(S_e + S_o)$, giving $(S_e + S_o) = \pi^2/6$ and $(S_o - S_e) = \pi^2/12$. Function $y(x)$ is even, $y(0) = 0$, $y(l) = l^2$; compare derivative with (8.24); $y(x) = x^2(-l \leqslant x \leqslant l)$.

▶16. Intermediate value $\hat{f}(\mu) = \dfrac{2A}{(2\pi)^{1/2}\mu} \dfrac{\sin (\mu a/2)}{\exp (i3\mu a/2)}$

$$\times \frac{\exp (2iN\mu a)[1 - \exp (-4iN\mu a)]}{1 - \exp (-2i\mu a)} .$$

▶17. Integrate by parts.

▶18. Integrate by parts.

▶22. Use (8.8) and its complex conjugate, and $\int_{-\pi}^{\pi} \exp [i(n - m)\theta] \, d\theta = 2\pi \, \delta_{nm}$.

▶23. Write $\sin (\omega_0 t) = \dfrac{-i}{2} [\exp (i\omega_0 t) - \exp (-i\omega_0 t)]$.

▶24. Change variable of integration to $u = z - x$.

▶26. $FT[\exp (i\omega_c t)] = (2\pi)^{1/2} \delta(\omega - \omega_c)$.

Section 8.11

1. $A_n = (2V/\pi n) \sin (n\pi/3)$,

$$f(t) = \frac{V}{3} + \frac{V\sqrt{3}}{\pi} \left[\cos \left(\frac{2\pi t}{T}\right) + \frac{1}{2} \cos \left(\frac{4\pi t}{T}\right) - \frac{1}{4} \cos \left(\frac{8\pi t}{T}\right) \right.$$

$$\left. - \frac{1}{5} \cos \left(\frac{10\pi t}{T}\right) + \frac{1}{7} \cos \left(\frac{14\pi t}{T}\right) + \cdots \right].$$

2.   Show that

$$\int_0^{2\pi/\omega} (f(t) - f_N(t))^2 \, dt = \frac{\pi}{\omega} \left[\sum_{n=0}^{N} (A_{Nn}^2 + B_{Nn}^2) \right.$$

$$\left. - 2A_{Nn}A_n - 2B_{Nn}B_n) \right] + \text{constant},$$

and then minimize with respect to A_{Nn} and B_{Nn}.

3.

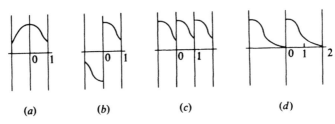

(a)   (b)   (c)   (d)

Fig. S.2 Continuations of $\exp(-x^2)$ in $0 \leqslant x \leqslant 1$ to give: (a) cosines only; (b) sines only; (c) period 1; (d) period 2.

(a) (iii) $\frac{1}{2}(1 + e^{-1})$, (iv) $\frac{1}{2}(1 + e^{-4})$; (b) (iii) $\frac{1}{2}(1 + e^{-1})$, (iv) e^{-1}.

4. $f(t) = \dfrac{1}{\pi} + \dfrac{1}{2}\cos(\omega t) - \dfrac{2}{\pi}\displaystyle\sum_{n=1}^{\infty}\dfrac{(-1)^n}{4n^2 - 1}\cos(2n\omega t).$

Approximate fluctuation amplitude $= (\pi/20)$.

5. $u(r, \theta) = \dfrac{1}{\pi} + \dfrac{r}{2}\sin\theta - \dfrac{2}{\pi}\displaystyle\sum_{n=1}^{\infty}\dfrac{r^{2n}}{4n^2 - 1}\cos(2n\theta).$

6. $C_{\pm(2m+1)} = \mp i2/[\pi(2m + 1)]$; $\sum|C_n|^2 = (4/\pi^2)\cdot 2\cdot(\pi^2/8)$, the values $n = \pm 1, \pm 3$ contributing more than 90 per cent of the total.

7. The t integral is

$$\int_{-\infty}^{\infty}\exp\left[i(E_1' + E_2' - E_1 - E_2)\right]dt$$

yielding $\delta(E_1' + E_2' - E_1 - E_2)$. Similarly the **R** integral yields $\delta(\mathbf{p}_1 + \mathbf{p}_2 - (\mathbf{p}_1' + \mathbf{p}_2'))$.

8. With **k** as the polar axis,

$$\tilde{V}(\mathbf{k}) \propto 2\pi\int_0^{\infty}dr\, r\exp(-\mu r)\int_0^{\pi}\exp(-ikr\cos\theta)\sin\theta\, d\theta$$
$$= (-2\pi/ik)\int_0^{\infty}\exp(-\mu r)[\exp(ikr) - \exp(-ikr)]\, dr$$
$$= -4\pi(\mu^2 + k^2)^{-1}.$$

9. Overshoot has limiting value of 0.562.

CHAPTER 9

▶2. Substitute $p - 2y$ for x^2 everywhere.

(a) (i) $p^2 - 4p - 4$, yes, (ii) $(p - y)^2$, no, (iii) $(p^2 + 4)(2p^2 + p)^{-1}$, yes; (b) necessarily as in (a).

▶4. (i) (a) $p - 1$, (b) $4p - 4$, (c) 0; (ii) (a) $3 + p - p^2$, (b) $4p - p^2$, (c) $4 - p^2$.

▶5. (i) $p = x^2 + y^2$, $u = \sin(p^{1/2}) + 1$; (ii) $p = 3x + iy$, $u = \frac{1}{2}p^2$; (iii) $p = \sin x \cos y$, $u = 2p - 1$; (iv) $p = y - x^2$, $u = p + 2$; (v) $p = x^2 + y^2$, PI is $u = -3y$, (a) $x^2 + y^2 - 3y$, (b) $2x^2 + 2y^2 - 3y + g(x^2 + y^2)$ where $g(1) = 0$; (vi) $g(x^3 - y^3) + \frac{1}{3}x^3y^3$.

▶6.

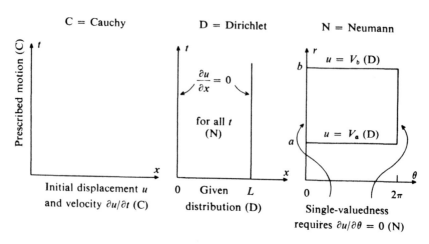

Fig. S.3   Boundary conditions for the situations of ▶6.

▶7. $u(x, y) = f(p) = c(ax + by) + k$.

▶8. Suppose A has a dependence $x^\sigma y^\nu$, and then consider the implications of $\lambda_1 + \lambda_2$ and $\lambda_1 - \lambda_2$ not being functions of x and y; $\lambda_1 = -2$, $\lambda_2 = 3x[y + (3/2)x]^{-1}$.

▶10. $\dfrac{\partial u}{\partial x} = gf_2' + \dfrac{\partial g}{\partial x}f_2$, $\dfrac{\partial u}{\partial y} = \lambda g f_2' + \dfrac{\partial g}{\partial y}f_2$ with $\lambda = -(B/C)$;

similarly the required second derivatives are obtained.

▶11. $yf(x + \lambda y) = \lambda^{-1}(x + \lambda y)f(x + \lambda y) - \lambda^{-1}xf(x + \lambda y)$
       $= G(x + \lambda y) + xF(x + \lambda y)$.

▶13. Recall that $\lambda_1 + \lambda_2 = -(2B/C)$ and $\lambda_1\lambda_2 = A/C$.

▶14. Recall that $A + 2B\lambda + C\lambda^2 = 0$ and $B^2 = AC$.

▶15. In the notation of (9.61) the boundary conditions are $f(p) + g(p) = 1$, $f'(p) + \frac{1}{2}g'(p) = 1$, for all p; these lead to $u(x, y) = 2(x + y) - 2(x + \frac{1}{2}y) + 1 = y + 1$ and $u(0, 1) = 2$.

▶16. Parabolic equation with $\lambda = -1$; general solution $u(x, y) = f(x - y) + xg(x - y)$; boundary conditions yield $f(p) \equiv 0$, $g(p) = p + 1$; $u(x, y) = x^2 - xy + x$.

▶17. Writing $\xi = x + iy$ and $\eta = x - iy$, $y^3 - 3yx^2 = \frac{1}{2}i(\xi^3 - \eta^3)$; $g(p) = \frac{1}{2}ip^3$, $h(p) = -\frac{1}{2}ip^3$.

▶18. Multiply through by t and write η for x^2/Kt.

▶19.

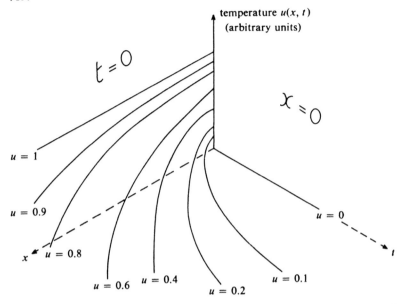

temperature $u(x, t)$
(arbitrary units)

$t = 0$

$x = 0$

$u = 1$

$u = 0.9$

x $u = 0.8$

$u = 0$

$u = 0.6$ $u = 0.4$

$u = 0.2$

$u = 0.1$

t

Fig. S.4 Sketch of the (three-dimensional) contours of equal temperature u given by equation (9.85). The height of the surface containing all the contours at any particular pair of values (x, t) gives the value of the temperature at point x after time t (relative to the value unity everywhere at time $t = 0$). Projections of the contours onto the $u = 0$ plane would be parabolas $t = cx^2$.

▶20. (i) After differentiation and substitution, both sides contain a factor $\alpha(r^2/2\beta t^3 - 1/t^2) \exp(-r^2/2\beta t)$.

▶24. See the footnote to section 9.9.

▶25. Substitute from (9.91).

▶28. $R \triangleq \Omega L^{-1}$, $C \triangleq T\Omega^{-1}L^{-1}$, $(RC)^{-1} \triangleq L^2 T^{-1}$.

▶29. The terms of (9.98) are sums and differences of terms such as $\exp(\pm \lambda x)$ and $\exp(\pm i\lambda x)$.

Section 9.14

1. (i) $u(x, y) = (y^2 - x^2)^{1/2}$; (ii) $u(x, y) = 1 + f(x^2 - y^2)$ where $f(0) = 0$.

2. PI $u = y$, CF $u = f(y - \ln \sin x)$, (i) $u = \ln \sin x$; (ii) $u = (y - \ln \sin x)^2 + y$.

3. $u = f(3x - 2y) + 2(x + y), f(p) = 3 + 2p, u = 8x - 2y + 3, u(2, 4) = 11$.

4. $\dfrac{\partial^2 u}{\partial x^2} = \dfrac{\partial^2 u}{\partial \xi^2}\left(\dfrac{\partial \xi}{\partial x}\right)^2 + 2\,\dfrac{\partial^2 u}{\partial \xi\, \partial \eta}\left(\dfrac{\partial \xi}{\partial x}\right)\left(\dfrac{\partial \eta}{\partial x}\right) + \dfrac{\partial^2 u}{\partial \eta^2}\left(\dfrac{\partial \eta}{\partial x}\right)^2$

$\qquad\qquad\qquad\qquad + \dfrac{\partial^2 \xi}{\partial x^2}\dfrac{\partial u}{\partial \xi} + \dfrac{\partial^2 \eta}{\partial x^2}\dfrac{\partial u}{\partial \eta},\ \text{etc.},$

$A' = A\left(\dfrac{\partial \xi}{\partial x}\right)^2 + 2B\,\dfrac{\partial \xi}{\partial x}\dfrac{\partial \xi}{\partial y} + C\left(\dfrac{\partial \xi}{\partial y}\right)^2,$

$B' = A\,\dfrac{\partial \xi}{\partial x}\dfrac{\partial \eta}{\partial x} + B\left(\dfrac{\partial \xi}{\partial x}\dfrac{\partial \eta}{\partial y} + \dfrac{\partial \xi}{\partial y}\dfrac{\partial \eta}{\partial x}\right) + C\,\dfrac{\partial \xi}{\partial y}\dfrac{\partial \eta}{\partial y},$

$C' = A\left(\dfrac{\partial \eta}{\partial x}\right)^2 + 2B\,\dfrac{\partial \eta}{\partial x}\dfrac{\partial \eta}{\partial y} + C\left(\dfrac{\partial \eta}{\partial y}\right)^2.$

5.   Similar triangles by a common angle and corresponding sides in proportion.

(i) $\dfrac{e^2 ab}{4\pi\epsilon_0(b^2 - a^2)^2}$;   (ii) $\dfrac{e^2 ab}{4\pi\epsilon_0}\left[\dfrac{1}{(b^2 - a^2)^2} - \dfrac{1}{b^4}\right].$

Obtain (ii) from (i) by adding a further image charge $+ea/b$ at O, to give zero net electrostatic flux from the sphere but maintain its equipotential property.

6. $\dfrac{\partial}{\partial x} = \dfrac{\partial}{\partial \xi} + \dfrac{\partial}{\partial \eta},\quad \dfrac{\partial}{\partial y} = i\,\dfrac{\partial}{\partial \xi} - i\,\dfrac{\partial}{\partial \eta}.$

7.

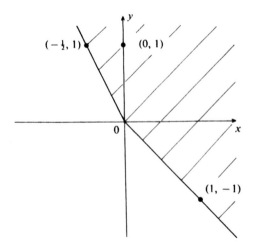

Fig. S.5   Region in which the solution to ▶15 is valid, under the conditions of example 7, is shown shaded.

Solution $u = 2(x + y) - 2(x + \frac{1}{2}y) + 1$ is only valid if $x + y > 0$ and $x + \frac{1}{2}y > 0$, i.e. in the shaded area of fig. S.5.

8. $u(x, y) = f(x + 2y) + g(x + 3y) + x^2 + y^2$ leading to $f(p) = 2p + 1$, $g(p) = -p^2$ and $u = 1 + 2x + 4y - 6xy - 8y^2$.

9. Either $\lambda = -3, 0$, or integrate with respect to y and obtain first-order equation; $u(x, y) = f(y - 3x) + F(x) + \frac{1}{2}x^2 y^2$.

10. $c = v \pm \alpha$ where $\alpha^2 = T/\rho A$; $u(x, 0) = a \cos kx$, $\partial u(x, 0)/\partial t = 0$; $u(x, t) = a \cos [k(x - vt)] \cos (k\alpha t) - (v a/\alpha) \sin [k(x - vt)] \sin (k\alpha t)$.

11. $CF = f(x + iy) + g(x - iy)$, PI are (i) $(x^4 + y^4)/12$; (ii) $\frac{1}{2}x^2 y^2$; for these two functions, their difference $= [(x + iy)^4 + (x - iy)^4]/24$.

12. $u(x, y) = f(x + iy) + g(x - iy) + (x^4/12)[y^2 - (1/15)x^2]$. In the last term x and y may be interchanged with equal validity.

13. $\left(\dfrac{\partial}{\partial x} - \dfrac{1}{c}\dfrac{\partial}{\partial t}\right)\left(\dfrac{\partial}{\partial x} + \dfrac{1}{c}\dfrac{\partial}{\partial t}\right) u = 0,$

then use that, at constant u, $\quad \dfrac{\partial x}{\partial t}\bigg|_u = -\left(\dfrac{\partial u}{\partial t}\right)\bigg/\left(\dfrac{\partial u}{\partial x}\right).$

14. $\dfrac{4\partial^2 u}{\partial \xi\, \partial \eta} = 3[(1 - i)\xi + (1 + i)\eta],$

$u = \dfrac{3\xi\eta}{8}[(\xi + \eta) + i(\eta - \xi)] = \frac{3}{4}(x^2 + y^2)(x + y).$

15. $-k\, \partial u(0)/\partial x = J_1$, $k\, \partial u(w)/\partial x = J_2$ determine g and α in (9.79).

16. $V_0[1 - (2/\sqrt{\pi}) \int^{\frac{1}{2}x(CR/t)^{1/2}} \exp (-v^2)\, dv]$; consider as V_0 applied at $t = 0$ and continued and $-V_0$ at $t = T$ and continued.

$$V(x, t) = \frac{2V_0}{\sqrt{\pi}} \int_{\frac{1}{2}x(CR/t)^{1/2}}^{\frac{1}{2}x(CR/(t - T))^{1/2}} \exp (-v^2)\, dv.$$

For $t \gg T$, maximum at $x = (2t/CR)^{1/2}$ with value $[V_0 T \exp (-\frac{1}{2})]/(2\pi)^{1/2} t$.

17. $\lambda = -\mu = (\omega/2k)^{1/2}$, where k is the diffusion constant; ratio of depths, $x_Y = (365)^{1/2} x_D$; only yearly variation significant, phase is $\mu_Y x_Y$ behind surface, this is $\ln 20$; coldest at 25 January $+ (\ln 20)/2\pi$ years $\approx$ 16 July.

19. Using form (9.98),

$A = -C, \qquad B = -D,$
$A \sin \lambda l + B \cos \lambda l + C \sinh \lambda l + D \cosh \lambda l = 0,$
$A \cos \lambda l - B \sin \lambda l + C \cosh \lambda l + D \sinh \lambda l = 0,$

consistency yields the given condition.

20. $U(x, s) = F(s) \exp (-sx/c)$, where $F(s)$ is the transform of $f(t)$; $u(x, t) = f(t - (x/c))U(t - (x/c))$; $u(x, t) = 0$ for $x \geq ct$, $= vt - vx/c$ for $x \leq ct \leq x + hc/v$, $= h$ for $x \leq ct - hc/v$.

21. $(\partial^2 V/\partial x^2) - LCs^2 V = 0$, $V(0, s) = E/s$, $V(l, s) = 0$, leading to

$$V(x, s) = E\{\exp (sx/c) - \exp [(2l - x)s/c]\}/\{s[1 - \exp (2sl/c)]\}.$$

Writing denominator as

$$-s \exp (2sl/c)[1 - \exp (-2sl/c)]$$

and expanding gives

$$V(x, s) = (E/s)\{\exp (-sx/c)$$
$$- \exp [(x - 2l)s/c] + \exp [-(2l + x)s/c] - \cdots\}.$$

CHAPTER 10

▶1. $u(0, y, t) = u(a, y, t) = u(x, 0, t) = u(x, b, t) = 0$ are the required boundary conditions leading to

$$u(x, y, t) = \sin (n\pi x/a) \sin (m\pi y/b)(A \sin \omega t + B \cos \omega t)$$

where n, m, are integers.

▶2. $\dfrac{-\hbar^2}{2m_0} \dfrac{X''}{X} = \dfrac{p_x^2}{2m_0}$ etc., $\dfrac{i\hbar T'}{T} = E$.

▶3. As in ▶2, but with solutions $X = A \sin (p_x x/\hbar)$, etc. where $p_x a/\hbar = n_x \pi$.

▶5. $u(x, y) = AC \exp (\mu x + i\mu y)$.

▶6. $\partial /\partial r = \exp (-t) \partial /\partial t$, and the equation becomes $d^2 R/dt^2 - n^2 R = 0$.

▶8. See ▶6.

▶9. $-\dfrac{(1 - \mu^2)}{M} \dfrac{d}{d\mu} \left[-(1 - \mu^2) \dfrac{dM}{d\mu} \right] + l(l + 1)(1 - \mu^2) = m^2$.

▶10. (i) $6u/r^2$, $-6u/r^2$, 0, $l = 2$, $m = 0$; (ii) $2u/r^2$, $(\cot^2 \theta - 1)u/r^2$, $-u/(r^2 \sin^2 \theta)$, $l = 1$, $m = 1$.

▶13. $Y_0^0 = (4\pi)^{-1/2}$,

$Y_1^0 = (3/4\pi)^{1/2} \cos \theta$,

$Y_1^{\pm 1} = \mp (3/8\pi)^{1/2} \sin \theta \exp (\pm i\phi)$,

$Y_2^0 = (5/16\pi)^{1/2}(3 \cos^2 \theta - 1)$,

$Y_2^{\pm 1} = \mp (15/8\pi)^{1/2} \sin \theta \cos \theta \exp (\pm i\phi)$,

$Y_2^{\pm 2} = (15/32\pi)^{1/2} \sin^2 \theta \exp (\pm 2i\phi)$.

$$\iint Y_2^{1*} Y_2^1 \, d\phi \, d(\cos \theta)$$

$$= \int_0^\pi \int_0^{2\pi} \frac{15}{8\pi} \sin^2 \theta \cos^2 \theta \, e^{-i\phi} e^{i\phi} \, d\phi \sin \theta \, d\theta = 1.$$

$$\iint Y_2^{1*} Y_1^1 \, d\phi \, d(\cos \theta) = \int_0^\pi \int_0^{2\pi} \left[-\left(\frac{15}{8\pi}\right)^{1/2} \right]$$

$$\times \left[-\left(\frac{3}{8\pi}\right)^{1/2} \right] \sin \theta \cos \theta \, e^{-i\phi} \sin \theta \, e^{i\phi} \, d\phi \sin \theta \, d\theta = 0.$$

▶14.

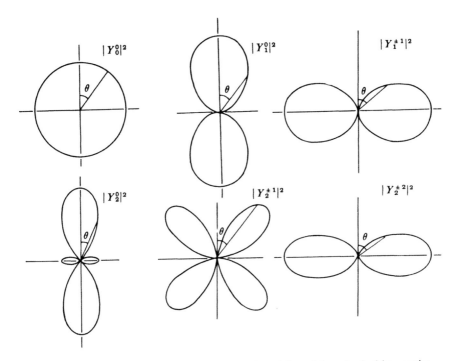

Fig. S.6 Sections of polar plots of the squared modulus of the spherical harmonics $Y_l^m(\theta, \phi)$ up to $l = 2$. There is no ϕ dependence and the three-dimensional plots would be obtained by rotating the curves shown about the lines $\theta = 0$.

▶15. $r^2(\frac{3}{4}r^{-5/2}S - r^{-3/2}S' + r^{-1/2}S'') + 2r(-\frac{1}{2}r^{-3/2}S + r^{-1/2}S')$

$$+ [k^2r^2 - l(l + 1) - \frac{1}{4}](r^{-1/2}S) = 0.$$

▶16. $\dfrac{1}{r^2}\dfrac{r}{\sin kr}\dfrac{\partial}{\partial r}(-\sin kr + kr \cos kr) = -k^2.$

▶17. Use equations (8.11 b) and (8.19).

▶19. (i) $\simeq GMr^{-1}$, GMa^{-1}.

Section 10.8

1. (i) $u(x, y) = C \exp[\frac{1}{2}\lambda(x^2 + 2y)]$; (ii) $u(x, y) = C(x^2y)^{\lambda/2}$.

2.

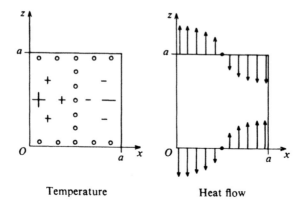

Temperature	Heat flow

Fig. S.7   The temperature and heat flow distributions in the cube of example 2. There is no variation of either with y.

3. $u(x, t) = (A \cos mx + B \sin mx + C \cosh mx + D \sinh mx) \cos(\omega t + \epsilon)$ where $m^4 a^4 = \omega^2$.

5.   The angular part of the wave function for an electron in an unpolarized atom has the form $\psi = \sum_{m=-l}^{l} a_m Y_l^m(\theta, \phi)$, where the a_m all have the same magnitude and random phases; $|\psi|^2$ is thus proportional to the given expression when averaged over many atoms.

6.   First term can only contain $Y_1^{\pm 1}$, $Y_2^{\pm 1}$, the second only Y_0^0, Y_1^0, Y_0^0; $f(\theta, \phi) = (\pi)^{1/2} [Y_0^0 - 3^{-1/2} Y_1^0 - (2/3)^{1/2} Y_1^1 - (2/15)^{1/2} Y_2^{-1}]$.

7. $u = \frac{1}{2} A_0 + \sum_{n=1}^{\infty} r^n (A_n \cos n\phi + B_n \sin n\phi)$ to give the stated function when $r = 1$, leading to

$$u(r, \phi) = \frac{1}{\pi} + \frac{r}{2} \sin \phi - \sum_{n=1}^{\infty} r^{2n} \frac{2 \cos 2n\phi}{\pi(4n^2 - 1)}.$$

8. $u = A \ln r + B + \sum_{n=1}^{\infty} (A_n r^n + B_n r^{-n}) \cos n\phi$;

at $r = b$: $B_n = -b^{2n} A_n$, $B = -A \ln b$,

at $r = a$: $c|\phi| = A \ln(a/b) + \sum_{1}^{\infty} A_n (a^{2n} - b^{2n}) a^{-n} \cos n\phi$, giving on Fourier analysis that

$$u(r, \phi) = \frac{c\pi}{2} \frac{\ln(b/r)}{\ln(b/a)}$$

$$- \frac{4c}{\pi} \sum_{m \, odd} \frac{a^m}{m^2(b^{2m} - a^{2m})} \left(\frac{b^{2m}}{r^m} - r^m \right) \cos m\phi.$$

9. $E_n = \frac{16\rho A^2 c^2}{l(2n + 1)^2 \pi^2}$; $E = \frac{2\rho c^2 A^2}{l} = \int_0^l \frac{2Tv \, dv}{(l/2)}$.

10. (i) $\int_0^1$ L.H.S. $d\mu = t^{-1}[(1 + t^2)^{1/2} - (1 - t)]$, then equate powers of t.

11. $u(r, \phi) = \dfrac{4V}{\pi} \sum_{n=0}^{\infty} \dfrac{(-1)^n}{2n + 1} \left(\dfrac{r}{b}\right)^{2n+1} \cos(2n + 1)\phi.$

12. Write $C(x, t) = C_0 + \sum_1^{\infty} A_n \sin(n\pi x/L) f_n(t)$ with $f_n(t) \rightarrow 0$ as $t \rightarrow \infty$; Initial condition $C(x, 0) = 0$ determines A_n;

$$C(x, t) = C_0 - \frac{4C_0}{\pi} \sum_{n=0}^{\infty} \frac{1}{(2n + 1)}$$

$$\times \sin\left[\frac{(2n + 1)\pi x}{L}\right] \exp\left[-\frac{K(2n + 1)^2}{L^2} \pi^2 t\right].$$

13. Since there is no heat flow at $x = \pm a$, use a series with period $4a$, $u(x, 0) = 100$ for $0 < x \leqslant 2a$, $= 0$ for $-2a \leqslant x < 0$.

$$u(x, t) = 50 + \frac{200}{\pi} \sum_{n=0}^{\infty} \frac{1}{(2n + 1)}$$

$$\times \sin\left[\frac{(2n + 1)\pi x}{2a}\right] \exp\left[-\frac{k(2n + 1)^2 \pi^2 t}{4a^2 s \rho}\right].$$

$$\text{Heat} = \int_0^{\infty} dt\, k\, \frac{\partial u(0, t)}{\partial x} = 50 s \rho \text{ per unit cross-sectional area.}$$

Take $n = 0$ term only giving $t \approx 2300$ s.

14. (i) $u(z, 0, \phi) = 2\pi G \rho[(a^2 + z^2)^{1/2} - z]$; $M = \pi a^2 \rho$;
(ii) for $z < 0$, the factor in [ ] is $(a^2 + z^2)^{1/2} + z$;
(iii) field is $\partial u/\partial r$ at $\theta = 0$ for $r < a$, then let $a \rightarrow \infty$ giving $2GM/a^2 = 2\pi G \rho$.

CHAPTER 11

▶2. Equation (11.16) is the N–R formula for $f(x) = x^2 - X = 0$.
▶3. There are $N/2$ pairs of intervals to be added.
▶4. Second difference must be constant ($= 2$). This leads to $f(3) = 4$, $f(9) = 58$, and $f(8)$ should be 44.
▶5. $y = (1 - x)^{-2}$.

Section 11.12

1. 5.370.
2. 6.951 after two iterations.
3. $(1/5)(2x^2 + 3)^{-4/5} \neq 0$ at the root.
4. $f(N) = (N - 3x^2)^{-1} \approx -\frac{1}{3}N^{-1}$, 2.645 741 1, accurate value 2.645 751 3.
5. Use differences, the third difference should be the same everywhere; $p(0.5) = 0.175$, $p(1.0) = 0.200$, $p(1.1) = 0.121$, $p(1.2) = 0.000$.

6. $y(x) = 1 - \frac{1}{2}x^2 + \frac{1}{8}x^4 - \cdots$. Actual solution, $y(x) = \exp(-x^2/2)$.

x	0.0	0.1	0.2	0.3	0.4	0.5
y (approx.)	1	0.9950	0.9802	0.9560	0.9232	0.8828
y (true)	1	0.9950	0.9802	0.9560	0.9231	0.8825

CHAPTER 12

▶3. $L = (b - a)(1 - k^2)^{-1/2}$; k is found from (12.8) at the end-points.

▶4. $k_1(1 + y'^2)^{1/2} = 1$.

▶7. $A = \dfrac{l}{\pi} \displaystyle\int_0^{1/2} \sin^2 \dfrac{2\pi s}{l}\, ds$.

▶8. $k\rho' = (\rho^2 - k^2)^{1/2}$.

Section 12.7

1. Same steps as example 12.3.
2. Minimizing curve is $x^2 + y^2 = 2$.
3. (a) x is a double root of $\lambda x = \cosh x$;
    (c) from (12.11), $S = 2\pi k \int_{-b}^{b} \cosh^2(z/k)\, dz$;
    (d) $k = a - b^2/2a - 7b^4/24a^3$.
4. $I = \int n(r)[r^2 + (dr/d\theta)^2]^{1/2}\, d\theta$.
    Take axes so that $\theta = 0$ when $r = \infty$. If $\phi = \frac{1}{2}(\pi - \text{deviation})$, $\phi = \theta$ at $r = a$. Equation reduces to

    $$\frac{\phi}{(a^2 + \alpha^2)^{1/2}} = \int_\infty^a \frac{dr}{r(r^2 - a^2)^{1/2}},$$

    which can be evaluated by putting $r = \frac{1}{2}a(y + y^{-1})$, or successively $r = a \cosh \psi$, $y = \exp \psi$, to yield a deviation $\pi[(a^2 + \alpha^2)^{1/2} - a]/a$.
5. Apply (12.18) to obtain $\nabla^2\phi - \partial^2\phi/\partial t^2 = \mu^2\phi$.
6. (a) $\partial x/\partial t = 0$ and so $\dot{x} = \dot{q}_i\, \partial x/\partial q_i$;

    (b) use $\dot{q}_i \dfrac{d}{dt}\left(\dfrac{\partial T}{\partial \dot{q}_i}\right) = \dfrac{d}{dt}(2T) - \ddot{q}_i \dfrac{\partial T}{\partial \dot{q}_i}$.

7. $\phi(x, t) = m(\frac{1}{2}v^2 t + vx)$.
8. $r^2\dot{\theta} = k$, $\ddot{r} - r\dot{\theta}^2 + dV/dr = 0$; $V = -k^2a^2/2r^4$.
9. $-\lambda y'(1 - y'^2)^{-1/2} = 2gP(s)$, where $y = y(s)$ and $P(s) = \int_0^s \rho(s')\, ds'$; $y = -a\cos(s/a)$ and $2P(\pi a/4) = M$ give $\lambda = -gM$; $\rho(s) = (M/2a)\sec^2(s/a)$.

CHAPTER 13

▶6. Reduce the numerator until it contains the same integral as the denominator. There is no need to evaluate I_0.

▶7. Integrate by parts.

Section 13.6

1. In S–L equation $p = 1$, $q = x$, $\rho = x^2$. Try $y = x(1 - x)$; $I = \frac{21}{60}$, $J = \frac{2}{210}$.

2. $\dfrac{d^2Z}{dr^2} + \dfrac{1}{r}\dfrac{dZ}{dr} + \dfrac{\omega^2}{c^2}Z = 0$, with $Z(a) = 0$ and $Z'(0) = 0$;

   S–L equation with $p = r$, $q = 0$, $\rho = r/c^2$, $\lambda = \omega^2$;

   estimate $\lambda = (c^2v/2a^2)\left(\dfrac{1}{2} - \dfrac{2}{v + 2} + \dfrac{1}{2v + 2}\right)^{-1}$, which minimizes to

   $c^2(2 + \sqrt{2})^2/2a^2 = 5.83c^2/a^2$, when $v = \sqrt{2}$.

3. (ii) As in example 13.2, estimate is $\hbar^2\alpha/2m + k/8\alpha$ [kinetic energy + potential energy] and minimum occurs at that α which makes the two terms equal.

4. $H = -(\hbar^2/2m)\nabla^2 - e^2/r$; writing J_n for $\int_0^\infty r^n e^{-2\beta r}\,dr$, $\int \psi^* H\psi\,dv = 4\pi(\hbar^2\beta/m - e^2)J_1 - (4\pi\hbar^2\beta^2/2m)J_2$, $\int \psi^*\psi\,dv = 4\pi J_2$; minimum estimate, when $\beta = me^2/\hbar^2$, or $-me^4/2\hbar^2$. Wave function is

   $(m^3e^6/\pi\hbar^6)^{1/2}\exp(-me^2r/\hbar^2)$.

   In SI units e^2 is replaced by $e^2/4\pi\epsilon_0$.

5. In S–L equation extended to two independent variables, $p = 1$, $q = 0$, $\rho = 1$, and

   $$k_0^2 \leqslant \frac{\iint [1 \times (\nabla u)^2 + 0 \times u^2]\,dx\,dy}{\iint 1 \times u^2\,dx\,dy}.$$

   Use as the trial function $u(x, y) = x(4 - x)y(1 - y)$. Numerator $= (64 \times 17)/90$. Denominator $= 512/450$; direct solution $k^2 = 17\pi^2/16$.

6. $E_1 \leqslant \dfrac{\hbar}{2}\left(\dfrac{k}{m}\right)^{1/2}\dfrac{(14n^2 + 7n - 3)}{(4n - 1)}$ which has a minimum value of $\dfrac{3\hbar}{2}\left(\dfrac{k}{m}\right)^{1/2}$

   when integer $n = 0$.

7. $-m\ddot{y}_1 = Ta^{-1}[y_1 - (y_2 - y_1)]$ and similarly for the other masses. Putting $y_i = x_i \cos \omega t$ gives required equation; estimates for $ma\omega^2/T$, $\frac{10}{17}$, $\frac{58}{17}$; true values $2 - \sqrt{2}$, $2 + \sqrt{2}$.

8. Take angles of links with vertical as θ_i;

   $$V = \frac{mga}{2}\sum_{i=1}^{n}\left(\sum_{j=1}^{i-1}\theta_j^2 + \tfrac{1}{2}\theta_i^2\right),$$

   $$T = \frac{ma^2}{24}\sum_{i=1}^{n}\dot\theta_i^2 + \frac{ma^2}{2}\sum_{i=1}^{n}\left(\sum_{j=1}^{i-1}\dot\theta_j + \tfrac{1}{2}\dot\theta_i\right)^2.$$

   Trial (i) $\theta_1 = 1$, $\theta_{i \neq 1} = 0$; $\omega^2 \leqslant (6n - 3)g/(6n - 4)a \simeq ng/l$ for $n \gg 1$:

   (ii) $\theta_n = 1$, $\theta_{i \neq n} = 0$; $\omega^2 \leqslant 3ng/2l$;

   (iii) $\theta_i = 1$ all i; $\omega^2 \leqslant 6n^2g/4n^3a \simeq 3g/2l$, for $n \gg 1$.

   Trial (iii) is, as expected, the best guess and can be compared with a solid rod.

CHAPTER 14

▶1. (i) $3412 \to 1432 \to 1234$, i.e. even number (two) of changes;
(ii) $321 \to 123$, odd; (iii) not permutation of 1234.

▶3. $2 \cdot 4 \cdot 1 - 2 \cdot 0 \cdot (-2) - 1 \cdot 3 \cdot 1 + 1 \cdot 0 \cdot 1 + (-3) \cdot 3 \cdot (-2) - (-3) \cdot 4 \cdot 1 = 35.$
$\underset{123}{} \qquad \underset{132}{} \qquad \underset{213}{} \qquad \underset{231}{} \qquad \underset{312}{} \qquad \underset{321}{}$

▶4. (i) $-(3)\begin{vmatrix} 1 & -3 \\ -2 & 1 \end{vmatrix} + (4)\begin{vmatrix} 2 & -3 \\ 1 & 1 \end{vmatrix} - (0)\begin{vmatrix} 2 & 1 \\ 1 & -2 \end{vmatrix}$

$$= -3(-5) + 4(5) - 0 = 35;$$

(ii) $(-3)\begin{vmatrix} 3 & 4 \\ 1 & -2 \end{vmatrix} - (0)\begin{vmatrix} 2 & 1 \\ 1 & -2 \end{vmatrix} + (1)\begin{vmatrix} 2 & 1 \\ 3 & 4 \end{vmatrix}$

$$= -3(-10) - 0 + 1(5) = 35.$$

▶5. (i) $abc + 2fgh - af^2 - bg^2 - ch^2$; (ii) 0.

▶6. $ab(ab - cd)$.

▶8. (i) $|A| = -24$, no; (ii) $\begin{vmatrix} 2 & -b & -b \\ 1 & -2a & 2a \\ 1 & -6a+b & 6a+b \end{vmatrix} = 0,$

$$\text{yes,} \quad \frac{x}{4ab} = \frac{y}{4a+b} = \frac{z}{4a-b}.$$

▶10. $|A| = 61$, $|B| = 1$, $|D| = |E| = 61$.

▶11. (i)$AB = O$; $BA = \begin{pmatrix} 0 & 0 \\ 3 & 0 \end{pmatrix}$.

▶12. $(AA^{-1})_{kl} = (A)_{km}(A^{-1})_{ml} = a_{km}\dfrac{C_{lm}}{|A|} = \dfrac{\delta_{kl}}{|A|}|A| = \delta_{kl}.$

▶13. (i) $(AI)_{ij} = a_{ik}(I)_{kj} = a_{ik}\delta_{kj} = a_{ij}$;
(ii) $(IA)_{ij} = \delta_{ik}a_{kj} = a_{ij} = (AI)_{ij}$;
(iii) $(II)_{ij} = \delta_{ik}\delta_{kj} = \delta_{ij}.$

▶14. (i) consider $(\widetilde{AA^{-1}}) = \tilde{I}$;
(ii) start from $AB(AB)^{-1} = I$ and apply (14.36 a) twice;
(iii) build up from (ii).

▶16. Replace B by BC in (14.46) then $\widetilde{ABC} = \widetilde{BC}\tilde{A} = \tilde{C}\tilde{B}\tilde{A}$, then replace C by CD etc.

▶18. $r_{kj} + is_{kj} = r_{jk} - is_{jk}$ since $H = \tilde{H}^*$, where r_{kj} and s_{kj} are the real and imaginary parts of h_{kj}.

▶19. (i) B is (c), (d), (f); (ii) C is (a), (e), (h).

▶22. $\lambda^3 - 7\lambda^2 + 6 = 0$; $\lambda = 1, 3 \pm \sqrt{15}$; $(1, 1, 3)$,
$(5 + \sqrt{15}, 7 + 2\sqrt{15}, -4 - \sqrt{15})$, $(5 - \sqrt{15}, 7 - 2\sqrt{15}, -4 + \sqrt{15})$.

▶26. (i) Show $\tilde{U}U = I$.

▶27. Eigenvalues are -4, $\mu_1 = 6$, $\mu_2 = -6$; $y_1 = 3^{-1/2}(x_1 + x_2 + x_3)$,
$y_2 = 6^{-1/2}(x_1 - 2x_2 + x_3)$, $y_3 = 2^{-1/2}(-x_1 + x_3)$.

▶28. (a), (b), (e), (f).

▶29. $(\vec{\nabla}\phi) = (2a_{1t}x_t, 2a_{2j}x_j, 2a_{3k}x_k) = \lambda(x_1, x_2, x_3)$.

▶31. $x_1:x_2 = 5 + \sqrt{19}: -2(4 + \sqrt{19})$.

Section 14.13

1. (a) -1; (b) $(3!)(\pm 1)^2 = 6$; (c) $\epsilon_{ijk}\epsilon_{jik} = -\epsilon_{ijk}\epsilon_{ijk} = -6$; (d) write as $\frac{1}{2}(\epsilon_{ijk}a_{ij} + \epsilon_{ijk}a_{ji}) = \frac{1}{2}(\epsilon_{ijk} + \epsilon_{jik})a_{ij} = 0$.

2. $(\mathbf{a} \wedge \mathbf{b}) \cdot (\mathbf{c} \wedge \mathbf{d}) = (\mathbf{a} \cdot \mathbf{c})(\mathbf{b} \cdot \mathbf{d}) - (\mathbf{d} \cdot \mathbf{a})(\mathbf{b} \cdot \mathbf{c})$.

3. $\lambda^3 - \lambda^2(a + b + c) - \lambda(b^2 + c^2 + a^2 - ab - ac - bc) - 3abc + a^3 + b^3 + c^3 = 0$; invariant under cyclic interchange of a, b, c; $BC = CB$ implies $cb + ac + ab = a^2 + b^2 + c^2$ and $3abc = a^3 + b^3 + c^3$, equation becomes $\lambda^3 - \lambda^2(a + b + c) = 0$.

4. (i) x only appears on the leading diagonal, so equation must be a cubic; $x = a$ or b or c each make two rows or columns identical or simple multiples of each other and therefore must be the 3 required solutions;
(ii) the equation is linear in x as can be seen by imagining the subtraction of one row from each of the other two and then one column from each of the other two, leaving x in only one element. Adding all rows of the original matrix together gives $3x + 3$ as a common factor. Thus the solution is $x = -1$.

5. Determinant of left-hand side coefficients equals zero. Eliminate x, y and z to obtain $\eta^2 - 3\eta + 2 = 0$; for $\eta = 1$, $x = 1 + 2z$, $y = -3z$; for $\eta = 2$, $x = 2z$, $y = 1 - 3z$.

6. Inverse matrix $= \dfrac{1}{2}\begin{bmatrix} 1 & 1 & -2 \\ -7 & 1 & 4 \\ 5 & -1 & -2 \end{bmatrix}$;
$\quad\begin{array}{l} x_1 = -3/2, \\ x_2 = 7/2, \\ x_3 = -3/2. \end{array}$

7. Inverse matrix $= \begin{bmatrix} 11 & -31 & 2 \\ -5 & 15 & -1 \\ -6 & 17 & -1 \end{bmatrix}$;
$\quad\begin{array}{l} x = 3, \\ y = 1, \\ z = 2. \end{array}$

8. (i) $(\widetilde{U^{-1}AU})^* = \tilde{U}^*\tilde{A}^*\tilde{U}^{-1*} = U^{-1}AU$;
(ii) $(iA)^* = -iA^* = i\tilde{A} = (\widetilde{iA})$;
(iii) If $AB = BA$, $(AB)^* = A^*B^* = \tilde{A}\tilde{B} = \widetilde{BA} = \widetilde{AB}$;
if $(AB)^* = (\widetilde{AB})$, $BA = \tilde{B}^*\tilde{A}^* = (\widetilde{AB})^* = AB$.
(iv) $(\widetilde{I \pm S}) = I \mp S$, $\tilde{A} = [(\widetilde{I + S})^{-1}](\widetilde{I - S}) = (I + S)^{-1}(I + S) = (I - S)^{-1}(I + S)$, it then follows that $\tilde{A}A = I$;

$$S = \begin{bmatrix} 0 & -\tan(\theta/2) \\ \tan(\theta/2) & 0 \end{bmatrix}.$$

9. $1, 1 + (\alpha\beta)^{1/2}, 1 - (\alpha\beta)^{1/2}$; $(0, 0, 1)$, $(\alpha^{1/2}, \beta^{1/2}, 0)$, $(\alpha^{1/2}, -\beta^{1/2}, 0)$; $\alpha\beta$ real and > 0, $|\alpha| = |\beta|$; express α and β as $|\alpha| \exp(i\theta_\alpha)$ and $|\beta| \exp(i\theta_\beta)$.

10. $2^{-1/2}(0, 0, 1, 1)$, $6^{-1/2}(2, 0, -1, 1)$, $(39)^{-1/2}(-1, 6, -1, 1)$, $(13)^{-1/2}(2, 1, 2, -2)$.

11. For A: $(1, 0 - 1)$, $(1, \alpha_1, 1)$, $(1, \alpha_2, 1)$; for B: $(1, 1, 1)$, $(\beta_1, \gamma_1, -\beta_1 - \gamma_1)$, $(\beta_2, \gamma_2, -\beta_2 - \gamma_2)$; simultaneous and orthogonal $(1, 0, -1)$, $(1, 1, 1)$, $(1, -2, 1)$.

12. Max. $= 6$ at $\pm 6^{-1/2} (2, 1, 1)$; min. $= 3$ at $\pm 3^{-1/2} (1, -1, -1)$; [other eigenvalue $= 4$].

13. (i) Use the mutual orthogonality of eigenvectors, $\pm 2^{-1/2} (0, 1, -1)$;
    (ii) $Q = \tilde{y} \Lambda y$, where $x = Uy$, and is the required expression; explicitly

$$Q = 6 \left(\frac{2x + y + z}{\sqrt 6} \right)^2 + 3 \left(\frac{x - y - z}{\sqrt 3} \right)^2 + 4 \left(\frac{y - z}{\sqrt 2} \right)^2,$$

which is as in question 12.

14. Divide through by 100, then eigenvalues are $\frac{1}{4}$ and 1, thus semi-axes are 2 and 1; major axis makes arctan $(-4/3)$ with the positive x-axis.

15. (i) Longest axis corresponds to smallest eigenvalue whose eigenvector is $(1, 1, 1)$;
    (ii) divide through by 3, then eigenvalues are -1, -1, 9; eigenvector of the non-repeated root is the axis of symmetry and is $(1, 1, -1)$.

16.

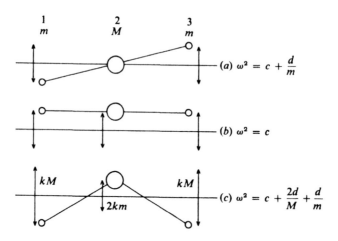

Fig. S.8 The normal modes of the coupled pendulums in example 16.

As $d \to 0$ all swing independently with frequency $c^{1/2}$; as $d \to \infty$ only mode (b) can be excited and pendulums swing as a rigid body with frequency $c^{1/2}$.

17. (i) obtain $(\lambda^n - \lambda^m) \tilde{u}^m A u^n = 0$; (iii) $c_{ij} = (\tilde{P} A P)_{ij} = (\tilde{P})_{ik} A_{kl} P_{lj} = u_k^i A_{kl} u_l^j = \tilde{u}^i A u^j = 0$ for $i \neq j$.

18. Values of λ are 2, -1, and -3; required vectors $(0, 1, 1)$, $(-1, 1, 1)$, $(-1, 1, 2)$; $x = -\eta - \chi$, $y = \xi + \eta + \chi$, $z = \xi + \eta + 2\chi$; $2\xi^2 - \eta^2 + 3\chi^2$ and $\xi^2 + \eta^2 - \chi^2$ [or any scaling of these]. If the roles of the quadratics are reversed the values of λ are $-\frac{1}{3}$, -1, and $\frac{1}{2}$.

19. $\lambda = -1, 2,$ and 4; $x = 2\xi - 2\eta + 2\chi$, $y = \xi + \eta + \chi$, $z = -3\xi + \eta - \chi$.

20. $A = \dfrac{1}{2}m \begin{bmatrix} 1 & 0 & 0 \\ 0 & 1 & 0 \\ 0 & 0 & 1 \end{bmatrix}$, $B = \dfrac{S}{a} \begin{bmatrix} 1 & -\frac{1}{2} & 0 \\ -\frac{1}{2} & 1 & -\frac{1}{2} \\ 0 & -\frac{1}{2} & 1 \end{bmatrix}$,

$$\frac{ma\omega^2}{S} = 2 \text{ or } 2 + \sqrt{2} \text{ or } 2 - \sqrt{2},$$

with corresponding amplitudes $(1, 0, -1)$, $(1, -\sqrt{2}, 1)$, $(1, \sqrt{2}, 1)$.

21. (i) Show $\mathbf{f}_i \cdot \mathbf{e}_j = \delta_{ij}$; show $\mathbf{f}_i \cdot \mathbf{f}_j = \cdots = h_{ik}g_{kl}h_{jl} = \cdots = h_{ij}$, using that H is symmetric;

(ii) $(u_i g_{ij} u_j)^{1/2}$, $(v_i h_{ij} v_j)^{1/2}$, $u_i v_i$;

(iii) $H = a^{-2} \begin{bmatrix} 3/2 & -1/2 & -1/2 \\ -1/2 & 3/2 & -1/2 \\ -1/2 & -1/2 & 3/2 \end{bmatrix}$;

(iv) Normal to the plane is $\mathbf{n} = v_i \mathbf{f}_i$ where $v_1 = p$, $v_2 = q$, and $v_3 = r$. Length of this is $(v_i h_{ij} v_j)^{1/2} = a^{-1}[\frac{3}{2}(p^2 + q^2 + r^2) - qr - pr - pq]^{1/2} = M$ (say).

(a) $p^{-1}\mathbf{e}_1 \cdot \mathbf{n} = M^{-1}$, (b) $\arccos (p/Ma)$.

CHAPTER 15

▶1. (i) Follow example 15.1, $u_1' = x_1 \cos (\phi - \theta) - x_2 \sin (\phi - \theta)$, $u_2' = x_1 \sin (\phi - \theta) + x_2 \cos (\phi - \theta)$; (ii) $\phi = -\pi/2$.

▶2. $\left(\dfrac{\partial u_i}{\partial x_i}\right)' = \dfrac{\partial u_i'}{\partial x_i'} = \dfrac{\partial (a_{kl} u_k)}{\partial x_i} \dfrac{\partial x_l}{\partial x_i'} = a_{kl} \dfrac{\partial u_k}{\partial x_l}$ $a_{ll} = \delta_{lk} \dfrac{\partial u_k}{\partial x_l} = \dfrac{\partial u_k}{\partial x_k}$.

▶3. $u_{11}' = s^2 x_1^2 - 2sc x_1 x_2 + c^2 x_2^2 \neq c^2 x_2^2 + cs x_1 x_2 + sc x_1 x_2 + s^2 x_1^2$.

▶4. Start with (15.13 b), interchange i and j, interchange dummy subscripts p and q.

▶5. (ii) $w_{kk}' u_{pi}' = a_{qp} a_{ji} v_{qj} = a_{qp} a_{ji} (w_{il} u_{qj}) = w_{il} u_{pi}'$.

▶6. (a) $x_i x_j$ is a second-order tensor and the value of the form at any particular point in space is fixed; contraction with $x_i x_j$ gives (b) 0, (c) $4x_1^2 x_2^2$ which is not invariant.

▶7. (i) $\delta_{ij} a_i b_j$; (ii) $\delta_{ij} \dfrac{\partial^2 \phi}{\partial x_i \partial x_j}$; (iii) $\epsilon_{ijk} \dfrac{\partial u_k}{\partial x_j}$;

(iv) $\delta_{ij} \dfrac{\partial^2 u_i}{\partial x_k \partial x_j}$; (v) $\epsilon_{ijk} \epsilon_{klm} \dfrac{\partial^2 u_m}{\partial x_j \partial x_i}$;

(vi) $\delta_{ij} \epsilon_{jkl} c_i a_k b_l = \epsilon_{ikl} c_i a_k b_l$.

▶10. Use exactly the procedures employed for δ_{ij} and ϵ_{ijk}.

▶11. Polar, **I**; axial, **H**, **B**, **μ**, **C**. Consider (a) e**v**; (b) $\mu_0 \mathbf{I} \wedge \mathbf{r}$; (c) $\mu_0 \mathbf{H}$; (d) $-\mathbf{\mu} \cdot \mathbf{B}$; (e) $\mathbf{\mu} \wedge \mathbf{B}$.

▶12. Using p for polar, a for axial, and s for scalar; (a) $p = p \wedge a$; (b) $p = p \wedge a$; (c) $s = p{\cdot}p$; (d) $p = p + a \wedge p + a \wedge p + a \wedge (a \wedge p)$.

▶13. (Scalar × isotropic tensor) − (outer product of two vectors).

▶14. $T = \frac{1}{2} \sum_\alpha m^{(\alpha)} \epsilon_{ijk} \omega_j x_k^{(\alpha)} \epsilon_{ilm} \omega_l x_m^{(\alpha)}$.

▶15. $\partial u_1/\partial x_2 = \partial u_2/\partial x_1 = k$, $u = (kx_2, kx_1, 0)$.

▶16. Normal to quadric is $(2e_{1j}y_j, 2e_{2j}y_j, 2e_{3j}y_j) = 2(v_1, v_2, v_3)$; $e = c|y|^{-2} = cr^{-2}$.

▶17. $lx_1 + mx_2 + nx_3 = \pm(c/e)^{1/2}$; consider quadric referred to its principal axes as $eX_1^2 = c$ with X_1 axis having direction l, m, n, with respect to original axes.

Section 15.10

1. Writing $T_{ii} = T_0$, $U_{ij} = \frac{1}{2}(T_{ij} + T_{ji}) - \frac{1}{3}T_0\delta_{ij}$, $V_{ij} = \frac{1}{3}T_0\delta_{ij}$, $S_{ij} = \frac{1}{2}(T_{ij} - T_{ji})$.

2. Contract with the outer product of (x, y, z) with itself to obtain $-(x^2 + y^2 + z^2)^2$, i.e. an invariant.

3. (i) $W = \begin{bmatrix} 0 & w_{12} & -w_{31} \\ -w_{12} & 0 & w_{23} \\ w_{31} & -w_{23} & 0 \end{bmatrix}$;

   (ii) $v_1 = w_{23}$, $v_2 = w_{31}$, $v_3 = w_{12}$;

   (iv) $(\mathbf{v} \wedge \mathbf{b})_i = \epsilon_{ijk}v_j b_k = \frac{1}{2}\epsilon_{ijk}\epsilon_{jlm}w_{lm}b_k = \cdots = -w_{ik}b_k$.

4. (i) $\epsilon_{ijk}\epsilon_{jlm}u_l v_m w_k$ and use (15.29);
   (ii) $\epsilon_{ijk}\,\partial(\phi u_k)/\partial x_j$;
   (iii) $\partial(\epsilon_{ijk}u_j v_k)/\partial x_i$;
   (iv) $\epsilon_{ijk}\epsilon_{klm}\,\partial(u_l v_m)/\partial x_j$ and use (15.29);
   (v) start with $\mathbf{u} \wedge \operatorname{curl} \mathbf{u}$, $\epsilon_{ijk}u_j\epsilon_{klm}\,\partial u_m/\partial x_l = \cdots$
   $$= u_j(\partial u_j/\partial x_i) - u_j(\partial u_i/\partial x_j).$$

5. $\operatorname{curl} \mathbf{H} = \mathbf{J} + \dot{\mathbf{D}}$; $\operatorname{div} \mathbf{D} = \rho$; $\operatorname{curl} \mathbf{E} + \dot{\mathbf{B}} = 0$; $\operatorname{div} \mathbf{B} = 0$.

6. (ii) arrange the answer to (i) as $\frac{1}{2}\partial(A^2)/\partial x_i - \partial(A_iA_j)/\partial x_j + A_i(\partial A_j/\partial x_j)$;
   (iii) (a) $\operatorname{curl} \mathbf{A} = 0$; (b) $\operatorname{div} \mathbf{B} = 0$; see previous question.

7. $a_j = \frac{1}{2}\epsilon_{jik}\sigma_{ik}$, $s_{ik} = \frac{1}{2}(\sigma_{ik} + \sigma_{ki})$ where σ_{ik} is the conductivity tensor; $\mathbf{j}$, $\mathbf{E}$ and $s\mathbf{E}$ are all polar vectors, thus so is $(\mathbf{E} \wedge \mathbf{a})$, showing $\mathbf{a}$ is axial.

8. $I = ma^2 \begin{bmatrix} 20 & 0 & 2 \\ 0 & 20 & 4 \\ 2 & 4 & 20 \end{bmatrix}$, $x_1 = \begin{bmatrix} 2 \\ -1 \\ 0 \end{bmatrix}$,

   $$x_2 = \begin{bmatrix} 1 \\ 2 \\ \sqrt{5} \end{bmatrix}, \quad x_3 = \begin{bmatrix} 1 \\ 2 \\ -\sqrt{5} \end{bmatrix}.$$

9. $I = 8ma^2 \begin{bmatrix} 8 & -1 & 1 \\ -1 & 8 & 1 \\ 1 & 1 & 6 \end{bmatrix}$ with principal moments $40\,ma^2$, $64ma^2$ and $72ma^2$.

10. Principal susceptibilities and axes are

$$\lambda = 4, \ \pm(0, 2^{-1/2}, 2^{-1/2}),$$
$$\lambda = 2, \ \pm(2 + c_1^2)^{-1/2}(c_1, 1, -1) \quad \text{with } c_1 c_2 = -2.$$

(i) Lowest value of energy – equation (15.45) – when $(0, 1, 1)$ axis along field;

(ii) permitted settings are $(0, n_2, n_3)$, hence as in (i);

(iii) permitted settings are $(n_1, 0, n_3)$ giving energy $= -\frac{1}{2}\mu_0 k H^2 V \times$ $(2n_1^2 + 3n_3^2)$, which, subject to $n_1^2 + n_3^2 = 1$, minimizes with $(0, 0, 1)$ direction along field.

11. The principal permeability (in direction $(1, 1, 2)$) has value 0. Thus all nails lie in the plane to which this is the normal.

12. $j_i = \sigma_{ik} E_k$ gives $I \sin\theta \cos\phi = a\pi r^2 E_1$, $I \sin\theta \sin\phi = a\pi r^2 E_2$, $I \cos\theta = b\pi r^2 E_3$ and then $V/l = E_1 \sin\theta \cos\phi + E_2 \sin\theta \sin\phi + E_3 \cos\theta$; the current must flow *along* the wire; E is not along the wire.

13. Take $p_{11} = p_{22} = p_{33} = -p$, $p_{ij} = e_{ij} = 0$ for $i \neq j$ leading to $-p = (\lambda + \frac{2}{3}\mu)e_{kk}$. Fractional volume change $= e_{kk}$. λ and μ as in example 15.3.

14. $$\frac{KE(2 - \sigma)}{(1 + \sigma)(1 - 2\sigma)}, \quad \frac{KE(1 + 4\sigma)}{2(1 + \sigma)(1 - 2\sigma)} \text{ (twice), use (15.60).}$$

15. All wall forces normal, $p_{ij} = e_{ij} = 0$ for $i \neq j$, $u_2 = u_3 = 0$, $-S = \lambda e_{11} + 2\mu e_{11}$, modulus $= \lambda + 2\mu$. λ and μ as in example 15.3.

16. $$p_{ij} = 2a(\nabla \cdot \mathbf{u})\delta_{ij} + 2b\left(\frac{\partial u_i}{\partial x_j} + \frac{\partial u_j}{\partial x_i}\right);$$

form $\sum_j (\partial/\partial x_j)$ of this equation, substitute for $\partial p_{ij}/\partial x_j$ and then form $\sum_i (\partial/\partial x_i)$ of the result; wave velocity for $\nabla \cdot \mathbf{u}$ is $[(2a + 4b)/\rho]^{1/2}$;

$$\rho \frac{\partial^2}{\partial t^2}(\text{curl } \mathbf{u})_k = \epsilon_{kji}\frac{\partial}{\partial x_j}\left(\frac{\partial p_{il}}{\partial x_l}\right),$$

then use above equation and $\epsilon_{kji}\, \partial^2/\partial x_j\, \partial x_i = 0$; wave velocity for $\nabla \wedge \mathbf{u}$ is $[2b/\rho]^{1/2}$.

CHAPTER 16

▶2. (a), (b), (d) are analytic since everywhere they satisfy (16.5) and the partial derivatives are continuous; (c) is analytic only at the point $z = 0$.

▶3. Second and fourth quadrants, $\partial u/\partial x = \pm 1$ for $x \gtrless 0$, $\partial v/\partial y = \pm 1$ for $y \lessgtr 0$,

▶4. (a) $(3 - 2i)z$; (b) $\sin(z)$; (c) zz^*; (d) $z + z^{-1}$.

▶5. At $z = 1$ series diverges; at $z = -1$ series converges to $-\ln 2$.

▶6. (ii) Ratio $= (n + 1)|z|$, which is >1 for all $z \neq 0$ and sufficiently large

n; (iii) ratio $= n|z|/(n + 1)$, which is < 1 if $|z| < 1$, but is > 1 for sufficiently large n if $|z| > 1$.

▶7. $\dfrac{1}{r!s!} = \dbinom{r + s}{r} \dfrac{1}{(r + s)!}$.

▶8. (i) $|t| < 1$, except $t = 0$; (ii) $|t| > 1$; (iii) a positive increasing spiral starting at $t = 1$; (iv) a positive decreasing spiral starting at $t = 1$.

▶9. $t^{1/n} = \exp\left(\dfrac{1}{n}\log t + i\dfrac{2\pi k}{n}\right)$, $k = 0, 1, \ldots, n - 1$.

▶10. (a) $\exp(-2y)\cos(2x)$;
(b) $\frac{1}{2}\sin 2y \sinh 2x$;
(c) $\sqrt{2}\exp(\pi i/3)$ and $\sqrt{2}\exp(4\pi i/3)$;
(d) $\exp(2^{-1/2})$ or $\exp(-2^{-1/2})$;
(e) $\cos 1 - i\sin 1 = 0.540 - i\,0.841$;
(f) $8\sin(\ln 2) = 5.11$;
(g) $\exp(-\frac{1}{2}\pi - 2\pi n)$, n integral, $n = 0$ gives principal value;
(h) $\ln(8) + i(2n + \frac{1}{2})\pi$.

▶11. Use (16.22) and $dw/dz = (dz/dw)^{-1}$.

▶12. Write in modulus and argument form and use the formulae for $\cos\alpha$ and $\sin\alpha$ in terms of $\cos 2\alpha$.

▶13. In (ii) $dF/dz = \frac{1}{2}ikz^{-1/2}$, in (iii) $dF/dz = 2ikz$; (a) $|\mathbf{E}| = 2k|z|$; (b) at $y = 0$, $dF/dz = \frac{1}{2}ikx^{-1/2}$, hence $E_x = 0$, $E_y = \frac{1}{2}kx^{-1/2}$ and $\sigma \propto E_y$; (c) $|\mathbf{E}| = |\frac{1}{2}ikz^{-1/2}|$; (d) if θ is angle of radius vector, $\arg dF/dz = \theta + \pi/2$, direction of $\mathbf{E} = \pi - (\theta + \pi/2)$ by (16.34 b), hence result.

▶14. (a) See (16.47); differentiate (16.44) w.r.t. x and y and then (b) let $y \to 0$ with $x > 0$, and (c) square results and add.

▶15. (i) $f(t)$ is analytic except at $t = 0$, $\to Et$ as $t \to \infty$, and has no real part on $t = a\exp(i\theta)$; (ii) put $t = a\exp(i\theta)$ giving $z = \frac{3}{2}a\cos\theta + i\frac{1}{2}a\sin\theta$; (iii) $dz/dt = 0$ at $t = \pm 2^{-1/2}a$ and $=\infty$ at $t = 0$; (iv) express t in terms of z and substitute; (v) put $z = \frac{3}{2}a\cos\theta + i\frac{1}{2}a\sin\theta$ into $F(z)$.

▶16. $\int z^{-1}\,dz = \text{Log }z$.

▶17. (i) $\int_0^{2\pi} iR^3\exp(3i\theta)\,d\theta = 0$; (ii) $\int [(x^2 - y^2) + 2ixy](dx + i\,dy)$ for each of 4 sides; of the 16 integrals, 8 are trivially zero, 4 have value zero, and the other 4 cancel in pairs.

▶19. $\dfrac{1}{\xi - z} = \dfrac{1}{\xi - a}\displaystyle\sum_0^\infty \left(\dfrac{z - a}{\xi - a}\right)^n$.

▶21. (a) $f(z) = (2z^2)^{-1}(1 + \frac{1}{2}z + \cdots)$, integral $= 2\pi i \times \frac{1}{4} = \frac{1}{2}\pi i$; (b) the same; (c) by expansion or $\lim[(z - 2)f(z)]$ as $z \to 2$, residue at $z = 2$ is $-\frac{1}{4}$, total integral $= 0$.

▶22. $\Delta_{01} = 0$, $\Delta_{11} = \arctan(\frac{1}{2})$, $\Delta_{10} = -\arctan(\frac{1}{2})$, thus no root included.

▶24. Residue at $z = -a$ is $\beta a^{-\beta-1}\exp(-i\beta\pi)$; $I = \pi\beta a^{-\beta-1}\operatorname{cosec}(\pi\beta)$.

Section 16.16

1. $\partial u/\partial y = -\exp(x)(y \cos y + x \sin y + \sin y)$, $z \exp(z)$.

2. $f = \dfrac{\sin 2x - i \sinh 2y}{\cosh 2y - \cos 2x}$. Special case of z real, $= x$, shows $f(z) = \cot(z)$; poles at $z = n\pi$.

3. (a) ∞; (b) 1; (c) 1; (d) 1; (e) $\lim (1 + p/n)^n = e^p$, $R = e^{-p}$.

4. $|[(x + c) + iy]/[(x - c) + iy]| = \exp(u)$, which can be arranged as $(x - c \coth u)^2 + y^2 = c^2 \operatorname{cosech}^2 u$; $c \coth u_1 = -d$, $c \coth u_2 = d$, $|c \operatorname{cosech} u_i| = a$, $\cosh u_i = d/a$ with $u_1 = -u_2$, capacity $\propto (u_2 - u_1)^{-1}$.

5. $W(t) = (-iV/\pi) \log t$; $\qquad W'(z) = (-iV/\pi) \log [(z/a) + ((z^2/a^2) - 1)^{1/2}]$, potential $= (V/\pi) \times \arg [\ldots]$.

6. Ellipses $\dfrac{x^2}{(a + 1)^2} + \dfrac{y^2}{(a - 1)^2} = \dfrac{c^2}{4a^2}$; hyperbolae $\dfrac{x^2}{\cos^2 \alpha} - \dfrac{y^2}{\sin^2 \alpha} = c^2$;

   slit $-c \leqslant x \leqslant c$, $y = 0$; slits $\pm c$ to $\pm\infty$ for both cases; $\phi = 0$ on strip; if $P = \phi + i\psi = -k \operatorname{arcosh}(z/c)$, $dP/dz = -k/(c \sinh(-P/k))$.

7. $f(z) = \displaystyle\sum_0^x a_n z^n$

   with $a_{2n+1} = \dfrac{(-1)^{n+1} p^{2n+1}}{(2n + 1)!} \displaystyle\sum_{r=1}^x \dfrac{(-1)^r}{r^{2n+1}}$, $a_{2n} = 0$;

   $R^{-1} = \overline{\lim} [p \cdot 1 \cdot (2n + 1)^{-1}] = 0$ by the root test.

8. Zeros at $z = n\pi$, simple poles at $z = (n + \frac{1}{2})\pi$, essential singularity at $z = \infty$.

9. (i) $|z| = \frac{3}{8}$, $|8z^3 + z| \leqslant \frac{51}{64} < 1$; $|z| = \frac{5}{8}$, $|8z^3| = \frac{125}{64} > \frac{104}{64} \geqslant |z + 1|$;
   (ii) write as $8(z - \gamma)(z - \alpha - i\beta)(z - \alpha + i\beta) = 0$, γ is < 0, and then zero coefficient of z^2 shows $\alpha > 0$. Show $-\frac{3}{8} > \gamma > -\frac{1}{2}$ and use $-8\gamma(\alpha^2 + \beta^2) = 1$.

10. (i) $y^8 + 5 = 0$ has no real roots; (ii) one negative real zero and a conjugate pair in the second and third quadrants; zeros at $-\frac{3}{2}$, $-1 - i$, $-1 + i$.

11. $A = 3 - 12t^2 + t^4$, $\qquad B = -2t - 2t^3$; $\qquad \Delta_c[\arctan(B/A)] = 0$, $\Delta_c[\arg f(z)] = 4\pi$, hence 2 zeros inside $|z| = 1$.

12. Pole at $z = 1/a$, $\pi a^{-n}(a^2 - 1)^{-1}$.

13. Only pole inside unit circle at $z = ia - i(a^2 - 1)^{1/2}$, residue $-i\frac{1}{2}(a^2 - 1)^{-1/2}$; $2\pi[a(a^2 - 1)^{-1/2} - 1]$.

14. Follow example 16.5 and use Jordan's lemma, pole at $z = i$; $\pi\frac{1}{2}\exp(-\alpha)$.

15. Factorize denominator as $(2x + i)(2x - i)(x + i)(x - i)$, simple poles at $\frac{1}{2}i$ and i.

16. Use Jordan's lemma and a semicircular contour indented at $z = \pm a$.

17. (i) Only pole at origin with residue π^{-1}; integral $= 2i$;
    (ii) each is $O(\exp(-\pi R^2 - \pi R 2^{-1/2}))$;
    (iii) sum of integrals is $2i \int_{-R}^{R} \exp(-\pi r^2)\, dr$.

18. Follow example 16.7 (page 478); $\pi \operatorname{cosec}(\pi\alpha)$.

19. See example 16.7 (page 478).

20. Evaluate $\displaystyle\int \frac{\pi \cot \pi z}{(\frac{1}{2} + z)(\frac{1}{4} + z)}\, dz$

   around a very large circle; residue at $z = -\frac{1}{2}$ is 0, and at $z = -\frac{1}{4}$ is $4\pi \cot(-\frac{1}{4}\pi)$.

21. Behaviour of integrand for large $|z|$ is $|z|^{-2}\exp[(2\alpha - \pi)|z|]$. Residue at $z = m$ is $\sin^2(m\alpha)(-1)^m/(m\alpha)^2$ for $m = -\infty,\ldots, -1, 0, 1,\ldots, \infty$. Even function, so $\sum_1^\infty = \frac{1}{2}[\sum_{-\infty}^\infty - (m = 0\ \text{term})]$. No contribution from contour, leading to value $\frac{1}{2}$ for required summation.

22. Poles at (i) ib and $-ib$, (ii) $t = s - a = 0$, of order $n + 1$, (iii) $t = 0$ and $t = -2|a|$.

23. $\int_\Gamma$ and $\int_\gamma$ tend to 0 as $R \to \infty$ and $\rho \to 0$. Put $s = r\exp(i\pi)$ and $s = r\exp(-i\pi)$ on the two sides of the cut and use $\int_0^\infty \exp(-t^2 x)\, dt = \frac{1}{2}(\pi/x)^{1/2}$. No poles inside contour.

24. γ is traversed negatively, residue is 1.

Index

Where a topic is discussed on two consecutive pages, reference is made only to the more appropriate of these. For discussions spread over several pages the first and last page numbers are given; these references are usually to the major treatment of the corresponding topic.

Where a reference is made to an exercise or example, the information may be in the corresponding solution, rather than in the question itself.

Printed in the United States
22159LVS00001B/254